HANDBOOK OF

# Web Surveys

# HANDBOOK OF

# Web Surveys

*Second Edition*

## SILVIA BIFFIGNANDI
University of Bergamo
Italy

## JELKE BETHLEHEM
Faculty of Social and Behavioral Sciences,
Institute of Political Science
Leiden University
The Netherlands

contact:
biffisil@teletu.it
j.g.bethlehem@fsw.leidenuniv.nl

website:
**www.web-survey-handbook.com/**

This second edition first published 2021
© 2021 by John Wiley & Sons, Inc.

*Edition History*
John Wiley and Sons, Inc. (1e, 2012)

*Registered Office*
John Wiley & Sons, Inc., 111 River Street, Hoboken, NJ 07030, USA

*Editorial Office*
111 River Street, Hoboken, NJ 07030, USA

For details of our global editorial offices, customer services, and more information about Wiley products visit us at www.wiley.com.

Wiley also publishes its books in a variety of electronic formats and by print-on-demand. Some content that appears in standard print versions of this book may not be available in other formats.

*Library of Congress Cataloging-in-Publication Data*

Names: Biffignandi, Silvia, author. | Bethlehem, Jelke G., author. | John
   Wiley & Sons, Inc., publisher.
Title: Handbook of web surveys / Silvia Biffignandi,
   University of Bergamo, Italy
   [and] Jelke Bethlehem, Leiden University, Faculty of Social and
   Behavioral Sciences, Institute of Political Science, The Netherlands.
Description: Second edition. | Hoboken, NJ : Wiley, 2021. |
   Includes bibliographical references and index.
Identifiers: LCCN 2020028687 (print) | LCCN 2020028688 (ebook) | ISBN
   9781119371687 (hardback) | ISBN 9781119371694 (adobe pdf) | ISBN
   9781119764496 (epub)
Subjects: LCSH: Internet surveys. | Surveys–Methodology.
Classification: LCC HM538 .B55 2021  (print) | LCC HM538  (ebook) | DDC
   001.4/33–dc23
LC record available at https://lccn.loc.gov/2020028687
LC ebook record available at https://lccn.loc.gov/2020028688

Cover Design: Wiley
Cover Images: © Adisa/Getty Images

Set in 10.5/12pt AGaramond by SPi Global, Pondicherry, India

SKY10025786_032421

# Contents

# Preface

The last 10 years have witnessed a significant increase in Internet penetration. What is particular about this growth is that a number of generations are currently experiencing the contemporary and highly technological environment. Social media, constant connectivity, and on-demand entertainments are innovations that Millennials (aged between 23 and 38 in 2019) adapted to as they grew up. For those born after 1996, the so-called Generation Z (aged between 7 and 22 in 2019), these innovations are mostly taken for granted, having been part of their lives from the beginning. The iPhone was launched in 2007, when the oldest members of Generation Z were 10. By the time they are in their teens, young Americans access the Internet mainly via mobile devices, Wi-Fi, and high-bandwidth cellular services. Pre-Millennial generations play an important role in the general population, but for them, this environment based on technological communication is a new experience.

The implications of some population subgroups having adapted to the technological environment (Millennials and pre-Millennials) while others have lived in this "always on" technological environment all their lives are of relevance for survey-based research, particularly in the case of web surveys. The way that questionnaires are administered undoubtedly has an impact which differs according to population group. Furthermore, the behavior of the respondents while participating depends on their digital experience, their generational characteristics, and their attitude toward technology in their lives. Therefore, surveys—and in particular web and mobile web surveys—have to adopt a number of changes in their methodology to take into account any differences in the cultural backgrounds of potential survey participants and the characteristics of the eventual devices used.

Due to high Internet penetration and the relatively low cost of conducting web surveys compared with other methods, the number of surveys being conducted via the Internet has increased dramatically over recent years. The panorama of survey-based research has changed drastically over the last few decades.

First there was a change from traditional paper-and-pencil interviewing (PAPI) to computer-assisted interviewing (CAI). Since the 1990s, there has been a gradual replacing of face-to-face surveys (CAPI), telephone surveys (CATI), and mail surveys (CASI, CSAQ) with web-based surveys. With the relatively recent diffusion of smartphones and other mobile devices, it has become possible to run mobile web surveys, i.e., questionnaires sent to interviewees may be submitted and also completed via mobile devices. A web survey is a simple way to access a large group of potential respondents. Questionnaires can be distributed at very low cost. They require no interviewers, and there are no mailing or printing costs involved. Surveys can be launched rapidly, and little time is lost between the moment the questionnaire is ready and the moment that fieldwork begins. Web surveys also offer interesting new possibilities, such as the use of multimedia (images, sound, animation, and video). Panel surveys are also moving toward data collection via the web.

The recent trend toward the use of big data and the integration of data sources will not render the role of web surveys obsolete, although they may in the future have a different role.

At first sight, web surveys appear to have much in common with other types of survey, seeming to be just another way to collect data, with questions asked over the Internet instead of face-to-face, by telephone or via e-mail. There are however a number of factors that may render the results of web surveys unreliable. Some examples are under-coverage, self-selection, and measurement errors. These can cause estimates of population characteristics to be biased, thus leading to incorrect conclusions being drawn from the data collected.

Under-coverage occurs when the target population is wider than the number of people with Internet access. This leads to bias in estimates in the case of relevant differences between those with Internet access and those without.

Self-selection is when a questionnaire is simply made available via Internet to all, with individuals nominating themselves.

A respondent is therefore anyone who happens to have Internet access, visits the website, and decides to take part in the survey. These participants generally differ significantly from nonparticipants.

General-population surveys that aim to provide reliable and accurate statistics are traditionally carried out face-to-face or by telephone. Interviewers are used to persuade people to take part and to help respondents to provide the right answers. In web surveys, there is no interviewer assistance, a fact that can have serious impact on the quality of the data collected.

The diffusion of smartphones increases the possibility for interviewees to be reached via their mobile device and to have the questionnaire completed via the same device, resulting in the current trend in running mobile web surveys. Consequently, there are new risks for error in the survey due to device characteristics and the behavior of the user.

The researcher should have in mind that when a web survey is run a mobile web survey takes place, if  questionnaire is not blocked against mobile devices. Here, for simplicity the term web surveys is used, meaninig the  mobile web survey is included. Summing up, web surveys afford several challanges and need reseacher be conscious of the methodological issues for a good survey. At the time beeing, collecting data through web surveys is going to become a common practice both in market research, academic research and official statistics. Knowledge about how to manage a web survey, risks, errors and advantages is important.

This book provides an insight into the possible use of web surveys and mobile web surveys for data collection. Web surveys allow for lower data collection costs.

It is also expected that web surveys lead to increased response rates. Is this the case? What about the quality of the data collected? This book examines many theoretical and practical aspects of mobile web surveys and can therefore be considered as a handbook for those involved in practical survey research, including survey researchers working with official statistics (e.g., in national statistical institutes), academics, and commercial market research.

The book's two authors have widespread expertise in survey methodology. They come from two different countries (the Netherlands and Italy) and different research organizations (a national statistical institute and a university). They therefore provide a broad view on the various theoretical and practical aspects of mobile web surveying.

The second edition of the book involves a revision of each chapter with the following criteria:

(1) to maintain the existing text and content as much as possible, (2) to update the existing text and content with comments based on new literature and results, and (3) to add new paragraphs (if necessary) to cover new relevant topics (see the contents and chapter description below). A number of new examples have been provided, some of the existing examples have been updated or substituted, and some applications have been replaced. Updates have also been included to highlight new trends in web surveys and emerging solutions. There are two new chapters on topics concerning mobile web surveys: one presenting a flowchart to illustrate the steps involved in running a survey via web and the other examining adaptive design. It was therefore necessary to renumber the chapters in respect to the first edition.

The first two chapters of the book provide an introduction into web surveys. Chapter 1 provides a historic account of developments in survey research and shows how web surveys have become a tool for data collection. Section 1.3.1 examines the Blaise system, which has been around for more than 30 years. New developments have taken place over the last 10 years, but no papers have been written on this subject. The section looks at the history and recent developments regarding Blaise; it was written by Lon Hofman and Mark Pierzchala and is published for the first time here.

Chapter 2 is an overview of basic aspects of web surveys. It describes how and where they can be used. Official statistics departments, research institutions, market research companies, and private forums are all interested in web surveys studying both households/individuals and businesses.

Chapter 3 presents a flowchart illustrating the steps (and sub-steps) of web surveys, each accompanied by a short description. The flowchart is of potential use to both practitioners as a guideline for how the survey process should be carried out and to researchers in highlighting and explaining the positioning of their studies at the different steps of the survey process. It is also useful in discussing the errors that can occur in different steps. The chapter provides an introduction to the framework and its structure and discusses the relevance of bearing in mind the framework and the survey steps when considering web survey errors. It then goes on to describe the concept of the step and the structure of the flowchart, breaking down the web survey process into six main steps. These are analyzed in detail, and an overview of survey errors is provided.

Chapter 4 examines the aspects of sampling. It is stressed that valid population inference is possible only if some form of probability sampling is used and that a proper sampling frame is required for this. A number of sampling designs and estimation procedures useful for web surveys are discussed.

A researcher conducting a survey may encounter a number of practical problems, and Chapter 5 provides an overview of possible errors, with two types of error examined in further detail. The first concerns errors in measurement. These can be caused by specific issues in questionnaire design, as well as a number of other aspects such as technology, incorrect unit definition, and so on. The second type of error regards nonresponse. This is a phenomenon that can affect all surveys, but the specific aspects of nonresponse in web surveys require particular attention. The chapter provides advice on relationships between errors and information on the various types.

A web survey is just one form of data collection. There are others, such as face-to-face, telephone, mail, and mobile surveys. Chapter 6 compares these various methods with online data collection, discussing the advantages and disadvantages of each one.

As web surveys do not involve interviewers, the respondents complete the questionnaire on their own. Furthermore, when questionnaires are sent out, they may very well be received and even completed on a mobile device (such as smartphones, which are very widespread). This means that questionnaire design is of crucial importance. Questionnaires must be adapted in order to be suitable for mobile devices; otherwise they cannot be used for this purpose. Small imperfections in the questionnaire may have serious consequences in terms of data quality. Questionnaire design issues are addressed in Chapter 7.

Chapter 8 examines strategies for data collection with adaptive/responsive survey design. In this case, strategies are not defined in advance, but instead are adapted, if necessary, during fieldwork. These designs may contribute to countering growing problems of nonresponse. This chapter was written by Annamaria Bianchi and Barry Schouten, who applied their particular expertise in this field to the subject examined.

A web survey may not always be the best solution for providing reliable and accurate statistics, with quality being affected by problems of under-coverage and low response rates. An interesting alternative is to set up a mixed-mode survey, in which several data collection methods are combined either sequentially or concurrently. This approach is less expensive than a single-mode interviewer-assisted survey (conducted either face to face or by telephone) and solves under-coverage problems, but at the same time it poses other difficulties, known as mode effects, with one of the most significant of these being measurement error. Mixing modes is also of critical importance, as is the fact that in practice, a web survey is always mobile, unless questionnaire access via mobile device is restricted. All these aspects, as well as others concerning mixed-mode surveys, are discussed in Chapter 9.

Chapter 10 is devoted to the problem of under-coverage. This remains an important problem in many countries due to poor Internet coverage and the fact that Internet access is often unevenly distributed throughout the population. The chapter demonstrates how this can lead to survey estimates being biased. A number of techniques that may reduce under-coverage bias are discussed.

Chapter 11 examines self-selection. The correct and scientifically well-founded principle is to use probability sampling in order to select survey subjects and therefore allow reliable estimates regarding population characteristics to be calculated. Nowadays, it is easy to set up a web survey. Even those without any survey knowledge or experience can create one through dedicated websites. Many of the resulting web surveys do not apply probability sampling, but instead rely on self-selection of respondents. This causes serious problems with estimation. Self-selection and its consequences in terms of survey results are discussed in this chapter, demonstrating that correction techniques are not always effective, and there are many reasons why web-survey-based estimates are biased.

Nonresponse, under-coverage, and self-selection are typical examples, and adjustment weighting is often applied in surveys in order to reduce such biases. Chapter 12 describes various weighting techniques, such as post-stratification, generalized regression estimation and raking ratio estimation. The effectiveness of these techniques in reducing bias caused by under-coverage or self-selection is examined.

Chapter 13 introduces the concept of response probabilities, describing how they can be estimated through response propensities. If estimated accurately, response probabilities can be used to correct biased estimates. Here, two general approaches are described: response propensity weighting and response propensity stratification. The first attempts to adjust the original selection probabilities, while the second is a form of post-stratification.

Chapter 14 is devoted to web panels. There are many such panels, particularly in the field of commercial market research. One crucial aspect is how the panel members (households, individuals, companies, and shops) are recruited. This can be carried out via a proper probability sample, or through self-selection. There are consequences for the validity of the results of the specific surveys conducted with the panel members. The chapter discusses several quality indicators.

The accompanying website, www.web-survey-handbook.com, provides the survey data set for the general population survey (GPS), which has been used for many examples and applications in the book. The data set is available in SPSS (SPSS Corporation, Chicago, IL) format.

SILVIA BIFFIGNANDI

JELKE BETHLEHEM

The editors acknowledge the contributions of:

Lon Hofman (Manager Blaise, Statistics Netherlands) and Mark Pierzchala (owner of MMP Survey Services, Rockville, USA) who wrote Section 1.3.1.

Annamaria Bianchi (University of Bergamo) and Barry Schouten (Statistics Netherlands) who wrote Chapter 8.

# CHAPTER ONE

# The Road to Web Surveys

## 1.1 Introduction

Web surveys are a next step in the evolution process of survey data collection. Collecting data for compiling statistical overviews is already very old, almost as old as mankind. All through history, rulers of countries used statistics to take informed decisions. However, new developments in society always have had their impact on the way the data were collected for these statistics.

For a long period, until the year 1895, statistical data collection was based on complete enumeration of populations. The censuses were mostly conducted to establish the size of the population, to determine tax obligations of the people, and to measure the military strength of the country. The idea of sampling had not emerged yet.

The year 1895 marks a fundamental change. Populations had grown bigger and bigger. It was the period of industrialization. Centralized governments required more and more information. The time was ripe for sample surveys. The first ideas emerged around 1895. There was a lot of discussion between 1895 and 1934 about how to select samples: by means of probability sampling or some other sample selection technique.

By 1934, it was clear that only surveys based on probability sampling could provide reliable and accurate estimates. Such methods of data collection were accepted as a scientific. In the period from 1940s to the 1970s, most sample surveys were probability based. Questionnaires were on paper forms. They were completed in face-to-face, telephone, or mail.

*Handbook of Web Surveys*, Second Edition. Silvia Biffignandi and Jelke Bethlehem.
© 2021 John Wiley & Sons, Inc. Published 2021 by John Wiley & Sons, Inc.

Somewhere in the 1970s another significant development started. The fast development of microcomputers made it possible to introduce computer-assisted interviewing (CAI). This made survey data collection faster, cheaper, and easier and increased data quality. It was time in which acronyms like CATI (computer-assisted telephone interviewing) and CAPI (computer-assisted personal interviewing) emerged.

The next major development was the creation of the Internet around 1982. When more and more persons and companies got access to the Internet, it became possible to use this network for survey data collection. The first Internet surveys were e-mail surveys. In 1989 the World Wide Web was developed. This software allowed for friendly graphical user interfaces for Internet users. The first browsers emerged and the use of Internet exploded. In the middle of 1990s, the World Wide Web became widely available, and e-mail surveys were increasingly replaced by web surveys.

Web surveys are attractive because they have a number of advantages. They allow for simple, fast, and cheap access to large groups of potential respondents. Not surprisingly, the number of conducted web surveys has increased rapidly over time. There are, however, also potential methodological problems. There are ample examples of web surveys not based on probability sampling. Therefore, generalization of survey results to the population is questionable. The interviewed may access the Internet using various types of devices. Thus, web surveys can be completed and received not only on personal computer (PC) or laptop; it is highly probable the survey to be received in the mobile phone. The so-called mobile web surveys are fully part of web surveys. This implies some methodological problems to be considered, and further research on the impact of mobile is called for.

This chapter describes the historical developments that have led to the emergence of web surveys. As an illustration, Section 1.3 shows how these developments were implemented at Statistics Netherlands and led to new software for survey data collection.

## 1.2  Theory

### 1.2.1  THE EVERLASTING DEMAND FOR STATISTICAL INFORMATION

The history of data collection for statistics goes back in time for thousands of years. As far back as Babylonian era, a census of agriculture was carried out. This already took place shortly after the invention of the art of writing. The same thing happened in China. This empire counted its people to determine the revenues and the military strength of its provinces. There are also accounts of statistical overviews compiled by Egyptian rulers long before Christ. Rome regularly took censuses of people and of property. The collected data were used to establish the political status of citizens and to assess their military and tax obligations to the state.

Censuses were rare in the Middle Ages. The most famous one was the census of England taken by the order of William the Conqueror, King of England. The compilation of his *Domesday Book* started in the year 1086 AD. The book records a wealth of information about each manor and each village in the country. Collected information was about more than 13,000 places. More than 10,000 facts were recorded for each country.

To collect all this data, the country was divided into a number of regions. In each region, a group of commissioners was appointed from among the greater lords. Each county within a region was dealt with separately. Sessions were organized in each county town. The commissioners summoned all those required to appear before them. They had prepared a standard list of questions. For example, there were questions about the owner of the manor; the number of free man and slaves; the area of woodland, pasture, and meadow; the number of mills and fishponds, to the total value; and the prospects of getting more profit. The *Domesday Book* still exists, and many county data files are available on CD-ROM and the Internet.

Another interesting example of the history of official statistics is in the Inca Empire that existed between 1000 and 1500 AD. Each Inca tribe had its own statistician, called the *quipucamayoc*. This man kept records of the number of people, the number of houses, the number of llamas, the number of marriages, and the number of young men that could be recruited for the army. All these facts recorded on *quipus*, a system of knots in colored ropes. A decimal system was used for this. At regular intervals, couriers brought the quipus to Cusco, the capital of the kingdom, where all regional statistics were compiled into national statistics. The system of quipucamayocs and quipus worked remarkably well. The system vanished with the fall of the empire.

An early census also took place in Canada in 1666. Jean Talon, the intendant of New France, ordered an official census of the colony to measure the increase in population since the founding of Quebec in 1608. Name, age, sex, marital status, and occupation were recorded for every person. It turned out there lived 3,215 people in New France.

The first censuses in Europe took place in the Nordic countries. The first census in Sweden–Finland took place in 1749. Not everyone welcomed the idea of a census. Particularly religious people believed that people should not be counted. They referred to the census ordered by King David in biblical times, which was interrupted by a terrible plague and never completed. Others said that a population count would reveal the strengths and weaknesses of a country to foreign enemies. Nevertheless, censuses took place in more and more countries. The first census in Denmark–Norway has been in 1769. In 1795, at the time of the Batavian Republic under Napoleon's influence, the first count of the population of the Netherlands took place. The new centralized administration wanted to gather quantitative information to devise a new system of electoral constituencies (see Den Dulk and Van Maarseveen, 1990).

In the period until the late 1880s, there were some applications of *partial investigations*. They were statistical inquiries in which only part of a complete human population has been interviewed. The way the persons were selected from the population was generally unclear and undocumented.

In the second half of the 19th century, so-called monograph studies became popular. They were based on Quetelet's idea of the average man. According to Quetelet, many physical and moral data have a natural variability. This variability can be described by a normal distribution around a fixed, true value. He assumed the existence of something called the *true value*. Quetelet introduced the concept of *average man* ("l'homme moyenne") as a person of which all characteristics were equal to the true value (see Quetelet, 2010, 2012).

The period of the 18th and 19th centuries is called the era of the Industrial Revolution, too. It led to important changes in society, science, and technology. Among many other things, urbanization started from industrialization and democratization. All these developments created new statistical demands. The foundations for many principles of modern statistics were laid. Several central statistical bureaus, statistical societies, conferences, and journals, were established soon after this period. First ideas about survey sampling emerged in the world of official statistics. If a starting year must be chosen, 1895 would be a good candidate. Anders Kiaer, the founder and first director of Statistics Norway, started in this year a fundamental discussion about the use of sampling methods. This discussion led to the development, acceptance, and application of sampling as a scientific method.

Anders Kiaer (1838–1919) was the founder and advocate of the survey method that is now widely applied in official statistics and social research. With the first publication of his ideas in 1895, he started the process that ended in the development of modern survey sampling theory and methods. This process is described in more detail in Bethlehem (2009).

There have been earlier examples of scientific investigations based on samples, but they were lacking proper scientific foundations. The first known attempt of drawing conclusions about a population using only information about part of it was made by the English merchant John Graunt (1662). He estimated the size of the population of London. Graunt surveyed families in a sample of parishes where the registers were well kept. He found that on average there were three burials per year in 11 families. Assuming this ratio to be more or less constant for all parishes and knowing the total number of burials per year in London to be about 13,000, he concluded that the total number of families was approximately 48,000. Putting the average family size at 8, he estimated the population of London to be 384,000. Since this approach lacked a proper scientific foundation, John Graunt could not say how accurate his estimates were.

More than a century later, the French mathematician Pierre-Simon Laplace realized that it was important to have some indication of the accuracy of his estimate of the French population. Laplace (1812) implemented an approach that was more or less similar to that of John Graunt. He selected 30 departments distributed over the area of France in such a way that all types of climate were represented. Moreover, he selected departments in which accurate population records were kept. Using the central limit theorem, Laplace proved that his estimator had a normal distribution. Unfortunately, he disregarded the fact that sampling was purposively, and not at random. These problems made application of the central limit theorem at least doubtful.

In 1895 Anders Kiaer (1895, 1997), the founder and first director of Statistics Norway, proposed his *representative method*. It was a partial inquiry in which a large number of persons were questioned. Selection of persons was such that a "miniature" of the population was obtained. Anders Kiaer stressed the importance of *representativity*. He argued that if a sample was representative with respect to variables for which the population distribution was known, it would also be representative with respect to the other survey variables. Example 1.1 describes the Kiaer's experiment about the representative method.

---

▨ EXAMPLE 1.1 **The representative method of Anders Kiaer**

Anders Kiaer applied his representative method in Norway. His idea was to survey the population of Norway by selecting a sample of 120,000 people. Enumerators (hired only for this purpose) visited these people and filled in 120,000 forms. About 80,000 of the forms were collected by the representative method and 40,000 forms by a special (but analogue) method in areas where the working-class people lived.

For the first sample of 80,000 respondents, data from the 1891 census were used to divide the households in Norway into two strata. Approximately 20,000 people were selected from urban areas and the rest from rural areas.

There was a selection of 13 representative cities from the 61 cities in Norway. All five cities having more than 20,000 inhabitants were included, and eight cities representing the medium sized and small towns, too. The proportion of selected people in cities varied: in the middle-sized and small cities, the proportion was greater than in the big cities. Kiaer motivated this choice by the fact that the middle-sized and small cities did not represent only themselves but a larger number of similar cities.

In Kristiania (nowadays Oslo) the proportion was 1/16, in the medium-sized towns the proportion varied between 1/12 and 1/9, and in the small towns it was 1/4 or 1/3 of the population.

Based on the census, it was known how many people lived in each of the 400 streets of Kristiania, the capital of Norway. The sorting of the streets was in four categories according to the number of inhabitants. Then, there was the specification of a selection scheme for each category: the adult population enumeration was in 1 out of 20 for the smallest streets. In the second category, the adult population enumeration was in half of the houses in 1 out of 10 of streets. In the third category, the enumeration concerned one-fourth of the streets, and the enumeration was every fifth house; and in the last category of the biggest streets, the adult population enumeration was on half of the streets and in 1 out of 10 houses in them.

In selecting the streets their distribution over the city was considered to ensure the largest possible dispersion and the "representative character" of the enumerated areas.

In the medium-sized towns, the sample was selected using the same principles, though in a slightly simplified manner. In the smallest towns, the total adult population in three or four houses was enumerated.

The number of informants in each of the 18 counties in the rural part of Norway was decided considering census data. To obtain representativeness, municipalities in each country, it was used a classification according to their main industry, either as agricultural, forestry, industrial, seafaring, or fishing municipalities. In addition, the geographical distribution was considered.

The total number of the representative municipalities amounted to 109, which is six in each county on average. The total number of municipalities was 498.

The selection of people in a municipality was done in relation to the population in different parishes, and so all different municipalities were covered. The final step was to instruct enumerators to follow a specific path. In addition, instruction to the enumerators was to visit different houses situated close to each other. That is, they were supposed to visit not only middle-class houses but also well-to-do houses, poor-looking houses, and one-person houses.

Kiaer did not explain in his papers how he calculated estimates. The main reason probably was that the representative sample construction was as a miniature of the population. This made computations of estimates trivial: the sample mean is the estimate of the population mean, and the estimate of the population total could be attained simply by multiplying the sample total by the inverse of sampling fraction.

A basic problem of the representative method was that there was no way of establishing the precision of population estimates. The method lacked a formal theory of inference. It was Bowley (1906, 1926) who made the first steps in this direction. He showed that for large samples, selected at random from the population, estimates had an approximately normal distribution. From this moment on, there were two methods of sample selection:

- Kiaer's representative method, based on purposive selection, in which representativity played an essential role and for which no measure of the accuracy of the estimates could be obtained;

- Bowley's approach, based on simple random sampling, for which an indication of the accuracy of estimates could be computed.

Both methods existed side by side until 1934. In that year the Polish scientist Jerzy Neyman published his famous paper (see Neyman, 1934). Neyman developed a new theory based on the concept of the confidence interval. By using random selection instead of purposive selection, there was no need any more to make prior assumptions about the population. The contribution of Neyman was not only that he proposed the confidence interval as an indicator for the precision of estimates. He also conducted an empirical evaluation of Italian census data and proved that the representative method based on purposive sampling was not able to provide satisfactory estimates of population characteristics. He established the superiority of random sampling (also referred to as *probability sampling*) over purposive sampling. Consequently, use of purposive sampling was rejected as a scientific sampling method.

Gradually probability sampling found its way into official statistics. More and more national statistical institutes introduced probability sampling for official statistics. However, the process was slow. For example, a first test of a real sample survey using random selection was carried out by Statistics Netherlands only in 1941 (see CBS, 1948). Using a simple random sample of size 30,000 from the population of 1.75 million taxpayers, it was shown that estimates were accurate.

The history of opinion polls goes back to the 1820s, in which period American newspapers attempted to determine political preference of voters just before the presidential election. These early polls did not pay much attention to sampling. Therefore, it was difficult to establish accuracy of results. Such opinion polls were often called *straw polls*. This expression goes back to rural America. Farmers would throw a handful of straws into the air to see which way the wind was blowing.

It took until the 1920s before more attention was paid to sampling aspects. Lienhard (2003) describes how George Gallup worked out new ways to measure interest in newspaper articles. Gallup used *quota sampling*. The idea was to investigate a group of people that could be considered representative for the population. Hundreds of interviewers across the country visited people. Interviewers were given quota for different groups of respondents. They had to interview so many middle-class urban women, so many lower-class rural men, etc. In total, approximately 3,000 interviews were conducted out for a survey.

Gallup's approach was in great contrast with that of the *Literary Digest* magazine, which was at that time the leading polling organization. This magazine conducted regular "America Speaks" polls. It based its predictions on returned questionnaire forms that were sent to addresses taken from telephone directories books and automobile registration lists. The sample size for these polls was on the order of two million people. So the sample size was much larger than that of Gallup's polls.

The presidential election of 1936 turned out to be decisive for both methods. This is described by Utts (1999). Gallup correctly predicted Franklin Roosevelt to be the new president, whereas *Literary Digest* predicted that Alf Landon would beat Franklin Roosevelt. The prediction based on the very large sample size turned out to be wrong. The explanation was that the sampling technique of *Literary Digest* did not produce representative samples. In the 1930s, cars and telephones were typically owned by middle- and upper-class people. These people tended to vote Republican, whereas lower-class people were more inclined to vote Democrat. Consequently, Republicans were overrepresented in the *Literary Digest* sample.

As a result of this historic mistake, opinion researchers learned that they should rely on more scientific ways of sample selection. They also learned that the way a sample is selected is more important than the size of the sample.

The classical theory of survey sampling was more or less completed in 1952. Horvitz and Thompson (1952) developed a general theory for constructing unbiased estimates. Whatever the selection probabilities are, as long as they are known and positive, it is always possible to construct a useful estimate. Horvitz and Thompson completed the classical theory, and the random sampling approach was almost unanimously accepted. Most of the classical books about sampling were also published by then (Cochran, 1953; Deming, 1950; Hansen, Hurvitz, and Madow, 1953; Yates, 1949).

## 1.2.2 TRADITIONAL DATA COLLECTION

There were three modes of data collection in the early days of survey research: face-to-face interviewing, mail interviewing, and telephone interviewing. Each mode had its advantages and disadvantages.

*Face-to-face interviewing* was already used for the first censuses. Thus, it is not a surprise it was also used for surveys. Face-to-face interviewing means that interviewers visit the persons selected in the sample. Well-trained interviewers will be successful in persuading reluctant persons to participate in the survey. Therefore, response rates of face-to-face surveys are usually higher than surveys not involving interviewers (for example, mail surveys). Interviewers can also assist respondents in giving the right answers to the questions. This often results in better data. However, the presence of interviewers can also be a drawback. Research suggests that respondents are more inclined to answer sensitive questions properly if there are no interviewers present.

Survey agencies often send a letter announcing the visit of the interviewer. Such a letter can also give additional information about the survey, explain why it is important to participate, and assure that the collected information is treated confidentially. As a result, the respondents are not taken by surprise by the interviewers.

The response rate of a face-to-face survey is usually high and so is quality of the collected data. But a price has to be paid literally: face-to-face interviewing is much more expensive. A team of interviewers has to be trained and paid. They also have to travel, which costs time and money.

*Mail interviewing* is much less expensive than face-to-face interviewing. Paper questionnaires are sent by mail to persons selected in the sample. They are invited to answer the questions and to return the completed questionnaire to the survey agency. A mail survey is not interviewers based. Therefore, it is a cheaper mode of data collection than face-to-face survey. Data collection involve mailing costs (letters, postage, envelopes) both for sending the questionnaire and for delivering the questionnaire back; similar costs have to be considered for each reminder. Therefore, costs for stamps and questionnaire printing could be not completely irrelevant. Another advantage of mail survey is that the absence of interviewers can be experienced as less threatening for potential respondents. Therefore, respondents are more inclined to answer sensitive questions properly.

The absence of interviewers also has a number of disadvantages. There are no interviewers to explain questions or assist respondents in answering them. This may cause respondents to misinterpret questions, which has a negative impact on the quality of the collected data. Furthermore, it is not possible to use show cards. A *show card* is typically used for answering closed questions. Such a card contains the list of all possible answers to a question. Respondents can read through the list at their own pace and select the answer corresponding to their situation or opinion. Mail surveys put high demands on the design of the paper questionnaire. For example, it should be clear to all respondents how to navigate through the questionnaire and how to answer questions.

Since the persuasive power of the interviewers is absent, response rates of mail surveys tend to be low. Of course, reminder letters can be sent, but this is often not very successful to let the response rate become very high. More often survey questionnaire forms end up in the pile of old newspapers.

In summary, the costs of a mail survey are relatively low, but often a price has to be paid in terms of data quality: response rates tend to be low, and also the quality of the collected data is also often not very good. Dillman (2007) believes, however, that good results can be obtained by applying his *Tailored Design Method*. This is a set of guidelines for designing and formatting mail survey questionnaires. They pay attention to all aspects of the survey process that may affect response rates or data quality.

Face-to-face interviewing was preferred in the early days of survey interviewing in the Netherlands. The idea was in the 1940s that poor people had poor writing skills and they were not interested in the topics of the surveys. Therefore, they had a lower probability to complete mail questionnaires. People completing and returning questionnaire forms were assumed to be more interested in the survey topics because their intelligence and socioeconomic position was above average.

A third mode of data collection is *telephone interviewing*. Interviewers are needed to conduct a telephone survey, but not as many as for a face-to-face survey, since they do not have to travel from one respondent to the next. They can remain in the call center of the survey agency and conduct more interviews in the same amount of time. Therefore, interviewer costs are less high. An advantage of telephone interviewing over face-to-face interviewing is that respondents may be more inclined to answer sensitive questions, because the interviewer is not present in the room. A drawback in the first days of telephone surveys may be that telephone coverage in the population was low. Not every respondent could be contacted by telephone.

Telephone interviewing has some limitations. Interviews cannot last too long, and question answer may not be written. Obviously, no show cards can be used; lists can be presented by reading them out loud (by the interviewers).

This implies a possible recency effect in the answers. Another problem may be the lack of a proper sampling frame. Telephone directories may suffer from severe under-coverage because many people do not want their phone number to be listed in the directory. Another new development is that increasingly people replace their landline phone by a mobile phone. This fact increases under-coverage in the telephone directories. For example, according to Cobben and Bethlehem (2005), only between 60% and 70% of the Dutch population can be reached through a telephone dictionary. However it has to be advised that mobile phone numbers are not listed in directories in many countries. Thus, a problem arises.

Example 1.2 is about the first telephone survey in the Netherlands.

🔲  EXAMPLE 1.2 **The first telephone survey in the Netherlands**

The first telephone survey took place in the Netherlands on June 11, 1946. See NIPO (1946) for a detailed description. An interview to a few hundred owners of telephones in Amsterdam asked a few questions about listening to the radio. The call to the people was between 20:00 and 21:30 hours on a Tuesday night. Some results are in Table 1.1.

TABLE 1.1    **The first telephone survey in the Netherlands**

| Are you listening to the radio at this moment? | Percentage |
| --- | --- |
| Was listening | 24 |
| Was not listening | 38 |
| Line busy | 5 |
| No answer | 31 |
| Did not have a radio | 2 |

If people declared they were listening to the radio, the program they were listening to was asked. It turned out that 85% was listening the "Bonte Dinsdagavondtrein," a very famous radio show at that time.

Telephone interviewing has some limitations. Interviews cannot last too long, and no written answer is possible. Obviously, no show cards are used; lists are read loud (by the interviewers).

This implies a possible recency effect in the answers. Another problem may be the lack of a proper sampling frame. Telephone directories may suffer from severe under-coverage because many people do not want their phone number is in the directory. Another new development is that increasingly people replace their land-line phone by a mobile phone. This fact increases under-coverage in the telephone directories. For example, according to Cobben and Bethlehem (2005), only between 60% and 70% of the Dutch population had at that time a telephone dictionary.

A way to avoid the under-coverage problems of telephone directories is to apply *random digit dialing* (RDD) to generate random phone numbers. A computer algorithm computes valid random telephone numbers. Such an algorithm is able to generate both listed and unlisted numbers. Thus, there is complete coverage. An example of an algorithm used in the United Kingdom is to take a number from a directory and replace its last digit by a random digit. RDD also has drawbacks. In some country it is not clear what an unanswered number means. It can mean that the number is not in use. This is a case of over-coverage. No follow-up is needed. It can also mean that someone simply does not answer the phone, which is a case of nonresponse, which has to be followed up. Another draw-back of RDD is that there is no information at all about nonrespondents. This makes correction for nonresponse very difficult (Bethlehem, Cobben, and Schouten (2011), see also Chapter 12).

The choice of the mode of data collection is not any easy one. It is usually a compromise between quality and costs. In large countries (like the United States) or sparsely populated countries (like Sweden), it is almost impossible to collect survey data by means of face-to-face interviewing. It requires so many interviewers that have to do so much traveling that the costs would be very high. Therefore, it is not surprising that telephone interviewing emerged here as a major data collection mode. In a very small and densely populated country, like the Netherlands, face-to-face interviewing is much more attractive. Coverage problems of telephone directories and low response rates also play a role in the choice for face-to-face interviewing. More about data collection issues is in Couper et al. (1998).

### 1.2.3 THE ERA OF COMPUTER-ASSISTED INTERVIEWING

Collecting survey data can be a costly and time-consuming process, particularly if high-quality data are required, the sample is large, and the questionnaire is long and complex. Another problem of traditional data collection is that the completed paper questionnaire forms may contain many errors. Substantial resources must therefore be devoted to cleaning the data. Extensive data editing is required to obtain data of acceptable quality.

Rapid developments in information technology since the 1970s have made it possible to reduce these problems. By introducing microcomputers for data collection, important innovation in surveys took place. A computer program for asking questions and recording the answers replaced the paper questionnaire.

The computer took control of the interviewing process, and it checked answers to the questions. Thus, *computer-assisted interviewing* (CAI) emerged.

CAI comes in different modes of data collection. The first mode of data collection that emerged was *computer-assisted telephone interviewing* (CATI). Couper and Nicholls (1998) describe its development in the United States in the early 1970s. The first nationwide telephone facility for surveys was established in 1966. The idea at that time was not implementation of CAI but simplifying sample management. The initial systems evolved in subsequent years into full featured CATI systems. Particularly in the United States, there was a rapid growth of the use of these systems. CATI systems were little used in Europe until the early 1980s.

Interviewers in a CATI survey operate a computer running interview software. When instructed to do so by the software, they attempt to contact a selected person by telephone. If this is successful and the person is willing to participate in the survey, the interviewer starts the interviewing program. The first question appears on the screen. If correctly answered, the software proceeds to the next question on the route through the questionnaire.

Call management is an important component of the CATI systems. Its main function is to offer the right telephone number at the right moment to the right interviewer. This is particularly important in cases in which the interviewer has made an appointment with a respondent for a specific time and date. Such a call management system also has facilities to deal with special situations like a busy number (try again after a short while) or no answer (try again later). This all helps to increase the response rate. More about the use of CATI in the United States is in Nicholls and Groves (1986).

Small portable computers came on the market in the 1980s. This made it possible for the interviewers to take computers with them to the respondents. This is the computer-assisted form of face-to-face interviewing, called *computer-assisted personal interviewing* (CAPI). After interviewers have obtained cooperation of the respondents, they start the interviewing program. Questions display is one at a time. Only after the entering of the answer, the next question appeared on the screen.

At first, it was not completely clear whether this mode of data collection could use the computer. There were issues like the weight and size of the computer, the readability of the screen, battery capacity, and the size of keys on the keyboard. Experiments showed that CAPI was feasible. It became clear that CAI for data collection has three major advantages:

- It simplifies the work of interviewers. They do not have to pay attention any more to choosing the correct route through the questionnaire. The computer determines the next question to ask. Interviewers can concentrate more on asking questions and helping respondents giving the proper answers.

- It improves the quality of the collected data. Answers checking is by the software during the interview. Correction of the detected errors is automatic. The respondent is there to provide the proper information. This is much more effective than having to do data editing afterward in the survey agency and without the respondent.

- Data entering in the computer is immediate, during the interview. Straightaway checks are undertaken and detected errors corrected. Therefore, the record of a respondent is "clean" after completion of the interview. No more subsequent data entry and/or data editing is required. Compared with the old days of traditional data collection with paper forms, this considerably reduces time needed to process the survey data. Therefore, timeliness of the survey results is improved.

Or more information about CAPI in general, see Couper et al. (1998).

The computer-assisted mode of mail interviewing also emerged. It was called *computer-assisted self-interviewing* (CASI), or sometimes also *computer-assisted self-administered questionnaires* (CASAQ). The electronic questionnaire program is sent to the respondents. They run the software, which asks the questions and stores the answers. After the interview completion, the data are send back to the survey agency. Early CASI applications used diskettes or a telephone and modem to transmit the questionnaire and the answers to the question. Later it became common practice to use the Internet as a transport medium.

A CASI survey is only feasible if all respondents have a computer on which they can run the interview program. Since the use of computers was more widespread among companies than among households in the early days of CASI, the first CASI applications were business surveys. An example is the production of fire statistics in the Netherlands in the 1980s. Since all brigades had a microcomputer at that time, data for these statistics CASI were a mode of data collection. Diskettes were sent to the fire brigades. They ran the questionnaire on their MS-DOS computers. The answers were stored on the diskette. After having completed the questionnaire, the diskette was returned to Statistics Netherlands.

An early application in social surveys was the *Telepanel*, set up by Saris (1998). The Telepanel started in 1986. It was a panel of 2,000 households. They agreed on regularly completing questionnaires with the computer equipment provided to them by the survey organization. A home computer was installed in each household. It was connected to the telephone with a modem. It was connected to the television set in the household also. Then it was possible to use it as a monitor. After inserting the diskette into the home computer, it automatically established a connection with the survey agency to exchange information (downloading a new questionnaire or uploading answers of the current questionnaires). Panel members completed a questionnaire each weekend. The Telepanel was in essence very similar to the web panels, which are frequently used nowadays. The only difference was the Internet did not exist yet.

## 1.2.4 THE CONQUEST OF THE WEB

The development of the Internet started in the early 1970s. The first step was to create networks of computers. The U.S. Department of Defense decided to connect computers of research institutes. Computers were expensive. A network made it possible for these institutes to share each other's computer resources. The name of this first network was ARPANET.

ARPANET became a public network in 1972. Software to send messages over the network was developed. Thus, e-mail was born. Ray Tomlinson of ARPANET was sending the first e-mail in 1971.

The Internet was fairly chaotic in the first decade of its existence. There were many competing techniques and protocols. In 1982, the TCP/IP set of protocols was adopted as the standard for communication of connected networks. This can be seen as the real start of the Internet.

Tim Berners-Lee and scientists at CERN, the European Organization for Nuclear Research in Geneva, were interested in making it easier to retrieve research documentation over the Internet. This led in 1989 to the *hypertext* concept, a text containing references (hyperlinks) to other texts the reader can immediately access. To be able to view these text pages and navigate to other pages through the hyperlinks, Berners-Lee developed a computer software. He called this program a *browser*. The name of the first browser was *World Wide Web*. Now this name denotes the whole set of linked hypertext documents on the Internet.

In 1993, Marc Andreessen and his team at the National Center for Supercomputing Applications (NCSA) (Illinois, USA) developed the browser *Mosaic X*. It was easy to install and use. This browser had increased graphic capabilities. It already contained many features that are common in current browsers. It became a popular browser, which helped to spread the use of the World Wide Web across the world.

The rapid development of the Internet led to new modes of data collection. Already in the 1980s, prior to the widespread introduction of the World Wide Web, e-mail was explored as a new mode of survey data collection. Kiesler and Sproul (1986) describe an early experiment conducted in 1983. They compared an e-mail survey with a traditional mail survey. They showed that the costs of an e-mail survey were much less than those of a mail survey. The response rate of the e-mail survey was 67%, and this was somewhat smaller than the response rate of the mail survey (75%). The turnaround time of the e-mail survey was much shorter. There were less socially desirable answers and less incomplete answers. Kiesler and Sproul (1986) noted that limited Internet coverage restricted wide-scale use of e-mail surveys. In their view, this type of data collection was only useful for communities and organizations with access to and familiarity with computers. These were relatively well-educated, urban, white-collar, and technologically sophisticated people.

Schaefer and Dillman (1998) also compared e-mail surveys with mail surveys. They applied knowledge about mail surveys to e-mail surveys and developed an e-mail survey methodology. They also proposed mixed-mode surveys for populations with limited Internet coverage. They pointed out some advantages of e-mail surveys. In the first place, e-mail surveys could be conducted very fast, even faster than telephone surveys. This was particularly the case for large surveys, where the number of available telephones and interviewers may limit the number of cases that can be completed each day. In the second place, e-mail surveys were inexpensive, because there were no mailing, printing, and interviewer's costs.

The experiment of Schaefer and Dillman (1998) showed that response rates of e-mail and mail surveys were comparable, but the completed questionnaires of the e-mail survey were received much quicker. The answers to open questions were, on average, longer for e-mail surveys. This did not come as a surprise because of the relative ease of typing an answer on a computer compared to writing an answer on paper. There was lower item nonresponse for the e-mail survey. A possible explanation was that moving to a different question in an e-mail survey is much more difficult than moving to a different question on a paper form.

Couper, Blair, and Triplett (1999) found lower response rates for e-mail surveys in an experiment with a survey among employees of statistical agencies in the United States. They pointed out that nonresponse can partly be explained by delivery problems of the e-mails and not by refusal to participate in the survey. For example, if people do not check their e-mail or if the e-mail with the questionnaire does not pass a spam filter, people will not be aware of the invitation to participate in a survey.

Most e-mail surveys could not be seen as a form of CAI. It was merely the electronic analogue of a paper form. There was no automatic routing and no error checking. See Figure 1.1 for a simple example of an e-mail survey questionnaire. It is sent to the respondents. They are asked to reply to the original message. Then they answer the questions in the questionnaire in the reply message. For closed questions they do that by typing an X between the brackets of the option of their choice. The answer to an open question is typed between the corresponding brackets. After completion, they send the e-mail message to the survey agency.

```
1. What is your age?
   [ ]

2. Are you male or female?
   [ ] Male
   [ ] Female

3. What is your marital status?
   [ ] Married
   [ ] Not married

4. Do you have a job?
   [ ] Yes
   [ ] No

5. What kind of job do you have?
   [ write your job]

6. What is your yearly income?
   [ ] Less than 20,000
   [ ] Between 20,000 and 40,000
   [ ] More than 40,000
```

FIGURE 1.1 **Example of an e-mail survey questionnaire**

Use of e-mail imposes substantial restrictions on the layout. Example 1.3 describes a first way adopted to approach businesses for a web survey. Due to e-mail software of the respondent and the settings of the software, the questionnaire may look different to different respondents. For example, to avoid problems caused by line wrapping, Schaefer and Dillman (1998) advise a line length of at most 70 characters. Schaefer and Dillman (1998) also noted another potential problem of e-mail surveys: the lack of anonymity of e-mail. If respondents reply to the e-mail with the questionnaire, it is difficult to remove all identifying information. Some companies have the possibility to monitor the e-mails of their employees. If this is the case, it may become difficult to obtain high response rates and true answers to the questions asked.

Personalization may help to increase response rates in mail surveys. Therefore, this principle should also be applied to e-mail surveys. An e-mail to a long list of addresses does not help to create the impression of personal treatment. It is probably better to send a separate e-mail to each selected person individually.

### EXAMPLE 1.3 The first e-mail survey at Statistics Netherlands

The first test with an e-mail survey at Statistics Netherlands was carried out in 1998. At the time, Internet browsers and HTML were not sufficiently developed and used to make a web survey feasible.

Objective of the test was to explore to what extent e-mail could be used to collect data for the survey on short-term indicators. This was a noncompulsory panel survey, where companies answered a small number of questions about production expectations, order-books, and stocks.

The traditionally mode of data collection for this survey was a mail survey.

The test was conducted in one of the waves of the survey. 1,600 companies were asked to participate in the test. If they did, they had to provide their e-mail address. About 190 companies agreed to participate. These were mainly larger companies with a well-developed computer infrastructure.

A simple text form was sent to these companies by means of e-mail. After activating the reply option, respondents could fill in answers in the text. It was a software-independent and platform-independent solution, but rather primitive from a respondent's point of view.

The test was a success. The response rate among the participating companies was almost 90%. No technical problems were encountered. Overall, respondents were positive. However, they considered the text-based questionnaire old-fashioned, and not very user friendly.

More details about this first test with an e-mail survey at Statistics Netherlands can be found in Roos, Jaspers, and Snijkers (1999).

It should be noted that e-mail can also be used in a different way to send a questionnaire to a respondent. An electronic questionnaire can be offered as an executable file that is attached to the e-mail. The respondents download this interview program on their computers and run it. The advantage of this approach is that such a computer program can have a better graphical user interface. Such a program can also include routing instructions and checks. This way of data collection is sometimes called CASI. Example 1.4 describe an example of a CASI approach.

■ EXAMPLE 1.4 **The production statistics pilot at Statistics Netherlands**

In October 2004, Statistics Netherlands started a pilot to find out whether a CASI approach could be used to collect data for yearly production statistics.

One of the approaches tested is denoted by Electronic Data Reporting (EDR). It was a system for responding companies to manage interviewing programs (generated by the Blaise system) on their own computers. The EDR software was sent to respondents on CD-ROM, or respondents could download the software from the Internet.

After the installation of the software, new survey interviews could be sent to respondents by e-mail. These electronic questionnaires were automatically imported in the EDR environment. A simple click would start the interview. After offline completion of the interview, the entered data were automatically encrypted and sent to Statistics Netherlands.

The pilot made clear that downloading the software was feasible. It should be preferred over sending a CD-ROM because it was simpler to manage and less expensive, too. Some companies experienced problems with downloading and installing the software, because security settings of their computer systems and networks prevented them of doing so. User-friendliness and ease of navigation turned out to be important issues for respondents.

For more information about this pilot, see Snijkers, Tonglet, and Onat (2004, 2005).

This form of CASI also has disadvantages. It requires respondents to have computer skills. They should be able to download and run the interviewing program. Couper, Blair, and Triplett (1999) also note that problems may be caused by that fact that different users may have different operating systems on their computers or different versions of the same operating system. This may require different versions of the interviewing program, and it must be known in advance which operating system a respondent has. Moreover, the size of an executable file may be substantial, which may complicate sending it by e-mail.

E-mail surveys had the advantages of speed and low costs. Compared with CAI they had the disadvantages of a poor user interface and lack of adequate editing and navigation facilities. An e-mail questionnaire was just a paper question-naire in an e-mail. The Internet became more interesting for survey data collection after HTML 2.0 was introduced in 1995. HTML stands for Hypertext Markup Language. It is the markup language for web pages. The first version of HTML was developed by Tim Berners-Lee in 1991. Version 2 of HTML included support for forms. This made it possible to transfer data from a user to the web server. Web pages could contain questions, and the answers could be collected by the server. Example 1.5 shows some applicative aspects of the HTML questions.

 **EXAMPLE 1.5  Designing questions in HTML 2.0**

Version 2.0 of HTML made it possible to implement questions on a web page. The `<input>` tag can be used to define different types of questions. With `type=radio` this tag becomes a *radio button*. A *closed question* is defined by introducing a radio button of each possible answer. See Figure 1.2 for an example. Not more than one radio button can be selected. This corresponds to a closed question for which only one answer must be selected.

FIGURE 1.2  **A closed question in HTML**

Sometimes respondents must be offered the possibility to select more than one answer, like in Figure 1.3. Respondents are asked for their means of transport to work. Some people may use several transport means.

FIGURE 1.3  **A check-all-that-apply question in HTML**

For example, a person may first take a bicycle to the railway station and then continues by train. Such a closed question is sometimes also called a *check-all-that-apply* question. It can be implemented in HTML by means of a series of *checkboxes*. A checkbox is obtained by stetting the type of the <input> tag to checkbox.

Figure 1.4 shows the implementation of an open question. Any text can be entered in the input field. A limit may be set to the length of the text. An open question is defined with type=text for the <input> tag.

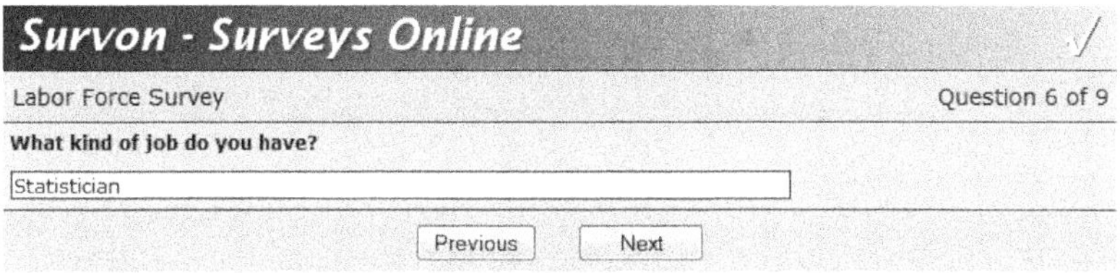

FIGURE 1.4  **An open question in HTML**

If an input field is preferred that allows for more lines of text to be answered, the <textarea> tag can be used for this.

There are no specific types of the <input> tag for other types of questions. However, most of these question types can be implemented with the input field of an open question. For example, Figure 1.5 shows a numeric question. The question is basically an open question, but extra checks on the answer only allow numbers to be entered within certain bounds.

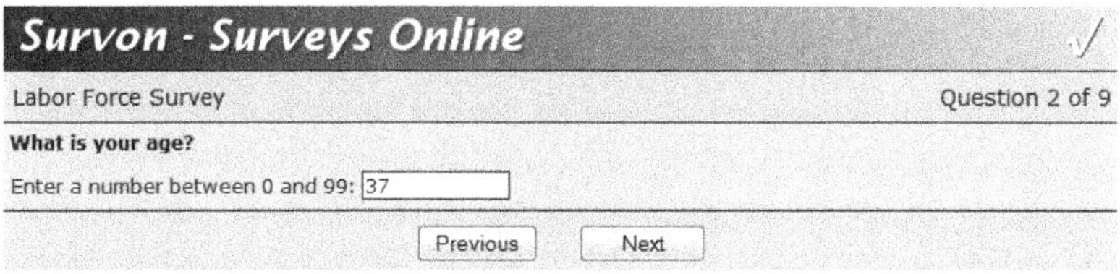

FIGURE 1.5  **A numeric question in HTML**

Date question can be specified as a set of three input fields: one for the day, one for the month, and one for the year.

In the first years of the World Wide Web, use of web surveys was limited by the low penetration of the Internet. Internet penetration was higher among establishments than among households. Therefore, it is not surprising that first experiments tested the use of web business surveys. Clayton and Werking (1998) describe a pilot carried out in 1996 for Current Employment Statistics (CES) program of the U.S. Bureau of Labor Statistics. They expected the web to offer a low-cost survey environment. Because it was a form of true online data collection, an immediate response to the answers of the respondents was possible. This could improve data quality. They also saw the great flexibility of web survey questionnaires. They could be offered in a form layout or in a question-by-question approach. The drawback was the limited number of respondents having access to the Internet. Only 11% of CES respondents had access to Internet and a compatible browser.

Roos and Wings (2000) conducted a test with Internet data collection at Statistics Netherlands for the construction industry. Respondents could choose between three modes:

- Completing a form offline. The form was sent as an HTML file that was attached to an e-mail. The form is downloaded, completed offline, and returned by e-mail.

- Completing a form online. The Internet address of an online web form was sent by e-mail. The form was completed online.

- Completing an e-mail form. An e-mail is sent containing the questionnaire in plain text. Respondents clicked the reply button, answered the questions, and sent the e-mail back.

A sample of 1,500 companies was invited to participate in the experiment. 188 companies were willing and able to participate. Of those, 149 could surf the Internet, and 39 only had e-mail. Questionnaire completion times of all three modes were similar to that of a paper form. Respondents preferred the form-based layout over the question-by-question layout. The conclusion of the experiment was that web surveys worked well.

General population web surveys were rare in the first period of existence of the Internet. This was due to the low Internet penetration among households. This prevented conducting representative surveys. However, there were polls on the Internet. Recruitment of respondents was based on self-selection and not on probability sampling. Users could even create their own polls on websites like *Survey Central, Open Debate,* and *Internet Voice* (see O'Connell, 1998).

Also in 1998, the *Survey2000* project was carried out. This was a large self-selection web survey on the website of the National Geographic Society. This was a survey on mobility, community, and cultural identity. In a period of two months, over 80,000 respondents completed the questionnaire. See Witte, Amoroso, and Howard (2000) for more details about this project.

It seems to be typical for this type of self-selection web surveys that they make it possible to collect data about a large number of respondents in a relatively short time. Other examples are given by Bethlehem and Stoop (2007). The survey *21minuten.nl* has been conducted a number of times in the Netherlands. This survey supposed to supply answers to questions about important problems in Dutch society. Within a period of six weeks in 2006, about 170,000 people completed the online questionnaires. A similar survey was conducted in Germany. It is called *Perspektive Deutschland.* More than 600,000 participated in this survey in 2005/2006.

It should be noted that these large sample sizes are no guarantee for proper statistical inference. Due to under-coverage (not everyone has access to the

Internet) and self-selection (no proper random sampling), estimates can be biased. This bias is independent of the sample size.

Internet penetration is still low in many countries, making it almost impossible to conduct a general population web survey. Since data collection costs can be reduced if the Internet is used, other approaches are sought. One such approach is *mixed-mode data collection*. A web survey is combined with one or more other modes of data collection, like a mail survey, a telephone survey, or a face-to-face survey. Researchers first attempt to collect as much data as possible with the cheapest mode of data collection (web). Then, the nonrespondents are re-approached in a different (next cheapest) mode. Example 1.6 describes a survey run using a mixed-mode approach.

---

EXAMPLE 1.6 **Experiment with a mixed-mode surveys**

Beukenhorst and Wetzels (2009) describe a mixed-mode experiment conducted by Statistics Netherlands. They used the Dutch Safety Monitor for this experiment. This survey asks questions about feelings of security, quality of life, and level of crime experienced. The sample for this survey was selected from the Dutch population register. All sampled persons received a letter in which they were asked to complete the survey questionnaire on the Internet. The letter also included a postcard that could be used to request a paper questionnaire. Two reminders were sent to those that did not respond by web or mail. If still no response was obtained, nonrespondents were approached by means of CATI, if a listed telephone number was available. If not, these nonrespondents were approached by CAPI.

To be able to compare this four-mode survey with a traditional survey, also a two-mode survey was conducted for an independent sample. Sampled persons were approached by CATI if their telephone number was listed in the directory, and otherwise they were approached by CAPI.

The response rate for four-mode survey turned out to be 59.7%. The response rate for the two-mode survey was higher. So, introducing more modes did not increase the overall response rate. However, more than half of the response (58%) in the four-mode survey was obtained with a self-administered mode of data collection (web or paper). Therefore, the costs of the survey were much lower. Interviewers were deployed in only 42% of the cases. For more detail, see Beukenhorst and Wetzels (2009) or Bethlehem, Cobben, and Schouten (2011).

---

A special case of mixed-mode data collection is related to the increasing diffusion of mobile phones and smartphones. When an invitation e-mail is sent, the questionnaire might be received either on a computer or a mobile phone or a smartphone. The interviewee could complete the web questionnaire using either the mobile device or the computer. Thus, it is better to talk about mobile web surveys rather than web survey. A recommendation is to run web surveys that are fully adapt for smartphones. Therefore, in presenting methods for web surveys, comments about the adaption for smartphone surveys will be discussed all along the chapters of this handbook. Some penetration data allow for understanding how the situation differs across the countries. The coverage of telephone directories, of Internet, and of mobile cells provides the feeling of the need to adopt a mixed-mode approach. A World Bank study reports that, in the 2018, Euro area fixed telephone subscription for 100 people is 44.4, mobile 122.6 and Internet 83.8. Table 1.2 shows the same indicators by country. Only some countries are shown in the table since the objective is just to evidence that there is a relevant difference across countries.

TABLE 1.2    **Penetration of fixed and mobile phone and of Internet (year 2018)**

| Country | Fixed telephone subscription (% of inhabitants) | Mobile cellular subscription (% of inhabitants) | % of individuals using the Internet |
|---|---|---|---|
| Austria | 42 | 125 | 88.0 |
| Denmark | 19 | 125 | 97.6 |
| Finland | 6 | 132 | 88.9 |
| France | 59 | 108 | 82.0 |
| Germany | 52 | 129 | 89.7 |
| Greece | 47 | 116 | 72.9 |
| Italy | 34 | 137 | 74.4 |
| The Netherlands | 35 | 124 | 94.7 |
| Norway | 11 | 107 | 96.5 |
| Portugal | 50 | 115 | 74.7 |
| Romania | 19 | 116 | 70.7 |
| Slovenia | 33 | 118 | 79.7 |
| Spain | 40 | 116 | 86.1 |
| Sweden | 24 | 125 | 92.1 |
| Switzerland | 39 | 130 | 90.0 |

Source: Data from International Telecommunication Union. World Telecommunication/ICT.

Note that Internet users are individuals who have used the Internet (from any location) in the last 12 months. Internet can be used via a computer, mobile phone, personal digital assistant, games machine, digital TV, etc. Fixed telephone subscriptions refer to the sum of active number of analogue fixed telephone lines, voice-over-IP (VoIP) subscriptions, fixed wireless local loop (WLL) subscriptions, ISDN voice-channel equivalents, and fixed public payphones. Mobile cellular telephone subscriptions are subscriptions to a public mobile telephone service that provide access to the PSTN using cellular technology.

## 1.2.5  WEB SURVEYS AND OTHER SOURCES

Current digital environment and technology trends are providing a huge amount of data about most phenomena. These data are available on the web. Often they are free of charge, if not protected for privacy.

Examples of data available in digital format are credit card transactions, tax data, social chatting, telephone use (calls details: time, location, length of the call, etc.), social security payments, GPS, videos. Using this type of data for statistical purposes is appealing and challenging. The term big data is currently used to characterize data with high volume, velocity, and variety. There is a debate on the definition and on the use of big data for statistical purposes.

Roughly speaking, big data are based on the automatic collection on everything that people do; they are not subject to statistical classification criteria and to statistical treatment for representativity. Also administrative data, i.e., information that are collected for registering units (people, businesses, sales, and so on) into an activity process, might be included into big data.

Practitioners and researchers are now wondering if big data could substitute web surveys to provide information for social and economic decision making. For a discussion about that, see Couper (2013).

It should be emphasized that conclusion is that big data and web surveys are complementary data sources, not competing data sources. The availability of big data to support research provides a new way to approach old questions as well as an ability to address some new questions that in the past were not considered.

Web surveys may be run as stand-alone surveys. However, source integration is a major trend for the future of the next 10 years of web surveys.

It is a new area of research to achieve the three following goals: (1) Minimize the cost associated with surveys. (2) Maximize the information, i.e., the findings based on big data generate more questions, and some of those questions could be best addressed by web surveys or other traditional survey methods. Moreover, information from one source could be useful for improving data to be estimated from a survey. (3) Minimize the respondent burden. Integration alleviates the burden of duplicating data gathering efforts and enables the extraction of information that would otherwise be impossible.

Therefore, it is necessary to work in the direction of using big data and integrating them with the survey results. It is important to face experimental applications having in mind the characteristics, nature, and the limitations of big data as statistical sources and the methodological soundness of the survey results.

At the time being, market research and private/public businesses have great interest in trying to use big data to investigate markets and individual behavior. The use of this data as exploratory source is the most plausible application, whereas using this data for statistical purposes and integration with web survey requires still a lot of effort around definitions, classifications, and estimation methodological problems.

Official statistics producers are investigating how to use other big data sources and how to produce estimates in a multisource framework. Some experiments have been already undertaken, with contrasting results. Most successful applications consist in the integration of web survey data and administrative data (i.e., administrative data could be considered a type of big data according to many authors). Administrative data have been used:

- To generate a survey frame or to supplement/update an existing frame. When surveys are run on the web, administrative data integration could help in applying the adaptive survey design (see Chapter 8) to improve the data collection process. An ultimate task could be the replacement of data collection (e.g., use of taxation data for small businesses instead of seeking survey data for them). In this case, however, data replacement would be about some basic data; surveys will be anyway useful for collecting data about specific topics and behaviors. For example, think of all the surveys with a focus on consumption. If big data assets are providing insights on consumption via passive observation, primary research via surveys will not have to collect this type of information, and it is finally possible to deliver on the vision of shorter surveys instead of simply providing complementary data to the desired information. Surveys can be short and focused on those variables that they are ideally suited for, resulting in better data quality.
- In editing and imputation.
- In estimation (e.g., as auxiliary information in calibration estimation, benchmarking, or calendarization).
- In comparing survey estimates with estimates from a related administrative program as well as other forms of survey evaluation have been experienced.

When using a multisource approach in web surveys, several aspects should be considered.

First of all is the heterogeneous nature of the sources with respect to the following basic characteristics: the aggregation level, the unit, the variables, the coverage, the time, the population, and the data type:

The aggregation level, i.e., some data sources consist of only microdata, some other data sources consist of a mix of microdata and aggregated data, whereas in some other cases data sources consist of only aggregated data. In some case, aggregated data are available besides microdata. There is still overlap between the sources, from which there arises the need to reconcile the statistics at some aggregated level. Of particular interest is when the aggregated data are estimates themselves. Otherwise, the conciliation can be achieved by means of calibration, a standard approach in survey sampling.

As regards the units, it has to be considered that sometimes there are no overlapping units in the data sources or only some units are overlapping. Also, as regards the variables, no overlapping variables in the data sources could occur, or only variables in the data sources could overlap.

Under-coverage versus there is no under-coverage has to be considered. The data sources are cross-sectional versus other data sources are longitudinal; thus, the researcher should take care of what type of data he is integrating. The set of population units from a population register could be known, or the population list is not known; this affects the possibility of generating a probability-based sample. In some cases, a data source contains a complete enumeration of its target population, or a data source is selected by means of probability sampling from its target population, or a data source is selected by non-probability sampling from its population. The database may be further split into two subcases depending on whether one of the data sources consists of sample data (and where the sampling aspects play an important role in the estimation process) or not. In the former case, specific methods should be used in the estimation process, for instance, taking the sampling weights into account and considering that sample data may include specific information that is not reported in register.

Another aspect is the configuration of the sources to be integrated. There are a few basic ways, most commonly encountered. However, in practice, a given situation may well involve several basic configurations at the same time.

The first and most basic configuration of the integration process of different sources is multiple cross-sectional data that together provide a complete data set with full coverage of the target population. Provided they are in an ideal error-free state, the different data sets, or data sources, are complementary to each other and can be simply "added" to each other in order to produce output statistics.

A second type of configuration is when there exists *overlap* between the different data sources. The overlap can concern the units, the measured variables, or both.

A third situation is when the combined data *entail under-coverage of the* target population in addition, even when the data are in an ideal error-free state.

A further configuration is when microdata and aggregated data are available. There is overlap between the sources, but there is the need to reconcile the statistics at some aggregated level. The conciliation can be achieved by means of calibration, which is a standard approach in survey sampling. Of particular interest is when the aggregated data are estimates themselves.

Finally, it is possible that multisource approach refers to longitudinal data. More questions arise; the most important issue is that of reconciling time series of different frequencies and qualities. For example, one source has monthly data and the other source has quarterly data.

Integration may occur between different types of sources: surveys (mainly web surveys), administrative data, other passive collected data, social network, and other unstructured data.

Integration of different configurations as well as of different types of data sources implies different methodological problems. For instance, integrating survey and administrative data through unit record linkage requires improving coherence across data collections, using standard classifications and questions, rationalizing content between surveys, and processes for combining separate sample surveys into one survey vehicle (Bycroft, 2010).

Example 1.7 discusses an integration of web scraped information, administrative data, and surveys, whereas Example 1.8 shows an application of integration between survey data and social network unstructured information.

As a result of the multisource integration, statistical output is based on complex combinations of sources. Its quality depends on the quality of the primary sources and the ways they are combined. Some studies are investigating the appropriateness of the current set of quality measures for multiple source statistics; they explain the need for improvement and outline directions for further work.

### EXAMPLE 1.7 Web scraping, administrative data, and surveys

Istat since 2015 has been experimenting web scraping, text mining, and machine learning techniques in order to obtain a subset of the estimates currently produced by the sampling survey on "Survey on ICT Usage and e-Commerce in Enterprises" yearly carried out on the web. Studies from Barcaroli et al., (2015, 2016), and Righi, Barcaroli, and Golini (2017) have focused in implementing the experiment and in evaluating data quality.

Trying to make the optimal use of all available information from the administrative sources to web scraping information, web survey estimates produced could tentatively be improved. The aim of the experiment is also to evaluate the possibility to use the sample of surveyed data as a training set in order to fit models to be applied to website information.

Recent and in progress steps are a further improvement to the performance of the models by adding explicative variables consisting not only of single terms, but the joint consideration of sequences of terms relevant for each characteristic object of interest. When a certain degree of quality of the resulting predictive models will be guaranteed, their application to the whole population of enterprises owning a website will be performed. A crucial task will be also the retrieval of the URLs related to the websites for the whole population of enterprises. Finally, once having predicted the values of the target variables for all reachable units in the population, the quality of estimates obtained will be analyzed and compared with the current sampling estimates obtained by the survey. In a simulation study, Righi, Barcaroli, and Golini (2017) found that the use of auxiliary variable coming from the Internet DB source highly correlated with the target variable does not guarantee enhancement of the quality of the estimates if selectivity affects the source. Bias may occur due to absence of some subgroups. Thus an analysis of the DB variable and the study of the relationship between populations covered or not by the DB source is a fundamental step to know how to use and which framework implement to assure high-quality output.

In conclusion, the approach that uses web scraping and administrative data together with the web survey looks to be promising; nevertheless quality results of the estimations are satisfactory only in some cases. The use of big data has to be carefully evaluated, especially if selectivity affects the source.

Example 1.8 focuses on an experiment of integration of social media data and surveys. Even if the study lacks statistical representativeness and indicators, it presents an interesting approach that should be deeply investigated and statistically formalized.

  EXAMPLE 1.8 **Social media and surveys**

Wells and Thorson (2015) introduce a novel method that combines a "big data" measurement of the content of individuals' Facebook (FB) news feeds with traditional survey measures to explore the antecedents and effects of exposure to news and politics content on the site. This hybrid approach is used to untangle distinct channels of public affairs content within respondents' FB news feeds.

The authors explore why respondents vary in the extent to which they encounter public affairs content on the website. Moreover, they examine whether the amount and type of public affairs content flows in one's FB are associated with political knowledge and participation above and beyond self-report measures of news media use.

To combine a survey with measurements of respondents' actual FB experiences, they created a FB application ("app") and embedded it within an online survey experience.

Respondents, undergraduates at a large Midwestern public university, visited a web page and gave two sets of permissions: they first consented to be participants in a research study—a form required by the institutional review board—and then they separately approved the app through their FB profile. Once they approved the app, they were returned to the survey to complete the questionnaire. While respondents completed the questionnaire, the app recorded specific elements of their FB experience (with respondents' permission), such as how many friends they had, what pages they followed, and what content appeared in their news feeds during the previous week. When respondents had completed the survey, the app had finished its work and automatically removed itself from respondents' profiles. This research was approved by a standard university institutional research board and was designed to comply with FB's Platform Policies and Statement of Rights and Responsibilities, each of which placed restrictions on the use and presentation of the data.

The resulting database offers an original combination of respondent's self-reported attitudes and media behaviors (including FB experience) with measure of part of their FB experience.

From the statistical point of view, the study has limitations (Beręsewicz et al., 2018; Biffignandi and Signorelli, 2016). The empirical study is run on a small sample of college volunteers. Thus, they have no claim of

representativeness. In addition they have considered only a single information platform (FB). Other limitations suggest to consider the results just as a
first experimental research. However, the approach proposed is in line with
interesting methodological innovations toward the combination of social
media trace with conventional methods. It opens the perspective to better
understand big data and then try to relate big data descriptive information
to socioeconomic theoretical hypotheses.

Obviously, it is underlined that the statistical perspective of representativeness of the results should be considered in future studies. No probability-based sample ad coverage limitations (partial coverage and
possibility of duplications) mine to the generalization of the results. New
methodological solutions need to be adopted for representativeness of these
interesting preliminary results.

### 1.2.6 HISTORIC SUMMARY

The history above shows that technology changes have impacted survey taking and
methods:

- Paper questionnaires were exclusively used for decades until the 1970s and
  1980s for both self-completion and by interviewers. Processing the data
  was expensive and focused on eliminating survey-taking mistakes.
- Computer questionnaires at first were used solely for interviewing, while
  paper questionnaires were still used for self-completion.
- The advent of the Internet meant that self-completion could now be computer based, but this was limited at first to browsers on PC.
- Computing advances in hardware, software, and connectivity enabled and
  forced changes in survey taking, processing, and methods.

### 1.2.7 PRESENT-DAY CHALLENGES AND OPPORTUNITIES

In the past 15 years, rapid technical and social changes have introduced a number
of challenges and opportunities. The following is a high-level list of challenges:

- The respondent is much more in charge of the survey including whether and
  how he/she will participate.
- There is such a vast proliferation of computing devices and platforms that survey takers cannot design and test for each possible platform.
- Modern-day surveys must be accessible to all self-respondents, including the
  blind, visually impaired, and the motor impaired.

- Few survey practitioners have all the skills needed to effectively design surveys for all platforms and to make them accessible at the same time.

  Pierzchala (2016) listed a number of technical challenges that face survey practitioners. This list was developed to communicate the magnitude of the challenges. The term *multis* refers to the multiple ways that surveys may have to adapt for a particular study:

- **Multicultural surveys**: There are differences in respondent understanding, values, and scale spacing due to various cultural norms. These can lead to different question formulation or response patterns.

- **Multi-device surveys**: There are differences in questionnaire appearance and function on desktops, laptops, tablets, and smartphones.

- **Multilingual surveys**: There are translations, system texts, alphabetic versus Asian scripts, left-to-right versus right-to-left scripts, and switching languages in the middle of the survey.

- **Multimode surveys**: There are interviewer- and self-administered surveys such as CATI and CAPI for interviewers and browser and paper self-completion modes (Pierzchala, 2006).

- **Multinational surveys**: There are differences in currency, flags and other images, names of institutions, links, differences in social programs, and data formats such as date display.

- **Multi-operable surveys**: These are differences in how the user interacts with the software and device including touch and gestures versus keyboards with function keys. Whether there is a physical keyboard or a virtual keyboard impacts screen space for question display.

- **Multi-platform surveys**: These are differences in computer operating systems, whether the user is connected or disconnected to/from the server, and settings such as for pop-up blockers.

- **Multi-structural surveys**: There can be differences in question structures due to visual versus aural presentation, memory demands on the respondent, and linear versus nonlinear cognitive processing.

- **Multi-version surveys**: In economic surveys, questionnaires can vary between industries. For example, an agricultural survey asks about different crops in different parts of the country, and different crops can have different questions.

These *multis* lead to changes in question wording, text-presentation standards, interviewer or respondent instructions, location of page breaks, number of questions on a page, question format, allowed responses, whether choices for *don't know* (DK) or *refusal* (R) are explicitly presented or are implied, and whether the user can advance without some kind of answer (even if *DK* or *RF*) or can just proceed at will to the next question or page.

There can be additional challenges. Governmental and scientific surveys can be long and complex. Surveys must be accessible and usable to the disabled.

Additionally, there are ever-tightening constraints including not enough time, not enough people or money, unclear and late and inconsistent specifications, last-minute changes, screens that are too small, and computers that are too slow.

## 1.2.8 CONCLUSIONS FROM MODERN-DAY CHALLENGES

The description of modern-day survey challenges leads to some conclusions:

- Modern-day surveys can be very hard.
- No single person has all the answers.
- New survey-producing methods are necessary to address all the challenges within ever-tightening constraints.
- Small screen sizes often lead to adaptations of survey instruments such as using fewer points in a scale question.
- With the proliferation of devices, it becomes harder to rely on *unimode* designs where all questions appear the same in all modes and devices (Dillman, Smyth, and Christian, 2014). Instead, the institute may strive for *cognitive equivalence* across all manifestations (de Leeuw, 2005).

## 1.2.9 THRIVING IN THE MODERN-DAY SURVEY WORLD

Updated survey design methods may give ways to handle and even thrive in the modern-day survey world. The idea is to use extremely powerful computer-based specification to replace document specification and manual programming. This idea is described in the following:

- Use a capable computer-based specification system to define the questionnaire. A drag-and-drop specification may be adequate for simpler surveys, but when you get to surveys that must handle more of the *multis* mentioned above, or when you get to thousands of questions, drag-and-drop becomes too onerous.
- Specification and survey methods research should use question structures (see below).
- The institute should define its question-presentation standards for each structure across all the *multis*. This requires some up-front work and decisions.
- When the specification is entered, the computer should generate the necessary source code and related configuration files for all *multis*. All these computer-generated outputs should conform to the institute's standards.
- Use a survey-taking system that has evolved to cope with the modern-day world.

# 1.3 Application

### 1.3.1 BLAISE

The historic developments with respect to surveys as described in the previous section took also place in the Netherlands. Particularly the rapid developments in computer technology have had a major impact on the way Statistics Netherlands collected its data. Efforts to improve the collection and processing of survey data in terms of costs, timeliness, and quality have led to a powerful software system called Blaise. This system emerged in the 1980s, and it has evolved over time so that it is now also able to conduct web surveys and mixed-mode surveys. The section gives an overview of the developments at Statistics Netherlands leading to Internet version of Blaise.

The advance of computer technology since the late 1940s led to many improvements at Statistics Netherlands for conducting its surveys. For example, from 1947 Statistics Netherlands started using probability samples to replace its complete enumerations for surveys on income statistics and agriculture. The implementation of sophisticated sampling techniques such as stratification and systematic sampling is much easier and less labor intensive on a computer than manual methods.

Collecting and processing statistical data was a time-consuming and expensive process. Data editing was an important component of this work. The aim of these data editing activities was to detect and correct errors in the individual records, questionnaires, or forms. This should improve the quality of the results of surveys. Since statistical offices attached much importance to this aspect of the survey process, a large part of human and computer resources were spent on it.

To obtain more insight into the effectiveness of data editing, Statistics Netherlands carried out a Data Editing Research Project in 1984. Bethlehem (1987) describes how survey data were processed. The overall process included manual inspection of paper forms, preparation of the forms for high-speed data entry including correcting obvious errors or following up with respondents, data entry, and further correction.

The Data Editing Research Project discovered a number of problems:

- Various people from different departments were involved. Many people dealt with the information: respondents, subject-matter specialists, data typists, and computer programmers.

- Transfer of material from one person/department to another could be a source of error, misunderstanding, and delay.

- Different computer systems were involved from mainframe to minicomputers to desktop computers under MS-DOS. Transfer of files from one system to another caused delay, and incorrect specification and documentation could produce errors.

- Not all activities were aimed at quality improvement. Time was also spent on just preparing forms for data entry, and not on correcting errors.
- The cycle of data entry, automatic checking, and manual correction was in many cases repeated three times or more. Due to these cycles, data processing was very time consuming.
- The structure of the data (the metadata) had to be specified in nearly every step of the data editing process. Although essentially the same, the "language" of this metadata specification could be completely different for every department or computer system involved.

The conclusions of the Data Editing Research Project led to general redesign of the survey processes of Statistics Netherlands. The idea was to improve the handling of paper questionnaire forms by integrating data entry and data editing tasks. The traditional batch-oriented data editing activities, in which the complete data set was processed as a whole, were replaced by a record-oriented process in which each record (form) was completely dealt with in one session.

More about the development of the Blaise system and its underlying philosophy can be found in Bethlehem and Hofman (2006).

The new group of activities was implemented in a so-called CADI system. CADI stands for *computer-assisted data input*. The CADI system was designed for use by the workers in the subject-matter departments. Data could be processed in two ways by this system:

- *Heads-up data entry.* Subject-matter employees worked through a pile of forms with a microcomputer, processing the forms one by one. First, they entered all data on a form, and then they activated the check option to test for all kinds of errors. Detected errors were reported on the screen. Errors could be corrected by consulting forms or by contacting the suppliers of the information. After elimination of all errors, a "clean" record was written to file. If employees could not produce a clean record, they could write the record to a separate file of "dirty" records to deal with later.
- *Heads-down data entry.* Data typists used the CADI system to enter data beforehand without much error checking. After completion, the CADI system checked in a batch run all records and flagged the incorrect ones. Then subject-matter specialists handled these dirty records one by one and correct the detected errors.

To be able to introduce CADI on a wide scale in the organization, a new standard package called Blaise was developed in 1986. The basis of the system was the Blaise language, which was used to create a formal specification of the structure and contents of the questionnaire.

The first version of the Blaise system ran on networks of microcomputers under MS-DOS. It was intended for use by the people of the subject-matter departments; therefore no computer expert knowledge was needed to use the Blaise system.

In the Blaise philosophy, the first step in carrying out a survey was to design a questionnaire in the Blaise language. Such a specification of the questionnaire contains more information than a traditional paper questionnaire. It did not only describe questions, possible answers, and conditions on the route through the questionnaire but also relationships between answers that had to be checked.

Figure 1.6 contains an example of a simple paper questionnaire. The questionnaire contains one route instruction: persons without job are instructed to skip the questions about the type of job and income.

Figure 1.7 contains the specification of this questionnaire in the Blaise system. The first part of the questionnaire specification is the *Fields section*. It contains the definition of all questions that can be asked. A question consists of an identifying name, the text of the question as presented to the respondents, and a specification of valid answers. For example, the question about age has the name *Age*, the text of the question is "*What is your age?*" and the answer must be a number between 0 and 99.

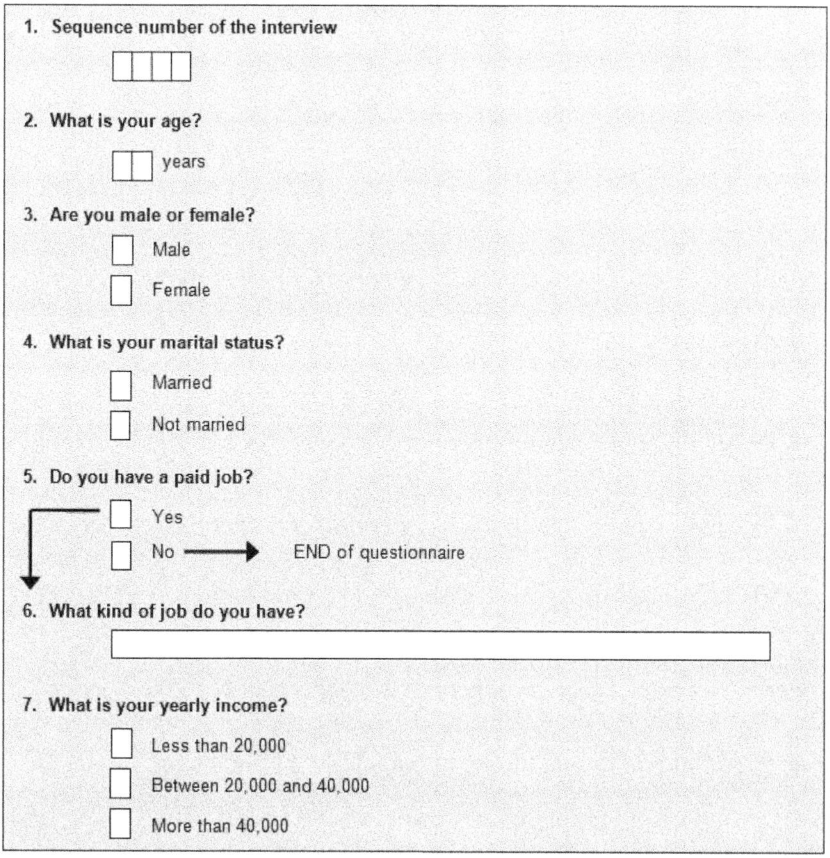

FIGURE 1.6 **A simple paper questionnaire**

```
DATAMODEL LFS "The Labour Force Survey";

FIELDS
    SeqNum   "Sequence number of the interview?": 1..1000
    Age      "What is your age?": 0..99
    Sex      "Are you male or female?": (Male, Female)
    MarStat "What is your marital status?":
               (Married "Married",
                NotMar  "Not married")
    Job      "Do you have a job?": (Yes, No)
    JobDes   "What kind of job do you have?": STRING[20]
    Income   "What is your yearly income?":
               (Less20  "Less than 20,000",
                Upto40  "Between 20,000 and 40,000",
                More40  "More than 40,000")

RULES
    SeqNum Age Sex MarStat Job
    IF Job = Yes THEN
       JobDes Income
    ENDIF

    IF Age < 15 "respondent is younger than 15" THEN
       MarStat = NotMar "he/she is too young to be married!"
    ENDIF

ENDMODEL
```

FIGURE 1.7 **A simple Blaise questionnaire specification**

The question *JobDes* requires a text not exceeding 20 characters. *Income* is a closed question. There are three possible answer options. Each option has a name (for example, *Less20*) and a text for the respondent (for example, "*Less than 20,000*").

The second part of the Blaise specification is the *Rules section*. Here, the order of the questions is specified and the conditions under which they are asked. According to the rules section in Figure 1.7, every respondent must answer the questions *SeqNum*, *Age*, *Sex*, *MarStat*, and *Job* in this order. Only persons with a job (*Job = Yes*) have to answer the questions *JobDes* and *Income*.

The rules section can also contain checks on the answers of the questions. Figure 1.7 contains such a check. If people are younger than 15 years (*Age <* 15), then their marital status can only be not married (*MarStat = NotMar*). The check also contains texts that are used to display the error message on the screen (*If respondent is younger than 15 then he/she is too young to be married!*).

The rules section may also contain computations. Such computations could be necessary in complex routing instructions or checks or to derive new variables.

The first version of Blaise used the questionnaire specification to generate a CADI program. Figure 1.8 shows what the computer screen of this MS-DOS program looked like for the Blaise questionnaire in Figure 1.7.

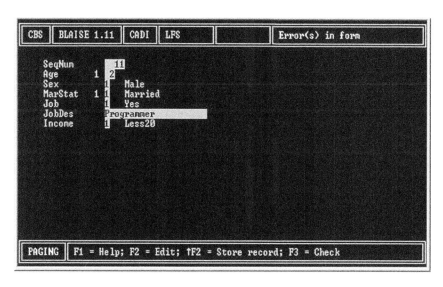

FIGURE 1.8 **A Blaise CADI program**

Since this program was used by subject-matter specialists, only question names are shown on the screen shown in Figure 1.8. Additional information could be displayed through special keys. Note that the input fields for the questions *Age* and *MarStat* contain error counters. These error indicators appeared because the answers of the questions *Age* (2) and *MarStat* (*Married*) did not pass the check.

After Blaise had been in use for a while, it was realized that such a system could be made much more powerful. The questionnaire specification in the Blaise system contained all knowledge about the questionnaire and the data needed for survey processing. Therefore, Blaise should be capable to handle CAI.

Implementing CAI means that the paper questionnaire is replaced by a computer program containing the questions to be asked. The computer takes control of the interviewing process. It performs two important activities:

- *Route control.* The computer program determines which question is to be asked next and displays that question on the screen. Such a decision may depend on the answers to previous questions. As a result, it is not possible anymore to make route errors.

- *Error checking.* The computer program checks the answers as data are entered. Range checks are carried out immediately, as well as consistency checks after entry of all relevant answers. If an error is detected, the program produces an error message, and data must be corrected.

Use of computer-assisted data collection has three major advantages. First, it simplifies the work of interviewer (for example, no more route control). Second, it improves the quality of the collected data. Third, data are entered in the computer during the interview resulting in a complete and clean record.

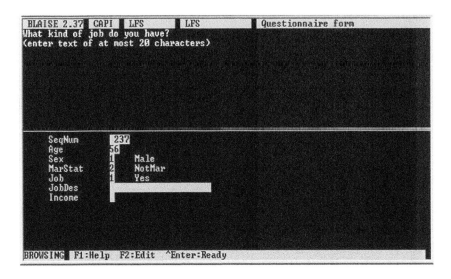

FIGURE 1.9  A Blaise CAPI program

Version 2 of Blaise was completed in 1988. It implemented CAPI. This is a form of face-to-face interviewing in which interviewers use a laptop computer to conduct the interview.

Figure 1.9 shows an example of a screen of a CAPI program generated by Blaise. The screen was divided in two parts. The upper part contains the current question to be answered (*What kind of a job do you have?*). After an answer had been entered, this question was replaced by the next question on the route.

Just displaying one question at the time gave the interviewers only limited feedback on where they are in the questionnaire. Therefore, the lower part of the screen displayed (in a very compact way) the current page of the questionnaire.

Statistics Netherlands started full-scale use of CAPI in regular survey in 1987. The first CAPI survey was the Labor Force Survey. Each month, about 400 interviewers equipped with laptops visited 12,000 addresses. After a day of interviewing, the laptop was connected to a telephone modem. The data were transmitted to the office at night. In return, new addresses were sent to the interviewers. The next morning the laptop was prepared for a new day of interviewing.

CATI was introduced in 1990 on desktop computers. Interviewers called respondents from a central unit (call center) and conducted interviews by telephone. The interviewing program for CATI was the same as that for CAPI. An important new tool for CATI was a call scheduling system. This system took care of proper delivering busy numbers (try again shortly), no answers (try again later), appointments, etc.

By the very early 1990s, nearly all household surveys of Statistics Netherlands had become CAPI or CATI surveys. Surveys using paper forms had almost become extinct. Table 1.3 lists all major and regular household surveys at that time together with their mode of interviewing.

TABLE 1.3    **Household surveys carried out by Statistics Netherlands in the early 1990s**

| Survey | Mode | Interviews per year |
|---|---|---|
| Survey on Quality of Life | CAPI | 7,500 |
| Health Survey | CAPI | 6,200 |
| Day Recreation Survey | CAPI | 36,000 |
| Crime Victimisation Survey | CAPI | 8,000 |
| Labour Force Survey | CAPI | 150,000 |
| Car Use Panel | CATI | 8,500 |
| Consumer Sentiments Survey | CATI | 24,000 |
| Social-Economic Panel | CATI | 5,500 |
| School Career Survey | CATI | 4,500 |
| Mobility Survey | CATI/CADI | 20,000 |
| Budget Survey | CADI | 2,000 |

In the middle of the 1990s, the MS-DOS operating system on microcomputers was replaced by Windows. This marked the start of the use of graphical user interfaces. Early versions of the Internet browser Internet Explorer were included in this operating system.

Blaise 4 was the first production version of Blaise for Windows released in 1998. When more and more people and companies were connected to the Internet, web surveys became a popular mode of data collection among researchers. The main reasons of this popularity were the high response speed, the possibility to provide feedback to respondents about the meaning of questions and possible errors, and the freedom for the respondents to choose their own moment to fill in the questionnaire.

The graphical user interface offered many more possibilities for screen layout. Figure 1.10 gives an example of a screen of the Blaise 4 CAPI program.

Since respondents are familiar with browsers from all their other activities on the Internet, there was no need to explain the graphical user interface.

The possibility to conduct web surveys was included in version 4.6 of Blaise released in 2003. The respondent completes the questionnaire online allowing continuous interaction between the computer of the respondent and the software on the Internet server.

The Internet questionnaire is divided into pages. Each page may contain one or more questions. After the respondent has answered all questions on a page, the answers are submitted to the Internet server. The answers are checked; a new page is returned to the respondent. The contents of this page may depend on the answers to previous questions.

Figures 1.11 and 1.12 show an example of the same page of a web survey when using Blaise 5. In this case, the page contains only one question. The first page will be displayed when using a tablet, and the second page will be displayed when using a smartphone.

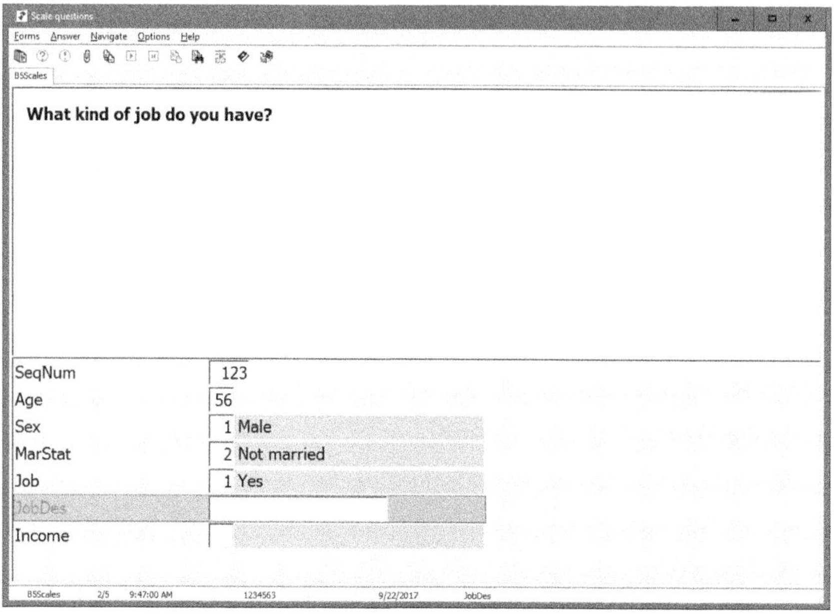

FIGURE 1.10  The screen of a CAPI program in Blaise 4

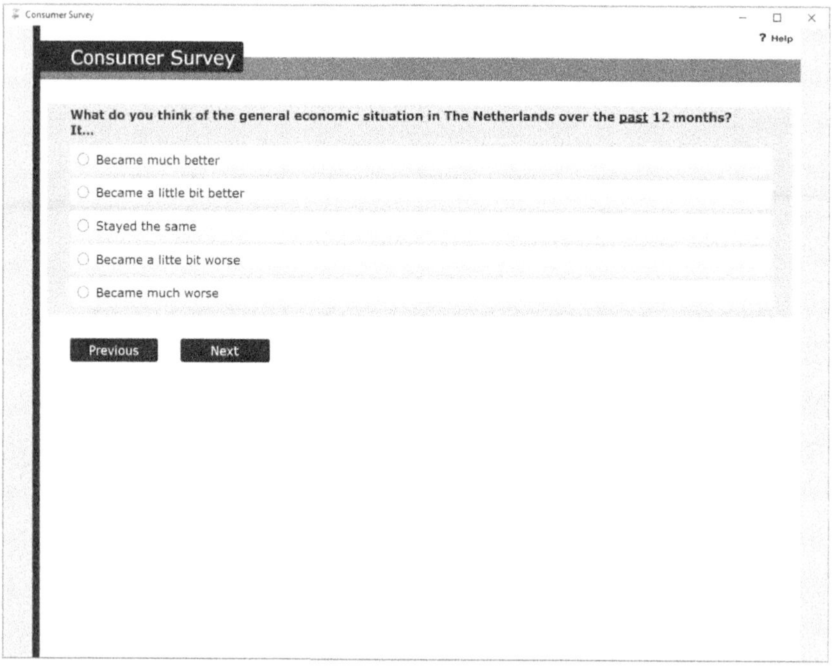

FIGURE 1.11  The screen of a Blaise 5 web survey on a tablet

FIGURE 1.12 **The screen of a Blaise 5 web survey on a smartphone**

The Blaise 5 system implements a number of source code features (*Languages, Modes, Roles, and SpecialAnswers*) that specifically address challenges listed above. It also implements a cross-platform layout designer, templates, and cross-platform settings that handle presentation and operability issues. Finally, Blaise 5 allows the institute to combine these features in as many ways as suits its survey program and population.

## 1.4 Summary

Web surveys are a next step in the evolution process of survey data collection. Collecting data for compiling statistical overviews is already very old, almost as old as mankind. All through history, statistics have been used by rulers of countries to take informed decisions. However, new developments in society always have had their impact on the way the data were collected for these statistics.

For a long period, until the year 1895, statistical data collection was based on complete enumeration of populations. The censuses were mostly conducted to establish the size of the population, to determine tax obligations of the people, and to measure the military strength of the country.

The first ideas about sampling emerged around 1895. There was a lot of discussion between 1895 and 1934 about how samples should be selected: by means of probability sampling or some other sample selection technique. By 1934 it was clear that only surveys based on probability sampling could provide reliable and accurate estimates. Such surveys were accepted as a scientific method of data collection.

Somewhere in the 1970s another significant development started. The fast development of microcomputers made it possible to introduce CAI. This made survey data collection faster, cheaper, and easier; it also increased data quality. It was time in which acronyms like CATI and CAPI emerged.

The next major development was the creation of the Internet around 1982. When more and more persons and companies got access to the Internet, it became possible to use this network for survey data collection. The first Internet surveys where e-mail surveys. In 1989 the World Wide Web was developed. In the middle of the 1990s, web surveys became popular.

Web surveys are attractive because they allow for simple, fast, and cheap access to large groups of potential respondents. There are, however, also potential methodological problems. There are ample examples of web surveys that are not based on probability sampling. It is not always easy to distinguish good from bad surveys. Attention to the methodological aspects is important both to run web surveys and to use web survey data.

The diffusion of mobile devices, especially smartphones, offers a recent attractive tool to reach interviewee for mobile web surveys, i.e., surveys where the contacted unit can receive and respond using either desk and portable computer or mobile devices. There are however methodological problems to be considered when applying mobile web surveys.

Current digital environment and technology trends are providing a huge amount of data about most phenomena. These data are available on the web and are based on the automatic collection on everything that people do; they are usually called big data; they are not subject to statistical classification criteria and to statistical treatment for representativity. However, they look like an attractive source of data to practitioners and researchers. They are wondering if big data could substitute web surveys to provide information for social and economic decision making. They will not substitute surveys; the message is that the two sources are complementary, but they require the researcher to consider the existing methodological problems. The big data offer a challenging opportunity to revise the role and questions faced from the surveys and to integrate web survey data with other data sources.

## KEY TERMS

**Blaise**: A software package for computer-assisted interviewing and survey processing developed by Statistics Netherlands.

**Census**: A way of gathering information about a population in which every element in the population has to complete a questionnaire form.

**Computer-assisted interviewing (CAI)**: A form of interviewing in which the questionnaire is not printed on paper. Questions are asked by a computer program.

**Computer-assisted personal interviewing (CAPI)**: A form of face-to-face interviewing in which interviewers use a laptop computer to ask the questions and to record the answers.

**Computer-assisted self-administered questionnaires (CASAQ)**: A form of data collection in which respondents complete the questionnaires on their own computer or device. See also CASI.

**Computer-assisted self-interviewing (CASI)**: A form of data collection in which respondents complete the questionnaires on their own computer or device. See also CASAQ.

**Computer-assisted telephone interviewing (CATI)**: A form of telephone interviewing in which interviewers use a computer to ask the questions and to record the answers.

**E-mail survey**: A form of data collection via the Internet in which respondents are sent a questionnaire that is part of the body text of an e-mail. The questionnaire is returned by e-mail after answering the questions in the text.

**Face-to-face interviewing**: A form of interviewing where interviewers visit the homes of the respondents (or another location convenient for the respondent). Together, the interviewer and the respondent complete the questionnaire.

**Mail survey**: A form of data collection where paper questionnaire forms are sent to the respondents. After completion of the questionnaires, they are returned to the research organization.

**Mobile web survey**: Self-administered surveys that can be conducted over mobile web-capable devices. They are similar to web surveys, but they have also unique features, such as administration on small screens and keyboards, different navigation, and reaching respondents in various situations, factors that can affect response processes.

**Probability sampling**: A form of sampling where selection of elements is a random process. Each element must have a positive and known probability of selection.

**Purposive sampling**: A form of non-probability sampling in which the selection of the sample is based on the judgment of the researcher as to which elements best fit the criteria of the study.

**Quota sampling**: A form of purposive sampling in which elements are selected from the population in such a way that the distribution of some auxiliary variables matches the population distribution of these variables

**Random digit dialing (RDD)**: A form of sample selection for a telephone survey where random telephone numbers are generated by some kind of computer algorithm.

**Representative method**: A methods proposed by Anders Kiaer in 1896 to select a sample from a population in such a way that it forms a "miniature" of the populations.

**Straw poll**: An informal survey conducted to measure a general feeling of a population. Sample selection is such that it usually does not allow concluding about the population as a whole.

**Survey**: A way of gathering information about a population in which only a sample of elements from the population has to complete a questionnaire form.

**Telephone interviewing**: A form of interviewing in which interviewers call selected persons by telephone. If contact is made with the proper person and this person wants to cooperate, the interview is started and conducted over the telephone.

**Web scraping**: Is data scraping; it is used for extracting data from websites. Note that data scraping is a technique in which a computer program extracts data from human-readable output coming from another program.

**Web survey**: A form of data collection via the Internet in which respondents complete the questionnaires on the World Wide Web. The questionnaire is accessed by means of a link to a web page.

## EXERCISES

**Exercise 1.1**   Which of the following options is not an advantage of computer-assisted interviewing (CAI) as compared with traditional modes of data collection?

**a.** Data quality is higher due to included checks.

**b.** The software is in charge of routing through the questionnaire.

**c.** CAI leads to higher response rates.

**d.** Data are processed quicker.

**Exercise 1.2**   What is an advantage of an e-mail survey over a traditional mail survey?

**a.** Data quality is higher due to included checks.

**b.** There is less under-coverage.

**c.** Response rates are higher.

**d.** It has better facilities for navigation through the questionnaire.

**Exercise 1.3**   Why were the first surveys on the Internet e-mail surveys and not web surveys?

**a.** E-mail surveys were cheaper.

**b.** The World Wide Web did not exist yet.

**c.** E-mail surveys are more user friendly.

**d.** E-mail surveys require less data communication over the Internet.

**Exercise 1.4**   When should the form-based approach be preferred over the question-by-question approach in a web survey?

**a.** The questionnaire is very long.

**b.** The questionnaire contains route instructions and edits.

**c.** All questions fit on one screen.

**d.** The survey is a business survey.

**Exercise 1.5**   Which of the four features is typically and advantage of web surveys?

**a.** There is no under-coverage.

**b.** The sample size is always large.

**c.** A survey can be designed and conducted very quickly.

**d.** Accurate estimates can always be computed.

**Exercise 1.6**   How to avoid the problem of under-coverage in a general population web survey?

**a.** Conduct a mixed-mode survey.

**b.** Increase the sample size.

**c.** Conduct a self-selection web survey.

**d.** Replace the web survey by an e-mail survey.

**Exercise 1.7**   Why source integration is an interesting perspective?

**a.** To optimize information.

**b.** To conduct mixed-mode surveys.

**c.** To totally avoid web surveys.

**d.** To run only paper surveys.

# REFERENCES

Barcaroli, G., Bianchi, G., Bruni, R., Nurra, A., Salamone, S., & Scarnò, M. (2016), Machine learning and statistical inference: the case of Istat survey on ICT. *Proceeding of the Italian Statistical Society Conference, SIS,* Salerno.

Barcaroli, G., Nurra, A., Salamone, S., Scannapieco, M., Scarnò, M., & Summa, D. (2015), Internet as a Data Source in the Istat Survey on ICT in Enterprises. *Austrian Journal of Statistics,* 44, pp. 31–43. doi:org/10.17713/ajs.v44i2.53.

Beręsewicz, M., Lehtonen, R., Reis, F., Di Consiglio, L., & Karlberg, M. (2018), *An Overview of Methods for Treating Selectivity in Big Data Sources.* Publication Office of the European Union, Luxembourg.

Bethlehem, J. G. (1987), The Data Editing Research Project of the Netherlands Central Bureau of Statistics. *Proceedings of the Third Annual Research Conference of the US Bureau of the Census,* U.S. Bureau of the Census, Washington, DC, pp. 194–203.

Bethlehem, J. G. (2009), *The Rise of Survey Sampling.* Discussion Paper 09015, Statistics Netherlands, The Hague/Heerlen, the Netherlands.

Bethlehem, J. G., Cobben, F., & Schouten, B. (2011), *Handbook on Nonresponse in Household Surveys.* Wiley, Hoboken, NJ.

Bethlehem, J. G. & Hofman, L. P. M. (2006), Blaise—Alive and Kicking for 20 Years. *Proceedings of the 10th International Blaise Users Conference,* Arnhem, the Netherlands, pp. 61–86.

Bethlehem, J. G. & Stoop, I. A. L. (2007), Online Panels—A Theft of Paradigm? The Challenges of a Changing World. *Proceedings of the Fifth International Conference of the Association of Survey Computing,* Southampton, U.K., pp. 113–132.

Beukenhorst, D. & Wetzels, W. (2009), *A Comparison of Two Mixed-mode Designs of the Dutch Safety Monitor: Mode Effects, Costs, Logistics.* Technical paper DMH 206546, Statistics Netherlands, Methodology Department, Heerlen, the Netherlands.

Biffignandi, S. & Signorelli, S. (2016), From Big Data to Information: Statistical Issues Through Examples, in Studies. In: Gaul, W., Vichi, M., & Weihs, C. (eds.), *Classification, Data Analysis, and Knowledge Organization.* Springer, Berlin.

Bowley, A. L. (1906), Address to the Economic Science and Statistics Section of the British Association for the Advancement of Science. *Journal of the Royal Statistical Society,* 69, pp. 548–557.

Bowley, A. L. (1926), Measurement of the Precision Attained in Sampling. *Bulletin of the International Statistical Institute,* XII, Book 1, pp. 6–62.

Bycroft, C. (2010), *Integrated Household Surveys: A Approach.* Statistics New Zealand, Wellington, New Zealand.

CBS, (1948), *Enige Beschouwingen over Steekproeven.* Reprint from: *Statistische en Economische Onderzoekingen* 3, Statistics Netherlands, The Hague, the Netherlands.

Clayton, R. L. & Werking, G. S. (1998), Business Surveys of the Future: The World Wide Web as a Data Collection Methodology. In: Couper, M. P., Baker, R. P., Bethlehem, J. G., Clark, C. Z. F., Martin, J., Nicholls, W. L., & O'Reilly, J. (eds.), *Computer Assisted Survey Information Collection.* Wiley, New York.

Cobben, F. & Bethlehem, J. G. (2005), *Adjusting Under-coverage and Non-response Bias in Telephone Surveys.* Discussion Paper 05006, Statistics Netherlands, Voorburg/Heerlen, the Netherlands.

Cochran, W. G. (1953), *Sampling Techniques.* Wiley, New York.

Couper, M. P. (2013), Is the Sky Falling? New Technology, Changing Media, and the Future of Surveys. *Survey Research Methods,* 7(3), pp. 145–156.

Couper, M. P., Baker, R. P., Bethlehem, J. G., Clark, C. Z. F., Martin, J., Nicholls II, W. L., & O'Reilly, J. M. (eds.) (1998), *Computer Assisted Survey Information Collection.* Wiley, New York.

Couper, M. P., Blair, J., & Triplett, T. (1999), A Comparison of Mail and E-mail for a Survey of Employees in U.S. Statistical Agencies. *Journal of Official Statistics*, 15, pp. 39–56.

Couper, M. P. & Nicholls, W. L. (1998), The History and Development of Computer Assisted Survey Information Collection Methods. In: Couper, M. P., Baker, R. P., Bethlehem, J. G., Clark, C. Z. F., Martin, J., Nicholls, W. L. & O'Reilly, J. (eds.), *Computer Assisted Survey Information Collection.* Wiley, New York.

de Leeuw, E. D. (2005). To Mix or Not to Mix Data Collection Modes in Surveys. *Journal of Official Statistics*, 21, pp. 233–255.

Deming, W. E. (1950), *Some Theory of Sampling.* Wiley, New York.

Den Dulk, K. & Van Maarseveen, J. (1990), The Population Censuses in The Netherlands. In: Maarseveen, J. V. & Gircour, M. (eds.), *A Century of Statistics, Counting, Accounting and Recounting in The Netherlands.* Statistics Netherlands, Voorburg, the Netherlands.

Dillman, D. A. (2007), *Mail and Internet Surveys: The Tailored Design Method.* Wiley, Hoboken, NJ.

Dillman, D. A., Smyth, J. D., & Christian, L. M. (2014), *Internet, Phone, Mail, and Mixed-Mode Surveys: The Tailored Design Method*, 4th Edition. Wiley, New York.

Graunt, J. (1662), *Natural and Political Observations upon the Bills of Mortality.* Martyn, London, U. K.

Hansen, M. H., Hurvitz, W. N., & Madow, W. G. (1953), *Survey Sampling Methods and Theory.* Wiley, New York.

Horvitz, D. G. & Thompson, D. J. (1952), A Generalization of Sampling Without Replacement from a Finite Universe. *Journal of the American Statistical Association*, 47, pp. 663–685.

Kiaer, A. N. (1895), Observations et Expériences Concernant des Dénombrements Représentatives. *Bulletin of the International Statistical Institute*, IX, Book 2, pp. 176–183.

Kiaer, A. N. (1997 reprint), Den Repräsentative Undersökelsesmetode. *Christiania Videnskabsselskabets Skrifter. II. Historiskfilosofiske klasse*, Nr 4 (1897). English translation: The Representative Method of Statistical Surveys, Statistics Norway, Oslo, Norway.

Kiesler, S. & Sproul, L. S. (1986), Response Effects in the Electronic Survey. *Public Opinion Quarterly*, 50, pp. 402–413.

Laplace, P. S. (1812), *Théorie Analytique des Probabilités. Oevres Complètes*, Vol. 7. Gauthier-Villar, Paris, France.

Lienhard, J. H. (2003), *The Engines of Our Ingenuity, An Engineer Looks at Technology and Culture.* Oxford University Press, Oxford, U.K.

Neyman, J. (1934), On the Two Different Aspects of the Representative Method: The Method of Stratified Sampling and the Method of Purposive Selection. *Journal of the Royal Statistical Society*, 97, pp. 558–606.

Nicholls, W. L. & Groves, R. M. (1986), The Status of Computer Assisted Telephone Interviewing. *Journal of Official Statistics*, 2, pp. 93–134.

NIPO (1946), Eerste Telefonische Enquête in Nederland verricht door NIPO. *De Publieke Opinie*, 1$^{e}$ jaargang, No. 4.

O'Connell, P. L. (1998), Personal Polls Help the Nosy Sate Curiosity. *New York Times*, June 18.

Pierzchala, M. (2006), *Disparate Modes and Their Effect on Instrument Design*. Paper presented at the 2006 International Blaise Users Conference, Papendal, the Netherlands (at http://www.blaiseusers.org/2006/Papers/207.pdf).

Pierzchala, M. (2016), *Blaise 5—Is Worth the Wait*. Paper presented at the 2016 International Blaise Users Conference, The Hague, the Netherlands (at http://blaiseusers.org/2016/papers/plen2_3.pdf).

Quetelet, L. A. J. (2010), *Lettre à S.A.R. le Duc Régant de Saxe Coburg et Gotha sur la Théorie des Probabilités, Appliquée aux Sciences Morales at Politiques (1846)*. Kessinger Pub Co., Brussels, Belgium.

Quetelet, L. A. J. (2012), *Sur l'Homme et le Développement de ses Facultés, Essai de Physique Sociale (Edit. 1835)*. Hachette Livre-BNF, Paris, France.

Righi, P., Barcaroli, G., & Golini, N. (2017), Quality Issues When Using Big Data in Official Statistics. In: Petrucci, A. & Verde, R. (eds.) *IProceedings of the Conference Statistics and Data Science: New Challenges, New Generations*, FUP Scientific Cloud for Books.

Roos, M., Jaspers, L., & Snijkers, G. (1999), *De Conjunctuurtest via Internet*. Report H4350-99-GWM, Statistics Netherlands, Data Collection Methodology Department, Heerlen, the Netherlands.

Roos, M. & Wings, H. (2000), Blaise Internet Services Put to the Test: Web-surveying the Construction Industry. *Proceedings of the 6th International Blaise Users Conference*, Kinsale, Ireland.

Saris, W. E. (1998), Ten Years of Interviewing Without Interviewers: The Telepanel. In: Couper, M. P., Baker, R. P., Bethlehem, J. G., Clark, C. Z. F., Martin, J., Nicholls II, W. L., & O'Reilly, J. M. (eds.), *Computer Assisted Survey Information Collection*. Wiley, New York, pp. 409–430.

Schaefer, D. R. & Dillman, D. A. (1998), Development of a Standard E-mail Methodology: Results of an Experiment. *Public Opinion Quarterly*, 62, pp. 378–397.

Snijkers, G., Tonglet, J., & Onat, E. (2004), *Projectplan Pilot e-PS*. Internal Report H3424-04-BOO, Development and Support Department, Division of Business Statistics, Statistics Netherlands, Heerlen, the Netherlands.

Snijkers, G., Tonglet, J., & Onat, E. (2005), *Naar een Elektronische Vragenlijst voor Productiestatistieken*. Internal Report, Development and Support Department, Division of Business Statistics, Statistics Netherlands, Heerlen, the Netherlands.

Utts, J. M. (1999), *Seeing Through Statistics*. Duxburry Press, Belmont, CA.

Wells, C. & Thorson, K. (2015), Combining Big Data and Survey Techniques to Model Effects of Political Content Flows in Facebook. *Social Science Computer Review*, 35, pp.1–20.

Witte, J. C., Amoroso, L. M., & Howard, P. E. N. (2000), Method and Representation in Internet-based Survey Tools. *Social Science Computer Review*, 18, pp. 179–195.

Yates, F. (1949), *Sampling Methods for Censuses and Surveys*. Charles Griffin & Co, London, U.K.

# CHAPTER TWO

# About Web Surveys

## 2.1 Introduction

The Internet is one of the data collection tools available for conducting surveys. It is a relatively new method. At first sight, it is an attractive mean of data collection because it offers a possibility to collect a large amount of data in a short time period at low cost. Therefore, web surveys have quickly become very popular. Due to the large diffusion of mobile devices connected to Internet, when sending a web survey, it is possible that the invitation and even the completion takes place on a mobile device like a tablet or, especially, a smartphone. Thus, in fact, a web survey becomes a mobile web survey. In this handbook we use the term web surveys, and web surveys only when we want to stress the importance of mobile devices we use the term mobile web survey.

The methodology of web surveys is not yet fully developed. Nevertheless, a lot of experiments and scientific discussion about the theory of web surveys are now in the literature. They are useful, because one can determine the advantages and the problems in doing a web survey. Therefore, the statistical studies described in this handbook are important for the future of web surveys and for those who are interested in learning about web surveys and how to run them.

*Handbook of Web Surveys*, Second Edition. Silvia Biffignandi and Jelke Bethlehem.
© 2021 John Wiley & Sons, Inc. Published 2021 by John Wiley & Sons, Inc.

Traditionally, surveys can use various modes of data collection:

- By mail using paper questionnaire forms.
- By telephone. The interviewer can use a paper form or a computer program for computer-assisted interviewing (computer-assisted telephone interviewing [CATI]).
- Face-to-face. The interviewer can use a paper form or a computer program for computer-assisted interviewing (computer-assisted personal interviewing [CAPI]).

Web surveys resemble mail surveys. Both modes of data collection rely on visual information transmission. Note that telephone surveys and face-to-face surveys use oral information transmission. Furthermore, there are no interviewers involved in data collection. Data collection is based on self-administered interviews.

Of course, a web survey is a computer-assisted form of data collection (like CAPI and CATI). Therefore, sometimes it is given name CAWI (computer-assisted web interviewing). Web survey questionnaires can include features like automatic routing through the questionnaire and automatic checking for inconsistencies. These features are not possible for mail survey questionnaires.

Like for any other survey, also web survey respondents have to be contacted first and invited to participate in the survey. In general, the following approaches are possible:

- Send an e-mail or an SMS (cell text message) with a link to the website containing the survey questionnaire. The link may include a unique identification code. The unique code ensures that a respondent will complete the questionnaire only once. It also ensures that only selected individuals complete the questionnaire. Example 2.1 describes a web survey that applied this approach. At the time being, e-mail and SMS can be read on mobile devices. Consequently, the questionnaire could be completed on mobile devices as well, or alternatively the contacted individual can postpone the completion when he will access to a laptop. This is a challenging situation because factors like the environment and contest where the invitation to the survey (i.e., the e-mail or the SMS) is received and the time lag between the invitation and the possibility to use a laptop for completion and other factors are requiring special attention.
- Send a letter by ordinary mail inviting the potential respondents to go to the survey website. The letter contains the address (URL) of the website and a unique code. Again this guarantees that only the proper individuals participate in the survey.
- Catch potential survey participants on the Internet when they are visiting a website. They are invited to click on a link or button to start the survey. They may be directed to a different website containing the survey, or the survey starts as a newly opened window (pop-up window) on the screen. The web survey may also be embedded in websites visited by the individual.

The last approach is a very simple way to conduct web surveys. No e-mails or letters have to be send. However, it has the disadvantage that no proper sampling procedure is used. This may lead to a response that lacks representativity. Sometimes such surveys may allow a respondent to complete the questionnaire more than once. Moreover, there is no guarantee that each respondent is a member of the intended survey population. Finally, technical aspects, like pop-up blockers, may prevent starting the survey questionnaire.

▓ EXAMPLE 2.1 **A web survey on technological communication and links between enterprises**

The survey is carried out within the survey research activities now provided from the CASI (Center for Statistical Analyses and Surveys/Interviewing, Centro Analisi Statistiche e Indagini) seated at the ex-Department of Mathematics, Statistics, and Informatics of the Faculty of Economics of the University of Bergamo. The survey topics were the use of e-commerce, the collaboration with other enterprises and/or the belonging to groups, the markets, and the employment. The questionnaire was kept simple (6 pages of 40 substantial items, 1 welcome page, and 1 final page) and asked mainly for qualitative answers or for percentage data.

The survey addressed to about 2,000 firms of the provinces of Bergamo (used as pilot province), Brescia, Lecco, Varese, and Mantova (each province is in the Lombardy region in Italy) in the manufacturing and building sector. E-mail addresses and stratification variables collection is from the administrative databases of the Chamber of Commerce for the Bergamo province and from a Unioncamere (Union of Chambers of Commerce) database for the other provinces.

The overall response rate was 21.9%, which is quite high considering that due to the quality of the list, 12% of the follow-up contacts were explicitly wrong e-mail addresses. Response rates by size of the firm, legal form, and economic activity did not differ very much. It is interesting to note that for small firms (less than 20 employees), the response rate was quite high compared with the other firm sizes.

Data collection took place, for the first wave, in spring 2000 and has been carried out according to the following steps:

- Invitation to participate in the survey sent by e-mail (survey presentation letter, a survey report as incentive, and other related advantages have been prospected).
- A link of the individual firm address for completion of the questionnaire was included in the presentation letter. Therefore no identification code (id) and no password was required.

> • Three e-mail reminders were sent (first reminder 14 days after the sur-
> vey follow-up and the two reminders with weekly periodicity). Mainly
> for research purposes a fourth e-mail remainder was sent after the end
> of the survey period. As described in literature, the three reminders
> were effective in improving the response rate; the forth reminder did
> not have an effect.

A web survey questionnaire consists of one or more web pages. Respondents
have to visit the website in order to answer the survey questions. Note that, at early
times of web surveys, the Internet was only deployed as a medium to transport the
empty questionnaire to the respondent and to transport the completed questionnaire back from the respondent to the survey agency. For example, a simple questionnaire form is implemented in an Excel spreadsheet. The respondents receive
the spreadsheet as an attachment of an e-mail. After they have downloaded this
electronic form, they fill it in on their own computer. Completed questionnaire
return is also by e-mail. This type of survey is called an *Internet survey*, because
it uses the Internet in a much broader sense, than just the HTML pages of the
World Wide Web.

This chapter describes the various forms of online data collection, from simple
e-mail surveys to advanced web surveys. It shows how to use web surveys for different target populations, for cross-sectional data collection, and for longitudinal
data collection (panels). It discusses the main reasons for online data collection, the
advantages and disadvantages, areas of application, and specific related problems.

## 2.2  Theory

Collecting data using a web survey has much in common with other modes of data
collection. There are the usual steps, like survey design, fieldwork, data processing,
analysis, and publication. At each step, however, consider the suitability of concepts and methods taken from traditional survey approaches (face-to-face, paper,
telephone). Where necessary, implication for eventual questionnaire reception and
completion on mobile devices needs to be borne in mind, and adequate adaptions
need to be applied. This handbook examines the most important practical and
methodological aspects of only web surveys and of mobile web surveys that need
careful consideration. They relate on the following questions:

• How to select the sample?
• How to contact potential respondents?
• How to construct a web questionnaire?
• How mobile web surveys differ or are similar to web surveys?
• How to make proper statistical inference based on a web survey data?

- What is the impact of sampling and non-sampling errors?
- What are the problems and solutions for mixed-mode surveys with web component?
- How to handle web panels?

Detailed answers to these questions are in the other chapters of this handbook. The current chapter gives a general overview of different approaches of conducting web surveys. For each approach and each situation, different problems may occur, and therefore different methodological solutions are required.

### 2.2.1 TYPICAL SURVEY SITUATIONS

In this section, identification of typical situations, in which to conduct a web survey, is presented. A number of different key aspects lead to different survey situations:

- *Target population.* There are general population surveys (among individuals or households), business surveys, specific population surveys (among specific populations like company employees, company customers, students at a university or school, or members of club), or open population surveys (among ill-defined populations like consumers of a product or service).
- *Survey administrator.* This can be a national statistical institute (NSI) or other official statistical government body, a commercial market research company, a university, or another research institute.
- *Cross-sectional versus longitudinal data collection.* A cross-sectional web survey measures the status of a population at one specific point in time, based on a sample selected for that purpose. A longitudinal web survey (or web panel) is recruited; it is maintained to allow measuring change over time. Also, surveys on specific topics can be selected from the web panel.
- *Technical implementation.* The questionnaire can be designed as a website on the World Wide Web. In this case, questionnaires are completed online. It is also possible to use the Internet as just a vehicle to transport a questionnaire form to the respondents. For example, a form in an Excel spreadsheet can be send as an attachment of an e-mail. In this case, the questionnaire is completed offline. This approach was used at early stages of Internet diffusion. Now, the questionnaire is mostly, even not exclusively, completed online. The challenge is now the choice of the device used for reception and completion of the questionnaire and/or on the choice of the mode, if a mixed-mode approach is adopted.

The choice of the *target population* for a web survey may have an impact on the magnitude of survey errors and particularly on non-sampling errors. Such errors may partly be caused by selection problems (for example, under-coverage) and partly by measurement errors (due to the lack of inter-viewer assistance).

If the target population is the general population (households or individuals), there is a problem with the sampling frame. Some countries have a population register. Such a register contains addresses. Therefore, use it as a sampling frame for a face-to-face or mail survey. Sometimes telephone numbers link to addresses, which makes it possible to use it as a sampling frame for telephone surveys. Unfortunately, these registers do not contain e-mail addresses, nor can e-mail addresses link to it.

Internet penetration varies greatly between countries (see Chapters 1 and 10 about under-coverage problems). Presently, Internet coverage is relatively high, say, between 60% and 90% in a number of European countries. These coverage rates seem to suggest that general population web surveys are possible in these countries and that they can compete with traditional data collection modes. However, note that a large Internet penetration does not imply high Internet use. Moreover, it also does not imply high quality if fast Internet connections are available. For example, not everyone with Internet access has broadband.

One should always bear in mind that not everyone has access to the Internet. One example is that not every employee of a company is allowed to use the Internet. Moreover, Internet access is substantially lower in many countries for specific subpopulations. For example, Hispanic blacks are underrepresented in the United States. Another example of underrepresented groups are people with low education and people living in rural areas. This situation exists in many countries. The elderly, also, are often underrepresented among Internet users. Under-coverage leads to web surveys that lack of representativity. Therefore, there is a risk to draw wrong conclusions from the survey results.

If the target population consists of businesses, it is quite probable, in most countries, that each business has Internet access and therefore has an e-mail address. Thus, the collection of business to sample for a web survey is rather close to the target population. However, obtaining a complete list of e-mail addresses for businesses may be a very difficult task. Partial lists sometimes exist, but complete lists are often lacking. NSIs regularly contact large enterprises for surveys. Therefore, they may have a complete list of e-mail addresses of certain economic branches or specific size classes of companies. In most cases, obtaining an e-mail population list for small enterprises and businesses could well be a difficult task for NSIs. Even if such a list is available, it requires a lot of effort to maintain it. In European countries, the maintenance of the business register—requested from Eurostat—allows for the updated list of businesses and their stratification variables and address (e-mail address, in most cases). NSIs are now going to run many surveys via web. Even census data collection is on the web, thus sharing many methodological problems with the web surveys.

If the target population is a closed population (employees of a company or students of university), often there is a sampling frame containing e-mail addresses of all members of the population. In such situations, there is no difference between the target population and the sampling frame. There are no coverage problems. This is the ideal case for a web survey.

With respect to the *survey administrator*, differences may occur with respect to the amount of information for setting up the surveys and the topics that are addressed in the web survey. NSIs and official statistics bodies probably have the largest amount of information available on the general population (households and individuals) and businesses (or institutions). They may have access to population registers, they may have census data, and they may manage demographic databases, the business register, and other sources of information. Therefore, although this huge amount of data may be insufficient for generating a sampling frame of e-mail addresses for the target population, they may well be in a rather good position to obtain this information. This is happening in the near future, since the trend in the diffusion of web surveys is gaining importance and the Internet penetration within the population continues to increase. Currently, the advantages of the NSIs are twofold:

- They often have a sampling frame for the general population of individuals of households based on addresses. This means that they are able to select a suitable probability based sample from the target population. They can contact the selected persons/households using an alternative mode (for example, by mail) while using the web mode as a second step in the data collection process.
- As regards businesses, they have a full population list, at least for large businesses, together with contact points and, probably, e-mail addresses. Therefore, a sampling frame adequate for web is available.

Example 2.2 describes how Statistics Netherlands has managed to use the available information to gain insights on the web participation of the population.

🖿 EXAMPLE 2.2 **The ICT survey pilot**

Statistics Netherlands carried out a pilot with the ICT survey to find out whether it was possible to use the web for data collection. This survey collects information on the use of computers and Internet in households and by individuals. The regular ICT survey was a CATI survey. It was rather expensive. It also suffered from under-coverage because the sample selection was from the telephone directory. It was not possible to select households with unlisted numbers and mobile-only households.

The sample selection for the pilot was from the population register. Therefore, there was no under-coverage. All persons in the sample received an invitation letter by mail. The letter contained the Internet address of the survey and a unique login code. Respondents had the possibility to complete the questionnaire on paper. To prevent those with Internet to respond by paper, the paper questionnaire was not included in the invitation letter.

People had to apply for the paper form by returning a stamped return postcard.

After one week, a postcard was sent to all nonrespondents with a reminder to complete the survey questionnaire, either by web or mail. Two weeks after receipt of the invitation letter, the remaining nonrespondents were approached again. Part of these nonrespondents received a reminder letter, and another part was called by telephone (if a telephone number was available). The telephone call was just to remind the nonrespondents and did not replace the paper/web questionnaire form.

It turned out the postcard reminders worked well. Each time they were sent, there was a substantial increase in response. The telephone reminder did not work as well as the postcard reminder. Of the people that promised by telephone to fill in the form, only 40% actually did so.

For the survey administrator, a favorable environment for running web surveys is a closed population. For example, businesses or institutions' administrative database records for each employee contain individual data and corporate e-mail address.

It is a similar situation for the universities. The university database records the student's e-mail address (the institutional one and—sometimes—the private one).

Business customer's databases record detailed information on each customer including contact references, among them e-mail addresses.

Other survey administrators, such as academic researchers, market research companies, or private businesses, may not have proper sampling frames available. One solution of this problem could be to let an NSI select the sample for them. Another could be to obtain a copy of the sampling frame after privacy-related information has been removed. Nevertheless, privacy-related laws may prevent NSIs to make available sampling frame information to third parties.

With respect to the topic of the survey, it should be borne in mind that NSIs and other government statistical bodies collect data primarily for policy decisions. There may be different surveys for different social and economic indicators. Many surveys are compulsory, which means that the contacted elements are obliged to respond. If they do not, they may be fined. Sometimes questionnaires are rather complex since many topics are covered. Surveys conducted by academic researchers, market research organizations, and other companies tend to be more heterogeneous, covering a number of different issues: product characteristics, customer satisfaction regarding products and services, employee satisfaction, trends in consumer preferences or behavior, health, use of technological products, and so on. Generally speaking, survey topics dealt with by this type of survey administrator are mainly devoted to a more or less traditionally defined target population, and therefore an appropriate survey frame definition becomes more difficult. Surveys carried out by this type of survey administrator could often make use of a simpler and shorter questionnaire.

With respect to the distinction between *cross-sectional* and *panel* data collection, it should be noted that cross-sectional surveys gather data about one moment in time, whereas panel surveys collect information at many successive points in time with the focus on investigating changes over a period of time. Panel surveys are discussed in Chapter 14. The main problem with panel surveys is the lack of representativity of the panel and of the samples selected from it. The real world is full of panelists recruited by means of self-selection and therefore not representative for the population. In this case, the panel is a large group of elements that is approximately as large as the population, but the large size doesn't imply representativeness. Large probability-based panels are representative, but more difficult to recruit.

If a panel is for longitudinal studies, all respondents are tracked back to the moment they entered the panel. Therefore, when doing a longitudinal analysis of survey results, DiSogra and Callegaro (2009) recommend computing cumulative standardized response rates (taking into account different recruitment waves), i.e., rates based on a multiple recruitment approach. This approach captures the dynamics of a panel member's history with regard to nonresponse and attrition, i.e., of loss of respondents of the recruited panel.

With regard to *technical implementation of the questionnaire*, there are two approaches possible i.e. online data collection and offline data collection:

- *Online data collection* is a way of data collection for which the respondents have to remain online during the process of answering the questions. The questionnaire is implemented as one or more web pages. The respondent has to surf to the survey website in order to start the questionnaire. The questionnaire can be question based or form based. *Question based* means that every web page contains a single question. After answering a question, the respondent proceeds to the next question that is on the next page. If the questionnaire contains routing instructions and consistency checks, we recommend the page-based approach. *Form based* means that there is a single web page containing all questions. This page looks like a form. Usually there are no routing instructions and no consistency checks. The questionnaire can be optimized for the mobile devices; in such a case, questions are presented in a user-friendly format for reading and completing in smartphones or other devices.

- *Offline data collection.* The electronic questionnaire form (an HTML page, an Excel spreadsheet, or another of interviewing software tool) is send to the respondent by e-mail, or the respondent can download it from the Internet. The respondent fills in the form or spreadsheet offline. After completing the questionnaire, it is returned (uploaded, send by e-mail) to the survey agency. Statistics Netherlands, for example, used this approach for a number of business surveys. A computer-assisted interviewing program was send to the selected businesses. The businesses run this program offline and answer the questions. After completion, contact is with the Internet again, and the data are uploaded to the survey agency.

In the case of an electronic form, the advantages of a printable questionnaire can be combined with those computer-assisted interviewing (routing and consistency checking). Note that it is also possible not to bother the respondents with consistency checking. This means no errors will be detected during form completion. However, errors may afterward be detected by the survey agency, possibly resulting in returning the form to the respondent for error correction. This is much less efficient.

## 2.2.2  WHY ONLINE DATA COLLECTION?

The challenges of the online surveys have been recently increasing due to the explosion in the Internet use and of mobile devices. Potential advantages of this data collection mode are interesting, but several disadvantages or at least methodological question are arising. Thus, the researcher needs to understand and know the methodological issues to be faced and the risks of errors. The following paragraphs are focusing on the advantages and disadvantages and related critical issues.

**2.2.2.1  Advantages.**  Survey participants are increasingly responding to web surveys on their smartphone as opposed to their personal computers. Web or mobile web surveys combine the advantages of the web with the advantages of mobile devices. They have several advantages. The three most important advantages are that mobile web surveys are faster, simpler, and cheaper.

With respect to the time required to conduct a mobile web survey, the following observations can be made:

- The time it takes to get in contact with the respondent can be considerably reduced if the invitation is sent by e-mail or text message (SMS).
- Follow-ups can be carried out very quickly by e-mail. The timing of reminders can be tailored to the respondents. A typical pattern for web surveys is that many completed questionnaires are returned almost immediately. The number of returns diminishes fast after a few days. Web surveys allow for a short time lag between request and reminder than mail surveys. Biffignandi and Pratesi (2002) showed that the time interval between the first contact and the first follow-up can be shorter than in mail surveys. The intervals between successive follow-ups can also be very short, and no more than three reminders are necessary. The fourth reminder is almost ineffective. The authors suggest that 10 days is an adequate time interval between first contact and first follow-up, whereas the time interval between successive reminders is around 1 week. Furthermore, Crawford, Couper, and Lamias (2001) showed that a quick reminder after two days works fairly well.
- The time it takes to deliver a complete questionnaire is also very short. As soon as it is completed, the questionnaire is submitted and delivered. Thus, there is no time lag between the moment the respondent returns the questionnaire and the moment it is received.

- The time it takes to store the collected data is dropped. Responses are instantly recorded into a database and prepared for analysis.

To sum up, the entire data collection period is significantly shortened. All data can be collected and processed in little more than a month. There are even opinion polls on the web for which design, data collection, analysis, and publication all take place in one day. Recently, mobile web surveys are applied for in-the-moment research. In-the-moment survey administration is via mobile device (especially smartphone) in the moment and in the location of the research matter. For example, customers contact is while using the product or consuming it, while shopping or while entering or exiting a location. In this case, the results are rather immediate; the representativeness of the results has to be evaluated, subjects to be interviewed have to be selected using probabilistic criteria, and data have to be adequately processed.

A second advantage of web and mobile web surveys is that they can be tailored to the situation. Therefore they may make life simpler for the respondents and the researcher. Here are some examples:

- Respondents may be allowed to save a partially completed form. At a later point in time, they can continue and complete the questionnaire. This is particularly important for business surveys where sometimes different departments of the business each have to complete a specific part of the questionnaire.

- The questionnaire may be filled with already available information. The respondents only have to check them for changes. This can be useful in web panels. Example of such preloaded data are address of the respondent and employment status.

- There can be a facility that automatically generates an e-mail message to the survey agency if the respondent indicates he has complaints about the questionnaire. Such information can help to improve surveys and avoid future problems.

- Response rates can be monitored over time. If the response is lower than expected, action can be undertaken. For example, customized e-mail reminders are sent. However, although there are no costs attached to sending reminders, a good rule is to send them at well-defined moments in time, should they be necessary. Do not overload slow respondents by reminders. The literature shows that this may lead to irritation and break-off or lower data quality. Three or four reminders are enough.

- The proper survey software can check that no respondent can complete the questionnaire more than once. Of course, this requires handing out unique identification codes to the individuals selected in the sample. Note that this does not work in case of self-selection surveys.

- Like in computer-assisted interviewing, web questionnaires may contain route instructions. These instructions monitor that respondents only answer relevant questions and skip irrelevant questions.

By applying *usability tests,* the web questionnaires may be improved. Usability refers to the ease of use of a software application for a web questionnaire. Measures of usability refer to the speed in performing a task, the frequency of errors in performing a task, and user satisfaction with regard to an application interface, in terms of being easy to understand and use. Two techniques are especially valuable in usability testing:

- *Qualitative interviews.* Usability tests apply to small groups of individuals. Using current or proposed standards for the interface, a fully functional web questionnaire creates. A group of people is invited considered typical of respondents. After completing the questionnaire, they get an in-person in-depth interview. The case study reported in Section 2.3 is an example of using qualitative interviews for questionnaire testing.

- Analysis of paradata. *Paradata* are data concerning the actual web questionnaire completion process. As users complete the questionnaire, the actions are collected. The value of paradata in web and mobile web questionnaire testing, and in an analysis of response behavior, is becoming more and more significant. Information on the characteristics of the respondent's technical environment, respondent response time, errors made, and navigation behavior help to detect and correct problems in the questionnaire. Device used in completion is registered. This is especially important in mobile web surveys.

Survey data collection is expensive. Businesses even if interested in understanding consumers' behavior have to face with budget constraint. Many statistical agencies of governments face budget cuts; also, other survey organizations attempt to reduce data collection costs. Thus, it is worthwhile to consider cheaper modes of data collection. Web surveys or mobile web surveys are substantially cheaper than other modes of data collections. Thus, they are an appealing tool for data collection.

Of course, a web survey requires initial investment in computers, servers, and software. Additionally, there are initial costs for the sampling design (if the researcher uses a probability sampling approach) and for web questionnaire design and implementation. Skilled and specialized personnel, with an understanding of usability and visual design, are necessary to design and implement a web survey. There is a need to use programs for mobile optimization of questionnaires.

After these survey steps are over, there are no further data collection costs other than the costs of help desk personnel. Such a help desk is important in order to answer respondents' questions or to solve their problems (Lozar Manfreda, and Vehovar, 2008).

Field data collection is relatively cost-free and not dependent on the number of questionnaires administered and completed. Automatically, the database with survey data generates, making data input costs irrelevant as well. No time and effort related to data entry and verification is required. For a comparison of the timing of return rates in mail and web surveys, see Dillman, Smith, and Christian (2014).

To sum up, large numbers of completed questionnaires can be collected in a very short time and at low costs.

Web surveys also have some other attractive properties worthwhile mentioning:

- The possibility of obtaining server-side and client-side information allows for easily monitoring response burden in web surveys. This makes it possible to record how much time respondents need to complete the questionnaire. Analysis may show how the response burden is related to the response rate.
- The use of short questionnaires reduces the response burden. It may help to split a large questionnaire into a number of small questionnaires. The administration of small questionnaires is possible at different moments in time. This does not increase the costs of the survey.
- Web surveys are less intrusive, and they suffer less from social desirability effects.
- Geographical boundaries are not a problem. Geography is not limiting web surveys in the same way as face-to-face interviews and mail and telephone surveys. Therefore, international target populations may be easily reached without special additional costs or time delays.

### 2.2.2.2 Disadvantages and Problems.

A major problem of web-based surveys is sample selection. For research applications, a random sample is desirable and often essential, and researchers may simply not have a comprehensive sampling frame of e-mail addresses for people who drink fruit juices or go to church. Despite the huge growth of the Internet, there are still many people who do not have access to, or choose not to use, the Internet. There are also wide disparities in Internet access among ethnic, socioeconomic, and demographic groups. A sampling frame, including e-mail addresses, of all members of the target population should be available to draw a random sample. In practice, this list is very rarely available. Therefore, large coverage problems arise, and this is the most relevant issue.

Sampling problems may particularly be an issue for general population surveys. For many specific populations, there are no problems. Examples are companies collecting customer satisfaction data, employers measuring job satisfaction, educators collecting course evaluations and conducting examinations, bloggers wanting to consult with their readers, and event organizers checking proposed attendance and meal and other preferences. While there is still a need for some caution, in terms of learning how to use the new technology with confidence, the use of web surveys has been growing rapidly and will clearly continue to grow. Inside the innovative contest web surveys present new methodological challenges, like the integration with other data sources.

A disappointing aspect of web surveys is that they do not contribute to solving the problem of decreasing response rates. It is widely recognized that they usually result in a rather low response rates. It should be noted that, despite low response rates, the use of server-side and client-side paradata can help to focus efforts on specific population that most need it.

## 2.2.3  AREAS OF APPLICATION

Web surveys may be used in any field of application provided that the elements in the population have Internet access and that they have some basic computer skills. In some cases, as described in detail in Chapter 12, a probability-based sample from the general population is selected. Then some people without a computer may receive one (with Internet access), together with basic instructions for use. This solution has typically been adopted for general population web panels.

If the web data collection is possible for all potential respondents, a web survey can be a very useful data collection tool, combining low costs and high quality.

Unfortunately, often not all elements in the target population have Internet access, or computers with adequate processing power to process questionnaires, or sufficient computer literacy. This problem holds true for general population surveys as well as for many other possible target populations. Even if Internet penetration is growing, differences may exist between countries and between groups within populations. Large differences in computer equipment, screen settings, and technical literacy may have a substantial impact. Thus, to carry out a good web survey, a statistical sound approach is needed, which attempts to minimize a possible bias as much as possible. And if a bias cannot be avoided, there should be statistical techniques applied to correct for this bias afterward.

In practice, despite the methodological challenges, many surveys (especially commercial surveys) are conducted on the web without properly taking into consideration the impact on the reliability of the results arising from the lack of Internet coverage and/or lack of computer literacy. Such surveys are administered exclusively via the web and therefore reach only one part of the target population. If the survey is not blocked, the web survey is a mobile web survey and will reach the participants also through the mobile device. Nevertheless, only who is using Internet is contacted. When using web survey results, one should be aware of potential problems. Therefore, it is important to assess the quality of the web survey by analyzing the methodological description in the documentation.

  EXAMPLE 2.3  **Reliable web surveys**

Certain surveys are not affected by the instrument bias a web survey may cause. When measuring job satisfaction among high-tech workers, the bias will be minimal. Getting feedback from employees on a benefit package can have a slightly higher bias if not all employees have computer access.

However, an attempt to determine what role of the United States should have been in the Libyan war of 2011 would probably produce highly biased estimates, because one would only obtain the opinions from computer-literate people with Internet access. They will not be representative for the whole population.

Web surveys may be conducted for profiling purposes. Examples are member surveys, audience profiling, and donor profiles. Web surveys may also be used for data collection by asking people to provide information about themselves. Other interesting applications are socioeconomic research, planning support, and social behavior studies. A web survey can also be used for attitude polls, opinion polls, program evaluation, and community cultural planning surveys. Other possible topics are economic aspects and performances, as well as market trends and customer/employee satisfaction.

**EXAMPLE 2.4  The Kauffman Firm Survey (KFS)**

The Kauffman Firm Survey (KFS) is a panel study of new businesses founded in 2004. They are tracked over their early years of operation. The survey focuses on the nature of new business formation activities, characteristics of the strategy, offerings, employment patterns of new businesses, the nature of the financial and organizational arrangements of these businesses, and the characteristics of their founders.

The KFS created the panel by using a random sample from a Dun & Bradstreet (D&B) database list of new businesses started in 2004. The list contained in total about 250,000 businesses.

The KFS oversampled these businesses based on the intensity of research and development employment in the businesses' primary industries. The KFS sought to create a panel that included new businesses created by a person or team of people, purchases of existing businesses by a new ownership team, and purchases of franchises. To this end, the KFS excluded D&B records for businesses that were wholly owned subsidiaries of existing businesses, businesses inherited from someone else, and not-for-profit organizations. Previous research on new businesses showed also variability in how business founders perceive the operation of their starting businesses. Therefore, a series of questions was asked to business owners about indicators of business activity and whether these were conducted for the first time in the reference year (2004). These indicators were payment of state unemployment (UI) taxes, payment of Federal Insurance Contributions Act (FICA) taxes, presence of a legal status for the business, use of an Employer Identification Number (EIN), and use of Schedule C to report business income on a personal tax return.

A self-administered web survey and *computer-assisted telephone interviewing* (CATI) were used for data collection.

Over time several changes in the original sample of businesses occurred and went for different reasons out of the panel. Due to panel attrition the number of units is becoming slightly smaller each year. Since only the 2004 cohort is under study, no continuous refreshment is planned. At the time being the Kauffman Firm Survey (KFS) is a panel study of 4,928 businesses founded in 2004 and tracked over their early years of operation. Until September 2020, the University of Chicago NORC Data Enclave has been managing a secure remote access to the KFS confidential microdata file access for researchers. Free access was allowed. After September 2020, access to the KFS is shifted towards a fee-based model.

Objective of the panel survey is to provide information about creation and development aspects on new businesses (especially of high-technology and women-owned businesses). Firm characteristics, revenue and expenses, profit and loss, owner characteristics, and, since 2007, information about predominant markets and Internet sales are also collected.

## 2.2.4 TRENDS IN WEB SURVEYS

Dillman, Smyth, and Christian (2014) describe some changes in the survey environment from the 1970 to 2000, focusing on the factors like human interaction, trust that the survey is legitimate, time involvement for each respondent, attention given to each respondent, respondent control over access, and respondent control over whether to respond. These observations indicate that during the 1990s human interaction and individual contact relevance are decreasing due to the use of IT (i.e., computer-aided and web surveys) and to massive use of e-mails. Trust on survey relevance and legacy is very low, and the possibility of refusal and of filtering against surveys (anti-spam, disclosure rules) is very high. These observations are in line with developments with respect to web surveys.

More recently, Biffignandi (2010a, 2010b) has been focusing on major trends with respect to web surveys. The author underlines that survey methodology has quite recently undergone a paradigm shift: the causes of survey errors and how to prevent them. This new paradigm stresses the total error concept, which includes sampling error (the central error in the traditional approach) and non-sampling errors. Non-sampling errors can be numerically larger than sampling errors.

The flowchart of Chapter 3 describes the survey process. Errors might occur at each step of the process. The new paradigm focuses on the following aspects:

- All kinds of events and behavior occurring during the survey process are considered.
- The overall response rate is only a very simple measure of survey quality, although it is frequently used as an indicator. To some extent this measure could be useful in identifying weak points in the process (for instance, a large amount of refusal might be due to a bad-contact process), but it fails to consider that people who have web access, or that are respondent in a web survey, could significantly differ from other units.
- Overall response rates do not give information regarding the response propensity of different respondent subgroups (late respondents versus early respondents, and sociodemographically different subgroups) or on respondent behavior.
- Response rates are anyway becoming low, and there is a need to investigate the reasons. Incentive is thought as a possible solution. Göritz (2006, 2010, 2015) and Brown et al. (2016) suggest a generally positive effect of incentives in web surveys. Singer and Ye (2013) conclude that in all survey modes, prepaid cash incentive is the most effective. In this case, it is required to use a mode rather than web to contact respondents. If the e-mail contact option is adopted, Dillman, Smyth, and Christian (2014) comment that electronic incentive sent to all sample member is likely the best option. With respect to sampling error, due to imperfect frames in web surveys, traditional probabilistic samples are in many cases not easy to implement. Therefore, it is not possible to compute the sampling error, as the theory of statistical inference does not apply.

As consequence of this new paradigm, attention is going at:

- How to face decreasing response rates. Possible solutions may be:
  - Keeping respondents focused on the relevant parts of the computer screen and keeping distraction to a minimum can help to get completed questionnaire. To accomplish this task, studies based on eye-tracking analysis are to be carried out.
  - An interesting strategy for improving response rates is to use mixed-mode surveys (see Chapters 3 and 9). However, new problems arise with the mixed approach, since mode effects are to be considered in analyzing survey results. Occurrence and treatment of mixed-mode effects need further investigation. Chapter 9 is about it.
- How to use paradata (i.e., data collected during the interviewing process). Increasing attention is going to be devoted to the analysis of this type of data. In particular, they help to identify typologies of response behavior explaining the potential variations in participation in web-based surveys and providing a

valuable insight into understanding nonresponse and various aspects of response behavior. From the methodological point of view, behavioral analyses rely to the Cognitive Aspects of Survey Methodology Movement (CASM), and, in many empirical studies, the theory of planned behavior (TPB) model is applied (Ajzen, 1991). The main objective is to obtain a more comprehensible picture on how intentions form. For example, based on the TPB, two alternative models were empirically tested, in which the roles of trust and innovativeness were theorized differently—either as moderators of the effects that perceived behavioral control and attitude have on participation intention (moderator model) or as direct determinants of the attitude, perceived behavioral control, and intention (direct effects model).

• How to get representative web surveys and/or panels? Many access panels consist of volunteers, and it is impossible to evaluate how well these volunteers represent the general population. In any case, they represent a non-probability sample. Recent research attempts to tackle the task of how to apply probabilistic recruitment to panels and how to draw inferences from them are present in recent literature. One approach to correct for a lack of representativity is, for example, to apply propensity score methodology (Steinmetz et al., 2014). Propensity scores serve to reweight web survey results.

Generally speaking the methodology and quality of data collected in area of socio-economics could greatly benefit by:

a. The development of suitable estimation methods aimed at capturing the bias and specific variance connected with the frame characteristics and participation process of this type of survey;

b. Research, principally based on experimental designs, allowing the effects of various factors to be tested (for example, the effects of different types of question structure, various contact modes, etc.);

c. Research, based on behavioral models, that allows response and participation processes to be analyzed and modeled in the context of the individual behavior of survey respondents.

# 2.3 Application

The Italcementi Group is a large Italian company. The following data are about the situation when the study took place. With an annual production capacity of approximately 75 million tons of cement, it is the world's fifth largest cement producer. The group had companies in 22 countries around the world. Italcementi regularly monitors, by means of a mixed-mode survey, the working conditions and working climate in the company. As regards the survey data is collected part by web and part by paper questionnaire forms.

The target population consists of all employees of all companies belonging to the Italcementi Group around the globe. With no sample selection, all employees are invited to participate in the survey. Thus, in principle it is a census and not a survey. There are no inference issues. Statistics computation is simply from the data.

Nonresponse can be a problem in surveys. Some people may fail to complete the survey questionnaire because they could not be contacted (despite several attempts), they refused to cooperate, or they were not able to answer the questions. This may lead to biased estimates of population characteristics. Nonresponse is not a big problem in the Italcementi survey. In many countries, almost everyone participates. See also Table 2.1.

The first step in the survey process is the design of the questionnaire. In fact, there are two questionnaires: one for blue-collar employees and one for white-collar employees. Italcementi Group as a whole, the company to which the employee belongs, and the specific location of the company cover the three main

TABLE 2.1   **Response rates of the Italcementi survey by country**

| Country | Overall response rate (%) | White-collar response rate (%) | Blue-collar response rate (%) |
|---|---|---|---|
| Albania | 91.9 | — | — |
| Bulgaria | 51.9 | 59.8 | 46.9 |
| Dubai | 100.0 | — | — |
| Egypt | 48.9 | 66.1 | 50.4 |
| France/Belgium | 44.7 (F) 51.7 (B) | 65.6 | 38.1 |
| Gambia | 90.7 | — | — |
| Greece | 52.3 | 79.7 | 31.2 |
| India | 94.1 | 94.3 | 94.0 |
| Italia | 48.6 | 65.2 | 30.5 |
| Kazakhstan | 66.7 | 61.8 | 68.2 |
| Morocco | 64.4 | 67.4 | 62.9 |
| Mauritania | 98.7 | — | — |
| North America | 43.3 | 76.3 | 21.9 |
| Singapore | 75.0 | — | — |
| Spain | 53.7 | 66.2 | 45.1 |
| Sri Lanka | 65.5 | — | — |
| Thailand | 78.2 | 84.1 | 72.7 |
| Turkey | 78.6 | 79.6 | 77.9 |
| Total | 54.7 | — | — |

sections of the questionnaire. In addition, a section with questions about personal characteristics is included. The paper questionnaire for white-collar employees is 8 pages long, while that for blue-collar employees is 6 pages.

The management committee of the enterprise group discusses at first the new survey. Next, by means of cognitive interviews, the questionnaire is tested. After approval of the questionnaire, fieldwork can start. A letter on the intranet announces the survey. The Enterprise Group CEO signs it. The letter distribution is also as part of pay packets. Moreover, the survey is announced on the company notice boards.

It is a mixed-mode survey including both a web mode and a paper mode. To complete the questionnaire on the web, invitation is to everyone with an e-mail address. Those without Internet can complete the paper questionnaire.

The enterprise has companies in many different countries. Thus, employees speak different languages. By translating the two questionnaires into the languages spoken in the countries in which the companies are located, language problems are overpassed. For individuals with literacy problems, data collection is via the collaboration of a private voluntary organization.

For the 2007 survey (the starting year of the survey), 22,276 employees were eligible for the survey. In countries with a large number of employees (France, Belgium, Italy, Egypt, Thailand, Morocco), two-thirds of the white-collar employees were invited to complete the questionnaire on the web, and one-third were asked to complete the paper questionnaire. In all remaining countries, all white-collar employees received a paper questionnaire. For all blue-collar workers, invitation and completion was paper mode. An exception was France, where blue-collar workers were receiving an Internet questionnaire.

Online data collection took place in the period from June to September. Data collection in the field ended in October. The results diffusion was in December. The total number of respondents was equal to 12,183. This comes down to a response rate of 54.7%. The response rate of the white-collar workers (71.5%) was much higher than the response rate of the blue-collar workers (45%). The response rate of the web respondents was higher than the response rate of the paper questionnaire respondents. Table 2.1 shows the response rates by country.

In 2007, the first survey was conducted. This survey is run in the subsequent years in a similar way. The survey turned out to be very informative with respect to the attitudes of employees about their jobs and the company they belong to.

Some examples of (translated) questions in the Italcementi survey are shown below. Figure 2.1 contains two questions that use a so-called Likert scale. They both use a 4-point scale. When responding to a Likert scale question, respondents specify their level of agreement to a statement. The scale name is after its inventor, *the psychologist Rensis Likert.*

Figure 2.2 shows a so-called matrix question. Such a question combines a number of single question all with the same type of answer.

**How satisfied are you with your job?**
**Select one answer only**

1. Very satisfied
2. Rather satisfied
3. Rather dissatisfied
4. Very dissatisfied

**Thinking about your last performance assessment, how useful would you say it was? Select one answer only**

1. Very useful
2. Somewhat useful
3. Not very useful
4. Not at all useful

FIGURE 2.1  Questions with Likert scales

For each of the following aspects of personnel management, please indicate whether you think the organizational activities are efficient or inefficient. Select one answer in each row

|  | Efficient | Inefficient |
|---|---|---|
| 1. Administrative aspects (payroll and benefits administration, holidays) | ○ | ○ |
| 2. Identification of training needs | ○ | ○ |
| 3. Organization of quality training courses | ○ | ○ |
| 4. Management of career development | ○ | ○ |
| 5. Transfer of knowledge and skills within the company | ○ | ○ |
| 6. Anticipating staff needs | ○ | ○ |
| 7. Controlling employee awareness of safety measures in the workplace | ○ | ○ |
| 8. Controlling staff costs | ○ | ○ |
| 9. Recruitment of new staff | ○ | ○ |
| 10. Integration of newly recruited staff | ○ | ○ |
| 11. Management of international careers | ○ | ○ |
| 12. Cooperation with operational managers | ○ | ○ |

FIGURE 2.2  A matrix question

# 2.4  Summary

A web survey is a relatively new data collection technique. The role of the web in conducting surveys is more important due to the spread of the Internet. At first sight, it is an attractive means of data collection because it has many advantages. Among them are costs, timeliness, and the possibility of improving survey quality. Web surveys allow for simple, fast, and cheap access to large groups of potential respondents. Web surveys have quickly become very popular. Recently, due to the high penetration of mobile devices connected to Internet, mobile web surveys are becoming an interesting data collection tool. In practice every web survey becomes a mobile web survey, if access to these devices is not blocked. For these reasons it is important to check how questionnaire works on mobile devices, especially on smartphones. Optimization procedures exist to adapt the questionnaire. Web survey and mobile web surveys deserve also methodological challenges like selection effects and measurement errors. There are ample examples of web surveys not based on probability sampling. It is not always easy to distinguish good surveys from bad. Mobile web surveys encompass even more challenges related to several technical problems like screen and browser characteristics, adaption of the format of the questionnaire, and behavioral issue like when is the questionnaire received. The respondent could be traveling, working, out for a walk, or sporting activity.

This chapter describes the various forms of online data collection, from simple e-mail surveys to advanced web surveys. It shows how to use web surveys for different target populations, for cross-sectional data collection, and for longitudinal data collection (panels). It discusses the main reasons for online data collection, the advantages and disadvantages, areas of application, and specific related problems.

## KEY TERMS

**Computer-assisted personal interviewing (CAPI)**: A form of face-to-face interviewing in which interviewers use a laptop computer to ask the questions and to record the answers.

**Computer-assisted telephone interviewing (CATI)**: A form of telephone interviewing in which interviewers use a telephone to ask the questions and to record the answers.

**Computer-assisted web interviewing (CAWI)**: A form of self-interviewing in which respondents complete the questionnaires on the Internet. CAWI is a synonym for web survey.

**Cross-sectional survey:** A survey that observes a sample from the target population at one point in time. Objective is to describe the state of the population on that moment in time.

**Internet survey**: A general term for various forms of data collection via the Internet. Examples are a web survey and e-mail surveys. Included are also forms of data collection that use the Internet just to transport the questionnaire and the collected data.

**Longitudinal survey**: A survey that observes the same sample from the target population at several points in time. Objective is to describe the changes of the population over time.

**Paradata**: Data about the process by which the survey data are collected.

**Qualitative interview**: An in-person in-depth interview with respondents that have completed a survey questionnaire. Such an interview aims at uncovering usability problems like difficult questions or cumbersome tasks.

**Self-selection survey**: A survey for which the sample has been recruited by means of self-selection. It is left to the persons themselves to decide to participate in a survey.

**Usability testing:** Conducting an experiment to check whether respondents find it easy to complete the web survey questionnaire. Aspects tested include the speed with which the survey task is carried, the number of errors made, and familiarity with the user interface.

**Web panel**: A survey in which the same individuals are interviewed via the web at different points in time.

**Web survey**: A form of data collection via the Internet in which respondents complete the questionnaires on the World Wide Web. The questionnaire is accessed by means of a link to a web page.

## EXERCISES

**Exercise 2.1**   Which of the following statements does not apply to web surveys?

a. The survey can be conducted faster.

b. The survey can be conducted cheaper.

c. The response rate is high.

d. The survey collects a large number of interviews.

**Exercise 2.2**   In what respect does a web survey resemble a mail survey?

a. They both rely on visual information transmission.

b. They both rely on oral information transmission.

c. They both use computer-assisted interviewing techniques.

d. They both cost the same amount of time to conduct.

**Exercise 2.3**  Which of the following phenomena is not a problem of self-selection web surveys?

**a.** A respondent can complete a questionnaire more than once.

**b.** Persons not belonging to the target population can complete the questionnaire.

**c.** The survey results show a lack of representativity.

**d.** It is difficult to get a large number of respondents.

**Exercise 2.4**  What is the difference between a cross-sectional survey and a longitudinal survey?

**a.** A cross-sectional survey measures changes over time, and a longitudinal survey measures the state of the population at one point in time.

**b.** A cross-sectional survey measures the state of a population at one moment in time, and a longitudinal survey measures time changes over time.

**c.** A cross-sectional survey mainly measures facts and behavior, and a longitudinal survey measures attitudes and opinions.

**d.** For cross-sectional surveys, any mode of data collection is adequate, whereas in longitudinal surveys only Internet is possible.

**Exercise 2.5**  What is offline data collection?

**a.** Any form of data collection that does not use the Internet.

**b.** A form of Internet data collection for which the questionnaire is not written in HTML.

**c.** A survey that uses e-mail to transfer information.

**d.** A survey that uses the Internet to transfer the electronic questionnaire to the respondents.

**Exercise 2.6**  What is the main reason national statistical institutes consider using web surveys?

**a.** It shows government also uses modern ICT.

**b.** It reduces nonresponse in surveys.

**c.** It improves the quality of the collected data.

**d.** It reduces data collection costs.

# REFERENCES

Ajzen, I. (1991), The Theory of Planned Behavior. *Organizational Behavior and Human Decision Processes*, 50, pp. 179–211.

Biffignandi, S. (2010a), Modeling Non-sampling Errors and Participation in Web Surveys. *Proceedings of the 45th SIS Scientific Meeting*, Padova, Italy.

Biffignandi, S. (2010b), Internet Survey Methodology—Recent Trends and Developments. In: Lovric, M. (ed.), *International Encyclopedia of Statistical Science*. Springer, Heidelberg, Germany.

Biffignandi, S. & Pratesi, M. (2002), Internet Surveys: The Role of Time in Italian Firms Response Behaviour. *Research in Official Statistics*, 5, pp. 19–33.

Brown, J. A., Serrato, C. A., Hugh, M., Kanter, M. H., Spritzer, K. L., & Hays, R. D. (2016), Effect of a Post-paid Incentive on Response Rates to a Web-Based Survey. *Survey Practice*, 9, 1. https://doi.org/10.29115/SP-2016-0001.

Crawford, S. D., Couper, M. P., & Lamias, M. J. (2001), Web Surveys. Perceptions of Burden. *Social Science Computer Review*, 19, pp. 146–162.

Dillman, D. A., Smyth, J. D., & Christian, L. M. (2014), *Internet, Mail and Mixed-Mode Surveys. The Tailored Design Methods*. 4th Edition, Wiley, Hoboken, NJ.

DiSogra, C. & Callegaro, M. (2009), Computing Response Rates for Probability-Based Web Panels. *Proceeding of Section on Survey Research of the American Statistical Association*, Washington, DC.

Göritz, A. (2006), Incentives in Web Studies: Methodological Issues and a Review. *International Journal of Internet Science*, 1, pp. 58–70.

Göritz, A. (2010), Using Lotteries, Loyalty Points, and Other Incentives to Increase Participant Response and Completion. In: Gosling, S. & Johnson, J. (eds.), *Advanced Methods for Conducting Online Behavioral Research*, pp. 219–233. American Psychological Association, Washington, DC. doi: https://doi.org/10.1037/12076-014.

Göritz, A. (2015), Incentive Effects. In: Engel, U., Jann, B., Lynn, P., Scherpenzeel, A., & Sturgis, P. (eds.), *Improving Survey Methods: Lessons from Recent Research*, pp. 339–350. Routledge, London.

Lozar Manfreda, K. & Vehovar, V. (2008), Internet Surveys. In: de Leeuw, E., Hox, J. J., & Dillman, D. A. (eds.), *International Handbook of Survey Methodology*. Lawrence Erlbaum Associates, New York.

Singer, E. & Ye, C. (2013), The Use of Incentives in Surveys. *Annals of the American Academy of Political and Social Sciences*, 645 (1), pp. 112–141.

Steinmetz, S., Bianchi, A., Tijdens, K., & Biffignandi, S. (2014), Improving Web Survey Quality—Potentials and Constraints of Propensity Score Weighting. In: Callegaro, M., Baker, R., Bethlehem, J., Göritz, A., Krosnick, J.A., & Lavrakas, P. J. (eds.), *Online Panel Research: A Data Quality Perspective*. Wiley, Chichester. pp. 273–298.

# CHAPTER THREE

# A Framework for Steps and Errors in Web Surveys

## 3.1 Introduction

The framework proposed in this chapter is a collection of interrelated ideas and concepts that provide guidance for researchers undertaking a web survey. To carry out a survey, one must conduct a process that involves many steps and decisions. The process-oriented approach divides the survey into steps that define the flow of the survey process, and this approach must be applied to web surveys as well. Steps differ in some respects from the traditional modes. Thus, the flow of the web survey process must be explicitly defined. Bethlehem and Biffignandi (2012) and Tourangeau, Conrad, and Couper (2013) pointed out the specifics of the web procedures without focusing on the whole process. Thorsdottir and Biffignandi[1] presented a flowchart for the steps of a probability-based web survey. This chapter follows the flowchart approach to illustrate the process, the decisions involved at the various steps of the process, and the related risks of errors. The flowchart

---

[1] Fanney Thorsdottir and Silvia Biffignandi presented and discussed their flowchart at a WG3 task force (TF10) meeting organized by the Webdatanet COST Action (IS1004) while chairing and co-chairing WG3 and leading TF10 on General Framework for Error Categorization in Internet Surveys. A final draft of the flowchart was presented at the Webdatanet conference (Salamanca, May 26–28, 2015). An extension for the mode selection has been presented at the Total Survey Error Conference (Baltimore, September 2015).

---

*Handbook of Web Surveys*, Second Edition. Silvia Biffignandi and Jelke Bethlehem.
© 2021 John Wiley & Sons, Inc. Published 2021 by John Wiley & Sons, Inc.

will help the surveyor[2] in properly planning a web survey and will help researchers highlight the steps of the survey process for their respective research. Starting from the flowchart presented by Thorsdottir and Biffignandi (Webdatanet Conference, 2015), the flowchart presented in this chapter is slightly different since mobile extension of the web survey is also considered (i.e., mobile web survey). Considering the wide diffusion of mobile devices (especially smartphones), mobile web survey is becoming a very frequent, say, mostly a standard, situation. When a web survey is administered if no disallow is activated, the survey may be received and completed not only using a PC but even from a mobile device. This situation is called mobile web survey, i.e., no prevention toward mobile devices in a web survey (for a discussion of the mobile surveys, see Chapter 2). For a detailed description of the choices for smartphone participation in web surveys, see Peterson et al. (2017). Mobile web survey implies several methodological challenges; nevertheless the survey process requires only a few differences with respect to the web-only survey.

The flowchart divides the mobile web (or the web-only survey) process into the following six major steps:

1. *Determine the survey objective.*

2. *Metadata definition.*

3. *Designing the mobile web or only web survey (deciding on the mode, the sampling frame, and the sampling approach, designing the questionnaire, designing the paradata methodology, and selecting the sample, software, or programming language).*

4. *Collecting the data.*

5. *Creating the database.*

6. *Processing data.*

The use of the flowchart brings a total survey error (TSE) perspective, which is an effective approach for understanding error sources in a comprehensive way (Weisberg, 2005). Biemer TSE approach aims at helping researchers make design decisions that maximize data quality within the constraints of a limited budget. TSE perspective disaggregates overall error into component, mainly distinguishing between sampling and non-sampling errors. The TSE paradigm (see, for example, Platek and Särndal, 2001 and the ensuing discussions) refers to the concept of optimally allocating the available survey resources to minimize TSE for key estimates. In principle, to apply the TSE paradigm, the major sources of error

---

[2] The flowchart is useful not only for specialized survey research organizations but also for smaller organizations or individual researchers. Any surveyor, even for simple and rather small surveys, no matter if he is in favor for more or less sophisticated methodological solutions, should refer to it. He should go across each step and decision to run a good survey. Large statistical organizations (like National Statistical Institutes) follow a more complex survey process because of the different context, laws, regulations, and objectives governing the survey process. Therefore, a specific flowchart exists in specialized documents for official statistics.

should be identified so that the survey resources can be allocated to reduce their errors, within specified constraints on costs. An overview of the TSE history and recent research results on the relationships between different types of errors is given in Biemer et al. (2017). The idea of the TSE is of optimally balancing the dimensions of survey quality within the survey budget and schedule. It is possible to distinguish between TSE and total survey quality. This considers the *fitness for use* of an estimate, that is, quality from both the producer and user perspectives. The "fitness for use" concept implies not only accuracy is considered; for instance, attributes such as the timeliness, accessibility, and comparability of the data have to be granted in the survey process, as well as various types of errors (see Chapter 5), such as measurement, nonresponse, and so forth (Groves, 1989).

The message in this chapter is that both TSE and total quality survey perspective embody the need to consider different error sources and that errors occur at every step of the survey process and, sometimes, they are interrelated.

In this chapter, a framework of the web survey process is proposed. The framework's main purpose is to provide a clear overview of the necessary decisions when organizing a mobile web survey or web-only survey and to create a shared overview of the survey process steps. Practitioners, as well as researchers, can refer to the flowchart and obtain a clear understanding of the procedure to follow, the choices to do, and the type of errors that might occur. Thus, there is a more complete, integrated perspective in studying and interpreting errors. Understanding them becomes more easily.

## 3.2 Theory

The mobile web survey or the web-only survey process differs in some respects from a survey based on traditional modes (paper and pencil, telephone, and fax). Errors arising in mobile web surveys and only mobile surveys have some specific characteristics due mainly to the coverage aspects of the target population and to various other possibilities (the availability of auxiliary variables, the time required to deliver the questionnaire and to receive the completed questionnaire, technical skills, the equipment needed, and many other situations). Therefore, a well-defined outline of the survey research steps and of the errors that might occur at each step is vital for the surveyor. Furthermore, in many cases, the errors at different steps have some relationships, even if the relationship is not clearly defined and formalized. Thus, decisions pertaining to the reduction of an error at a specific step could increase the errors occurring at other steps of the survey process. For instance, increasing response rate by stressing the interviewee with a high number of solicitations might decrease the quality of the data (more item nonresponse, less accuracy, and so forth). Therefore, the overall quality of the survey process could be either improved or deteriorated depending on the decision made.

The flowchart mentioned in the previous section is in Figure 3.1. It shows the main steps and sub-steps for a web-only or mobile web survey based on probability sample.

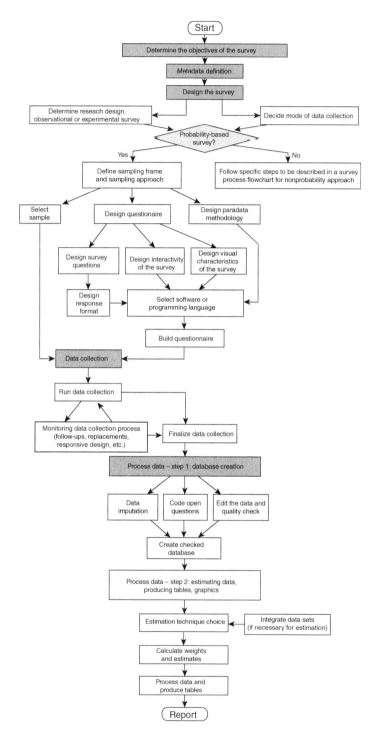

FIGURE 3.1  **Flowchart of a probability-based web-only or mobile web survey**

The six main steps (gray highlighted) and their related sub-steps (hell gray and white highlighted) are listed in the previous section. Hereunder, detailed comments of the flowchart steps are reported.

The first step is ***Determine the survey objectives***, and it is a preliminary point of every survey, independent from the mode. Objectives need to be clearly stated and the questions to be identified. Since this flowchart considers web mode (either web only or mobile web), it should be critically evaluated if the objectives and the structure of the survey are adequate for that mode. Web mode implies a self-completed questionnaire; thus, for example, self-completed surveys are better in capturing sensitive topics. If the questionnaire is long self-completion, especially using web mode, it is not in favor of high participation rates and smaller measurement error. This becomes especially critical in mobile web surveys, where the participation situation has a high probability to be disturbed from other activities or tasks the interviewee is doing.

The second step is ***Metadata definition***. Recent literature and empirical research brought significant attention to metadata, showing a set of information describing the survey (that is, elements for evaluating the data quality, like the list and description of the variables, the target population, the sample units, and so forth). The metadata refer to variables and activities occurring in every step of the process. The International Organization for Standardization (ISO) defines metadata as "data describing and defining other data in a certain context." In such a case, the contest is the condition under which the data collection and processing takes place. The definition of metadata is an important step, recommended for all surveys, independently from the mode. Details to consider in mobile web surveys (and in web-only surveys) differ from traditional surveys due to the characteristics of various steps and sub-steps. Metadata are mainly semantic based; a bad or imprecise metadata description affects the quality of the questionnaire and could cause errors in the answers (questions not correctly understood, questions that each interviewed could interpret in a different way). Metadata are useful for the selection of appropriate sampling techniques and for guiding and evaluating the procedures of the survey process. The so-called metadata database contains metadata, which complement the data database. See Example 3.1. The data user should look at the metadata to be conscious of the quality and meaning of the data he or she is going to use. On the other hand, a surveyor should take care to write down the metadata precisely because accuracy is crucial for undertaking the survey process in a correct and valuable way. For example, if a surveyor is drawing his probability-based sample from a largely incomplete sampling list, it will not be possible to obtain highly accurate results from the inference process. Thus, information about the sampling frames characteristics (coverage, reference time of the frame, i.e. is it updated or an old one?) is an important quality. Another example is a bad or insufficient description of one variable; in this case, each interviewed could assign to the variable a different meaning, sorting out in scarcely meaningful results. From the user point of view, this implies both the use of low-quality data and the impossibility to understand what is the effective meaning of the variable and how to interpret the data.


■ EXAMPLE 3.1  A metadata database: variables definitions

The Eurostat website is offering a metadata database that includes Euro-SDMX Metadata Structure (ESMS) (a set of international standards for the exchange of statistical information between organizations), classifications, legislation and methodology, concepts and definitions (CODED, Eurostat's Concepts and Definitions Database, and other online glossaries relating to survey statistics), glossary, national methodologies, and standard code lists.

A section reports the description of the variables in different sources. Variable descriptions are detailed explanations of the researcher's intended meaning of the variable in the questionnaire, and it is one example of basic metadata. For example, for the purposes of the Labour Force Survey, the following definition is used: "Employees are defined as persons who work for a public or private employer and who receive compensation in the form of wages, salaries, fees, gratuities, payment by results or payment in kind; non-conscripted members of the armed forces are also included."

In structural business statistics, employees are defined as "those persons who work for an employer and who have an employment contract and receive compensation in the form of wages, salaries, fees, gratuities, piece-work pay or remuneration in kind."

Furthermore, a worker is a wage or salary earner of a particular unit if he or she receives a wage or salary from the unit, regardless where he or she works (in or outside the production unit). A worker from a temporary employment agency is considered to be an employee of the temporary employment agency not of the unit (customer) in which they work. Metadata states that "employees include part-time workers, seasonal workers, person on strike or on short-term leave, but excludes those persons on long-term leave. Employees does not include voluntary worker."

If the variable is not precisely declared, the respondents could compile the questionnaire according different concepts; one could exclude part-time workers, whereas another could include them. Therefore, in such a survey, measurement error would arise, or a high number of nonresponses to the specific question (item nonresponse) would emerge due to the unclear variable definition.

Most statistical offices, both NSIs and various research bodies, present a section on metadata. Research institutes, marketing research societies, and every business or institution collecting survey data should provide a clear metadata definition and communicate it to the users.

The third step is the ***Designing the mobile web or web-only survey***; this may be broken down into sub-steps. Firstly, two basic sub-steps to consider are as follows: (1) decide if the study should be experimental or observational, and (2) decide the mode of data collection.

Regarding the sub-step *Decide if the study should be experimental or observational* (sub-step 1), it should be kept in mind that an experimental study tries to catch how different factors affect the results; thus, the task is to highlight relationships between factors and the results (or outputs). There is no special interest in estimating the values of the variables at the target population level. Observational studies, on the contrary, aim at estimating the values of the variables at the target population level. *Designing a survey for an experimental study* does not necessary require a probability-based sample, because the major task is getting a sort of case study for investigating causal relationships. For example, in-the-moment surveys, typically reaching the interviewee on the smartphone, are often lacking in probability sampling criteria; thus, mostly they have just a value of experimental studies capturing emotions and opinion when the individual is experimenting some event or action. Observational studies focus at the level of variables estimation; therefore, the probability-based sampling technique is crucial, and the sampling design is an important step. Socioeconomic surveys in general aim at the estimation of the whole target population estimates.

Sub-step B, *Deciding the mode of data collection,* is important because it verifies if organizing a survey only via the web (or mobile web) is feasible and effective. Criteria for mode selection are general and related to several aspects of the research environment and the specific issue. A mobile web survey or a web survey in general, because it is self-completed, fits extremely well for sensitive research questions and/or for short and simple questionnaires. Efficient implementation of complex questionnaires may be efficiently implemented; this happens particularly in official statistics. Web only or mobile web in this case is more problematic; mixed-mode is preferable. One relevant constraint in the use of a probability-based mobile web survey is the availability of an adequate sampling frame. Thus, the choice of the mode depends on many factors, and a critical one is the sampling frame availability. An inadequate mode choice might let many types of errors arise (coverage errors, extremely high unit nonresponse, and so forth) bringing about a poor-quality result. Due to the importance of an adequate mode selection for a probability-based mobile web survey, Thorsdottir and Biffignandi present a flowchart to show the major steps driving the mode choice. Figure 3.2 presents the actions and the decisions to be undertaken when choosing the mode of data collection.

Moving to Figure 3.2, when selecting the mode, the first problem is deciding if it is possible to draw a probability-based survey from target population under study, i.e., the question is if everyone does have an e-mail address. If everyone has an e-mail address and a complete list (good sampling frame) is available, then it is possible to proceed with the web-only survey. However, if the list of e-mail addresses is incomplete (bad sampling frame) or does not exist, the surveyor must

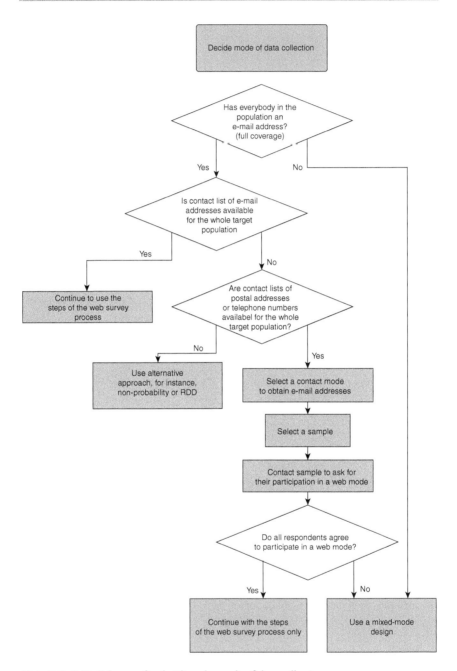

FIGURE 3.2  Sub-steps for deciding the mode of data collection

decide if an alternative sampling frame is available, for example, a sampling frame of telephone numbers or postal addresses. If the alternative sampling frame is available, a mixed-mode approach should be adopted. The surveyor should select a contact mode (telephone or mail) and approach the sampled interviewees to ask them to participate in the survey and if they can provide an e-mail address or not. If the researcher intends to conclude the survey via web, he can provide a personal computer and Internet access (with e-mail address) to those without Internet and e-mail address. In this case, the data collection takes place via a web or mobile web survey. Thus, from this step of Figure 3.2, it is possible to follow the decision steps of the flowchart in Figure 3.1. Whether the researcher don't want to provide Internet access or interviewees do not agree to participate via the web or if they do not provide an e-mail address, the interview should be administered using an alternative mode. In such a case, the surveyor must run a mixed-mode survey with a web component (see Chapter 9). In this case also, from this step of Figure 3.2, it is possible to follow the decision steps of the flowchart in Figure 3.1.

If no alternative sampling frame exists, only a non-probability approach is possible, or an RDD telephone contact could be undertaken.[3]

There are considerable advantages associated with web/mobile web survey compared with face-to-face, mainly in term of cost and timeliness. However, web mode does perform more poorly in terms of coverage and participation. For this reason, researchers sometimes consider using a mixed-mode approach including web, even if no problem exists in term of sampling frame (see experiments and comments in Jäckle, Lynn, and Burton, 2015).

If the conditions to a web survey are present, the surveyor will return to the steps and sub-steps of the Figure 3.1 flowchart.

At this stage, the sub-step to be faced is to *Define the sampling frame and the sampling approach*. This refers only to probability-based surveys. The identification of the sampling frame requires a complete e-mail list of the target population; thus, the entire target population should be online and listed (e-mail list) without under-coverage or duplications.[4] Once the sampling frame and the sampling *approach is decided, the sample selection takes place* (see Chapter 4 about sampling problems and methods). Sampling strategies do not differ from those of traditional surveys. Other sub-steps are the *Questionnaire design* and the *Design paradata methodology*. It is possible to work on in a parallel way. Paradata are usually "administrative data about the survey," and, during the data collection, they are gathered. Web survey paradata include how many times the respondent

---

[3] It should be advised that it is not possible to process non-probability-based survey data using statistical inference methods to estimate population. Comments about probability versus non-probability-based samples and the exclusive advantages of probability-based samples are in Chapter 14.

[4] As stated above, if some units of the target population are not on Internet and/or an adequate online sampling list is available, alternative sampling frames could be available (for instance, ordinary mail addresses), and the sample could be drawn from this sampling list. In this case, sampled units that are not on the Internet could receive the invitation letter either by e-mail or by ordinary mail.

had been accessing the questionnaire, item nonresponses, editing errors, and the time the questionnaire was completed. Thus, there are paradata about each *observation* in the survey. When considering mobile web surveys, a variety of devices could be used to complete the questionnaire; therefore paradata include the type of device, possibly the device where the survey contact/questionnaire is opened and the one from which the completed questionnaire is submitted. Paradata are useful for understanding problems in the questionnaire. For example, questions not clear enough or ambiguous are identified. Usually unclear questions present plenty of item nonresponse and higher response time, and/or they are inducing up and down navigation in the questionnaire. When monitoring survey participation, paradata are useful in order to allow for additional strategies in the solicitation process. For example, looking at the characteristics of the respondents during the data collection, it is possible to use adaptive design (see Chapter 8). This brings to higher response rate and limited costs. Paradata also provide insights into the under-coverage and over-coverage. In summary, paradata are an interesting information source that can improve the survey process and identify errors and their interrelations.

According to Callegaro (2013), in web surveys, it is possible to distinguish between device-type paradata and questionnaire navigation paradata. Device-type paradata provide information regarding the kind of device used to complete the interview (i.e., tablet or desktop). They provide information about the technical features of the device (browser, screen resolution, IP address, and several other characteristics). Questionnaire navigation paradata describe the full set of activities undertaken in completing the questionnaire, for example, mouse clicks, forward and backward movements along the questionnaire, number of error messages generated, time spent per question, and question answered before dropping out (if dropout exists). Other authors (for example, Heerwegh, 2011) distinguish between client-side paradata (they include click mouse and everything related to the activities of the respondent) and server-side paradata (they include everything collected from the server hosting the survey). The literature proposes also other classifications, and the technology evolution is going to offer new types of paradata. Capturing paradata is one of the main challenges. Software industry improves greatly and constantly. Traditionally most programs were collecting only a few server-side paradata. Not every program is registering client-side paradata. Technological innovation and commitment on this important task have contributed to enlarge the offer of programs collecting paradata and transmitting the often unintelligible strings into useful data sets. Due to high innovation in this field, software is fast over; some discussion about the topic is found in Olson and Parkhurst (2013) and in Kreuter (2013).

Currently, it is clear that paradata are useful data types for several different functions, such as monitoring nonresponse and measurement profiles, checking for measurement error and bias, improving questionnaire usability, and fixing many other problems. Due to their potential usefulness in helping to understand relationships between different errors, improving the data collection process, and the quality of results, paradata require a decisional sub-step to plan their structure and data collection.

Regarding the *Design questionnaire* sub-step, several decisions must be undertaken. First, decision about the *Questions' wording* and the *Response format* takes places. *Question wording* rules are similar to the other modes, except that the sentences should be especially simple and short; this recommendation is stronger for web surveys than for the other modes. As regards *Response format*, even if most of the basic criteria are like traditional modes, the general rules and response formats in web surveys are different and specific to a self-completed questionnaire and to the digital format used (see Chapter 7). The criteria for mobile web survey are similar to the ones for web surveys. Some specific requirements are due to the technical structure of the devices, especially to the characteristic of the small size of the screen, especially in smartphones. A poorly structured questionnaire and technically not adequate to mobile phones could critically affect the quality of the survey; errors could arise in terms of response rate, item nonresponse, and estimates. An important issue arises when a mixed-mode approach is adopted. In this case, the decision if the optimal approach for each mode or a unimode approach has to be adopted is debated (see Dillman, Smyth, and Christian 2014). Recently it has been suggested to achieve the balance between the basic presumption that survey questions should be as identical as possible between modes (unimode approach) and, at the same time, consider that mode effects might be reduced by optimizing each questionnaire for corresponding mode (mode-specific approach).

Another important sub-step related to the questionnaire is *Interactivity*. A web process provides the opportunity for automatically interacting with the respondent. For example, pop-up windows with a needed definition or other forms of metadata could appear. Some questions (one or more questions) are compulsory by design, meaning the blocking of the questionnaire's progress, until the questions are fulfilled (Example 3.2, Figure 3.3). Furthermore, the researcher could allow for going up and down throughout the questions or just a question-by-question (top-down) approach. In this case, when going from one page of the questionnaire to the successive page, an error message is given, and the uncompleted question appears again. In some cases, activation of the automatic check is necessary, for example, if the question asks for a percentage composition, the answer check is during the compilation and an error message appears (Example 3.3, Figure 3.4). The error correction takes place before continuing the questionnaire compilation.

### EXAMPLE 3.2 Compulsory question

In this case, the respondent is not allowed for skipping the answer to question 1. An error message appears and the question is submitted again.

## Nutrition & Health

⚠ There was an error on your page. Please correct any required fields and submit again. Go to the first error

⚠ This question is required

1. Do you think that the quality of food products has an impact on health? *

○ Yes

○ No

[ Next ]

FIGURE 3.3　**Error message for a compulsory question in the Nutrition & Health Survey**

Interactivity provides advantages in the quality of the web or mobile web surveys by reducing imputation errors and item nonresponses (see Example 3.4). The desired degree of interactivity is guiding the questionnaire implementation, since more sophisticated interactivity actions are supported only from some specific programs.

▣　EXAMPLE 3.3　**Automatic control in web questionnaire**

The question is asking for a percentage composition. Automatically is checked if the sum is equal to 100%.

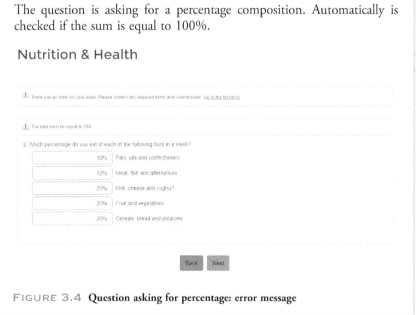

## Nutrition & Health

⚠ There was an error on your page. Please correct any required fields and submit again. Go to the first error

⚠ The total must be equal to 100

2. Which percentage do you eat of each of the following food in a week?

| | |
|---|---|
| 10% | Fats, oils and confectionery |
| 12% | Meat, fish and alternatives |
| 20% | Milk, cheese and yoghurt |
| 20% | Fruit and vegetables |
| 20% | Cereals, bread and potatoes |

[ Back ] [ Next ]

FIGURE 3.4　**Question asking for percentage: error message**

▣ EXAMPLE 3.4 **Paper versus web: errors comparison**

In the transition phase from paper to web for the business survey (structural business statistics [SBS] survey) conducted in Italy, errors in the paper and web questionnaire are compared. From Table 3.1, it is evident there is a general improvement in various types of errors: less need for checking (checks pending) in web data, a smaller average number of errors corrected, and fewer replacements.

TABLE 3.1 **Types of detected errors by response mode (averages on total responses: SBS survey)**

|       | Checks pending | Corrected errors | Replacements |
|-------|----------------|------------------|--------------|
| Paper | 0.89           | 4.74             | 4.89         |
| Web   | 0.85           | 2.88             | 3.39         |

The third important sub-step in questionnaire design is *Visualization*. This is a critical point in a web/mobile web survey. Colors, pictures, character formats, and the presence or absence of a progress bar are all factors affecting the interviewee's perception and could greatly improve or reduce response errors (response values elated to the interpretation of the content and of the questions, item nonresponses, decision to participate in the survey, etc.). Colors, for instance, affect the readability of the screen, possibly making completion less pleasant. Dark and highly contrasted colors are more difficult to read, as well as too hell colors. Formats and pictures have a different impact if presented on a PC screen rather than on a smartphone screen. On a small screen pictures are disturbing. Thus, the visual readability of the questionnaire is essential to enhance participation, not increase measurement errors (due to bad understanding of the questions or distraction due to not adequate—in the content and in the size—pictures).

In *Designing the questionnaire,* it is important considering that the questionnaire will be answered using different devices (desktop, laptop, tablet, smartphone) (see details in Chapter 5). The question with error message of Figure 3.5 shows how it looks like on a smartphone. If a mobile device accesses a website without a mobile version, the user will still be able to navigate the page. However, differences in screen size will usually require the user to perform gestures or scrolling in order to browse the content in its entirety. Therefore, if a smartphone accesses a nonmobile version of an online survey, then it is likely that the respondent will see only a portion of the content/question or question completion may require zooming first in order to select desired response. Some question types (e.g., multiple choice, lists) will not generally take advantage of the smartphone's native features. Peytchev and Hill (2010) reported the results for a series of experiments comparing various aspects of questionnaire design and layout, including horizontal

FIGURE 3.5 Message for asking answer to a compulsory question: smartphone device

scrolling, number of questions per screen, direction of response options, impact of embedded images, and the use of open-ended options, among others, using a smartphone. Couper and Mavletova (2014) explore the effect of scrolling versus paging design on the break-off rate, item nonresponse, and completion time in mobile web surveys. The scrolling design leads to significantly faster completion times, lower (though not statistically significant) break-off rates, fewer technical problems, and higher subjective ratings of the questionnaire. In general, web/mobile web survey design display should be accurately adapted on mobile phones, netbooks, and tablets as they do on desktop and laptop computers. Let's say a mixed devices approach should be adopted. When the questionnaire design is completed and the paradata decided, the software or the selection of the *programming language* to implement the web questionnaire is undergone, and the digital version of the questionnaire (including paradata and metadata) is created.

Following the implementation of the web questionnaire and the sample selection, the fourth main step of the survey process is the **Collecting data**. Conducting data collection implies sending the web questionnaire link to the sampled units along with an invitation letter for survey participation (see Chapter 7) and monitoring the data collection process, for example, by sending solicitations and, eventually, applying the responsive design (see Chapter 8) to better finalize the sampled units' participation.

Data processing takes place from the fifth step forward. ***Processing data—step 1*** is on *database creation*, where many error risks need identification and their corrections adopted. For instance, item nonresponse must be considered, the reasons for them evaluated, and, when necessary, imputation methods applied (sub-step *Data imputation*). Coding should also be considered a sub-step (sub-step *Code open questions*). Erroneous coding could cause misinterpretation of the survey results. Compilation of the questionnaire automatically generates the database and avoids the significant risk of errors connected with data transcription. It is a great advantage of web surveys; data errors are only due to incoherence or respondent little attention. Therefore, a data quality check is still necessary in web surveys, even if the risks for errors are smaller than in traditional paper-and-pencil interviews.

Once the checked database has been created, the second phase of data processing takes place (*Processing data—step 2*). Note that several computations and activities carried out in the second phase of data processing are undertaken according to decisions done at the *Designing the web survey* step. Application of the estimation and weighting techniques allows for the extension of the survey results to the target population and to present the survey results (sub-steps: *Estimation technique choice, Calculation of weights and estimators,* and *Process data and produce tables*; see Chapters 12 and 13). Estimation procedures are like to those applied in the traditional survey modes. However, several aspects encompass many advantages that are specific to the web. In many cases, it could be possible to link the survey microdata to another database providing individual data for auxiliary variables. A key variable useful for connecting the two databases at the individual level must be available though. Through database integration, modeling survey results, understanding participation behavior, and better profiling the respondents are possible. Integration with administrative databases, paradata databases, or other survey databases could be undergone.

Thus, in selecting the estimation techniques, the researcher needs to integrate the survey database (sub-step: *Integrate datasets*) with another database (e.g., an administrative database or a paradata database). This is useful when the researcher has planned to use auxiliary variables in the data processing, to compute weights, or to improve estimates. For instance, propensity scores technique (Chapter 13) is one approach that uses auxiliary variables for estimation purposes.

After deciding possible integration and applying the appropriate estimation and weighting technique, the data processing ends up; tables and figures are then produced to synthesize the survey results. Presenting the survey results includes offering quality indicators as well. AAPOR (2016) provides a detailed description of rates and indicators and emphasizes that all survey researchers should adopt the standardized final disposition for the outcome of the survey. For example, response rate calculation is in many different ways. Not to simply state "the response rate is $X$" is important. The researcher should exactly name the rate he is talking. In particular, there are two types of response rate:

1. The response rate type 1 (RR1), i.e., the number of complete interviews divided by the number of interviews plus the number of non-interviews (i.e. refusal, break-off plus non-contacts and others, plus all cases of unknown eligibility);

2. The response rate type 2 (RR2), where also partial interviews are considered at the numerator of the rate.

In web surveys, as discussed in Chapter 5, in regard to possible errors that might occur throughout the process, quality also uses specific indicators that take advantage of the paradata collected by the survey system during the survey process. For example, it is possible to compute the time spent in the completion of the questionnaire.

The final step, the *End of the survey process*, involves writing the conclusions. Usually, a final reporting is written. A clear and consistent reporting of the methods complements comments about substantive research results. This is a key component for a methodologically reliable survey. AAPOR (2016) is stressing this idea. A set of standardization of the outcome details and outcome rates has been proposed, and they should be available and part of a survey documentation.

## 3.3 Application

In the construction of the termed PAADEL (Agro-Food and Demographic Panel in Lombardy), the steps described in the flowchart (Figures 3.1 and 3.2) were followed. This section examines some problems that arise at the different steps and the interrelationship between errors. In particular, focus is on the step of choice of the mode and on the use of adaptive design during the step of the data collection.

Skipping the first two steps (Deciding the objective, Metadata description, and Designing the survey), which are related to the survey's subject matter and beyond the present study's scope, focus should shift to the mode choice. The panel's objective was to have a probability-based web data collection. Target population were Lombardy region inhabitants. Internet penetration in the population was not high, and an exhaustive list of e-mail addresses did not exist; thus, there was a need for a proxy for the population list. The proxy list available contained postal addresses and phone numbers. Therefore, a probability-based survey was not possible without some preliminary step to select a probability-based sample; the adoption of a mixed-mode approach to cover the part of the population not on the Internet was the solution. Thus, a *contact mode* had to be decided, considering that postal and telephone number codes were available. The selection of the survey mode for the sampled units took place using a mixed-contact mode, partially telephone and partially mail. The survey mode was a mixed-mode as well; only part of the sampled units had an e-mail address and accepted a web survey, while for some others the interview was by telephone or mail.

TABLE 3.2   **Responses to the survey by mode and percentage composition**

| Mode | % | Mode | % |
|---|---|---|---|
| Web | 68.5 | Web | 45.5 |
| Phone | 71.2 | Phone | 30.2 |
| Mail | 53.2 | Mail | 24.3 |
| **Total** | **63.5** | **Total** | **100.0** |

The data collected by mode are in Table 3.2. The response rate has been satisfactory. The web mode turned out to be the most important component of the mixed-mode approach.

To address the web component of the mixed-mode approach, the steps in Figure 3.1 were followed.

To show how to use, at the monitoring data collection step, adaptive design to increase the response rate in a web survey and how different error types are interrelated, Bianchi and Biffignandi (2014) applied experimental responsive design strategies, in retrospect, to the recruitment of the mixed-mode panel. Especially targeting the units contributing the most to nonresponse bias during data collection was useful. The identification of such units took place through indicators representing proxy measures for nonresponse bias. In their study, the authors adopted three strategies, and the results show that the method proves promising in reducing nonresponse bias. When evaluated in the TSE framework, i.e., considering how different errors relates to the adopted responsive strategies, the results are not uniform across variables. In general, there is a reduction of total errors.

# 3.4  Summary

A flowchart for the web/mobile web survey process of a probability-based survey is proposed and discussed. The flowchart will be useful for practitioners and researchers. Practitioners have a guide to follow when undertaking the survey in an efficient way, without forgetting or overlooking important decisional steps. Not considering the steps in the flowchart could compromise the survey's quality and increase the amount of errors. Because a detailed description of the most important flowchart steps is in the chapters of this book, surveyors have the opportunity to gain a deeper insight into different issues ad techniques.

When analyzing their empirical results and evaluating errors and the risk of errors, researchers can identify the steps or sub-steps to which the results are related and determine how the decisions on one step (or sub-step) could affect the results at other steps, thus improving the quality of the survey process.

## KEY TERMS

**Metadata**: Metadata is "data that provides information about other data." In short, it's data about data.

**Mixed-mode survey**: A survey in which various modes of data collection are combined. Modes can be used concurrently (different groups are approached by different modes) or sequentially (nonrespondents of a mode are re-approached by a different mode).

**Paradata**: Paradata of a web survey are data about the process by which the data were collected.

**Probability-based panel:** A panel for which members are recruited by means of probability sampling.

## EXERCISES

**Exercise 3.1**    Selecting the survey mode for a probability-based survey requires knowing:

**a.** The sampling frame list of the web population.

**b.** The sampling frame list of the whole population (both web and non-web).

**c.** Only the sampling frame list of the non-Internet population.

**d.** Only the phone number of the Internet population.

**Exercise 3.2**    When a bias error occurs?

**a.** When designing the survey.

**b.** When choosing the sampling technique.

**c.** When drawing the sample.

**d.** When estimating the model.

**Exercise 3.3**    Probability-based web survey can take place:

**a.** When the target frame list is available.

**b.** Whether people invitation to participate in a survey appears when accessing a website.

**c.** When an e-mail list of a few people of the target population is available.

**d.** When the list of e-mail addresses is available.

**Exercise 3.4**   The flowchart is describing:

**a.** The sampling methods.

**b.** Step of actions and decisions in a web survey.

**c.** When an e-mail list of a few people of the target population is available.

**d.** When the list of e-mail addresses is available.

**Exercise 3.5**   Total survey error approach is considering:

**a.** Only sampling errors.

**b.** The overall quality of the survey steps.

**c.** Only the measurement error.

**d.** A non-probability-based survey.

**Exercise 3.6**   The selection of an inadequate mode is affecting:

**a.** Only the coverage of the sampling frame.

**b.** Only the response rate.

**c.** The overall quality of the survey.

**d.** The response rate and the sampling frame.

## REFERENCES

Bethlehem, J. & Biffignandi, S. (2012), *Handbook of Web Surveys*, 1st edition. Wiley, Hoboken, NJ.

Bianchi, A. & Biffignandi, S. (2014), Responsive Design for Economic Data in Mixed-Mode Panels. In: Mecatti, F., Conti, P. L., & Ranalli, M. G. (eds.), *Contributions to Sampling Statistics*. Springer, Berlin, pp. 85–102.

Biemer, P., de Leeuw, E., Eckman, S., Edwards, B., Kreuter, F., Lyberg, L. E., Tucker, N., & West, B. T. (eds.). (2017), *Big Data in Practice*. Wiley, Hoboken, NJ.

Callegaro, M. (2013), Paradata in Web Surveys. In: Kreuter, F. (ed.), *Improving Surveys with Paradata: Analytic Uses of Process Information*. Wiley, Hoboken, NJ, pp. 261–280.

Couper, M. & Mavletova, A. (2014), Mobile Web Surveys: Scrolling versus Paging; SMS versus e-mail Invitations. *Journal of Survey Statistics Methodology*, 2, pp. 498–518.

Dillman, D., Smyth, J., & Christian, L. M. (2014), *Internet, Mail and Mixed Mode Surveys. The Tailored Design Method*. Wiley, Hoboken, NJ.

Groves, R. M. (1989), *Survey Errors and Survey Costs*. Wiley, New York.

Heerwegh, D. (2011), Internet Survey Paradata. In: Das, M., Ester, P., & Kaczmirek, L. (eds.), *Social and Behavioral Research and the Internet: Advances in Applied Methods and Research Strategies*. Taylor and Francis, Oxford.

Jäckle, A., Lynn, P., & Burton, J. (2015), Going Online with a Face-to-Face Household Panel: Effects of a Mixed Mode Design on Item and Unit Non-Response. *Survey Research Methods*, 9, 1, pp. 57–70.

Kreuter, F. (2013), *Improving Surveys with Paradata: Analytic Uses of Process Information.* Wiley, New Jersey.

Olson, K. & Parkhurst, B. (2013), Collecting Paradata for Measurement Error Evaluation. In: Kreuter, F. (ed.), *Improving Surveys with Paradata: Analytic Uses of Process Information.* Wiley, Hoboken, NJ, pp. 73–95.

Peterson, G., Griffin, J., La France, J., & Li, J. (2017), Smartphone Participation in Web Surveys. In: Biemer, P., de Leeuw, E., Eckman, S., Edwards, B., Kreuter, F., Lyberg, L. E., Tucker, N., & West, B. T. (eds.), *Total Survey Error in Practice.* Wiley, Hoboken, NJ.

Peytchev, A. & Hill, C. A. (2010), Experiments in Mobile Web Survey Design: Similarities to Other Modes and Unique Considerations. *Social Science Computer Review*, 28, pp. 319–333.

Platek, R. & Särndal, C. (2001), Can a Statistician Deliver. *Journal of Official Statistics*, 17, 1, pp. 1–20.

The American Statistical Association for Public Opinion Research (AAPOR). (2016), *Standardized Definitions, Final Disposition of Case Codes and Outcome Rates for Surveys*, 9th edition. AAPOR.

Tourangeau, R., Conrad, F., & Couper, M. (2013), *The Science of Web Surveys.* Oxford University Press, New York.

Weisberg, H. (2005), *The Total Survey Error Approach: A Guide to New Science of Survey Research.* Wiley, Hoboken, NJ.

# CHAPTER FOUR

# Sampling for Web Surveys

## 4.1 Introduction

A web survey is a survey in which data is collected using the World Wide Web. Like for all surveys, the aim of a web survey is to investigate a well-defined population. Such populations consist of concrete objects, like persons, households, or companies. It is typical for a survey that information is collected by means of asking questions of the representatives of the objects in the population. To do this in a uniform and consistent way, a questionnaire is used.

One way to obtain information about a population is to collect data about all its elements. Such an investigation is called a *census* or *complete enumeration*. Traditionally censuses are based on paper questionnaires and on face-to-face interviews. This approach has a number of disadvantages:

- It is very expensive. It involves a large amount of people (for example, interviewers) and other resources.

- A census is very time consuming. Collecting and processing a large amount of data take time. This affects the timeliness of the results. Less timely information is less useful.

- Large investigations increase the response burden of people. As many people are more frequently asked to participate, they will experience it more and more as a burden. Therefore, they will be less and less inclined to cooperate.

*Handbook of Web Surveys*, Second Edition. Silvia Biffignandi and Jelke Bethlehem.
© 2021 John Wiley & Sons, Inc. Published 2021 by John Wiley & Sons, Inc.

The last census for England and Wales took place in March 2011. It involved around 25 million households. It was possible to complete the questionnaire online. For the 2010 census in the United States, the Internet was not used. The reason was that the U.S. Census Bureau did not expect the use of the Internet would lead to lower costs or higher response rates. Moreover, there was concern about increased security risks. This decision resulted in over 120 million forms being mailed to every house in the address list.

In many other countries censuses have been conducted in 2010 or 2011. For example, the demographic census of Italy took place in October 2010. Experimentally online participation was an option. Paper questionnaires were sent to households. The households had the choice to complete either the paper form or a form on the Internet. They could also return the form directly to a municipality official. The Internet was also used for the 2010 agricultural survey. Farmers had the choice to complete a web questionnaire or to wait for a face-to-face interview. More than 66,000 questionnaires were completed on the web.

It is interesting to note that a special Facebook page was created for the Italian census. Individuals could comment on this page about the use of the Internet for data collection. Most posted comments were positive: "fast," "accurate," "clear," "efficient," and "absolutely necessary."

A *survey* collects information on only a small part of the population. This small part is obtained by taking a *sample*. The sample only provides information on the sampled elements of the population. No information will be obtained on the non-sampled elements. Still, if the sample is selected in a scientifically sound way, it is possible to make inference about the population as a whole. This principle of generalizing from the sample to the population is summarized in Figure 4.1.

"Scientifically sound" sample selection means that the sample is selected using *probability sampling*. If it is clear how this selection mechanism works and it is possible to compute the probabilities of selection in the sample, reliable and precise conclusions can be drawn about the population as a whole. The principles of probability sampling have been successfully applied in official and academic statistics since the 1940s and, to a lesser extent, in commercial market research, too. For an overview of the history of survey sampling, see, for example, Chapter 1 or Bethlehem (2009).

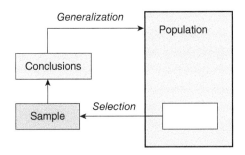

**FIGURE 4.1  Generalizing from a sample to a population**

There are various ways in which probability samples can be selected. Choice may depend on what is available with respect to sampling frames and auxiliary information. It may also depend on what the objective of the survey is. This chapter provides an overview of sampling techniques that are relevant to web surveys.

## 4.2 Theory

### 4.2.1 TARGET POPULATION

The first step in setting up a web survey is to define the *target population*. This is the population to investigate and to which the conclusions refer. Such a population need not necessarily consist of people, examples could be all IT companies in a certain state, all schools in a country, or all farms in a specific region. Example 4.1 is about target population identification.

The definition of the target population must be unambiguous. It must always be possible to determine in practice whether a certain element does or does not belong to the target population. Both failures to include relevant elements in the population and exclude irrelevant ones may affect the survey results.

EXAMPLE 4.1 **A survey about road pricing**

There was an intensive political discussion in the Netherlands in January 2010 about the introduction of a system of road pricing. An important participant in this discussion was the Dutch Automobile Association (ANWB). This organization conducted a poll on its website. It was a self-selection survey. No sample was selected. Everybody could participate. The poll was supposed to determine the opinion of the ANWB members about road pricing. The target population consisted of all members of the association. However, everyone could participate. One could participate even more than once. There was nothing preventing this. All respondents were asked whether they were a member. Consequently, nonmembers could be excluded from the analysis.

Within a period of a few weeks, the questionnaire was completed more than 400,000 times. About 50,000 respondents indicated they were not an ANWB member. One should take into account that people may not have answered the question about membership properly. There is always a risk of socially desirable answers.

Denote the target population by $U$. Assume it is finite and the number of elements $N$ known. Note that this is not always the case. Examples are the number of people having access to Internet or the number of foreign visitors in a country.

The elements of the target population must be *identifiable*. It means they can uniquely be assigned sequence numbers 1, 2, …, N. For each element encountered in practical situations, it must be possible to determine its sequence number. In some cases, this process is straightforward. An example is a population of persons identified by means of their social security number. Denote the target population by

(4.1)                                            $U = \{1, 2, ..., N\}.$

In the survey design phase, the researcher translates the objectives of the survey into concrete operational procedures. This involves defining the *target variables* of the survey. See Example 4.2. These variables measure the various aspects of the phenomena to investigate.

### EXAMPLE 4.2  Target variables

Suppose a research organization intends to carry out a General Election Survey. Objective of the survey is to measure voting behavior. Target variables could be whether one voted (yes/no), party one voted for (choice from a list), reasons for voting for this party (open question), etc.

A target variable is denoted by the letter $Y$. The values of this variable for all elements in the population $U$ are indicated by

(4.2)                                            $Y_1, Y_2, ..., Y_N.$

So, $Y_k$ is the value of variable $Y$ for element $k$, where $k = 1, 2, …, N$. Usually, an additional number of variables are measured in the survey. These variables are called *auxiliary variables*. These variables assist in differentiating the survey results for various subpopulations. They can also be very useful in improving estimates of population characteristics. Examples of auxiliary variables are demographic characteristics like sex, age, and marital status. Denote an auxiliary variable by the letter $X$. The values of variable $X$ for all elements in the population $U$ are indicated by

(4.3)                                            $X_1, X_2, ..., X_N.$

So, $X_k$ is the value of $X$ for element $k$, where $k = 1, 2, …, N$. Here, $Y_k$ and $X_k$ indicate single values. Both target variables and auxiliary variables can be of one of three types:

- *Continuous variables.* These variables measure quantities, amounts, sizes, or values. It is possible to carry out meaningful computations on these values, like calculating totals and averages. Examples of such variables are income and age of a person, the number of cars he owns, etc.
- *Categorical variables.* These variables divide the target population into subpopulations. The values denote labels of categories. Elements with the same label belong to the same category. It is not meaningful to carry out computations on the values of a categorical variable. Examples of categorical variables are race, religion, marital status, and region of residence.
- *Indicator variables.* Such a variable measures whether or not an element has a certain property. It can only assume two values 0 and 1. The value is 1 if an element has the property, and the value is 0 if it does not have it. An example of an indicator variable is employment. If a person has a job, the value of the variable is 1, and otherwise its value is 0.

The aim of a survey is to get information about the target population. This information is quantified in the form of *population parameters*. A population parameter is a function that only depends on the values in the population for one or more variables. These variables can be target variables as well as auxiliary variables.

One simple example of an often used population parameter for a continuous variable is the *population total*:

$$(4.4) \qquad Y_T = \sum_{k=1}^{N} Y_k = Y_1 + Y_2 + \cdots + Y_N.$$

Suppose the target population consist of all farms in a country and $Y$ denotes the number of cows a farm has, then the population total is the total number of cows in the country. Related to the population total is the *population mean*:

$$(4.5) \qquad \overline{Y} = \frac{1}{N} \sum_{k=1}^{N} Y_k = \frac{Y_1 + Y_2 + \cdots + Y_N}{N} = \frac{Y_T}{N}.$$

The population mean is simply obtained by dividing the population total by the population size. If the target population consists of all inhabitants of a town and $Y$ denotes the age of a person, the population mean is the mean age in this town. Another important population parameter is the *adjusted population variance*. It is defined by

$$(4.6) \qquad S^2 = \frac{1}{N-1} \sum_{k=1}^{N} \left( Y_k - \overline{Y} \right)^2.$$

This quantity gives an indication of the amount of variation in the values of the target variable. If all values of $Y$ are equal, the variance is 0. The more the values of $Y$ are apart, the larger the variance will be. The adjusted population variance also appears in formulas for the variance of estimators.

For indicator variables, the population total denotes the number of elements in the population having a certain property. The population mean is the fraction of elements with that property. The *population percentage* is defined by

$$(4.7) \quad P = 100\overline{Y} = \frac{100}{N} \sum_{k=1}^{N} Y_k = 100 \frac{Y_1 + Y_2 + \cdots + Y_N}{N} = 100 \frac{Y_T}{N}.$$

Note that for indicator variables, the adjusted population variance reduces to

$$(4.8) \qquad S^2 = \frac{1}{N-1} \sum_{k=1}^{N} \left(Y_k - \overline{Y}\right)^2 = \frac{P(100-P)}{N-1}.$$

There are no specific population parameters for categorical variables. Of course, totals, fractions, or percentages of elements in categories can be estimated. In fact, this comes down to replacing the categorical variable by a set of indicator variables, one for each category. The focus in this book is on estimating population means and population percentages.

## 4.2.2 SAMPLING FRAMES

To select a sample from a target population in a scientifically sound way, two ingredients are required: a sampling design based on some form of probability sampling and a sampling frame. Several sampling designs are described in subsequent sections. This section will discuss sampling frames.

A *sampling frame* is a complete listing of all elements in the target population. For every element in the list, there must be information on how to contact that element. Such contact information can consist of, for example, a name and address, a telephone number, or an e-mail address. Lists can exist on paper (a card-index box for the members of a club, a telephone directory) or in a computer (a database containing a register of all companies). If such lists are not available, detailed geographical maps are sometimes used. Example 4.3 discusses a postal list as sampling frame, whereas Example 4.4 is about the recruitment of the LISS panel.

Some countries, like the Netherlands and the Scandinavian countries, have a population register. The population register of the Netherlands is a decentralized system. Each municipality maintains its own register. Demographic changes related to their inhabitants are recorded. It contains information on gender, date of birth, marital status, and country of birth. Periodically, all municipal information is combined into one large register, which is used by Statistics Netherlands as a sampling frame for its surveys. Samples of persons can be selected from it.

In some countries, there are address lists available. These lists contain addresses of all houses in the countries. Such address lists may have been compiled as part of a census operation. If they are maintained well, they can be used to select a sample of households. In Italy, Istat accesses the LAC (Anagrafical List of the Municipalities) as a frame for surveys and for the census.

**EXAMPLE 4.3** **Postal Address Files**

An often used sampling frame is a *Postal Address File* (PAF). Postal service agencies in several countries maintain databases of all delivery points in the country. Examples are the Netherlands, the United Kingdom, Australia, and New Zealand. Such databases contain postal addresses of both private houses and companies. Typically, a PAF can be used to draw a sample of addresses and therefore also of households.

It is sometimes not clear whether addresses in a PAF belong to private houses or to companies. If the aim is to select a sample of households, there may be over-coverage caused by companies in the file.

The ideal sampling frame for a web survey would be a list of e-mail addresses of all members of the target population. Each element is represented by an e-mail address. A random sample of elements can be selected from this list. An e-mail is sent to all selected elements. This e-mail contains a link to the website of the online survey questionnaire.

It is important to control access to the questionnaire with a unique identification code for each sample element. This guarantees that only selected elements can get access to the questionnaire. The unique code can also help to prevent that someone completes a questionnaire more than once. The unique identification code should be included in the e-mail that is sent to the selected persons. Such codes are often part of the link to the survey website.

Sometimes, a sampling frame consisting of e-mail addresses is indeed available. An example is a large company where all employees have their own company e-mail address. Another example is an educational institution where all registered students have an e-mail address. For general population surveys, such a sampling frame is usually not available. Then a different sampling frame must be used. If a population register or address register is available, a letter can be send to a sample of persons or addresses. This letter invites selected persons to participate in the survey. It also contains a link to the survey questionnaire website. This approach is somewhat more cumbersome. If the link is mentioned in an e-mail, it is sufficient to click on it to start the questionnaire. If the link is mentioned in a letter, it must be typed in, which involves a risk of making typing errors.

■ EXAMPLE 4.4 **Sample selection for the LISS panel**

As part of an attempt to boost the Dutch knowledge economy, the government of the Netherlands granted funds in 2006 to establish a web panel consisting of approximately 5,000 households. This panel is called the *LISS panel*. LISS stands for Longitudinal Internet Studies for the Social Sciences. Universities are invited to submit research proposals that can be carried in this panel. See Scherpenzeel (2008) for a detailed description of this panel.

The panel has been constructed by selecting a random sample of households from the population register of the Netherlands. Selected households were recruited for this panel by means of a face-to-face interview (CAPI) or a telephone interview (CATI).

An initial sample of 10,150 addresses was selected from the population register. For each sample address, an attempt was made to find a corresponding landline telephone number. This was successful in 70% of the cases. These addresses were approached by telephone (CATI) for a recruitment interview. The other 30% of the cases were approached face-to-face (CAPI) for recruitment.

Unusable addresses were removed. These were, among others, non-existing addresses, non-inhabited addresses, business addresses, and addresses with people unable to participate (due too long-time illness or language problems). The remaining sample size was 9,944 addresses.

The recruitment interview was a 10 minute interview, in which some basic demographic questions were asked, as well as questions about an Internet connection at home, social integration, political interest, leisure activities, survey attitudes, loneliness, and personality. At the end of the interview, the respondent was invited to become a member of the web panel.

If the contacted household refused to participate in the short recruitment interview, it was asked to answer just three basis questions. These three questions were followed by the invitation to participate in the panel.

Table 4.1 shows the % of addresses at different phases of the recruitment process.

The response for the recruitment interview was reasonably high (75%). Of those respondents, 84% (corresponding to 64% of the addresses) expressed willingness to participate in the panel. Of those willing to participate, ultimately 75% did (corresponding to 48% of the starting cases).

TABLE 4.1    **The recruitment process for the LISS panel**

| Phase of the recruitment process | Remaining % of addresses |
| --- | --- |
| Start with 9,844 addresses | 100 |
| Completed CAPI or CATI recruitment or basic questions | 75 |
| Willing to participate in panel | 64 |
| Actually participated | 48 |

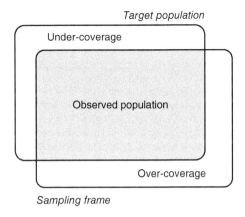

**FIGURE 4.2** Target population and sampling frame

The sampling frame should be an accurate representation of the population. There is a risk of drawing wrong conclusion from the survey if the sample is from a sampling frame that differs from this population. Figure 4.2 shows what can go wrong.

The first problem is *under-coverage*. This occurs if the target population contains elements that do not have a counterpart in the sampling frame. Such elements are never selected in the sample. An example of under-coverage is survey where the sample is selected from a population register. Illegal immigrants are part of the population, but will never be encountered in the sampling frame. Another example is a survey in which selection of the respondents is via the Internet. Then there will be under-coverage due to people without Internet access. Under-coverage can have serious consequences. If the elements outside the sampling frame systematically differ from the elements in the sampling frame, estimates of population parameters may be seriously biased. A complicating factor is that it is often not very easy to detect the existence of under-coverage.

The second sampling frame problem is *over-coverage*. This refers to the situation where the sampling frame contains elements that do not belong to the target population. If such elements end up in the sample and their data are used in the analysis, estimates of population parameters may be affected. It should be rather simple to detect over-coverage in the field. This should become clear from the answers to the questions asked.

Another example is given to describe coverage problems. Suppose a survey is carried out among the inhabitants of a town. It is decided to collect data by means of a web survey. Since there is no sampling frame containing e-mail addresses of the inhabitants, it is decided to recruit people for the survey by telephone. A sample of inhabitants is selected from the telephone directory of the town. Unfortunately, this sampling frame can have serious coverage problems. Under-coverage occurs because many people have unlisted numbers and some have no phone at all. Moreover, there is a rapid increase in the use of mobile phones.

In many countries, mobile phone numbers are not listed in directories. In the Netherlands not more than 60% of the people can be found in the telephone directory. A telephone directory also suffers from over-coverage, because it contains the telephone numbers of shops, companies, etc. Hence, it may happen that people are contacted that do not belong to the target population. Moreover, some people may have a higher contact probability than anticipated, because they can be contacted both at home and in the office.

A survey is often supposed to measure the status of a population at a specific moment in time. This is the so-called reference date. The sampling frame should reflect the status at this reference date. Since the sample will be selected from the sampling frame prior to the reference date, there may be discrepancies. The sampling frame may contain elements that do not exist any more at the reference date. People may have died or companies may have ceased to exist. These are cases of over-coverage. See Example 4.5. It may also happen that new elements have come into existence after the time of sample selection and before the reference date. For example, a person moves into town or a new company is created. These are cases of under-coverage.

---

**EXAMPLE 4.5 Over-coverage or nonresponse?**

Suppose a survey is carried in a town among the people of age 18 and older. Objective is to describe the situation at the reference date of May 1. The sample is selected in the design phase of the survey, say, on April 1. It is a large survey, so data collection cannot be completed in one day. Therefore, interviews are conducted in a period of two weeks, starting one week before the reference date and ending one week after the reference date.

Now suppose an interviewer contacts a selected person on April 29. It turns out the person has moved to another town. This is a case of over-coverage. What counts is the situation on May 1, and the person did not belong any more to the target population at the reference date. So, there is no problem. Since this is a case of over-coverage, it can be ignored.

The situation is different if an interviewer attempts to contact a person on May 5 and this person turns out to have moved on May 2. This person belonged to the target population at the reference date and therefore should have been interviewed. This is not a coverage problem, but a case of nonresponse. The person should be tracked down and interviewed.

---

Problems may also occur if the units in the sampling frame are different from those in the target population. Typical is the case in which the target population consists of persons and the sampling frame of addresses. This may happen if an address list (for example, a telephone directory) is used as a sampling frame.

Suppose persons have to be selected with equal probabilities. A naive way to do this would be to randomly select a sample of addresses and to draw one person at each selected address. At first sight, this is reasonable, but it ignores the fact that now not every person has the same selection probability: members in large families have a smaller probability of being selected than members of small families.

### 4.2.3 BASIC CONCEPTS OF SAMPLING

To be able to obtain reliable estimates of population parameters, a random sample is selected from the population. The elements in this sample are obtained by means of a random selection procedure. This procedure assigns to every element in the target population a fixed, positive, and known probability of selection. The most straightforward way to select a random sample is giving each element the same probability of selection. Such a random sample is called a *simple random sample.*

Samples can be selected with replacement or without replacement. *Sampling with replacement* means that a selected element is returned to the population (after its characteristics have been recorded) before the next element is drawn. It is possible to select an element more than once. *Sampling without replacement* means that a selected element is not returned to the population. Therefore, an element can only be selected at most once in a sample. Selecting an element more than once does not produce more information than selecting it once. Hence, selection without replacement is usually preferred.

Assume here that the sample selection is without replacement. It means that each element can appear at most once in the sample. Thus, the sample representation is by a set of indicators

$$(4.9) \qquad a = a_1, a_2, ..., a_N.$$

The indicator $a_k$ assumes the value 1 if element $k$ is selected in the sample, and otherwise it assumes the value 0, for $k = 1, 2, ..., N$. The expected value of $a_k$ is denoted by

$$(4.10) \qquad \pi_k = E(a_k).$$

The quantity $\pi_k$ is called the *first-order inclusion probability* of element $k$ (for $k = 1, 2, ..., N$). For deriving variance formulas, also second-order inclusion probabilities are required. The *second-order inclusion probability* of elements $k$ and $l$ (with $k \neq l$) is equal to

$$(4.11) \qquad \pi_{kl} = E(a_k a_l),$$

and by definition $\pi_{kk} = \pi_k$. The *sample size*, i.e., the number of selected elements, is denoted by $n$. Since the indicators $a_k$ have the value 1 for all elements in the sample and the value 0 for all other elements, the sample size is equal to the sum of the values of the indicators:

$$(4.12) \qquad n = \sum_{k=1}^{N} a_k.$$

The values of the target variable are observed for the sampled elements. These values are available to estimate population characteristics. The recipe to compute such an estimate is called an *estimator*, and the result of this computation is called the *estimate*. In order to be useful, an estimator must have a number of properties:

- The estimator must be *unbiased*. This means that average value of the estimate over all possible samples must be equal to the value of the (unknown) population parameter to estimate. On average, the estimator will result in the correct value. It will never underestimate or overestimate the population value in a systematic way. Consequently, the expected value $E(z)$ of an estimator $z$ must be equal to the value of the population parameter to be estimated: $E(z) = Z$.
- The estimator must be *precise*. It means that the variation in possible outcomes must be small. Consequently, the variance $V(z)$ of an estimator $z$ of a population parameter $Z$ must be small over all possible samples.
- *For reasons of simplicity*, linear estimators are preferred. It means an estimate is computed as a linear combination of the observed values of the target variable.

Imposing the conditions of unbiasedness and linearity leads to the estimator introduced by Horvitz and Thompson (1952). This estimator for the population mean is defined as

$$(4.13) \qquad \bar{y}_{HT} = \frac{1}{N} \sum_{k=1}^{N} a_k \frac{Y_k}{\pi_k}.$$

The indicators $a_k$ see to it that only the available sample values of the target variable are used in the computation of the estimate. Note that each value $Y_k$ is weighted with the inverse selection probability $\pi_k$. Thus, the estimator corrects for the fact that elements with a large inclusion probability are overrepresented in the sample.

The Horvitz–Thompson estimator (4.13) is an unbiased estimator of the population mean. The variance of this estimator is equal to

$$(4.14) \qquad V\left(\bar{y}_{HT}\right) = \frac{1}{N^2} \sum_{k=1}^{N} \sum_{l=1}^{N} (\pi_{kl} - \pi_k \pi_l) \frac{Y_k}{\pi_k} \frac{Y_l}{\pi_l}.$$

For without replacement samples of fixed size $n$, the variance can be rewritten in the form

$$(4.15) \qquad V(\bar{y}_{HT}) = \frac{1}{2N^2} \sum_{k=1}^{N} \sum_{l=1}^{N} (\pi_k \pi_l - \pi_{kl}) \left( \frac{Y_k}{\pi_k} - \frac{Y_l}{\pi_l} \right)^2 .$$

This expression shows that the variance can be reduced by taking the first-order inclusion probabilities proportional to the values of the target variable.

To quantify the precision of an estimate, a confidence interval can be computed. The *confidence interval* is a range of possible values of the population parameter. The interval encompasses the true value of the population mean with a high probability if an estimator is unbiased. This probability is called the *confidence level*. It is denoted by $(1 - \alpha)$ where $\alpha$ is a small probability. Often the value $\alpha = 0.05$ is used. The confidence level is then 95%.

The 95% confidence interval of the Horvitz–Thompson estimator is equal to

$$(4.16) \qquad (\bar{y}_{HT} - 1.96 \times S(\bar{y}_{HT}); \bar{y}_{HT} + 1.96 \times S(\bar{y}_{HT})),$$

where

$$(4.17) \qquad S(\bar{y}_{HT}) = \sqrt{V(\bar{y}_{HT})}$$

is the *standard error* of the estimator. For a 99% confidence interval, the value 1.96 must be replaced by 2.58.

A problem is that usually the value of the variance or standard error is not known. It can only be computed if all values of the target variable in the population are available. The way out is that an unbiased estimate of the standard error can be computed just using the sample data. This leads to an estimated confidence interval

$$(4.18) \qquad ((\bar{y}_{HT} - 1.96 \times s(\bar{y}_{HT}); \bar{y}_{HT} + 1.96 \times s(\bar{y}_{HT})),$$

where $s(\bar{y}_{HT})$ is the estimate for $S(\bar{y}_{HT})$.

The quantity $1.96 \times S(\bar{y}_{HT})$ is often called the *margin of error*. It is the maximum difference between an estimate and the true value. Thus, it shows how close the estimate is to the value to be estimated.

### 4.2.4 SIMPLE RANDOM SAMPLING

The best known and probably most often used sampling design is a *simple random sample* without replacement. This is a sample design in which all elements have the same probability of being selected. First-order inclusion probabilities of all elements are equal. It can be shown for without replacement sampling that all first-order inclusion probabilities always sum up to $n$. Therefore $\pi_k = n/N$, for $k = 1, 2, \ldots, N$. Furthermore, all second-order inclusion probabilities sum up to $n(n-1)$. Therefore $\pi_{kl} = n(n-1)/N(N-1)$, for $k, l = 1, 2, \ldots, N$ and $k \neq l$.

Suppose objective of the survey is to estimate the population mean of a continuous target variable $Y$. Substitution of the values of the first-order inclusion probabilities in expression (4.13) results in a simple estimator, the *sample mean*:

$$(4.19) \qquad \bar{y} = \frac{1}{n} \sum_{k=1}^{N} a_k Y_k = \frac{1}{n} \sum_{i=1}^{n} y_i,$$

where $y_1, y_2, \ldots, y_n$ denote the $n$ observations that have become available in the sample. This is an unbiased estimator with variance

$$(4.20) \qquad V(\bar{y}) = \frac{1-f}{n} S^2,$$

where $f = n/N$ is the *sampling fraction* and $S^2$ the population variance. Expression (4.20) shows that an increased sample size leads to more precise estimators. The standard error of the sample mean is equal to

$$(4.21) \qquad S(\bar{y}) = \sqrt{V(\bar{y})} = \sqrt{\frac{1-f}{n} S^2}.$$

To compute an estimated 95% confidence interval, an unbiased estimator for $S^2$ is required. The sample variance

$$(4.22) \qquad s^2 = \frac{1}{n-1} \sum_{i=1}^{n} \left( y_i - \bar{y} \right)^2$$

can be used for this. See Example 4.6 for a simulation experiment on the estimator precision.

The effect of sample size on the precision of an estimator can be shown by means of a simulation experiment. The fictitious population of the country of Samplonia has been constructed. The population consists of 1,000 people. The working population of Samplonia consists of 341 people.

One thousand simple random samples without replacement of size 20 have been selected from the working population. For each sample, the mean income is computed as an estimate of the mean income in the population. The distribution of these 1,000 estimates is displayed in the graph on the left in Figure 4.3.

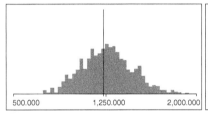 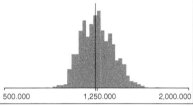

| 500.000 | 1,250.000 | 2,000.000 | 500.000 | 1,250.000 | 2,000.000 |

**FIGURE 4.3** Random samples from the working population of Samplonia

There is a lot of variation of the estimates around the population mean indicated by the vertical line. This variation can be reduced by increasing the sample size. The graph on the right shows the distribution of 1,000 estimates based on sample of the size 40. Indeed, doubling the sample size reduces the variance.

Objective of a survey also often is estimating a population percentage. Typical examples are the percentage of people voting for a political party, the percentage of households having an Internet connection, and the unemployment percentage. The theory for estimating percentages does not essentially differ from the theory of estimating means. In fact, percentages are just population means multiplied by 100 where the target variable $Y$ is an indicator variable, i.e., it only assumes the value 1 (if the element has the specific property) or 0 (if the element does not have the property). Because of this restriction on the possible values, formulas become even much simpler.

If $Y$ only assumes the values 1 and 0, the population mean $\overline{Y}$ is equal to the proportion of elements having a specific property. The population percentage $P$ is therefore equal to

(4.23)                                    $$P = 100\overline{Y}.$$

Estimation of a population percentage comes down to first estimating the population mean. The sample mean is an unbiased estimator for this quantity. Multiplication of the sample mean by 100 produces the sample percentage. This estimator is denoted by

(4.24)                                    $$p = 100\overline{y}.$$

Since the sample mean is an unbiased estimator for the population mean, the sample percentage is an unbiased estimator of population percentage.

The variance of this estimator can be found by working out the term $S^2$ in variance formula (4.25) for a population in which a percentage $P$ of the elements has a specific characteristic and a percentage $100 - P$ does not have this characteristic. This results in the simple expression

(4.25)                        $$V(p) = \frac{1-f}{n}\frac{N}{N-1}P(100 - P).$$

This variance can be estimated using the sample data. If $p$ denotes the sample percentage, then

(4.26)                        $$v(p) = \frac{1-f}{n-1}P(100 - p)$$

is an unbiased estimator of the variance (4.25). The estimated variance is used to obtain an estimated confidence interval. See Example 4.7 for an empirical estimation for the variance of the percentage estimation.

### EXAMPLE 4.7 Estimating a percentage

An election poll was conducted in June 2010 in the Netherlands. It was a web survey. The sample size was 1,000 persons. The population consisted of 12 million voters. The results showed that in the sample 22.1% would vote for the Liberal Party and 20.5% for the Social Democrats.

The estimated variance of the estimate for the Liberal Party is

$$v(p) = \frac{1-f}{n-1}P(100 - p) = \frac{1 - 1,000/12,000,000}{999}22.1 \times 77.9 = 1.723170.$$

The standard error of the estimate is therefore

$$s(p) = \sqrt{v(p)} = \sqrt{1.723170} = 1.312670.$$

The 95% confidence interval becomes

$$(p - 1.96 \times s(p); p + 1.96 \times s(p)) = (22.1 - 1.96 \times 1.312670; 22.1$$
$$+ 1.96 \times 1.312670) = (19.5; 24.7).$$

The conclusion can be that with 95% confidence interval the percentage of voters in the population will be between 19.5% and 24.7%.

The 95% confidence interval for the percentage of voters for the Social Democrats turns out to run between 18.0% and 23.0%. The intervals for both parties have a considerable overlap. So it cannot be concluded that in the population the Liberals will get more votes than the Social Democrats.

## 4.2.5 DETERMINING THE SAMPLE SIZE

A decision is needed in the survey design phase is the size of the sample to be selected. This is an important decision. If, on the one hand, the sample is larger than necessary, a lot of time and money may be wasted. If, on the other hand, the sample is too small, the required precision will not be achieved, making the survey results less useful.

There exists a relationship between the precision of an estimator and the sample size: the larger the sample is, the more precise the estimator will be. Therefore, the question about the sample size can only be answered if it is clear how precise the estimator must be. Once the precision has been specified, the sample size can be computed. A very high precision requires a large sample. If the costs per interview are high, this will make the survey expensive. Once a web survey questionnaire has been prepared on the Internet and a sample of e-mail addresses is available, the costs per interview can be very low. However, it should be realized that recruitment can be expensive if it is done by means of CAPI or CATI. In practice, the sample size will always be a compromise between costs and precision.

Here, some formulas for the size of a simple random without replacement are presented. The first considered situation is that for estimating population percentages. Then the case of estimating population means will be described.

**4.2.5.1 The Sample Size for Estimating a Percentage.** The starting point is that the researcher gives some indication of how large the *margin of error* at most may be. The margin is defined as distance between the estimate and the lower or upper bound of the confidence interval. Formulas are given for the sample size that is at least required to achieve this margin of error. In case of a 95% confidence interval, the margin of error is equal to

$$(4.27) \qquad\qquad 1.96 \times S(p).$$

For a 99% confidence interval, the value of 1.96 must be replaced by 2.58. Suppose the margin of error may not exceed a value $M$. In the case of a 95% confidence interval, rewriting this condition leads to

$$(4.28) \qquad\qquad S(p) \leq \frac{M}{1.96}.$$

The variance of the estimator for a population percentage can be found in expression (4.25). Substituting this expression in inequality (4.28) leads to the condition

$$(4.29) \qquad\qquad \sqrt{\frac{1-f}{n}\frac{N}{N-1}P(100-P)} \leq \frac{M}{1.96}.$$

The lower bound for the sample size can now be computed by solving $n$ from this equality. However, there is a problem, because expression contains an unknown quantity, and that is population percentage $P$. There are two ways to solve this problem:

- There is some indication of the value of $P$. Maybe there was a previous survey in which this quantity was estimated. Or may be a subject-matter expert may provide an educated guess. Such an indication can be substituted in expression (4.29), after which it can be solved.
- Nothing at all is known about the value of $P$. Now $P(100 - P)$ is a quadratic function that assumes its minimum value 0 in the interval [0, 100] for $P = 0$ and $P = 100$. Exactly in the middle, for $P = 50$, the function assumes its maximum value. This implies that the upper bound for the variance can be computed by filling in the value $P = 50$. So the worst case for the variance is obtained for this value of $P$. For any other value of $P$, the variance is smaller. If the value is determined so that the worst-case variance is not exceeded, the true variance will certainly be smaller. It should be noted that for values of $P$ between, say, 30% and 70%, the true variance will not differ much from the maximum variance.

Solving $n$ from inequality (4.29) leads to a lower bound of $n$ equal to

(4.30)
$$n \geq \frac{1}{\frac{N-1}{N}\left(\frac{M}{1.96}\right)^2 \frac{1}{P(100-P)} + \frac{1}{N}}.$$

A simple approximation can be obtained if the population size $N$ is very large. Then $(N-1)/N$ can be approximated by 1, and the value of $1/N$ can be ignored. This implies that expression (4.30) is reduced to

(4.31)
$$n \geq \left(\frac{1.96}{M}\right)^2 P(100-P).$$

Example 4.8 shows the computation of the sample size for percentage estimation.

---

**EXAMPLE 4.8** **The sample size for an opinion poll**

Suppose that in an earlier opinion poll 21% of the respondents indicated to vote for a specific party. A new poll will be conducted to measure the current support for this party. No dramatic changes are expected. Therefore, it is not unreasonable to fill in a value of 21 for $P$ in expression (4.31). Furthermore, the margin of error should not exceed $M = 3\%$. Substitution in expression (4.31) results in

$$n \geq \left(\frac{1.96}{3}\right)^2 21 \times 79 = 708.1.$$

So, the sample size must be at least equal to 709. The confidence level is 95%. For a confidence level of 99%, the value of 1.96 must be replaced by 2.58, leading to a minimum sample size of 1,227.

---

#### 4.2.5.2 The Sample Size for Estimating a Mean.
Expression (4.28) is also the starting point for the computation of the sample size if the objective of the survey is to estimate the mean of a continuous target variable. However, there is no simple expression for the standard error available. Expression (4.28) can be rewritten as

(4.32)
$$\sqrt{\left(\frac{1}{n} - \frac{1}{N}\right)S^2} \leq \frac{M}{1.96},$$

in which $S^2$ is the adjusted population variance. The problem is that usually this variance is unknown. Sometimes a rough estimate is available from a previous survey. Or maybe some indication can be obtained from a test survey. In these situations, the approximate value can be substituted in expression (4.32). Rewriting the inequality leads to

$$(4.33) \qquad\qquad n > \frac{1}{\left(\frac{M}{1.96S}\right)^2 + \frac{1}{N}}$$

The quantity $1/N$ can be ignored for large values of $N$. This produces the somewhat simpler expression

$$(4.34) \qquad\qquad n \geq \left(\frac{1.96S}{M}\right)^2.$$

## 4.2.6 SOME OTHER SAMPLING DESIGNS

An overview of four other sampling designs is given in this section. These sampling designs are stratified sampling, sampling with unequal probabilities, cluster sampling, and two-stage sampling. For more on sampling designs, see, for example, Cochran (1977) and Bethlehem (2009).

**4.2.6.1 Stratified Sampling.** To select a stratified sample, the population is divided into a number of subpopulations. These subpopulations are called *strata*. A sample is selected in each stratum. So, for each stratum, an unbiased estimate of the stratum mean or percentage can be computed. Then, these stratum estimators can be combined into an unbiased estimator of the population mean or percentages. There are various reasons to apply stratified sampling:

• If the strata are homogeneous, i.e., all elements within strata resemble each other, the variance of estimators will be small. So, estimators based on stratified sampling will be more precise than estimators based on simple random sampling;

• There may be situations in which not only estimates are required for the population as a whole but also for specific subpopulations. By using these subpopulations as strata in stratified sampling, the researcher can see that a sufficient number of observations become available in each subpopulation;

• By applying stratified sampling with the same fraction of observations in each stratum, the sample becomes at least representative with respect to these strata.

Stratified sampling can be implemented only if a proper sampling frame is available. There must be a separate sampling frame for each subpopulation. This condition sometimes prevents application of stratified sampling. For example, it is usually not possible to stratify a sample for a general population survey by level of education. The reason is there is no separate sampling frame for each level of education. However, it is possible in countries like the Netherlands to stratify a sample by region, because each municipality has its own population register.

Another example could be stratifying individuals by Internet access or not. This information is not available in a sampling frame. Therefore, it is not possible to stratify with respect to this characteristic.

To apply stratified sampling, the target population $U$ is divided into $L$ sub-populations (strata) $U_1, U_2, \ldots, U_L$ of sizes $N_1, N_2, \ldots, N_L$, respectively. The strata are nonoverlapping and together cover the whole population. This implies that

$$(4.35) \qquad \sum_{h=1}^{L} N_h = N_1 + N_2 + \cdots + N_L = N.$$

The $N_h$ values of the target variable $Y$ in stratum $h$ are denoted by

$$(4.36) \qquad Y_1^{(h)}, Y_2^{(h)}, \ldots, Y_{N_h}^{(h)}.$$

The mean in stratum $h$ can be written as

$$(4.37) \qquad \overline{Y}^{(h)} = \frac{1}{N_h} \sum_{k=1}^{N_h} Y_k^{(h)}$$

and the population mean can be written as

$$(4.38) \qquad \overline{Y} = \frac{1}{N} \sum_{h=1}^{L} N_h \overline{Y}^{(h)}.$$

So the population mean is a weighted average of the stratum means. The (adjusted) variance in stratum $h$ is equal to

$$(4.39) \qquad S_h^2 = \frac{1}{N_h - 1} \sum_{k=1}^{N_h} \left( Y_k^{(h)} - \overline{Y}^{(h)} \right)^2.$$

A stratified sample of size $n$ is selected from this population by selecting $L$ subsamples of sizes $n_1, n_2, \ldots, n_L$, respectively, where $n_h$ is the sample size in stratum $h$, for $h = 1, 2, \ldots, L$. In principle any sampling design can be applied within the strata, but usually simple random samples without replacement are used. If the $n_h$ observations in stratum $h$ are denoted by

$$(4.40) \qquad y_1^{(h)}, y_2^{(h)}, \ldots, y_{n_h}^{(h)},$$

the sample mean

$$(4.41) \qquad \overline{y}^{(h)} = \frac{1}{n_h} \sum_{i=1}^{n_h} y_i^{(h)}$$

in stratum $h$ is an unbiased estimator of the population mean in stratum $h$. Now, the stratum estimators can be combined into an estimator for the population mean of $Y$. Using expression (4.38), it can be shown that

$$(4.42) \qquad \bar{y}_S = \frac{1}{N} \sum_{h=1}^{L} N_h \bar{y}^{(h)}$$

is an unbiased estimator of the population mean. The variance of the sample mean in stratum $h$ is equal to

$$(4.43) \qquad V\left(\bar{y}^{(h)}\right) = \frac{1-f_h}{n_h} S_h^2$$

in which $f_h = n_h/N_h$. Since the subsamples are selected independently, it can be shown that the variance of estimator (4.42) is equal to

$$(4.44) \qquad V(\bar{y}_S) = \frac{1}{N^2} \sum_{h=1}^{L} N_h^2 \frac{1-f_h}{n_h} S_h^2.$$

This variance is small if the stratum variances $S_h^2$ are small. This is the case if there is little variation in the values of the target variable within strata, i.e., if the strata are homogeneous with respect to the target variable.

The variance of the estimator is influenced by the sample sizes $n_1, n_2, \ldots, n_L$ in the strata, i.e., the sample *allocation*. The variance is minimal for the so-called optimal allocation (also called *Neyman allocation*) (see, e.g., Cochran, 1977). This is the allocation where the $n_h$ are proportional to $N_h \times S_h$. Applying this allocation requires stratum variances to be known. If this is not the case and there are no estimates available, another option is to use *proportional allocation*. This is a sample allocation where the $n_h$ are proportional to $N_h$. As a result every element in the target population has the same selection probability. See Example 4.9 about selection of the sample in a business survey.

**EXAMPLE 4.9  Business surveys in the Netherlands**

Statistics Netherlands maintains a general business register. This is a database containing general information about all companies in the Netherlands. Variables included in the database are name and address, type economic activity (according to the so-called international NACE classification), and size class (in terms of number of employees).

For its business surveys, Statistics Netherlands draws stratified samples from this business register. Samples stratification is by type of economic activity and size class. This implies that a stratum contains companies with the same activity and of the same size. Therefore, the strata are homogeneous. This results in precise stratum estimates.

The sample allocation is often such that large companies have a larger selection probability than small companies. Since the values of many target variables (for example, turnover, profit, or investments) relate to the size of the companies, this also improves the precision of estimates.

Many of these surveys employ data collection using the Internet. There are two approaches. The first approach is a web survey. Companies can complete questionnaires on the Internet directly. The other is that they receive the interview software by e-mail or CD. They downloaded and installed this software. They run the software, answer the questions, and return the answers by e-mail.

**4.2.6.2 Unequal Probability Sampling.** It is an interesting property of the Horvitz–Thompson estimator that its variance is small if the first-order inclusion probabilities are more or less proportional to the values of target variable, i.e., $Y_k/\pi_k$ is approximately constant for all $k$. This is difficult to realize in practice because it requires all values of the target variable in the population to be known. If this is the case, there are no reason for carrying out a survey. Example 4.10 is about the selection of sample with unequal probability. However, sometimes the values in the population of an auxiliary variable $X$ are known. So, first-order inclusion probabilities could be taken proportional to the values of this variable. If there is a strong correlation between the target variable $Y$ and the auxiliary variable $X$, then the result will be a precise estimator. An example is a shoplifting survey where shops are sampled according to their floor size, assuming there is more shoplifting in larger shops than in smaller shops.

■ EXAMPLE 4.10 Sampling addresses for a web survey: a case of unequal probabilities

Suppose a web survey is conducted where the sampling frame is a list of addresses. A simple random sample of $n$ addresses is selected. For each selected address, one randomly selected person is asked to complete the questionnaire. This could, for example, be implemented by selecting the person with the next birthday.

The resulting sampling design is one in which not every person has the same probability of selection. Let $N$ denote the total number of persons in the population. This population is divided over $M$ addresses. There live $N_h$

persons at address $h$, where $N_1 + N_2 + \cdots + N_M = N$. The inclusion probability of a person $k$ at address $h$ is now equal to

$$(4.45) \qquad \pi_k^{(h)} = \frac{n}{M} \times \frac{1}{N_h}.$$

This expression is obtained by multiplying the inclusion probability of an address ($n/M$) by the probability of selecting a person at this address ($1/N_h$). Substitution of this expression in the Horvitz–Thomson estimator results in the estimator

$$(4.46) \qquad \bar{y}_{HT} = \frac{1}{N} \sum_{h=1}^{M} a_k \sum_{k=1}^{N_h} b_{hk} Y_{hk} \frac{MN_h}{n},$$

where the indicator $a_h$ indicates whether or not address $h$ is selected and the indicator $b_{hk}$ indicates whether or not element $k$ at address $h$ is selected. This estimator is not equal to the simple sample mean. It is a weighted mean, where the value $Y_{hk}$ for each respondent with the number of people $N_h$ at the address.

### 4.2.6.3 Cluster Sampling.

Another sampling design is *cluster sampling*. This type of sampling can be applied if there is no sampling frame for the elements in the population, but there is one for clusters of elements. In this situation a sample of clusters can be selected, and all elements in each selected cluster can be observed. A typical example of a cluster sample is an address sample where all people at a selected address are invited to participate in the survey.

Cluster sampling does not necessarily produce precise estimators. To the contrary, the more elements within clusters resemble each other, the less efficient the estimator will be. Another disadvantage of cluster sampling is that there is no control over the sample size. It simply depends upon the numbers of elements in the selected clusters.

### 4.2.6.4 Two-Stage Sampling.

One way to get more control is to select a *two-stage sample*. See Example 4.11 for a case study. First, a sample of clusters is selected, and then a sample of elements is drawn from each selected cluster. The reasons for applying this sampling design are more practical. Again, this procedure will not produce very accurate estimates, but it may be necessary to do this due to the lack of a proper sampling frame. The reason to apply a two-stage sampling design can also be reduction of costs. This only applies if interviewers are used for data collection. Interviewers have to travel less for a face-to-face survey if the addresses of selected persons are concentrated in clusters.

**▣ EXAMPLE 4.11 The Safety Monitor of Statistics Netherlands**

National statistical institutes are always under pressure to reduce data collection costs. This has led to considering new ways of data collection, in which web surveys play an important role. The most far-reaching change is to replace an expensive CAPI and CATI survey by a web survey. Another option is to introduce a mixed-mode survey. Statistics Netherlands has experimented with mixed-mode data collection for the Safety Monitor.

This survey measures how the Dutch feel with respect to security. Questions were asked, among others, about feelings of security, quality of life, and level of crime experienced.

The target population for this survey consisted of all persons of age 15 and older. The sample for this survey was selected from the Dutch population register. It was a stratified sample. Strata were constructed by crossing interview regions of Statistics Netherlands by the 25 police regions in which the country is divided. Within each stratum a two-stage sample was selected. In the first stage, municipalities were selected. In the second stage, persons were drawn in the selected municipalities.

All sample persons received a letter in which they were asked to complete the survey questionnaire on the Internet. The letter also included a postcard that could be used to request a paper questionnaire.

Two reminders were sent to those that did not respond by web or mail. If still no response was obtained, nonrespondents were approached by means of CATI, if a listed telephone number was available. If not, these nonrespondents were approached by CAPI. This four-mode survey is denoted by SM4.

To be able to compare this four-mode survey with a traditional survey, also a two-mode survey was conducted for an independent sample. Sampled persons were approached by CATI if there was a listed telephone number, and otherwise they were approached by CAPI. The two-mode survey is denoted by SM2.

The response rate for SM4 turned out to be 59.7%. The response rate for SM2 was 63.5%. So there was not a large difference. More than half of the response (58%) was obtained in the SM4 with a self-administered mode of data collection (CAWI or PAPI).

The conclusion was drawn that the four-mode survey did not increase the response. The costs of the survey were, however, much lower, because interviewers were deployed in only 42% of the cases. Focusing on just interviewer costs and ignoring all other costs (which are much lower), Beukenhorst and Wetzels (2009) found that the costs of SM4 were only 60% of the costs of SM2.

## 4.2.7 ESTIMATION PROCEDURES

The precision of an estimator may be improved by using auxiliary information. A good example is sampling with inclusion probabilities proportional to the values of an auxiliary variable. If there is a strong correlation between target variable and auxiliary variable, the variance of the Horvitz–Thompson estimator will be small. Another example is stratified sampling. This comes down to using categorical auxiliary variables that divide the population in homogeneous groups.

In the examples above, auxiliary information is used in the sampling design. The use of auxiliary information can also be in a different way, i.e., in the estimator itself. Some examples of improved estimation procedures are described here. It is assumed that simple random sampling without replacement is applied.

**4.2.7.1 The Ratio Estimator.** The *ratio estimator* assumes that a continuous auxiliary variable $X$ is available, the values of which are more or less proportional to the values of the target variable, i.e.,

$$(4.47) \qquad Y_k \approx BX_k$$

for some constant $B$. The ratio estimator is defined by

$$(4.48) \qquad \bar{y}_{RAT} = \frac{\bar{y}}{\bar{x}}\overline{X}$$

where $\bar{x}$ and $\bar{y}$ are the sample means of $X$ and $Y$ and $\overline{X}$ is the population mean of $X$. The estimator is asymptotically unbiased, and its variance is approximately equal to

$$(4.49) \qquad V(\bar{y}_{RAT}) \approx \frac{1-f}{n}\frac{1}{N-1}\sum_{k=1}^{N}\left(Y_k - \frac{\overline{Y}}{\overline{X}}X_k\right)^2.$$

It can be shown that the smaller this variance is, the better condition (4.47) is satisfied. See, for example, Cochran (1977).

**4.2.7.2 The Regression Estimator.** Example 4.12 is about this estimator. An even better estimator is the *regression estimator*. It assumes a linear relationship

$$(4.50) \qquad Y_k \approx A + BX_k$$

between the values of the target variable and the auxiliary variable. $A$ and $B$ are constants that have to be estimated using the sample data. This can be the done with ordinary least squares. The regression estimator is defined by

$$(4.51) \qquad \bar{y}_{REG} = \bar{y} - b(\bar{x} - \overline{X}),$$

where

$$(4.52) \qquad b = \frac{\sum\limits_{i=1}^{n} (x_i - \bar{x})(y_i - \bar{y})}{\sum\limits_{i=1}^{n} (x_i - \bar{x})^2}$$

and $x_1, x_2, \ldots, x_n$ denote the $n$ observations for $X$ that have become available in the sample. The estimator is asymptotically unbiased, and its variance is approximately equal to

$$(4.53) \qquad V(\bar{y}_{REG}) \approx \frac{1-f}{n} S^2 (1 - R_{XY}^2),$$

where $R_{XY}$ is the correlation between $X$ and $Y$ in the population. It is clear from expression (4.53) that the variance of the regression estimator is never larger than that of the simple sample mean. The stronger the correlation, the smaller the variance will be.

### EXAMPLE 4.12 A dairy farm survey

The target population of a survey consists of 200 dairy farms in the rural part of the fictitious country of Samplonia. Objective of the web survey is to estimate the average daily milk production per farm. Since all farms are connected to the Internet, a probability sample can be selected. A simple random sample is drawn.

Two estimators are compared: the simple sample mean and the regression estimator. The regression estimator uses the number of cows per farm as the auxiliary variable. This seems not unreasonable as one may expect milk production per farm to be more or less linearly related to the number of cows per farm.

The selection of a sample of size 40 and the computation of the estimator have been repeated 500 times for both estimators. This gives 500 values of each estimator. Figure 4.4 contains the distribution of the values of both estimators. The histogram on the left shows the distribution of the sample mean. The distribution of the regression estimator is shown on the right.

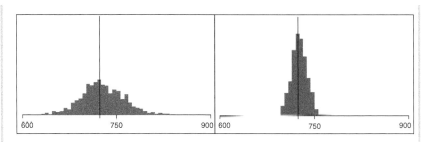

**FIGURE 4.4** Comparing the sample mean with the regression estimator

The regression estimator performs better than the direct estimator. The distribution of its values concentrates much more around the true value. The standard error of the sample mean is equal to 35.6, whereas it is 12.4 for the regression estimator. So, a more precise estimator can be obtained with the same sample size if the proper auxiliary information is available.

**4.2.7.3 The Post-stratification Estimator.** The ratio estimator and the regression estimator both use a continuous auxiliary variable. It is also possible to use a categorical auxiliary variable. A well-known example of this is the *post-stratification estimator*. Suppose this auxiliary variable divides the target population $U$ into $L$ subpopulations (strata) $U_1, U_2, \ldots, U_L$ of sizes $N_1, N_2, \ldots, N_L$, respectively. After a simple random sample has been selected, the sample mean can be computed in each stratum, after which these stratum estimates can be combined into an estimate

$$(4.54) \qquad \bar{y}_{PS} = \frac{1}{N} \sum_{h=1}^{L} N_h \bar{y}^{(h)}$$

for the population mean. Note that this expression is identical to expression (4.42) for stratified sampling. However, estimator (4.54) has different statistical properties, because the underlying selection mechanism is different. It can be shown that the post-stratification estimator is approximately unbiased and that its variance is equal to

$$(4.55) \qquad V\left(\bar{y}_{PS}\right) = \frac{1-f}{n} \sum_{h=1}^{L} W_h S_h^2 + \frac{1}{n^2} \sum_{h=1}^{L} (1 - W_h) S_h^2,$$

in which $W_h = N_h/N$ and $S_h^2$ is the population variance of the target variable in stratum $h$. If the strata are homogeneous with respect to the values of the target variable (i.e., there is little variation within strata), this variance will be small. Note that when $n$ becomes larger, expression (4.55) will be closer to expression (4.44) as the value of the second term in (4.55) quickly becomes smaller.

**4.2.7.4 The Generalized Regression Estimator.** A more general estimator can be defined, of which the regression estimator and the post-stratification estimator are special cases. This *generalized regression estimator* is introduced here, because it is used in nonresponse correction techniques.

Suppose there are $p$ auxiliary variables available. The $p$-vector of values of these variables for element $k$ is denoted by

$$(4.56) \qquad X_k = \left( X_{k1}, X_{k2}, ..., X_{kp} \right)'.$$

The symbol $'$ denotes transposition of a matrix or vector. Let $Y$ be the $N$-vector of all values of the target variable, and let $X$ be the $N \times p$ matrix of all values of the auxiliary variables. The vector of population means of the $p$ auxiliary variables is defined by

$$(4.57) \qquad \overline{X} = \left( \overline{X}_1, \overline{X}_2, ..., \overline{X}_p \right)'.$$

If the auxiliary variables are correlated with the target variable, then for a suitably chosen vector $B = (B_1, B_2, ..., B_p)'$ of regression coefficients for a best fit of $Y$ on $X$, the residuals $E = (E_1, E_2, ..., E_N)'$, defined by

$$(4.58) \qquad E = Y - XB,$$

vary less than the values of the target variable itself. Application of ordinary least squares results in

$$(4.59) \qquad B = (X'X)^{-1} X Y' = \left( \sum_{k=1}^{N} X_k X_k' \right)^{-1} \left( \sum_{k=1}^{N} X_k Y_k \right).$$

For any sampling design, the vector $B$ can be estimated by

$$(4.60) \qquad b = \left( \sum_{k=1}^{N} a_k \frac{X_k X_k'}{\pi_k} \right)^{-1} \left( \sum_{k=1}^{N} a_k \frac{X_k Y_k}{\pi_k} \right).$$

The estimator $b$ is an asymptotically design unbiased (ADU) estimator of $B$. It means the bias vanishes for large samples. Using expression (4.60), the generalized regression estimator is defined by

(4.61) $$\bar{y}_{GR} = \bar{y}_{HT} + \left(\overline{X} - \overline{x}_{HT}\right)' b,$$

where $\overline{x}_{HT}$ and $\bar{y}_{HT}$ are the Horvitz–Thompson estimators for the population means of $X$ and $Y$, respectively. The generalized regression estimator is an ADU estimator of the population mean of the target variable. If there exists a $p$-vector $c$ of fixed numbers such that $Xc = J$, where $J$ is a vector consisting of 1's, the generalized regression estimator can also be written as

(4.62) $$\bar{y}_{GR} = \overline{X}' b.$$

This situation occurs if $X$ contains an intercept term or a set of dummy variables corresponding to all categories of a categorical variable. It can be shown that the variance of the generalized regression estimator can be approximated by

(4.63) $$V\left(\bar{y}_{GR}\right) = \frac{1}{N^2} \sum_{k=1}^{N} \sum_{l=1}^{N} (\pi_{kl} - \pi_k \pi_l) \frac{E_k}{\pi_k} \frac{E_l}{\pi_l}.$$

This is the variance of the Horvitz–Thompson estimator, but with the values $Y_k$ replaced by the residuals $E_k$. This variance will be small if the residual values $E_k$ are small. Hence, use of auxiliary variables that can explain the behavior of the target variable will result in a precise estimator.

Given simple random sampling without replacement and use of just one continuous auxiliary variable, the generalized regression estimator reduces to the regression estimator defined in (4.51).

Suppose a categorical auxiliary variable is available with $p$ categories. Then this variable can be replaced by $p$ dummy variables. Associated with each element $k$ is a vector $X = (X_1, X_2, \ldots, X_p)'$ of dummy values. The $h$th dummy $X_{kh}$ assumes the value 1 if element $k$ belongs to stratum $h$, and it assumes the value 0 if it belongs to another stratum. In this case, $B$ turns out to be equal to

(4.64) $$B = \left(\overline{Y}^{(1)}, \overline{Y}^{(2)}, \ldots, \overline{Y}^{(L)}\right)',$$

and this vector can be estimated unbiasedly by the vector

(4.65) $$b = \left(\bar{y}_{HT}^{(1)}, \bar{y}_{HT}^{(2)}, \ldots, \bar{y}_{HT}^{(L)}\right)'$$

of Horvitz–Thompson estimators of the stratum means. The vector of population means of the $L$ auxiliary variables turns out to be equal to

(4.66) $$\overline{X} = (W_1, W_2, \ldots, W_L)',$$

where $W_h = N_h/N$. If we substitute these quantities in expression (4.62), the result is

$$(4.67) \qquad \bar{y}_{PS} = \frac{1}{N} \sum_{h=1}^{L} N_h \bar{y}_{HT}^{(h)}.$$

This is the post-stratification estimator (4.54) but written down for arbitrary sampling designs. So, the post-stratification estimator is indeed a special case of the generalized regression estimator.

## 4.3 Application

Some sampling concepts introduced in this chapter are illustrated using a fictitious target population. This population consists of 8,000 shops in a large town. Objective of a web survey is estimation of the average yearly value of theft of goods in a shop. Target variable of the survey is the value of the goods stolen in a shop in a specific year. There is also an auxiliary variable, and this is the floor space of the shop.

The population has been generated such that there is an approximate linear relationship between value of stolen goods and floor space of a shop. Figure 4.5 shows the scatter plot of these two variables. There is a clear relationship. Note

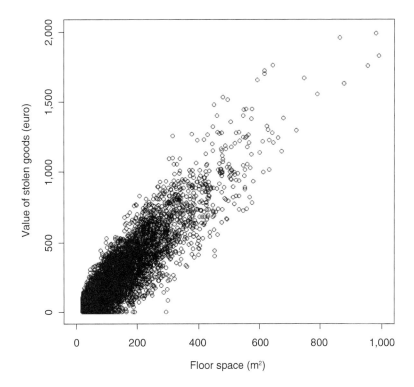

FIGURE 4.5 The relationship between the value of stolen goods and floor space

that the variation of the value of the stolen goods increases as the size of the shop increases. Also note that the distribution of floor space is very skew. There are many small shops and only a few big shops.

A business register is available containing (among other address information) the e-mail address of a contact person of each business. Therefore, a web survey can be conducted for which the sample is drawn by means of a probability sample.

Suppose a simple random of size $n = 100$ is selected without replacement from this population of shops and the sample mean of the value of the stolen goods is used as an estimator of the population mean of the value of stolen goods. Insight in the distribution of the estimator can be obtained by repeating the selection of the sample and the computation of the estimate for a large number of times. Figure 4.6 shows the results of this experiment for 1,000 repetitions. The dotted line represents the population mean to be estimated (223.61). As proven by Bowley (1906, 1926), the sample mean has approximately a normal distribution. The distribution is symmetric around the population value, implying that the estimator is unbiased.

The standard error of the sample mean is equal to 23.00. Hence the margin of error of the 95% confidence interval $= 1.96 \times 23.00 = 45.08$. Estimates will therefore not differ more than 45.08 from the true population mean (in 95% of the cases).

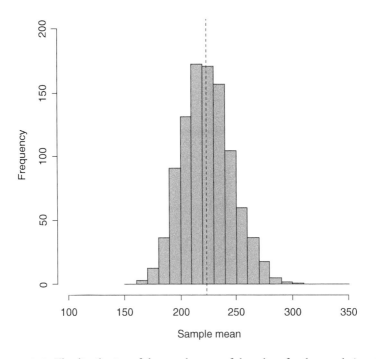

FIGURE 4.6 **The distribution of the sample mean of the value of stolen goods ($n = 100$)**

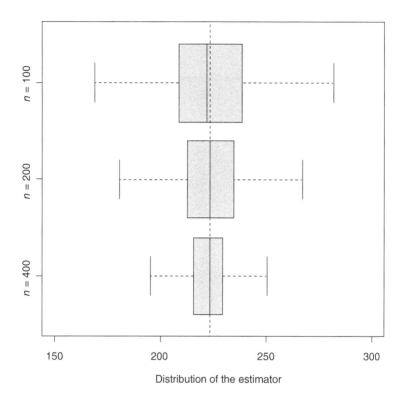

FIGURE 4.7 **Distribution of the estimator of the value of stolen goods for simple random sampling with various sample sizes**

The precision of estimates is for a large part determined by the sample size: the precision will be larger as the sample size increases. This is shown in Figure 4.7. The distribution of three different estimators is displayed by means of box plots. The distribution of each estimator was obtained by drawing 1,000 samples. In all three cases the sample design was simple random sampling without replacement, but the sample sizes were 100, 200, and 400. All three distributions are symmetric around the population mean 223.61. Therefore the estimators are unbiased.

It is clear that the variation of the estimates decreases as the sample size increases. The margin of error for a sample of size 100 is 45.08, and for a sample of size $n = 400$, it is 21.91. It is a rule of thumb that the precision is proportional to the square root of the sample size.

Business registers often contain additional information on companies, like the type of economic activity and size of the company (number of employees). It is assumed in this application that floor size is included in this register. This makes it possible to draw a stratified sample. The shops are divided into four (floor) size groups. The characteristics of the four strata are summarized in Table 4.2.

TABLE 4.2   Characteristics of the four size strata of the shops

| Stratum | Number of elements | Mean | Variance | Optimal allocation | Proportional allocation |
|---|---|---|---|---|---|
| Less than 50 | 2,160 | 51.785 | 51.044 | 12 | 27 |
| 50–100 | 2,315 | 127.963 | 83.697 | 21 | 29 |
| 100–250 | 2,733 | 301.444 | 150.686 | 44 | 34 |
| 250 and more | 792 | 703.215 | 277.652 | 24 | 10 |
| Total | 8,000 | 223.610 | 231.459 | 101 | 100 |

The variance of the shoplifting value is in the first three strata much smaller than in the population as a whole. Apparently these three strata are fairly homogeneous. This makes stratified sampling a promising approach.

The last two columns contain the allocation of a sample of size 100 for optimal and proportional allocation. Note that in the case of optimal allocation, most elements (44) are drawn from the third stratum. The reason is that this stratum is large and not very homogeneous. Only 12 elements are selected from the first stratum, because it is so homogeneous. Proportional allocation leads to a different distribution of the sample elements over the strata as the homogeneity of the strata is not taken into account. Stratified sampling is compared with simple random sampling in Figure 4.8. All distributions are based on 1,000 samples of size $n = 100$.

The distribution of the estimators for stratified sampling is shown for both optimal allocation and proportional allocation. They can be compared with the distribution of the sample mean. The precision of the stratification estimators is much higher than that of the sample mean. The margin of error of the sample mean is 45.08. In the case of stratification with optimal allocation, the margin of error is only 22.96. This is the highest precision that can be obtained with a stratified sampling. In the case of proportional allocation, the margin of error is only slightly larger (26.27).

If the individual values of an auxiliary variable $X$ are available for each element in the population and all these values are positive, sampling with unequal probabilities can be considered. This only improves the precision of an estimator if there is a correlation between the target variable and the auxiliary variable. In the case of the shoplifting survey, the correlation coefficient of the value of the stolen goods in the shop and the floor space of the shop is 0.892, which is a strong correlation. The margin of error is only 19.97 for a sample of size 100. Hence, it pays (in terms of precision) the used floor space as an auxiliary variable in the sampling design.

The precision of an estimator may be improved by using auxiliary information. Some examples of improved estimators were described in Section 4.2.7. Three estimators are applied to the fictitious shoplifting example: the ratio

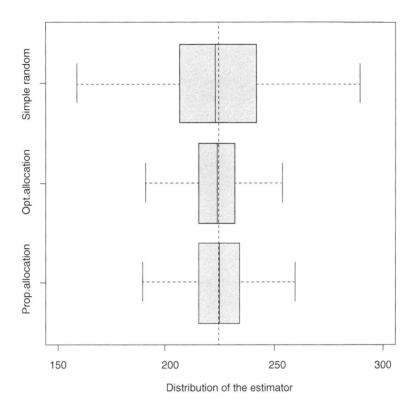

FIGURE 4.8 **Distribution of the estimator of the value of stolen goods for simple random sampling and different types of stratified sampling**

estimator, the regression estimator, and the post-stratification estimator. All three estimators require an auxiliary variable for which the population distribution is known.

The ratio estimator is most effective if the values of the target variable and the auxiliary are proportional. The regression estimator is the estimator of choice if there is a general linear relationship between the values of the target variable and the auxiliary variable. The post-stratification estimator can be used if the auxiliary variable is categorical and it divides the population into homogeneous strata.

Figure 4.9 shows the distribution of these three estimators for the shoplifting example. Floor space is used as the auxiliary variable for the ratio estimator and the regression estimator. This auxiliary variable is transformed into a categorical variable with four categories (see Table 4.2) for the post-stratification estimator. For reasons of comparison, also the distribution of the simple sample mean is shown.

All three estimators perform better than the simple sample mean. The regression estimator has the highest precision, closely followed by the ratio estimator. This is not surprising as there is a linear relationship between the value of stolen goods and the floor space of the shop. The ratio estimator assumes the constant

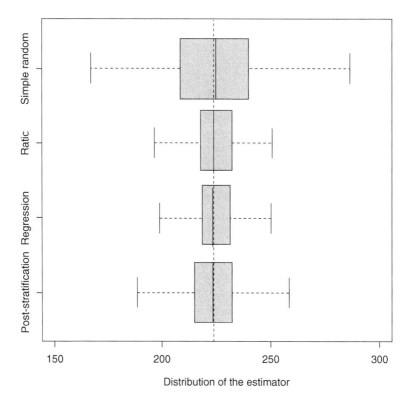

**FIGURE 4.9**  **Distribution of the ratio, regression, and post-stratification estimator**

term of the regression line to be 0. This is not the case here. Therefore, the ratio estimator does not perform as well as the regression estimator.

The post-stratification estimator has approximately the same behavior as the estimator for stratified sampling with proportional allocation. The explanation is that post-stratification results (on average) in numbers of observations per stratum that are close to proportional allocations.

The conclusion of this application is that it pays to use auxiliary information, either in the sampling design or in the estimation procedure.

## 4.4  Summary

A web survey is an instrument to collect information about a specific population. Typically, not all elements are investigated in a survey, but just a sample. This sample must have been selected by means of a probability sample where every element in the population has positive probability of selection.

There are various ways to do this. Common sampling designs are simple random sampling, stratified sampling, sampling with unequal probabilities, cluster sampling, and two-stage sampling.

If every element in the population has a known and positive probability of selection, it is always possible to define an unbiased estimator. This is the so-called Horvitz–Thompson estimator. This estimator can be improved by taking advantage of available auxiliary variables. Examples of such estimators are the ratio estimator, the regression estimator, and the post-stratification estimator.

## KEY TERMS

**Allocation**: The distribution of the sample over the strata in stratified sampling. Optimal (or Neyman) allocation results in the highest precision of the estimator. It requires knowledge of the variances in the strata. If this is not the case, proportion allocation can be used.

**Cluster sampling**: A sampling design in which the population has been divided into a number of clusters. A sample of clusters is drawn, and all elements in the selected clusters are included in the sample.

**Horvitz–Thompson estimator:** An unbiased estimator that can be computed if the selection probabilities of all elements are known and positive.

**Post-stratification estimator**: An estimator that can be computed if the sizes of all strata in a stratified population are available. It is a precise estimator if the strata are homogeneous.

**Ratio estimator**: An estimator that can be computed if the population mean of an auxiliary variable is available. It is a precise estimator if the target variable and auxiliary variable are (approximately) proportional.

**Regression estimator**: An estimator that can be computed if the population mean of an auxiliary variable is available. It is a precise estimator if there is a (approximate) linear relationship between target variable and auxiliary variable.

**Simple random sampling**: A sampling design in which elements are selected with equal probabilities.

**Stratified sampling**: A sampling design in which the population is divided into a number of strata and where a random sample is drawn from each stratum.

**Two-stage sampling**: A sampling design in which the population is divided into a number of clusters. A sample of clusters is drawn. From each selected cluster a sample of elements is drawn.

**Unequal probability sampling**: A sampling design in which elements are selected with probabilities that are proportional to the values of some auxiliary variable.

## EXERCISES

**Exercise 4.1**    Suppose a simple random sample of size 100 is selected from a population of size 1,000. What is the value of the second order inclusion probability for every pair of elements?

**a.** 0.1

**b.** 0.09

**c.** 0.01

**d.** 0.0099.

**Exercise 4.2**    Suppose a simple random sample of size 1,000 is selected from a population of size 16,000,000. What would happen to the variance of the sample mean if the sample size was doubled to 2,000?

**a.** It would be twice as small.

**b.** It would be approximately twice as small.

**c.** It would be twice as big.

**d.** It would be approximately twice as big.

**Exercise 4.3**    Under which condition will stratified sampling lead to more precise estimates than the simple sample mean?

**a.** The values of the target variable vary little within strata.

**b.** The stratum means vary little.

**c.** The strata are all of approximately the same size.

**d.** The subsamples in all strata are of the same size.

**Exercise 4.4**    Which of the statements below does not apply to cluster sampling?

**a.** It is not clear beforehand how large the sample size will be.

**b.** It can reduce travel costs of interviewers.

**c.** No sampling frame at all is necessary.

**d.** Generally, it will lead to more precise estimators than simple random sampling.

**Exercise 4.5**    Under which condition is the variance of the regression estimator smaller than the variance of the simple sample mean?

**a.** This is always the case.

**b.** Only if the correlation between target variable and auxiliary variable is greater than 0.

**c.** Only if the correlation between target variable and auxiliary variable is equal to 1.

**d.** Only if the squared correlation between target variable and auxiliary variable is greater than 0.

**Exercise 4.6** What is the effect of an increasing sample size on the value of the confidence level of the confidence interval?

**a.** It remains 0.95.

**b.** It increases in size and approaches 1.00.

**c.** It decreases in size and approaches 0.00.

**d.** It remains 0.05.

## REFERENCES

Bethlehem, J. G. (2009), *Applied Survey Methods, a Statistical Perspective.* Wiley, Hoboken, NJ.

Beukenhorst, D. & Wetzels, W. (2009), *A Comparison of Two Mixed-mode Designs of the Dutch Safety Monitor: Mode Effects, Costs, Logistics.* Technical paper DMH 206546, Statistics Netherlands, Methodology Department, Heerlen, the Netherlands.

Bowley, A. L. (1906), Address to the Economic Science and Statistics Section of the British Association for the Advancement of Science. *Journal of the Royal Statistical Society,* 69, pp. 548–557.

Bowley, A. L. (1926), Measurement of the Precision Attained in Sampling. *Bulletin of the International Statistical Institute,* XII, Book 1, pp. 6–62.

Cochran, W. G. (1977), *Sampling Techniques,* 3rd edition. Wiley, New York.

Horvitz, D. G. & Thompson, D. J. (1952), A Generalization of Sampling Without Replacement from a Finite Universe. *Journal of the American Statistical Association,* 47, pp. 663–685.

Scherpenzeel, A. (2008), An Online Panel as a Platform for Multi-disciplinary Research. In: Stoop, I. & Wittenberg, M. (eds.), *Access Panels and Online Research, Panacea or Pitfall?* Aksant, Amsterdam, pp. 101–106.

# CHAPTER FIVE

# Errors in Web Surveys

## 5.1 Introduction

Survey researchers have control over several different aspects of a survey. With the proper choice of a sampling frame, a sampling design, and an estimation procedure, they can obtain precise estimators of population characteristics. Unfortunately, not everything is under control. Survey researchers have to confront with various phenomena that may have a negative impact on the quality and therefore the reliability of the survey outcomes. Some of these disturbances are almost impossible to prevent. Efforts will then have to aim at reducing their impact as much as possible. Nevertheless, notwithstanding all efforts to eliminate or reduce problems, final estimates of population parameters are biased. Estimates differ from the true value. This difference is the *total error* of the estimate.

Errors may occur in surveys whatever the mode of data collection, but some errors are more likely to occur in some types of surveys. For an example, it makes a difference whether interviewers conduct interviews or the respondents complete the questionnaires themselves. The focus in this chapter is on errors in web surveys. Sometimes the impact of a specific type of error in a web survey is compared with other types of surveys.

Sources of error increase the uncertainty with respect to the correctness of estimates. This uncertainty can manifest itself in the distribution of an estimator in two ways: (1) it can lead to a systematic deviation (bias) from the true population value, or (2) it can increase the variation around the true value of the population parameter.

*Handbook of Web Surveys*, Second Edition. Silvia Biffignandi and Jelke Bethlehem.
© 2021 John Wiley & Sons, Inc. Published 2021 by John Wiley & Sons, Inc.

Let $\bar{y}_E$ be an estimator for the population mean $\bar{Y}$. Chapter 4 discusses the properties of a good estimator. One is that an estimator must be *unbiased*. This means its average value over all possible outcomes must be equal to the population mean to be estimated:

$$(5.1) \qquad E\left(\bar{y}_E\right) = \bar{Y}.$$

An estimator may be biased due to survey errors. Suppose one of the objectives of a survey is to estimate the average amount of time per day people spend on the Internet. If a web survey is conducted for this, people without Internet access will not be in the sample. Since these people do not spend time on the Internet, the estimate will be too high. The estimator has an upward bias. Denote this bias of the estimator $\bar{y}_E$ by

$$(5.2) \qquad B\left(\bar{y}_E\right) = E\left(\bar{y}_E\right) - \bar{Y}.$$

Another desirable property of an estimator is that its variance is as small as possible. This means that

$$(5.3) \qquad V\left(\bar{y}_E\right) = E\left(\bar{y}_E - E\left(\bar{y}_E\right)\right)^2$$

must be small. An estimator with a small variance is *precise*. An estimator can be made more precise by increasing the sample size or by using auxiliary information. This issue is discussed in more detail in Chapter 4.

A precise estimator may still be biased. Therefore, just the value of the variance itself is not a good indicator of how close estimates are to the true value. A better indicator is the *mean square error* (MSE). The definition of this quantity is

$$(5.4) \qquad M\left(\bar{y}_E\right) = E\left(\bar{y}_E - \bar{Y}\right)^2.$$

It is the expected value of the squared difference of the estimator from the value to be estimated. Writing out this definition leads to a different expression for the MSE:

$$(5.5) \qquad M\left(\bar{y}_E\right) = V\left(\bar{y}_E\right) + B^2\left(\bar{y}_E\right).$$

The MSE contains both sources of uncertainty: a variance component and a bias component. The MSE of an estimator is equal to its variance if it is unbiased. Only if both the variance and the bias are small, a small MSE is achieved. Figure 5.1 distinguishes four different situations encountered in practice.

The vertical line in each graph represents the population mean. The distribution in the upper-left corner shows the ideal situation for an estimator: it is precise

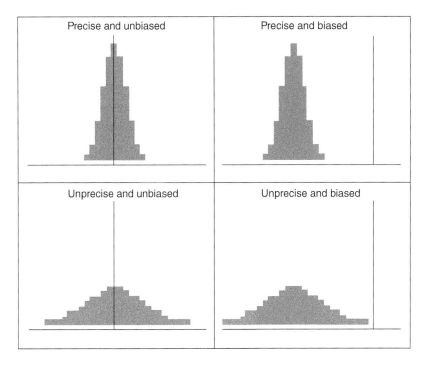

FIGURE 5.1  **The bias and precision of an estimator**

and unbiased. All possible outcomes are close to the true value, and there is no systematic overestimation or underestimation. The situation in the lower-left corner is less attractive. The estimator is still unbiased but has a substantial variance. Confidence intervals will be wider. Reliability is not affected. For the confidence level of a 95% confidence, interval remains 95%. The only difference is that these intervals are wider. It is still possible to draw the correct conclusions, but they are less precise.

The situation is completely different for the graph in the upper-right corner. The estimator is precise but has a substantial bias. The confidence interval computed using the survey data will almost certainly not contain the true value. The confidence level is seriously affected. Estimates are unreliable. Most likely, wrong conclusions are drawn. The graph in the lower-right corner offers the highest level of uncertainty. The estimator is biased. Moreover, it is also not precise. In this situation, the MSE has its largest value.

Survey estimates are never exactly equal to the population characteristics they intend to estimate. There will always be some error. A classification of possible causes is shown in Figure 5.2. This classification is an extended version of the one described by Kish (1967).

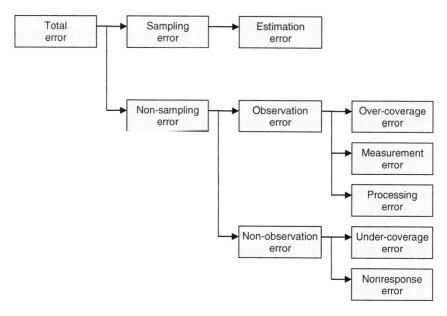

FIGURE 5.2 A classification of survey errors

The ultimate result of all these errors is a discrepancy between the survey estimate and the population parameter. Two broad categories of phenomena contributing to this total error are sampling errors and non-sampling errors.

*Sampling errors* are introduced by the sampling design. They are due to the estimates that are based on a sample from the population and not on a complete enumeration of the population. Sampling errors vanish when observing the complete population. Since only a sample is available for computing population characteristics, and not the complete data set, one has to rely on estimates. In general, the sampling errors decrease as the sample size increases, whereas non-sampling error increases as the sample size increases.

The *estimation error* denotes the effect caused by using a probability sample. Every new selection of a sample will result in a different set of elements and thus a different value of the estimator. Through the sampling design, control over the estimation takes place. For example, by increasing the sample size or by taking selection probabilities proportional to the values of some selected auxiliary variable, there is a reduction in the estimation error. Sampling errors are unrelated to the mode of data collection. Provided to correct selection probabilities, they are known and used in the estimator, it is always possible to construct an unbiased estimator (see Chapter 4). Problems in the sampling frame might cause differences between anticipated and true selection probabilities.

📱 EXAMPLE 5.1 **Selection probabilities in an address sample**

A researcher intends to select a simple random sample of persons. He does this by drawing a simple random sample of addresses and picking one person at random at each selected address. If he assumes equal selection probabilities and consequently uses the sample mean as an estimator for the population mean, there will be a selection error.

The true selection probabilities are not equal. They depend on the number of persons living at an address. For example, persons in large households have a smaller selection probability than single persons.

Members of large households will be underrepresented in the sample. If the target variable of the survey relates to the size of the household, there will be bias in the sample mean. For example, if the target variable would be the number of hours the members of the household spend on the Internet, the value of sample mean will have a substantial downward bias.

If web survey respondents are self-selected, a special type of selection error occurs. Since recruitment is by self-selection, the true selection probabilities are unknown. Usually the researcher assumes all selection probabilities are equal, so they use the sample mean as an estimator of the population mean. Unfortunately, self-selection probabilities tend to depend on the characteristics of the respondents and therefore may vary substantially. Thus, true selection probabilities differ from anticipated probabilities, resulting in a selection error.

True selection probabilities may differ from anticipated selection probabilities because of problems in the sampling frame. For example, if elements have multiple occurrences in the sampling frame, their selection probabilities will be larger. Detecting and avoiding this type of selection error is possible only by thorough investigation of the sampling frame.

The *non-sampling errors* arise because of the factors other than the inductive process of inferring about the population from a sample. In some situations, the non-sampling errors may be large and deserve greater attention than the sampling error. In any survey, it is assumed that the value of the characteristic to be measured has been defined precisely for every population unit. Such a value exists and is unique. This is called the true value of the characteristic for the population value. In practical applications, data collected on the selected units are called survey values, and they differ from the true values. Such difference between the true and observed values is termed as the observational error or response error. Such an error arises mainly from the lack of precision in measurement techniques and variability in the performance of the investigators. One or more of the following reasons may give rise to non-sampling errors or indicate its presence.

For example, the data specification may be inadequate and inconsistent with the objectives of the survey, the boundaries of area units may be incomplete or units wrongly identified, the questionnaire definitions and instructions may be ambiguous, units may be faulty selected, and so forth. These sources are not exhaustive but surely indicate the possible source of errors.

*Non-sampling errors* depend on problems that can occur even if the investigation of the whole population (instead of a sample) is undertaken. They denote errors made during the process of obtaining answers to questions asked. The non-sampling errors are unavoidable in census and surveys. The data collected by complete enumeration in census is free from sampling error but would not remain free from non-sampling errors.

One or more of the following reasons may give raise to non-sampling errors or indicate its presence: the data specification may be inadequate or inconsistent with the objectives of the survey or census; the questionnaire definitions and instructions may be ambiguous; the recall errors may pose difficulty in reporting the true data; the coding of the data may be erroneous.

In a sample survey, the non-sampling errors arise also due to defective frames and faulty selection of sampling units. The abovementioned sources of errors are not exhaustive but surely indicate several important possible sources of errors.

Non-sampling errors are observation errors and non-observation errors. *Observation errors* refer to errors made during the process of obtaining answers from respondents and recording and further processing these answers. It is possible to distinguish three types of observation errors: over-coverage errors, measurement errors, and processing errors.

An *over-coverage error* is due to elements that are included in the survey and that do not belong to the target population. Such elements should not be included in the survey. It is recommended to ignore them. If they are included, they may lead to errors in estimators. Example 5.2 provides a situation where over-coverage is registered in a tentative full population survey.

---

**EXAMPLE 5.2 A survey about road pricing**

There was an intensive political discussion in the Netherlands in January 2010 about the introduction of a system of road pricing. An important participant in this discussion was the Dutch Automobile Association (ANWB). This organization conducted a poll on its website. It was a self-selection survey. The poll was supposed to determine the opinion of the ANWB members about road pricing. Thus, the target population consisted of all members of the association. However, everyone with Internet could participate. Children could complete the questionnaire and people in other countries, too. There was no restriction. Everyone could complete the questionnaire even more than once. There was nothing preventing this.

Fortunately, by asking all respondents whether they were a member of the ANWB, it was possible to identify units not belonging to the sampling frame. Consequently, it was possible to remove the nonmember respondents (over-coverage) from the analysis, thus avoiding over-coverage error.

Within a period of a few weeks, more than 400,000 times the questionnaire completion took place. About 50,000 respondents indicated they were not an ANWB member. Thus, there was substantial over-coverage.

Note that people may not have answered the question about membership properly. There is always a risk of socially desirable answers. Hence, there could still be hidden over-coverage.

A *measurement error* occurs if the answers given by respondents differ from the true answer. They may not understand a question, do not know the true answer, or do not want to give the true answer. This is caused, for example, by interview effects; question wording effects and memory effects belong to this group of errors. Measurement errors are an important source of errors in web surveys. Therefore, Section 5.2.1 is devoted to this problem.

A *processing* error occurs if errors take place during the phase of recording and processing the collected data. In case of surveys with paper questionnaire forms, respondents or interviewers can make errors in writing down the answer. Such problems cannot occur in web surveys. However, mistakes occur also if the questionnaire is on the Internet. It is easy to click on the wrong answer, especially in mobile web surveys.

It is not uncommon that new variables are derived from those that have been measured in the survey. For example, one of the key variables of the Dutch Labor Force Survey is employment status in three categories: "employed," "unemployed," and "not in labor force." People are considered unemployed if they are working less than 12 hours a week and are actively seeking and available for one or more jobs with a total of more than 12 hours a week. Putting people in the right category requires the answers to a set of questions. Deriving new variables from existing ones usually is through computer programs. The algorithms for this can sometimes be complex, particularly if they must cope properly with every possible practical situation. Programming errors can be the cause of processing errors.

*Non-observation errors* are the second cause of non-sampling errors. These errors occur because it is impossible to obtain intended measurements. There are two types of non-observation errors: under-coverage errors and nonresponse errors.

*Under-coverage* occurs if elements in the target population do not have a corresponding entry in the sampling frame. These elements can and will never be selected for the survey. Under-coverage can be a problem in web surveys if the target population is wider than just those with access to the Internet. This typically happens if general population surveys take place by means of the Internet. The under-coverage errors can be substantial. Therefore, Chapter 10 is devoted to it.

Another type of non-observation error relates to *nonresponse*. Nonresponse is the phenomenon that elements selected in the sample do not provide the required information or that the collected information is useless. Nonresponse occurs in almost every survey, whatever the mode of data collection. Some specific groups respond better than other groups. Thus, some groups are overrepresented and other groups underrepresented. This leads to biased estimates. The nonresponse error may occur due to refusal by respondents to give information, or the sampling units may be inaccessible. This error arises because the set of units getting excluded may have characteristic so different from the set of units actually surveyed as to make the results biased. This error is termed as nonresponse error since it arises from the exclusion of some of the anticipated units in the sample or population. One way of dealing with the problem of nonresponse is to make all the efforts to collect information from a subsample of the units not responding in the first attempt. Section 5.2.2.1 describes the nonresponse problem in more detail.

The classification above makes clear that many things can go wrong during the process of collecting survey data, and usually it does. Taking preventive measures at the design stage might avoid some errors. However, any error will remain. Therefore, checking collected data for errors, and where possible correcting them, is very important. This activity is *data editing*. Data editing procedures are not able to handle every type of survey error. They are most suitable for detecting and correcting measurement errors, processing errors, and possibly over-coverage. Phenomena like selection errors, under-coverage, and nonresponse require a different approach.

Be advised that there are relationships between various types of errors. Several error trade-offs exist. For example, Eckman and Kreuter (2017) put in evidence the nonresponse–coverage trade-offs in web surveys. The impact on quality indicators is discussed, too. Although most web surveys under-cover those who do not have access to the Internet, some overcome this issue by providing Internet access, a computer, or an offline alternative so that these persons can participate. One such survey is the Longitudinal Internet Studies for the Social Sciences (LISS) panel in the Netherlands. A computer or Internet access is provided to non-Internet households, thus expanding its coverage to include these households. During panel recruitment, however, the LISS study observed lower recruitment rates from non-Internet households than Internet households. That is, increasing units of the survey to cover non-Internet households reduced the response rate to the recruitment survey (Leenheer and Scherpenzeel, 2013).

There are also trade-offs between errors and some constrains, like time, costs, and technology. Survey methodology literature has put forward responsive and adaptive survey designs (see Chapter 8) as means to make efficient trade-offs between nonresponse and costs.

There are also various survey-related features that impact on errors, e.g., the survey mode, the type of questionnaire (long or condensed), and the type of reporting hit the measurement error.

Mixed mode is thought to be useful in improving response to web and web/mail. Millar and Dillman (2011) conducted two experiments designed to evaluate several strategies for improving response to web and web/mail mixed-mode surveys to determine the best ways to maximize web response rates in a highly Internet-literate population with full Internet access.

Findings were that offering the different response modes sequentially (web is offered first and a mail follow-up option in the final contact) improves web response rates. Also, delivering a token cash incentive in advance using a combination of both postal and e-mail contacts improves web response rates. No evidence of impact on improving rates if a simultaneous choice of response modes is allowed.

It is advised to adopt a sequential mode with web first to tentatively improve response rates. However, the most adequate strategy to move from one single mode to mixed mode with web component is not yet clearly defined. Targeting the web invitation to a subset of the sample is likely to be more effective than inviting all sample members to the web (for more details and references about errors in mixed mode, see Chapter 9).

Also, trade-offs seem to differ across different modes (see Tourangeau, 2017). For example, trade-offs exist when deciding to include or not to include mobile phone numbers in a survey. An ever-increasing portion of the population can be reached only via mobile device. Thus, surveys that exclude mobile device numbers will miss these people. However, decisions about using mobile or landline numbers is critical for response rates. Literature shows that people reached via landline phone participate at higher rates than those reached via mobile devices (American Association for Public Opinion Research).

The flowchart presented in Chapter 3 (Figure 3.1) shows the steps of a survey process. At each step, errors might occur; errors at one step can affect errors at another step. The total survey error perspective is stressing that deciding to try to reduce a specific type of error needs the researcher to be careful. There is a risk that another type of error increases. It is difficult to measure the impact of the relationships between different errors, but attention on the possible impact and caution in decisions should be a rule when deciding how to proceed at different steps.

"Total" MSE shows the cumulative effects of sampling and non-sampling error sources on the survey estimate. Each error source contributes a variable error, a systematic error, or both. Variable errors are reflected in the variance of the estimate, while systematic errors are reflected in the bias squared component. The bias and variance components may be further decomposed into process-level and even subprocess-level components to further pinpoint specific error sources and, hopefully, their root causes. Such decompositions can be quite helpful for designing surveys and targeting and controlling the major error sources during survey implementation. For error evaluations, the major components of the total MSE are estimated and combined according to the decomposition formulas to form an estimate of the total MSE. Total survey error perspective is discussed, for instance, in Biemer (2010) and in Couper, Antoun, and Mavletova (2017).

When planning a survey, there are several types of constrains, and the limits in terms of money are one of them. There are trade-offs between costs and errors. Cost efficiency, i.e., the relationship between errors and costs, is a relevant task, There are costs from the user side (for example, response burden) and costs from the researcher side (coverage problems, questionnaire implementation, response rate, and so on). Snijkers et al. (2013) focus on costs business respondents are facing when completing the questionnaire (mainly time of personnel devoted to that task). Some other authors focus on costs from the research budget point of view and from quality point perspective. For example (Couper, Antoun, and Mavletova 2017; Peterson et al., 2017), allowing participation by completing the questionnaire using mobile devices could improve coverage of the population and reduce errors. However, complex web questionnaires and their adaption to mobile devices could increase the effort and costs in the questionnaire implementation. In some case allowing questionnaire access by mobile devices increases break-offs (Callegaro, 2010). Therefore, response rate does not improve. Concluding a balance between errors and costs (various types of costs) should be evaluated from the survey researcher taking the context into consideration.

# 5.2 Theory

### 5.2.1 MEASUREMENT ERRORS

Measurement error is a general concept also used outside the field of survey research. In general term, a definition is as the difference between the actual value of a quantity and the value obtained by a measurement. The measuring instrument can have a *random error* if repeated measurements produce values that vary around the true value. This may be caused by the limited precision of the instrument. It can also have a *systematic error* if repeated measurement produces values that are systematically too high or too low. This may be, for example, because of incorrect calibration of the instrument.

The measuring instrument of a survey is the questionnaire. This is not a perfect instrument. In the physical environment, for measuring someone's length, a measuring staff is used; by a weighing scale, the weight of a person can be determined. These physical measuring devices are generally very accurate. The situation is different for a questionnaire. It only indirectly measures someone's behavior or attitude. Schwarz et al. (2008) describe the tasks involved in answering a survey question:

- Step 1: The respondents need to understand the question. They have to determine the information to provide for the asked questions. If they do not understand the question, they may decide to rephrase the questions and answer this question.

- Step 2: They need to retrieve the relevant information from their memory. In the case of a nonfactual question (for example, an opinion question), they will not have this information readily available. Instead, they have to form an opinion on the spot using whatever information that comes to mind. In the case of a factual question (for example, a question about behavior), they have to retrieve from their memory information about events in the proper time period.

- Step 3: They have to translate the relevant information in a format fit for answering the questions. Although they may have an answer in mind, it is often not in the proper format. For example, they may have to reformat their answer so that it fits one of the answer options of a closed question.

- Step 4: Respondents may hesitate to give their answer. If the question is about a sensitive topic, they may decide to refuse to give an answer. If an answer is socially undesirable, they may change their answer, too.

The process of answering questions is a complex one. Several things can go wrong in this process, leading to measurement errors. Problems with the survey questions will affect the quality of the collected data and consequently also the survey results. It is of the utmost importance to carefully design and test the survey questionnaire. Several authors are focusing on the questionnaire construction, as Callegaro, Manfreda, and Vehovar (2015), Dillman, Smyth, Christian (2014). It is sometimes said that questionnaire design is an art and not a skill. Nevertheless, long years of experience have leaded to a number of useful principles. This section describes a number of issues.

Kalton and Schuman (1982) distinguish factual and nonfactual questions. *Factual questions* are to obtain information about facts and behavior. There is always an individual true value. This true value could also be determined, at least in theory, by some other means than asking a question of the respondent. Examples of factual questions are as follows: "What is your regular hourly rate of pay on this job?" "Do you own or rent your place of residence?" "Do you have an Internet connection in your home?"

The fact to be measured by a factual question must be precisely defined. Even a small difference in the question text may lead to a substantially different answer, as shown in several studies. As an example, a question about the number of rooms in the household can cause substantial problems if it is not clear what constitutes a room and what not. Should a kitchen, a bathroom, a hall, and a landing be included?

*Nonfactual questions* ask about attitudes and opinions. An *opinion* usually reflects views on a specific topic, like voting behavior in the next general elections. An *attitude* is a more general concept, reflecting views about a wider, often more complex issue. With opinions and attitudes, there is no such thing as a true value. They measure a subjective state of the respondent not observable by another means. The attitude only exists in the mind of the respondent.

There are various theories explaining how respondents determine their answer to an opinion question. One such theory is the *online processing model* described by

Lodge, Steenbergen, and Brau (1995). According to this theory, people maintain an overall impression of ideas, events, and persons. Every time they confront themselves with new information, this summary view spontaneously updates. When they have to answer an opinion question, their response considers this overall impression. The online processing model should typically be applicable to opinions about politicians and political parties.

There are situations in which people do not have formed an opinion about a specific issue. They just start to think about it when confronted with the question. According to the *memory-based model* of Zaller (1992), people collect all kinds of information from the media and in contacts with other people. Much of this information is stored in memory without paying attention to it. If asked to answer an opinion question, respondents may recall some of the relevant information stored in memory. Due to the limitations of the human memory, only part of the information is used. This is the information that immediately comes to mind when the question is asked. This is often information that only recently has been stored in memory. Therefore, the memory-based model is able to explain why people seem to be unstable in their opinions. Their answer may easily be determined by the way the media recently covered this issue.

The reminder of this section is devoted to a description of a number of effects that can lead to measurement error. Where possible, indication of whether web surveys in particular are vulnerable is given.

### 5.2.1.1 Satisficing.

When persons participate in a survey, they often will have to answer many questions. To do this properly requires a substantial cognitive effort. Although they may initially be motivated to do so, they are likely to become fatigued during the course of the interview. Interest in answering the questions will fade away. If the interview takes long to finish, they become impatient and distracted. As a result, they will devote less energy to answering the questions. As Krosnick (1991) describes it, respondents are less thoughtful about the meaning of the questions, they search their memories less thorough, they integrate information more carelessly, and they may select an answer option more haphazardly. The first more or less acceptable answer that comes into mind is given. This phenomenon is called *satisficing*.

Holbrook, Green, and Krosnick (2003) argue that satisficing occurs more in telephone survey than in face-to-face surveys. Respondents of telephone survey may be more distracted because they might be engaged in different activities while they answer questions. This is like a form of multitasking. Heerwegh and Loosveldt (2008) suggest that satisficing is even more a problem in web surveys. While respondents are answering for questions, they can also be involved in other activities on their computer, like answering e-mail or visiting other websites. They also note that the cognitive demands of answering web survey questions are higher than those of interviewer-assisted surveys. Contrasting results are obtained when considering the length of open-ended question as a satisficing indicator. Some experimental studies report mobile completed questionnaire providing shorter answers and suggest satisficing effect (Mavletova, 2013; Wells, Bailey, and Link, 2014). However, other authors do not identify differences between personal computer (PC) and mobile with respect to satisficing (Buskirk and Andrus, 2014; Toepoel and Lugtig, 2014).

Krosnick (1991) distinguished two forms of satisficing: weak satisficing and strong satisficing. *Weak satisficing* denotes the situation in which respondents still go through the four steps of answering a question, but do it less thoroughly. They put less effort in attempting to understand the meaning of the question, they search their memory less well for relevant information, they may integrate the retrieved information more carelessly, and they will pick more easily an arbitrary answer option. *Strong satisficing* denotes the situation in which respondents simplify the answer process even more. They interpret each question only superficially by skipping steps 2 and 3 (retrieval and processing of information). They just pick an answer that seems reasonable.

Satisficing can come in many forms. Two forms of weak satisficing (response order effects and acquiescence) and four forms of strong satisficing (status quo endorsement, non-differentiation, answering "don't know," and arbitrary answers) are described here.

**5.2.1.2 Response Order Effects.** *Response order effects* can occur if respondents are answering closed questions. They have to pick the proper answer from a list of possible answer options, which sometimes is long. Instead of thinking carefully about which option is appropriate, they chose the first reasonable option. In case of an interviewer-assisted survey (face-to-face or by telephone), the interviewer reads aloud the answer options. It is difficult for respondents to remember all options. Since only the last few options are still in their short-term memory, they restrict their judgment to these options. As a result, there is a preference for options near the end of the list. This is called a *recency effect*.

Self-administered surveys (web, mail) suffer, by contrast, from a *primacy effect*. This is the tendency to pick an answer early in the list of options. Reading to a list of possible options and considering each option require a considerable effort. Therefore, respondents may stop at the first reasonable option.

The description of response order effects is in more detail given by Krosnick and Alwin (1987). According to the underlying theory, web surveys suffer from primacy effects. This was indeed the case in an experiment with the Dutch Safety Monitor, described by Kraan et al. (2010). Schwarz, Hippler, and Noelle-Neumann (1992), Sudman, Bradburn, and Schwarz (1996), and Couper et al. (2004) were also showing the effect.

In web surveys not only the order of the response options matters but also the format of the presentation to the respondents of the response options. For a closed question, the HTML language offers various ways of displaying the possible answer options and selecting one of them. Figures 5.3–5.6 show some examples. Figure 5.3 shows the use of *radio buttons*. Clicking on the corresponding radio button, the respondent selects an option. The advantage of this technique is that it is always possible to select only one answer. Selecting a new answer deselects the currently selected answer. Once an answer is given, it is not possible to erase it and let the question remain unanswered.

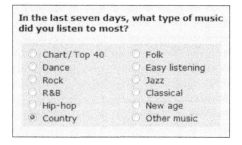

FIGURE 5.3  A closed question with radio buttons

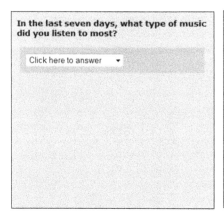

FIGURE 5.4  A closed question with two columns of radio buttons

FIGURE 5.5  A closed question with a drop-down box

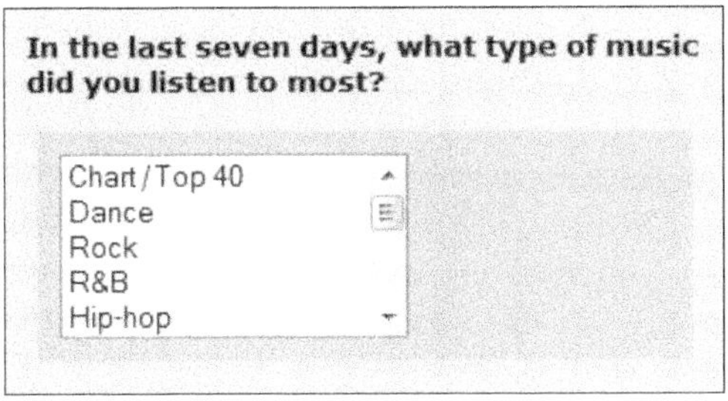

**In the last seven days, what type of music did you listen to most?**

Chart / Top 40
Dance
Rock
R&B
Hip-hop

FIGURE 5.6  A drop-down box with a fixed number of items

Comparing web respondents and mobile respondents, some empirical works show that most satisficing behaviors, like primacy effects or missing data, do not change with a switch from a PC to a mobile phone (Lugtig and Toepoel, 2016), nor the rate of on-substantive responses and primacy effects differ between mobile phone and PC respondents. Tourangeau et al. (2017) examined whether the differences across devices might moderate the effects of response order and other measurement errors due to devices (device effect). Like many of the prior studies examining mobile devices, response order affected the answers respondents gave, but findings showed few device effects.

Displaying radio buttons when the list of possible answers is long requires special attention. Possibly avoid it. All question information should be visible on the screen and not require any scrolling. Therefore, it is better to split a long list of radio buttons over a number of columns. Figure 5.4 gives an example. Here the distribution of the answers is over two columns. To give a visual clue that these two lists belong together, Dillman, Tortora, and Bowker (1998) suggest putting them in a kind of box. Figure 5.4 shows how a gray background helps in presenting the list in a compact way.

The HTML language also offers a different technique to select an item from a list. It is a *drop-down box*. Figure 5.5 shows an example. On the left, there is the initial state of the box. Only one option is visible: the first one or the selected one. To select an answer, the respondent must click on the drop-down box, after which the list of possible answers becomes visible. In Figure 5.5, on the right, the list appears to the respondent. If this list is very long, it only becomes partially visible. It depends on the browser used how long this list is. For example, 20 items are shown in Figure 5.6. If the list is longer, scroll bars to make other items visible are used. The respondent selects an answer by clicking on it in the list.

Drop-down boxes have a number of disadvantages. In the first place, respondents have to do more work to select an answer (as compared with radio buttons); this is still more heavy in smartphone device. They have to perform three actions: clicking the box, scrolling to the right answer, and clicking this answer. In the second place, there can be serious primacy effects when displaying only part of the list. In the third place, it is unclear how much space the question requires on the screen.

It is possible to modify the behavior of the drop-down box, so it always shows a fixed number of items in the list. Figure 5.6 shows an example in which this number is set to 5. The amount of space needed for such a question is now fixed and small. However, it suffers from a serious primacy effect. Couper et al. (2004) have shown that this effect is particularly large for the format in Figure 5.6. There-fore, where possible, radio buttons should be preferred.

The advantages and disadvantages of various answer formats of closed questions are also discussed by Couper (1999), Heerwegh and Loosveldt (2002), and Dillman (2007).

**5.2.1.3 Acquiescence.** A second form of weak satisficing is *acquiescence*. This is the fact that respondents tend to agree with statements in questions, regardless of their content. They simply answer "yes." Holbrook, Green, and Krosnick (2003) and Krosnick (1999) suggest that acquiescence occurs partly because respondents only superficially think about the statement offered in the question. This will result in a confirmatory answer.

Acquiescence is typically a problem for agree/disagree, true/false, or yes/no questions. Respondents tend to answer agree, true, or yes irrespective of the topic of the question. Krosnick (1999) estimates the bias due to acquiescence to be on the order of 10%. The literature suggests that acquiescence is more common among respondents with a lower socioeconomic status. Example 5.3 shows the results of an experiment about the effect of acquiescence.

▦  EXAMPLE 5.3  **Bias caused by acquiescence**

Schuman and Presser (1981) describe an experiment showing the effect of acquiescence. Respondents broken down into two groups were randomly selected. The first group was asked to respond to the statement "Individuals are more to blame than social conditions for crime and lawlessness in this country." There were two possible answers: agree or disagree. The statement for the second group was reversed: "Social conditions are more to blame than individuals for crime and lawlessness in this country." Table 5.1 shows the results of this experiment.

TABLE 5.1    **Are individuals or social conditions to blame?**

| Statement | Agree (%) | Disagree (%) |
|---|---|---|
| Individuals are more to blame | 59.6 | 40.4 |
| Social conditions are more to blame | 56.8 | 43.2 |

The percentages of respondents agreeing with the statement do not dif-fer much. The differences are within the margin of error. These percentages are always higher than the percentages of people disagreeing, whatever the statement. Reversing the statement does not seem to have an effect on the percentages.

According to de Leeuw (1992), there is less acquiescence in self-administered surveys than in interviewer-assisted surveys. Respondents tend to agree more with statements made in questions if interviewers are present. Without interviewers, respondents may feel more anonymous and therefore will be more inclined to answer sensitive questions honestly.

This suggests that acquiescence will be less of a problem in web surveys and more in face-to-face and telephone surveys.

**5.2.1.4 Endorsing the Status Quo.** Surveys sometimes ask respondents to give their opinion about changes. A typical example is a question whether government should change its policy with respect to a specific issue. Here are some examples of such questions:

- Should the defense budget of the United States be increased or decreased?
- Should gun control laws in the United States become stricter or less strict?
- Should the monarchy in the Netherlands be abandoned or not?
- Should same-sex marriages be recognized legally or should they be prohibited?
- Should new nuclear power plants be built in the country?

The easiest way to answer such a question without thinking is to select the option to keep everything the same. If the option of no change is not explicitly offered, not many respondents will insist on giving this answer. However, if this option is available, the number of people selecting it will substantially increase. According to Krosnick (1991), the percentage selecting "no change" will increase by 10%–40%.

The tendency to endorse the status quo raises the question whether or not to include a middle category (representing "no change") in a closed question. On the one hand, some researchers think it should be included. If it is not there, respondents with a neutral view remain forced to give an answer that does not correspond to their attitude or opinion. On the other hand, there are researchers who think to exclude the middle option is better. If it is there, it will be too easy for respondents to avoid giving a clear opinion.

**EXAMPLE 5.4 Including or excluding a middle option**

Kalton, Roberts, and Holt (1980) describe an experiment where the effect of including a middle option is determined. There were two random samples of approximately 800 persons each. One sample was offered a question with a middle category:

Do you think that drinking alcohol in moderation is good for your health or bad for your health, or do you think it makes no difference to your health?

The other sample had to answer the question without the middle category:

Do you think that drinking alcohol in moderation is good for your health or bad?

The result is shown in Table 5.2. Note that even if there was no middle option, 19.6% of the people gave this answer. This was coded by the interviewers as such under the option "Other." Offering the middle option increased the percentage of respondents in this category from 19.6% to 56.0%.

TABLE 5.2   **The effect of offering a middle option**

| Response | With middle option (%) | Without middle option (%) |
| --- | --- | --- |
| Good for health | 26.4 | 51.9 |
| Bad for health | 12.2 | 16.0 |
| Other | 5.4 | 12.5 |
| Makes no difference | 56.0 | 19.6 |
| Total | 100.0 | 100.0 |

Note that the middle category was the last option offered. Therefore, also a recency effect may contribute somewhat.

The good/bad ratio for the question with the middle option is 2.2, whereas without the middle option, it is 3.2. Apparently, this ratio is affected by excluding a middle option: respondents do not spread proportionally over the categories Good and Bad.

Similar experiments as described in Example 5.4 have been conducted by other researchers. Bishop (1987) showed that just mentioning the middle option in the question text (and not offering it explicitly as an answer option) also increases selection of the middle option. Furthermore, he showed that it makes a difference in a telephone survey whether the middle option is really placed in the middle of the set of answer option or at the end of it. In the latter case, the recency effect will cause even more respondents to choose for this option.

Heerwegh and Loosveldt (2008) compared a face-to-face survey with a web survey. They found that in the web survey (among students), more respondents selected the middle response option. They conclude the data collected by means of the web survey was of lower quality due to satisficing.

Note there will be no recency effect in a web survey. Therefore, putting the middle response option at the end of the list of answer options will not increase selection of this option even more. Tourangeau, Couper, and Conrad (2004) conducted experiments showing a preference for the visual middle of the set of answer options although this option did not correspond to the conceptual middle of the options.

**5.2.1.5 Non-differentiation.** Non-differentiation is a form of satisficing that typically occurs if respondents have to answer a series of questions with the same response options. The original idea was that this would make it easier for respondents to answer the questions. Changing the response options from question to question would increase the cognitive burden of respondents.

In the course of time, it has become clear that this approach is problematic because satisficing respondents tend to select the same answer for all these questions irrespective of the question content. For example, Heerwegh and Loosveldt (2008) compared a face-to-face survey with a web survey. They showed that respondents in the web surveys used less different scale values. So, there was more non-differentiation.

Combining series of questions with the same set of answer options into a *grid question* or a *matrix question*, more information are simultaneously collected. Each row of a matrix question represents a single question, and each column corresponds to an answer option. An example is in Figure 5.7.

At first sight, grid questions seem to have some advantages. A grid question takes less space on the questionnaire form than a set of single questions, and it provides respondents with more oversight. Therefore, it can reduce the time it takes to answer questions. Couper, Traugott, and Lamias (2001) indeed found that a matrix question takes less time to answer than a set of single questions.

According to Dillman, Smyth, and Christian (2009, 2014), answering a matrix question is a complex cognitive task. It is not always easy for respondents to link a single question in a row to the proper answer in the column. Moreover, respondents can navigate through the matrix in several ways: row-wise, column-wise, or a mix. This increases the risk of missing answers to questions, resulting in a higher item nonresponse.

Dillman, Smyth, Christian (2009, 2014) advise to limit the use of matrix questions as much as possible. If they are used, they should not be too wide or to too long. Preferably, the whole matrix should fit on a single screen. This is not so easy to realize as different respondents may have set different screen resolutions on their computer screens. If respondents have to scroll, either horizontally or vertically, they may easily get confused, leading to wrong or missed answers.

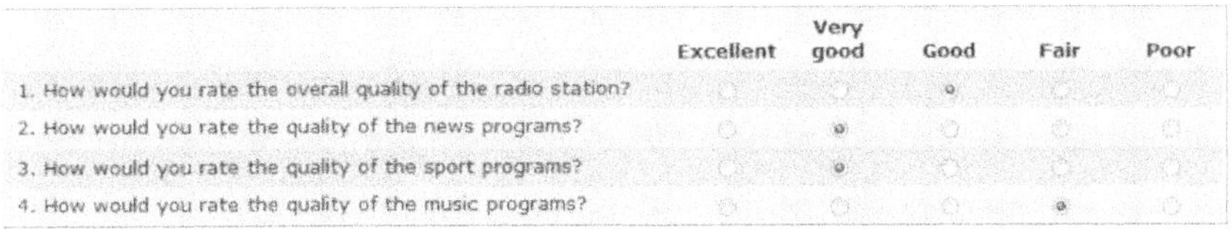

| | Excellent | Very good | Good | Fair | Poor |
|---|---|---|---|---|---|
| 1. How would you rate the overall quality of the radio station? | | | ● | | |
| 2. How would you rate the quality of the news programs? | | ● | | | |
| 3. How would you rate the quality of the sport programs? | | ● | | | |
| 4. How would you rate the quality of the music programs? | | | | ● | |

FIGURE 5.7 **An example of a matrix question**

Fricker et al. (2005) investigated differences between a web survey and a CATI survey. They showed that the respondents in the web survey gave less differentiated answers to attitude questions in matrix format.

Several authors, see, for example, Krosnick (1991) and Tourangeau, Couper, and Conrad (2004), express concern about a phenomenon that is sometimes called *straight-lining*. Respondents give the same answer to all questions in the matrix. The simply check all radio buttons in the same column. Often this is the column corresponding to the middle response option. Figure 5.8 shows an example.

Straight-lining the middle response option can be seen as a form of endorsing the status quo. It is also a form of non-differentiation. It is a means of quickly answering a series of questions without thinking. It manifests itself in very short response times. Thus, short response times for matrix questions (when compared with a series of single questions) are not always a positive message. It can mean that there are measurement errors caused by satisficing.

If a matrix question is used, much attention should be paid to its visual layout. For example, a type of shading as in Figure 5.8 reduces confusion and therefore reduces item nonresponse, too. *Dynamic shading* may even help more. Kaczmirek (2010) distinguishes pre-selection shading and post-selection shading. *Pre-selection shading* comes down to changing the background color of a cell or row of the matrix question if the cursor is moved over it by the respondent. Pre-selection shading helps the respondent to locate the proper answer to the proper question. It is active before clicking the answer. Post-selection shading means shading of a cell or row in the matrix after the answer has been selected. This feedback informs the respondent that answer to which question was selected. Kaczmirek (2010) concludes that pre-selection and post-selection shading of complete rows improves the quality of the answers. However, pre-selection shading of just the cell reduces the quality of the answers. That is because there was more non-differentiation.

Galesic et al. (2007) also experimented with post-selection shading by changing the font color or the background color immediately after respondents answered a question in the matrix. This helped respondents to navigate and therefore improved the quality of the data.

The disadvantages of grid questions seem to be more critical in mobile web surveys. Grids are critical when respondent uses a smartphone. Straight-lining (i.e., same response in a matrix question) is more likely if the questionnaire is answered from a smartphone rather than from a computer (Kolbas, 2014; Stern et al., 2015).

Thus, be advised that data quality can be different across the devices. For example, comparing matrix and item-by-item questions with 2, 3, 4, 5, 7, 9,

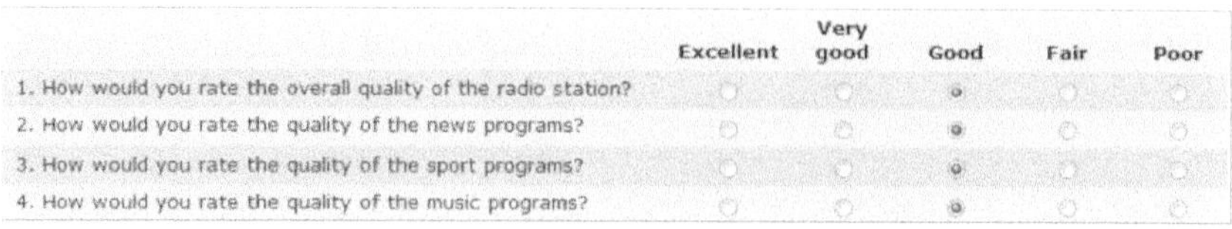

| | Excellent | Very good | Good | Fair | Poor |
|---|---|---|---|---|---|
| 1. How would you rate the overall quality of the radio station? | ○ | ○ | ● | ○ | ○ |
| 2. How would you rate the quality of the news programs? | ○ | ○ | ● | ○ | ○ |
| 3. How would you rate the quality of the sport programs? | ○ | ○ | ● | ○ | ○ |
| 4. How would you rate the quality of the music programs? | ○ | ○ | ● | ○ | ○ |

FIGURE 5.8 An example of a matrix question with straight-lining

and 11 response options in a web survey experiment, the impact of the device used to complete the survey is the following: straight-lining and response time are similar between the two question types across all response lengths, but item nonresponse tends to be higher for matrix than item-by-item question, especially among mobile respondents (Liu and Cernat, 2018).

Using matrix carefully is a critical point especially for business surveys. Business questionnaires are usually matrix designed to evaluate different items. Errors in business surveys are extensively discussed in Snijkers et al. (2013).

**5.2.1.6 Don't Know.** Questions are asked in a survey to collect information about respondents. However, respondents are sometimes not able to provide this information. They simply do not know the answer. If an opinion question is asked, it is usually assumed that respondents have an opinion with respect to the specific issue. This need not be the case. They may simply not have an opinion. Also, if a factual question is asked, respondents may lack the knowledge to answer it.

The question is how to ask a question in such a way that always a true answer is given: "don't know" if the respondent really does not know the answer and a "real" answer if the respondent has one. There are various approaches, and the most effective approach may depend on the type of survey.

On the one end of the spectrum, one could offer "don't know" just as one of the answer options. This may lead to a form of satisficing where respondents choose this option to avoid having to think about a real answer. On the other end of the spectrum, one can decide not to present "don't know" as an answer at all. So respondents are forced to give a real answer, even if they do not have one. It is clear that both approaches may lead to measurement errors.

Besides the two extremes mentioned, there are more ways to handle "don't know." A number of approaches are presented here.

*5.2.1.6.1 Offer "don't know" explicitly*
Offering "don't know" explicitly as one of the answer options has the advantage that respondents not knowing the answer can answer so. This approach accepts the existence of a group of persons that cannot answer the questions, and thus "don't know" is seen as a substantive answer.

This approach may suffer from satisficing. People not wanting to think about an answer or not wanting to give an answer have an escape by answering "don't know." Several authors have shown that explicitly offering "don't know" substantially increases the percentage of respondents choosing this option. See, for example, Sudman and Bradburn (1982).

*5.2.1.6.2 Offer "don't know" explicitly, but less obvious*
To make it less easy for respondents to choose the option "don't know," one could decide to offer this option but at the same time attempt to make it less obvious. The option could be placed elsewhere on the screen or shown in a smaller or less bright font.

Tourangeau, Couper, and Conrad (2004) experimented with questions where the "don't know" option was visually separated from the substantive options by a dividing line. This was counterproductive as it caused more respondents to select the "don't know" options, as more attention was drawn to this option.

Derouvray and Couper (2002) experimented with questions where the "don't know" option was displayed in a smaller and lighter font so that its visual prominence was reduced. This did not affect the number of respondents selecting this option.

Vis-Visschers et al. (2008) offered "don't know" as a special button at the bottom of the screen. It turned out that many respondents overlooked this option and complained they could not answer "don't know."

### 5.2.1.6.3 Offer "don't know" implicitly

To make "don't know" an even less obvious option, one could decide to not offer it explicitly on the screen. This is common in CAPI/CATI software. Only a list of substantive answer options is shown on the screen. If respondents insist they do not know the answer, the interviewer can record this by using a special key combination. For example, the option "don't know" is always by default available in the Blaise system for computer-assisted interviewing by pressing <Ctrl-K> (see Statistics Netherlands (2002)).

A special key combination may work for experienced interviewers, but not for respondents not experienced in web surveys. There are, however, different ways to offer "don't know" implicitly. Vis-Visschers et al. (2008) investigated an approach whereby respondents were offered questions with only substantive answer options. If they did not select an option and attempted to skip the question, the question was offered again, but then "don't know" was included in the list. It turned out that some respondents did not understand this mechanism, as they complained that they could not chose the answer "don't know." The other respondents less frequently selected "don't know."

Derouvray and Couper (2002) experimented with a similar approach. The answer option "don't know" was not offered for the question. If respondents attempted to skip the question without answering it, a new screen appeared offering two choices: (1) go back and answer the question, and (2) record the answer as "don't know" and proceed to the next question. This approach resulted in the lowest "don't know" rates. It also resembles to CAPI/CATI approach.

### 5.2.1.6.4 Do not offer "don't know"

To avoid satisficing, one may decide to not offer the option "don't know." This implies that respondents always have to provide a substantive answer, even if they do not know the answer. According to Couper (2008), this violates the norm of voluntary participation. Respondents should have the possibility to not answer a question. Forcing respondents to answer may frustrate respondents resulting in break-off. Dillman (2007) also strongly recommends not forcing respondents to answer. He warns for detrimental effects on respondent motivation, on data quality, and on the risk of break-off.

The treatment of "don't know" and the effects this can have on the collected data depend on the type of survey. Heerwegh and Loosveldt (2008) compared the use of "don't know" in a face-to-face survey with a web survey. There was a set of opinion questions about estimating parent's views of immigrants. "Don't know" was visually offered as the last response option. The respondents of the face-to-face survey were informed that it was possible to answer "don't know." The respondents of the web survey could see "don't know" as one of the options on the screen. The option "don't know" was selected much more frequent in the web survey. A possible explanation is that people without an opinion do not want to admit their ignorance to interviewers. They may feel foolish. Therefore, they feel forced to give a substantive answer although they lack relevant information for formulating a relevant judgment.

If there is a risk that "don't know" is avoided to prevent embarrassment, one may consider using a filter question. See Krosnick (1991) and Schuman and Presser (1981). This filter question asks whether respondents have an opinion about a specific issue. Only if they say they have, a next question asks what their opinion really is. Example 5.5 shows the effect of a filter question.

🖥 EXAMPLE 5.5 **Using a filter question for "don't know"**

Dutch parliament discussed in 2007 a possible change in the electricity law. The main purpose of the change was to make the law consistent with European directives. This was not a controversial or otherwise interesting issue. Therefore, it was not taken up by the media, and in fact the general public were not aware of the change in law.

Tiemeijer (2008) conducted an experiment in which he measured the effect of a filter question for "don't know." The question asked is shown in Figure 5.9.

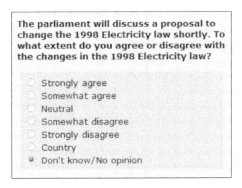

FIGURE 5.9 **Question about the 1998 electricity law**

The sample size was 395. The sample was divided randomly into three groups. The first group just answered the question in Figure 5.9. The second group also answered this question, but it was preceded by a weak filter question (*Do you have an opinion about the changes in the 1998 electricity law, or not?*). For the third group, there also was a filter question, but it was a stronger question (*Do you have heard or read enough about this proposal to be able to form an opinion about it?*). Respondents answering *No* to the filter question were classified as "Don't know/No opinion."

Table 5.3 contains the results. By asking no filter question, the percentage of "don't know" was 55%. This is a high percentage, but lower than expected since the topic of the question was completely unknown to the respondents. This percentage increases to 79% for the weak filter question and even to 86% for the strong filter question. Apparently, a filter question makes it less embarrassing for respondents to admit they do not know the answer.

TABLE 5.3    **The effect of offering a middle option**

| Response | No filter question (%) | Weak filter question (%) | Strong filter question (%) |
| --- | --- | --- | --- |
| Don't know/no opinion | 55 | 79 | 86 |
| Somewhat/strongly agree | 7 | 4 | 3 |
| Neutral | 23 | 3 | 2 |
| Somewhat/strongly disagree | 15 | 14 | 9 |
| Total | 100 | 100 | 100 |

### 5.2.1.7 Arbitrary Answer.

Selecting "don't know" as an answer is one way for respondents to avoid having to think about a proper answer. If giving this answer is considered undesirable, respondents may also decide to just pick an arbitrary answer. Converse (1964) already describes the problem of random answers. Krosnick (1991) calls this behavior "metal coin flipping."

This type of satisficing can also occur for a special type of question called the check-all-that-apply question. An example of such a question is shown in Figure 5.10. It is a closed question for which more than one answer can be selected. It is common practice for web surveys to use square *checkboxes* (instead of round *radio buttons*) for check-all-that-apply questions.

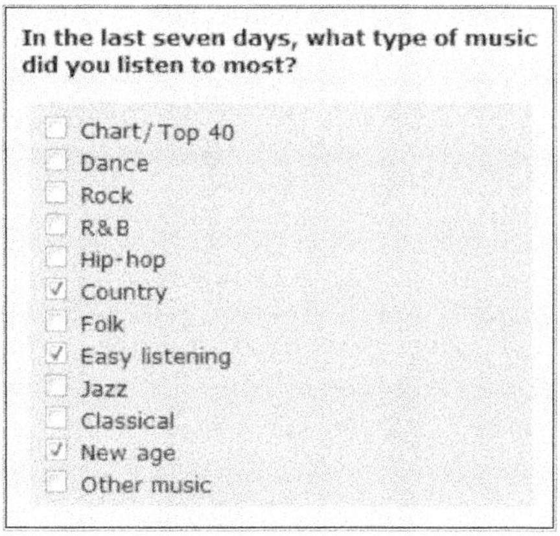

FIGURE 5.10   A check-all-that-apply question

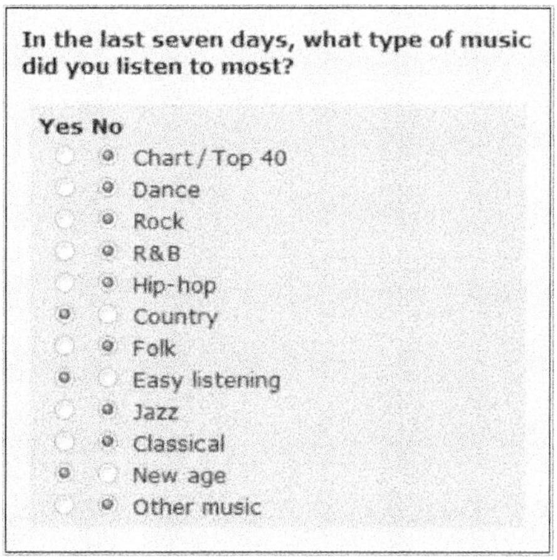

FIGURE 5.11   A check-all-that-apply question with forced choice

A check-all-that-apply question asks respondents to check all appropriate items from a (sometimes long) list of answer options. This can be a lot of work. Instead of checking all relevant answers, they may just check some arbitrary answers and stop when they think they have checked enough answers. Moreover, satisficing respondents tend to read only the first part of the list, and not the complete list. This causes a bias toward answers in the first part of the list.

It is common practice not to use check-all-that-apply questions in telephone surveys. Instead, respondents have to answer either yes (applies) or no (does not apply) for each item in the list. This raises the question whether in web surveys check-all-that-apply questions should be replaced by sets of these forced choice questions, too. This would mean changing the question format as in Figure 5.10 by a format as in Figure 5.11. Smyth et al. (2006) have shown the format in Figure 5.11 leads to more selected options and also respondents take more time to answer the questions. This is an indication that the format as in Figure 5.10 may cause satisficing.

It should be noted that completing the answer to the forced choice questions requires more work, and this may frustrate respondents. It may lead to straight-lining, a form of satisficing described in Section 5.2.1.5.

**5.2.1.8 Socially Desirable Answers.** With respect to data collection, there is a substantial difference between interviewer-assisted surveys (e.g., CAPI and CATI) on the one hand and self-administered surveys (web, mail) on the other. Inter-viewers carry out the fieldwork in a CAPI or CATI survey. There are no inter-viewers, however, in a web survey. It is a self-administered survey. Therefore, quality of collected data may be lower due to higher nonresponse rates and more errors in the answers to the questions.

de Leeuw (2008) and Dillman, Smyth, Christian (2009, 2014) discuss differ-ences between various modes of data collection. They observe that a positive effect of the presence of interviewers is that they are in control of the interview. They lead the respondent through the interview. They see to it that the right question is asked at the right moment. If necessary, they can explain the meaning of a ques-tion. They can assist respondents in getting the right answers to the question. Interviewers can motivate respondents, answers questions for clarification, provide additional information, and remove causes for misunderstanding. All this will increase the quality of the collected data.

The presence of interviewers can also have a negative effect. It will lead to more socially desirable answers for questions about potentially sensitive topics. Giving *socially desirable answers* is the tendency that respondents give answers that will be viewed as more favorable by others. This particularly happens for sensitive questions about topic like sexual behavior and use of drugs. If a true answer would not make the respondents look good, they will refuse to answer or give a different answer. A meta-analysis by de Leeuw (1992) shows that the effects of socially desirable answers are stronger in interviewer-assisted surveys. Respondents tend to give more truthful answers in self-administered surveys.

Holbrook and Krosnick (2010) describe an example of socially desirable answers in an election survey. Voting is seen as admirable and valued civic behav-ior. Therefore, nonvoters may be reluctant to admit they did not vote. This effect will occur particularly if the respondent is asked to report something embarrassing directly and explicitly. If the respondent can answer more anonymously and con-fidentially, a more truthful answer will be given. Holbrook and Krosnick (2010) find that socially desirable answer led to a bias telephone surveys but not in web surveys.

Kreuter, Presser, and Tourangeau (2008) conducted an experiment in which they compared the effects of socially desirable answers in a web survey, a CATI survey, and an IVR survey. IVR stands for interactive voice recognition. It is a tel-ephone survey without interviewers. Questions are asked by software, and respon-dents answer by keypad of the telephone or by giving a verbal answer. IVR can be placed somewhere between CATI and web in the spectrum from interviewer-assisted to self-administered. Kreuter, Presser, and Tourangeau (2008) had administrative information available that enabled them to compare the given answers with the true answers. They showed that the amount of correctly reported answers to sensitive questions in web surveys is higher than in the CATI and IVR surveys,

Kraan et al. (2010) describe an experiment in the Netherlands, which compares two different data collection modes for the Safety Monitor. Their findings show that respondents in a web survey are more critical about the performance of the police with respect to respondents in a CAPI or CATI survey. Apparently, respondents in the interviewer-assisted surveys are less inclined to make critical remarks about the police.

**5.2.1.9  Some Web Survey Design Issues.** Two different ways to offer the questionnaire in a web survey are possible: the question-based approach and the form-based approach. Figure 5.12 shows an example of the *question-based approach*. There is always only one question on the screen. After entering the answer to the question, the respondent clicks on a button to go to the next question. Alternatively, the respondent can go back to previous question to change an answer.

If the questionnaire contains routing instructions, the question-based approach is used. If answering the next question depends on the answers to one or more previous questions, the web server processes the answers to these previous questions before the next question is shown on the screen.

The question-based approach is also used if consistency checks are carried out immediately after an answer has been entered. This might result in showing a warning or error message, after which the question reappears on the screen in order to correct the answer.

The question-based approach is typically used for large and complex questionnaire where respondents remain online while answering questions. This approach is more or less similar to the approaches used for CAPI or CATI.

Figure 5.13 shows an example of the *form-based approach*. It more or less mimics the paper questionnaire on the screen. This approach does not allow routing instructions in the questionnaire. It is also not possible to have checks before the end of the page is reached.

Respondents need not be online while they answer the questions. It is possible to download the form, go offline, answer the questions, go online again, and upload the complete questionnaire. The form-based approach is generally recommended for small and simple questionnaires.

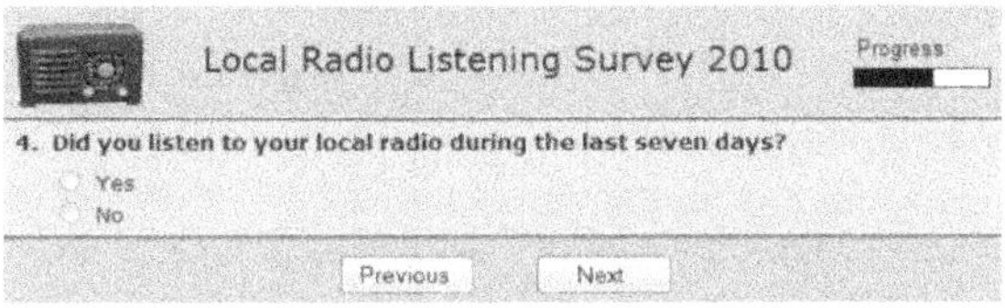

FIGURE 5.12  The question-based approach

The form-based approach

It is also possible to apply some kind of hybrid approach whereby a questionnaire is composed of a number of blocks of questions. Each block is presented as a form on the screen. There are no routing instructions, and checks are within blocks, but there might be between blocks. More about the advantages and disadvantages of both approaches can be found in Couper (2008).

One of the advantages of the form-based approach is that respondents see how long the questionnaire is. This may encourage them to finish the questionnaire (but also discourage them if it is very long). The question-based approach lacks this overview as respondents only see one question at the time. To help respondents, it is sometimes advised to include some kind of progress indicator. Such an indicator informs respondents about where they are in the questionnaire. The example in Figure 5.12 contains a progress indicator. Such indicators can take many different aspects, and research results are mixed about their usefulness. Progress indicators seem to help if they show quick progress, but they may have the opposite effect if progress is slow. It may be wise not to use progress indicators if the questionnaire has a complex routing structure. Then the number of questions to be answered may be different for different respondents. It may even depend on the answers to questions that have not yet been asked. Therefore, it is impossible to indicate progress. See Couper (2008); Heerwegh (2004); Kaczmirek et al. (2004); and Villar, Callegaro, and Yang (2013) for more on progress indicators.

The advantage of the question-based approach is that a single question always fits on the computer screen. This is not the case for the form-based approach. If the form contains many questions, it may only be partially visible. Consequently, the respondent has to scroll to see other parts of the questionnaire. Dillman, Smyth,

Christian (2009, 2014) warn against scrolling as it may increase the risk that respondents miss questions. Scrolling should be especially avoided if the question-naire is not access blocked to mobile devices.

It is not uncommon in CAPI and CATI surveys to have consistency checks in the questionnaire. These checks are carried out immediately after the relevant questions have been asked. If an inconsistency is detected, an error message is dis-played on the screen. The interviewer discusses the problem with the respondent, which may result in correcting the answers to one or more questions. Research shows that these checks improve the quality of the collected data.

Checks may also be included in web survey, but one should be careful as there are no interviewers who can explain the problem to the respondents. Couper (2008) suggests that error messages should at least be polite, illuminating, and help-ful. They should not contain threatening words and should not blame the respond-ent. Dillman, Smyth, Christian (2009, 2014) also warn that unfriendly, unspecific, unclear error messages may cause respondents to break off the interview.

The lack of interviewer assistance in a web survey makes it even more impor-tant to make clear to the respondents what is expected from them when they answer a question. The visual design of the question is of vital importance. Two examples illustrate this. The first example relates to asking for dates. Dates can be formatted in many different ways. If a researcher wants dates to be entered in a specific format, he should give as much as possible guidance to the respondents.

Figure 5.14 shows three different ways in which one could ask for the month and year in which a student started his/her university studies. The first question has the format of an open question. Answer is any text. The computer attempts to extract a date form the input. If this is not possible, an error message will appear. The second question has two input fields, one for the month and one for the year. It is still not clear whether to enter the date as words or as numbers. The third question also has two input fields. Now there is an indication both in the text of the question and below the input fields that digits are expected. Christian, Dill-man, and Smyth (2007) conducted an experiment in which they compared, among others, the format of the second question with the format of the third

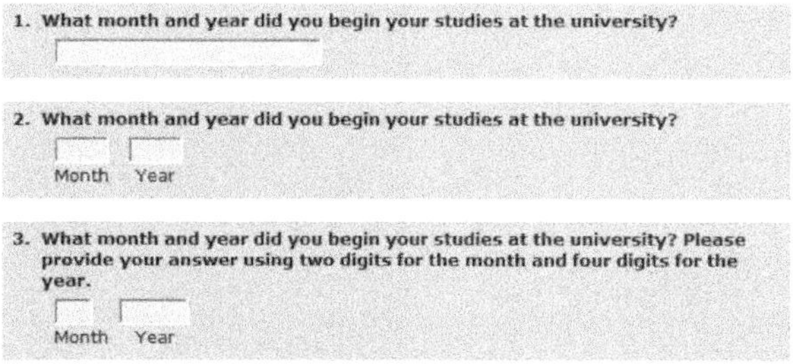

FIGURE 5.14 Formatting a date question

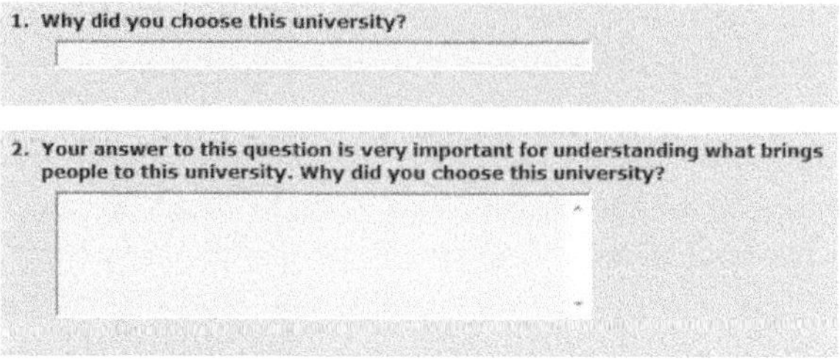

FIGURE 5.15  Formatting an open question

question. Only 44% of the respondents entered their answer properly for the second format. The percentage of right answers was 94% for the third format.

A second example relates to the format of open questions. The size of the input box should reflect the amount of information that is expected. Figure 5.15 shows two open questions. The input field of the first question suggests that only one line of text is sufficient. The input field of the second question suggests that respondents can enter several lines of text. The scroll bars even suggest that the text can be longer than the size of the box. Research shows that indeed the second format leads to longer answers than the first format.

**5.2.1.10  Other Measurement Errors.**  Section 5.2.1.9 discusses types of measurement errors that may be specific to web surveys. Other types of measurement errors also may occur irrespective of the mode of data collection. In this section some of them are mentioned.

Questions requiring respondents to recall events that have happened in the past are a source of errors. The reason is that people make *memory errors*. They tend to forget events, particularly when they happened a long time ago or when they are not too salient. Important events, more interesting events, and more frequently happening events will be remembered better than other events. The effects of recall errors are more severe as the length of the reference period is longer.

The question in Figure 5.16 is a simple question to ask, but for many people difficult to answer. Recall errors may even occur for shorter time. In the 1981 Health Survey of Statistics Netherlands, respondents had to report contacts with their family doctor over the last three months. Sikkel (1983) investigated memory effects. It turned out that the percentage of not reported contacts increased linearly in time. The longer ago an event took place, the more likely it is to forget it. The percentage of unreported events for this question increased on average with almost

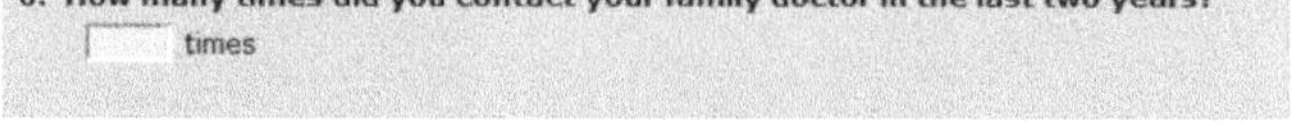

FIGURE 5.16  A recall question

4% per week. Over the total period of three months, about one-quarter of the contacts with the family doctor was not reported.

Recall questions may also suffer from *telescoping*. This occurs if respondents report events as having occurred either earlier or later than they actually did. As a result, within the reference period, an event not correctly or incorrectly reported is excluded from the reference period. Bradburn, Sudman, and Wansink (2004) note that telescoping leads more often to overstating than to understating a number of events. Particularly, for short reference periods, telescoping may lead to substantial errors in estimates.

Question order can affect the results in two ways. One is that mentioning something (an idea, an issue, a brand) in one question can make people think of it while they answer a later question, when they might not have thought of it if it had not been previously mentioned. In some cases, this problem may be reduced by randomizing the order of related questions. Separating related questions by unrelated ones might also reduce this problem, though neither technique will completely eliminate it.

Tiemeijer (2008) mentions an example where the answers to a specific question were affected by a previous question. The Eurobarometer (www.europa.eu/public_opinion) is an opinion survey in all member states of the European Union (EU) held since 1973. The European Commission uses this survey to monitor the evolution of public opinion in the member states. This may help in making policy decision. The content of the surveys and trend questions varies from time to time. The 2007 Eurobarometer contained the question as shown in Figure 5.17.

It turned out that 69% of the respondents were of the opinion that the country had benefited from the EU. A similar question was included at the same time in a Dutch opinion poll (*Peil.nl*). However, before the request of the opinion about the benefits had from their country, another question was administered, asking respondents to select the most important disadvantages of being a member of the EU. Among the items in the list, there were the too fast extension EU, the possibility of Turkey becoming a member state, the introduction of the Euro, the waste of money by the European Commission, the loss of identity of the member states, the lack of democratic rights of citizens, veto rights of member states, and possible interference of the European Commission with national issues. As a result, only 43% of the respondents considered membership of the EU beneficial.

Opinion questions may address topics about which respondents may not yet have made up their mind. They may even lack sufficient information for a balanced judgment. Questionnaire designers may sometimes provide additional information in the question text. Such information should be objective and neutral and should not influence respondents in a specific direction. Saris (1997)

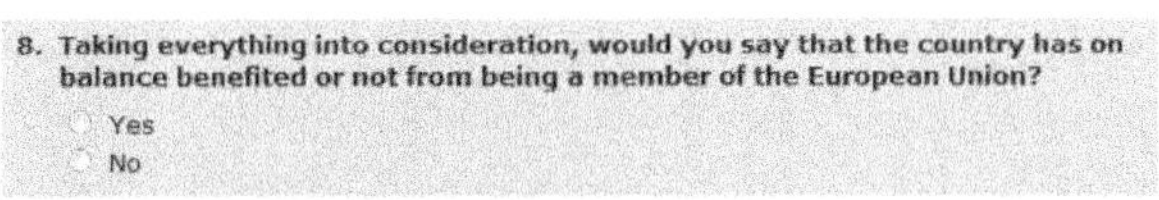

8. Taking everything into consideration, would you say that the country has on balance benefited or not from being a member of the European Union?

    Yes
    No

FIGURE 5.17 **A context-sensitive question**

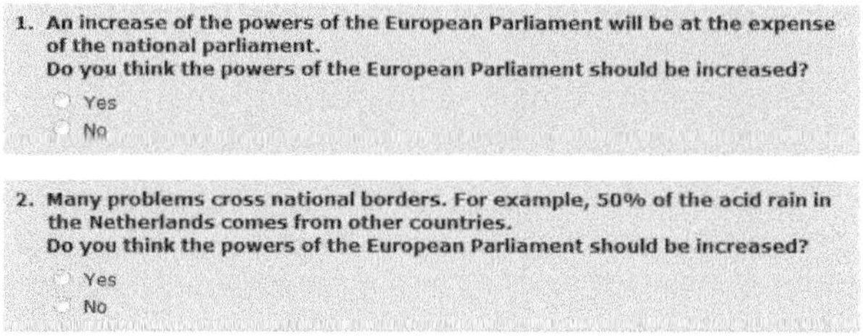

FIGURE 5.18   Questions containing additional information

performed an experiment to show the dangers of changes in the question text. He measured the opinion of the Dutch about increasing the power of the European Parliament. Respondents were randomly assigned one of the two questions in Figure 5.18.

In case respondents were offered the first question, 33% answered "yes" and 42% answered "no." In case respondents were offered the second question, 53% answered "yes" and only 23% answered "no." These substantial differences are not surprising, as the explanatory text in the first question stresses a negative aspect and the text in the second question stresses a positive aspect.

## 5.2.2  NONRESPONSE

### 5.2.2.1  The Nonresponse Problem.
Nonresponse occurs when elements (persons, companies) in the selected sample, and that also are eligible for the sample, do not provide the requested information, or that the provided information is not usable. The problem of nonresponse is that the researcher does not have control any more over the sample selection mechanism. A situation of nonresponse is described in Example 5.6. It becomes impossible to compute unbiased estimates of population characteristics. Validity of inference about the population is at stake.

There are two types of nonresponse: unit nonresponse and item nonresponse. *Unit nonresponse* occurs when a selected element does not provide any information at all, i.e., the questionnaire form remains empty. *Item nonresponse* occurs when some questions have been answered but no answer is obtained for some other, possibly sensitive, questions. So, the questionnaire form has been partially completed. This section focuses on unit nonresponse.

It is consequence of unit nonresponse that the realized sample size is smaller than planned. This decreases the precision of estimates. However, if there are no other effects, valid estimates can still be obtained, because computed confidence intervals still have the proper confidence level. To avoid realized samples that are too small, the initial sample size can be taken larger. For example, if a sample of 1,000 elements is required and the expected response rate is around 60%, the initial sample size should be approximately $1,000/0.6 = 1,667$.

The main problem of nonresponse is that estimates of population characteristics may be biased. This situation occurs if, due to nonresponse, some groups in the population are over- or underrepresented in the sample and these groups behave differently with respect to the characteristics to be investigated. Then nonresponse is defined to be *selective*.

It is likely that survey estimates are biased unless very convincing evidence to the contrary is provided. Bethlehem (2009) gives examples of a number of Dutch surveys where nonresponse is selective. A follow-up study of the Dutch Victimization Survey showed that people, who are afraid to be home alone at night, are less inclined to participate in the survey. In the Dutch Housing Demand Survey, it turned out that people who refused to participate have lesser housing demands than people who responded. And for the Survey of Mobility of the Dutch Population, it was obvious that the more mobile people were underrepresented among the respondents.

Nonresponse is a problem in almost every survey, whatever the mode of data collection. The magnitude and effect of nonresponse may differ from mode to mode (see Bethlehem, Cobben, and Schouten, 2011). This section gives a short overview of the nonresponse problem and describes aspects that are specific to web surveys.

It will be shown that the amount of nonresponse is one of the factors determining the magnitude of the bias of estimates. The higher the nonresponse rate, the higher and larger the bias will be.

 EXAMPLE 5.6 **The Dutch Housing Demand Survey nonresponse**

The effect of nonresponse is shown using data from the Dutch Housing Demand Survey. Statistics Netherlands carried out this survey in 1981. The initial sample size was 82,849. The number of respondents was 58,972, which comes down to a response rate of 71.2%.

To obtain more insight in the nonresponse, a follow-up survey was carried out among the nonrespondents. One of the questions asked was whether they intended to move within two years. Table 5.4 shows the results.

TABLE 5.4 **Nonresponse in the Dutch Housing Demand Survey 1981**

| Do you intend to move within two years? | Response | Nonresponse | Total |
| --- | --- | --- | --- |
| Yes | 17,515 | 3,056 | 20,571 |
| No | 41,457 | 20,821 | 62,278 |
| Total | 58,972 | 23,877 | 82,849 |

The percentage of people with the intention to move within two years is $100 \times 17{,}515/58{,}972 = 29.7\%$ if just the response data are used. This percentage is much lower for the complete sample (response and nonresponse): $100 \times 20{,}571/82{,}849 = 24.8\%$. The reason is clear: there is a substantial difference between respondents and nonrespondents with respect to the intention to move within two years. For nonrespondents, this percentage is only $100 \times 3{,}056/23{,}877 = 12.8\%$.

**5.2.2.2 Causes of Nonresponse.** Nonresponse can have many causes. It is important to distinguish these causes. To be able to reduce nonresponse in the field, one must know what caused it. Moreover, different types of nonresponse can have different effects on estimates and therefore may require different treatment. Main causes of nonresponse are noncontact, refusal, and not-able.

Nonresponse due to *noncontact* occurs if it is impossible to get into contact with the respondent. This can happen in a face-to-face survey if the respondent is not at home. Noncontact occurs in a telephone survey if the telephone is not answered. Various forms of noncontact are possible in web surveys. It depends on the way in which sample persons are selected. If the sampling frame is a list with e-mail addresses, noncontact occurs if the e-mail with the invitation to participate in the survey does not reach a selected person. The e-mail address may be wrong, or the e-mail may be blocked by a spam filter. If the sampling frame is the list of postal addresses and letters with an Internet address are sent to selected persons, noncontact may be caused by persons not receiving the letter. If recruitment for a web survey takes place by means of a face-to-face or telephone survey, noncontact can be due to respondents being not at home or not answering the telephone.

Nonresponse due to *refusal* can occur after contact has been established with a sample person. Refusal to cooperate can have many reasons: people may not be interested, they may consider it an intrusion of their privacy, they may have no time, etc. Sometimes a refusal can be temporary. In this case it may be attempted to make an appointment for another day and/or time. But often a refusal is permanent.

If sample persons for a web survey are contacted by an e-mail or a letter, they may postpone and forget to complete the questionnaire form. This can be seen as a weak form of refusal. Sending a reminder helps to reduce this form of refusal.

Nonresponse due to *not-able* is a type of nonresponse where respondents may be willing to respond but are not able to do so. Reasons for this type of nonresponse can be, for example, illness, hearing problems, or language problems. If a letter with an Internet address of a web questionnaire is sent to a sample person, this person receives the letter, and he/she wants to participate in the web survey, but does not have access to the Internet, this can also be seen as a form of nonresponse due to not-able.

It should be noted that lack of Internet access should sometimes be qualified as under-coverage instead of nonresponse. If the target population of a survey is wider than just those with Internet and the sample is selected using the Internet, people without Internet have zero selection probability. They will never be selected in the surveys. This is under-coverage. Nonresponse due to not-able occurs if people have been selected in the survey but are not able to complete the questionnaire form (on the Internet).

Response rates are usually more visible than coverage rates and are often taken as indicators of data quality. The risk is that survey researchers pursue high response rates at the expense of low coverage rates.

For example, Eckman and Kreuter (2017) through a simulation model explore how the relationship between coverage and response propensities affects response and coverage rates and under-coverage and nonresponse bias.

When we evaluate the overall quality of a survey, the distinction between under-coverage and nonresponse is not important. The American Association for Public Opinion Research (AAPOR) mandates reporting of coverage rates alongside response rates. AAPOR's new code (revised as of May 2015) requires explicit disclosure of "the sampling frame(s) and its coverage of the target population, including mention of each segment of the target population that is not covered by the design."

**5.2.2.3 Response Rate.** Due to the negative impact nonresponse may have on the quality of survey results, the *response rate* is considered to be an important indicator of the quality of a survey. Response rates are frequently used to compare the quality of surveys and also to explore the quality of a survey that is repeated over time. Bethlehem (2009) defines the response rate as

$$(5.6) \qquad \text{Response rate} = \frac{n_R}{n_E} = \frac{n_R}{n_R + n_{NR}},$$

where $n_R$ is the number of eligible respondents, $n_E$ is the total number of eligible persons in the sample, and $n_{NR}$ is the number of eligible nonrespondents. Eligible persons are persons that belong to the target population and have been selected in the sample. In practice it may be difficult to compute $n_E$ as it is not always possible to determine whether noncontacted persons are eligible or not.

Another complication concerns web and mail surveys. These are self-administered surveys. If there are no interviewers for recruiting respondents, it is not possible to establish eligibility or the cause of nonresponse. There are only two possibilities: the questionnaire form is completed or not. This may hinder analysis of the effects of nonresponse on the survey results.

**EXAMPLE 5.7 Response rates in the LISS panel**

The LISS panel (Longitudinal Internet Studies for the Social Sciences) is a web panel consisting of approximately of 5,000 households. This panel was set up in 2006 by CentERdata, a research institute in the Netherlands. Objective of the panel is to provide a laboratory for the development and testing of new, innovative research techniques.

The panel is based on a true probability sample of households drawn from the population register by Statistics Netherlands. The initial sample consisted of 10,150 households. Telephone numbers were added to the selected names and addresses. This was only possible for registered numbers. Households with a registered telephone were contacted by means of CATI. Addresses without a registered number and those who could not be contacted by telephone were visited by the interviewers (CAPI).

During the recruitment phase, persons in sampled households were first asked to participate in a short interview. General background questions about the respondent were asked. At the end of the interview, respondents were told about the panel and asked if they would like to participate. Households without access to the Internet, or who were worried that an Internet survey might be too complicated for them, were told about the simple-to-operate computer with Internet access that could be installed in their homes for free for the duration of the panel. To demonstrate the use of this computer, they were shown a demonstration video. The response process could be split into a number of steps:

Contact: Those who could be contacted for the recruitment interview.

Primary response: Those who participated in the recruitment interview.

Secondary response: Those who agreed to participate in the panel.

Tertiary response: Those who actually did participate in the panel.

The response rates are shown in Table 5.5. The column "Response rate" shows the response percentage in each separate step of the data collection process. The column "Cumulative" shows the cumulative effect of the current and previous steps.

TABLE 5.5 **Response rates in the recruitment phase of the LISS panel**

| Step | Response rate (%) | Cumulative (%) |
| --- | --- | --- |
| Contact | 85.7 | 85.7 |
| Primary response | 80.9 | 69.3 |
| Secondary response | 74.1 | 51.4 |
| Tertiary response | 89.4 | 45.9 |

The column "Response rate" in Table 5.5 shows the response rate in each of the four steps. The secondary response is lowest. Only three out of four participants in the recruitment interview also agreed to participate in the panel. Main reasons for this refusal were (1) that these persons considered the burden of participating in a panel too high and (2) they did not have Internet access and could not be persuaded to use it by offering a simple-to-use computer.

The tertiary response is highest: 9 out of 10 of those who agreed to participate in the panel also do so. Apparently, people kept their promise.

The cumulative column shows that ultimately only 45.9% of the sample persons became an active member of the LISS panel.

Note that noncontact rate is quite high. In 14% of the cases, it was not possible (by CAPI or CATI) to establish contact with the sampled households.

More in the response in the LISS panel can be found in Scherpenzeel (2009).

Like any other survey, web surveys suffer from nonresponse. A web survey is a self-administered survey. Therefore, web surveys have a potential of high nonresponse rates. An additional source of nonresponse are technical problems that may be encountered by respondents having to interact with the Internet, see, e.g., Couper (2000), Dillman and Bowker (2001), Fricker and Schonlau (2002), and Heerwegh and Loosveldt (2002). Slow modem speeds, unreliable connections, high connection costs, low-end browsers, and unclear navigation instructions may frustrate respondents. This often results in respondents breaking off completion of the questionnaire. In order to keep the survey response up to an acceptable level, every measure must be taken to avoid these problems. This requires a careful design of web survey questionnaire instruments.

Mixed mode is thought to be useful in improving response to web and web/ mail. Millar and Dillman (2011) conducted two experiments designed to evaluate several strategies for improving response to web and web/mail mixed-mode surveys to determine the best ways to maximize web response rates in a highly Internet-literate population with full Internet access.

Findings were that offering the different response modes sequentially (web is offered first and a mail follow-up option in the final contact) improves web response rates. Also, delivering a token cash incentive in advance using a combination of both postal and e-mail contacts improves web response rates. No evidence of impact in improving rates if a simultaneous choice of response modes is provided.

Be advised to adopt a sequential mode with web first to tentatively improve response rates. However, the most adequate strategy to move from one single mode to mixed mode with web component is not yet clearly defined. Targeting the web invitation to a subset of the sample is likely to be more effective than inviting all sample members to the web (for more details and references about errors in mixed mode, see Chapter 9).

### 5.2.2.4 The Effect of Nonresponse.

Under the *random response model*, it is possible to investigate the possible impact of nonresponse on estimators of population characteristics. This model assumes every element $k$ in the population to have an (unknown) response probability $\rho_k$. If element $k$ is selected in the sample, a random mechanism is activated that results with probability $\rho_k$ in response and with probability $1 - \rho_k$ in nonresponse. Under this model, a set of response indicators

$$(5.7) \qquad\qquad R_1, R_2, ..., R_N$$

is introduced, where $R_k = 1$ if the corresponding element $k$ responds and where $R_k = 0$ otherwise. So, $P(R_k = 1) = \rho_k$, and $P(R_k = 0) = 1 - \rho_k$.

Now suppose a simple random sample without replacement of size $n$ is selected from this population. This sample is denoted by the set of indicators $a_1, a_2, ..., a_N$, where $a_k = 1$ means that element $k$ is selected in the sample and otherwise $a_k = 0$. The response only consists of those elements $k$ for which $a_k = 1$ and $R_k = 1$. Hence, the number of available cases is equal to

$$(5.8) \qquad n_R = \sum_{k=1}^{N} a_k R_k.$$

Note that this realized sample size is a random variable. The number of nonrespondents is equal to

$$(5.9) \qquad n_{NR} = \sum_{k=1}^{N} a_k (1 - R_k),$$

where $n = n_R + n_{NR}$.

The values of the target variable become only available for the $n_R$ responding elements. The mean of these values is denoted by

$$(5.10) \qquad \bar{y}_R = \frac{1}{n_R} \sum_{k=1}^{N} a_k R_k Y_k.$$

It can be shown, see Bethlehem (2009), that the expected value of the response mean is approximately equal to

$$(5.11) \qquad E(\bar{y}_R) \approx \tilde{Y},$$

where

$$(5.12) \qquad \tilde{Y} = \frac{1}{N} \sum_{k=1}^{N} \frac{\rho_k}{\bar{\rho}} Y_k$$

and

$$(5.13) \qquad \bar{\rho} = \frac{1}{N} \sum_{k=1}^{N} \rho_k$$

is the mean of all response probabilities in the population. From expression (5.11) it is clear that, generally, the expected value of the response mean is unequal to the population mean to be estimated. Therefore, this estimator is biased. This bias is approximately equal to

$$(5.14) \qquad B(\bar{y}_R) = \widetilde{Y} - \vec{Y} = \frac{S_{\rho Y}}{\bar{\rho}} = \frac{R_{\rho Y} S_\rho S_Y}{\bar{\rho}},$$

where $S_{\rho Y}$ is the covariance between the values of the target variable and the response probabilities, $R_{\rho Y}$ is the corresponding correlation coefficient, $S_Y$ is the standard deviation of the variable $Y$, and $S_\rho$ is the standard deviation of the response probabilities. From this expression of the bias, a number of conclusions can be drawn:

- The bias vanishes if there is no relationship between the target variable and response behavior. This implies $R_{\rho Y} = 0$. The stronger the relationship between target variable and response behavior, the larger the bias will be.
- The bias vanishes if all response probabilities are equal. Then $S_\rho = 0$. Indeed, in this situation, the nonresponse is not selective. It just reduces the sample size.
- The magnitude of the bias increases as the mean of the response probabilities decreases. Translated in practical terms, this means that lower response rates will lead to larger biases.

Currently, response rates for general population surveys that are carried out as web surveys have a lower response rate than comparable CAPI or CATI surveys. Beukenhorst and Giesen (2010) report response rates for some web surveys of Statistics Netherlands: 21% for the Safety Monitor, 26% for the Mobility Survey, and 35% for the Health Interview Survey. Holmberg (2010) describes an experiment where respondents could choose between mail and web. Of the sample persons only 11.8% selected the web questionnaire, and 58.1% the mail questionnaire. The nonresponse was 30.1%. Those selecting the mail questionnaire did this because it was immediately available. They did not have to go to their computer.

### 5.2.2.5 Analysis and Correction of Nonresponse.

It is important to carry out a nonresponse analysis on the data that have been collected in a survey. Such an analysis should make clear whether or not response is selective, and if so, which technique should be applied to correct for a possible bias. Unfortunately, the available data with respect to the target variables will not be of much use. There are only data for the respondents and not for the nonrespondents. So it is not possible to establish whether respondents and nonrespondents differ with respect to these variables. The way out of this problem is to use auxiliary variables (see Figure 5.19).

An auxiliary variable is in this context a variable that has been measured in the survey and for which the distribution in the population (or in the complete sample) is available. So it is possible to establish whether there is a relationship between this variable and the response behavior.

Three different response mechanisms are distinguished. The first one is *Missing Completely at Random* (MCAR). The occurrence of nonresponse ($R$) is completely independent of both the target variable ($Y$) and the auxiliary variable ($X$). The response is not selective. Estimates are not biased. There is no problem.

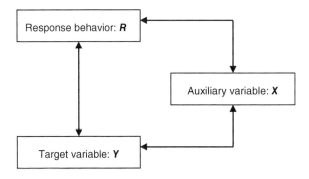

FIGURE 5.19 **Relationship between target variable, response behavior, and auxiliary variable**

In case of MCAR the response behavior ($R$) and any auxiliary variable ($X$) are unrelated. If it is also known that there is a strong relationship between the target variable ($Y$) and the auxiliary variable ($X$), this is an indication there is no strong relation between target variable ($Y$) and response behavior ($R$) and thus that estimators do not have a severe bias.

It should be noted that if there is no strong relationship between the auxiliary variable ($X$) and the target variable ($Y$), analysis of the relationship between the auxiliary variable ($X$) and the response behavior will provide no information about a possible bias of estimates.

The second response mechanism is *Missing at Random* (MAR). This situation occurs when there is no direct relation between the target variable ($Y$) and the response behavior ($R$), but there is a relation between the auxiliary variable ($X$) and the response behavior ($R$). The response will be selective, but this can be cured by applying a weighting technique using the auxiliary variable. Chapter 12 is devoted to such weighting techniques.

In case of MAR, response behavior ($R$) and the corresponding auxiliary variable ($X$) will turn out to be related. If it is also known that there is a strong relationship between the target variable ($Y$) and the auxiliary variable ($X$), this is an indication there is an (indirect) relation between target variable ($Y$) and response behavior ($R$) and thus that estimators may be biased.

The third response mechanism is *Not Missing at Random* (NMAR). There is a direct relationship between the target variable ($Y$) and the response behavior ($R$), and this relationship cannot be accounted for by an auxiliary variable. Estimators are biased. Correction techniques based on use of auxiliary variables will be able to reduce such a bias.

All this indicates that the relationship between auxiliary variables and response behavior should be analyzed. If such a relationship exists and it is known there is also a relationship between the target variables and auxiliary variables, there is a serious risk of biased estimates. So, application of nonresponse correction techniques should be considered.

■ EXAMPLE 5.8 **Nonresponse in the LISS panel**

The LISS panel (Longitudinal Internet Studies for the Social Sciences) is a web panel consisting of approximately of 5,000 households. More details can be found in Example 5.7.

The panel is based on a true probability sample of households drawn from the population register by Statistics Netherlands. The initial sample consisted of approximately 10,000 households. The response process could be split into a number of steps:

Contact: Those who could be contacted for the recruitment interview.

Primary response: Those who participated in the recruitment interview.

Secondary response: Those who agreed to participate in the panel.

Tertiary response: Those who actually did participate in the panel.

Since the sample was selected from the population register of the Netherlands, the age distribution of persons in the response can be compared with the age distribution in the complete sample. The mosaic plot in Figure 5.20 does this in a graphical way.

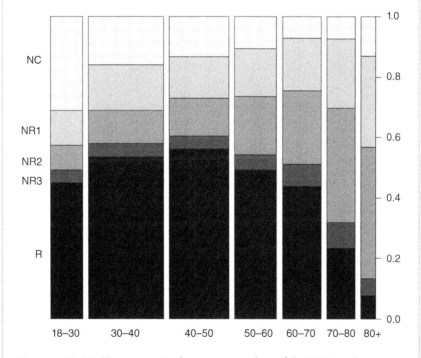

FIGURE 5.20 **Nonresponse in the recruitment phase of the LISS panel**

> The vertical bars represent the age categories. The width of these bars is proportional to the number of persons in the corresponding age groups. The shades of gray represent the steps in the response process: NC = noncontact, NR1 = primary nonresponse, NR2 = secondary nonresponse, NR3 = tertiary nonresponse, and R = ultimate response.
>
> The graph shows a clear relationship between age and response behavior. Striking is the low response for people of age 70 and over. This is mainly tertiary nonresponse. The elderly participate in the recruitment survey but do not want to become a member of the panel. The response is also low for young people (18–30 years). The main problem here is noncontact.

There is ample evidence that nonresponse often causes population estimates to be biased. This means that something has to be done in order to prevent wrong conclusions to be drawn from the survey data. There are several correcting approaches possible. The most frequently used one is *adjustment weighting*. It assigns weights to observed elements. These weights are computed in such a way that overrepresented groups get a smaller weight than underrepresented groups. Adjustment weighting has many aspects. Chapter 12 is completely dedicated to this approach.

# 5.3 Application

## 5.3.1 THE SAFETY MONITOR

Many national statistical institutes face a challenge of reducing data collection costs. This has led to considering new ways of data collecting, in which web surveys play an important role. The most far-reaching change would be to replace traditional, expensive, interviewer-assisted CAPI and CATI surveys by self-administered, and therefore cheaper, web surveys. Another option is to introduce mixed-mode surveys.

Statistics Netherlands has conducted experiments to determine whether a mixed-mode survey can replace a CAPI or CATI survey without affecting the quality of the results. One such experiment with the Dutch Safety Monitor is described in this section. More detailed account of these experiments is given by Beukenhorst and Wetzels (2009) and Kraan et al. (2010).

The Dutch Safety Monitor is an annual survey of Statistics Netherlands. It measures actual and perceived safety of the people in the country. Respondents are asked questions about feelings of safety, quality of life, and level of crime experienced. The sample for this survey is selected from the Dutch population register. Only persons of age 15 and older are selected.

Until 2008 the sample size was about 20,000. Local authorities also collected data on the same topics at the regional level. These surveys were conducted parallel to the Safety Monitor. This resulted in inconsistent estimates for safety feelings and crime victimization. Therefore it was decided to integrate the national and the regional surveys into a new Integrated Safety Monitor. This new survey had a sequential mixed-mode design with the web as one of the modes. To assess the effects of the change in survey design, the old Safety Monitor was carried out in parallel with the new Integrated Safety Monitor in 2008.

The old Safety Monitor applied two modes of data collection. If sampled persons had a known telephone number, they were approach by CATI. If this was not the case, they were approached by CAPI. The sample size was in 2008 approximately equal to 6,000 persons. This survey will be denoted by SM2.

The new Integrated Safety Monitor had four modes of data collection. All sample persons received a letter in which they were asked to complete the survey questionnaire on the Internet. The letter also included a postcard that could be used to request a paper questionnaire. Two reminders were sent to those that did not respond by web or mail. If still no response was obtained, nonrespondents were approached by means of CATI if a listed telephone number was available. If not, these nonrespondents were approached by CAPI. This four-mode survey will be denoted by SM4.

## 5.3.2 MEASUREMENT ERRORS

Objective of the Safety Monitor is to measure a number of indicators. Each indicator is based on a number of underlying questions. All indicators assume a value on a scale from 0 to 10. Three indicators are considered here:

- Harassment in the neighborhood (0 = No harassment, 10 = Harassment occurs frequently)
- Police performance (0 = Negative, 10 = Positive)
- Degradation of the neighborhood (0 = No degradation, 10 = Degradation occurs frequently)

Table 5.6 contains the average scores for these three indicators in both surveys. All differences are significant. Apparently, respondents in SM4 have a more negative attitude toward these aspects than respondents in SM2. This conclusion gives rise to questions what the cause of these differences is.

TABLE 5.6   **Average value of indicators in SM4 and SM2**

| Indicator | SM4 | SM2 | Difference |
|---|---|---|---|
| Harassment in neighborhood | 1.65 | 1.34 | +0.31 |
| Police performance | 5.50 | 5.88 | −0.38 |
| Degradation of neighborhood | 3.64 | 2.97 | +0.67 |

A next step in the analysis is establishing whether the differences can be attributed to mode effects. As an example, Table 5.7 contains the response distribution by mode of a question in SM4 that asks about the perceived occurrence of graffiti in the neighborhood.

There is striking difference for the interviewer-assisted modes on the one hand and the self-administered modes on the other. For the mail and web mode, there is a substantial shift of respondents from "Occurs (almost) never" to "Occurs sometimes." There are several possible explanations for this shift·

- It is case of non-differentiation. Respondents in the self-administered modes simply choose for the middle category.

- There is a recency effect in the interviewer-assisted modes. This would result in respondents selecting more the last answer option read ("Occurs (almost) never").

Figure 5.21 shows the analysis for an attitude question. It asks respondents for their judgment of the statement "I feel at home with the people here in the neighborhood."

TABLE 5.7    Response distribution of "perceived graffiti" by mode in SM4

| Answer | CAPI (%) | CATI (%) | Mail (%) | Web (%) |
|---|---|---|---|---|
| Occurs frequently | 12 | 13 | 12 | 11 |
| Occurs sometimes | 23 | 23 | 31 | 33 |
| Occurs (almost) never | 61 | 65 | 45 | 48 |
| Refuses to answer | 0 | 0 | 1 | 0 |
| Don't know | 4 | 1 | 12 | 7 |
| Total | 100 | 100 | 100 | 100 |

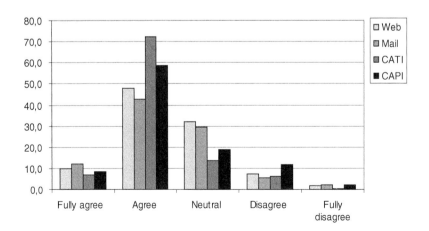

FIGURE 5.21    Response in SM4 to the question "I feel at home with the people here in the neighborhood"

There are two striking phenomena in this graph. The first one is that much more people agree with this statement in the CATI and CAPI mode. This may be caused by acquiescence. It is the tendency to agree with the statements made by the interviewer. Apparently people avoid contradicting the interviewer. CATI and CAPI are both interviewer-assisted modes of data collection. Therefore one can expect that acquiescence occurs only for these modes, and not for the web and mail modes.

Another striking phenomenon is that much less people choose for the neutral category in CAPI and CATI. This may be caused by the tendency of people to give more socially desirable answers when interviewers are present. Not having a clear opinion is not considered socially desirable. Therefore they may express an opinion they may not have.

It should be noted that the differences between SM4 and SM2 may be caused by other effects than mode effects. There can also be *selection effects*. This is the phenomenon that different subpopulations respond in different modes. For example, SM4 has a web mode that was not available in SM2. Since Internet coverage is high among young people, this may cause more young people to be in the SM4 survey. Therefore it is important to check composition of the response in the different modes. Kraan et al. (2010) did this for the variable age. It turned out there were no large differences in the age distributions by mode.

Paradata may provide some error indicators. For example, Kreuter, Presser, and Tourangeau (2008) suggest using unusually long or unusually short response times as proxy indicators for measurement error, controlling for other factors that influence response times.

## 5.3.3 NONRESPONSE

The response rate of SM4 turned out to be 59.7%. The response rate for SM2 was 63.5%. So there is not much difference. Table 5.8 shows the composition of the response for both surveys.

More than half of the response (58%) was obtained in the SM4 with a self-administered mode of data collection (web or mail). The conclusion can be drawn

TABLE 5.8   **Composition of the response of both safety monitors**

| Data collection mode | SM4 (%) | SM2 (%) |
| --- | --- | --- |
| Web | 41.8 | |
| Mail | 16.2 | |
| CATI | 30.5 | 71.6 |
| CAPI | 11.5 | 28.4 |
| Total | 100.0 | 100.0 |

that the four-mode survey did not increase the response. The costs of the survey were, however, much lower, because interviewers were deployed in only 42% of the cases. Focusing on just interviewer costs and ignoring all other costs (which are much lower), Beukenhorst and Wetzels (2009) found that the costs of SM4 were only 60% of the costs of SM2.

The quality of the response is not only determined by the response rate. The composition of the response is more important. If the response composition differs substantially from the sample composition, estimates may be biased. To get insight in possible composition differences, response rates in the categories of auxiliary variables can be compared. Table 5.9 shows the result.

The table shows there is little difference in the response rates. Only for *age* and *degree of urbanization* the range seems to be smaller for SM4. This is an indication that the composition may have improved.

A phenomenon that repeats itself in many general population surveys is the low response rate of specific groups. Well-known examples are people in highly urbanized areas and ethnic minority groups. Beukenhorst and Wetzels (2009) investigated whether the four-mode survey reduces this problem. This turned out not to be the case for the urbanized areas, because response was also very low here for the web and mail mode. They draw the same conclusion for ethnic groups.

The results of this experiment show that, in this case, the mixed-mode approach (with four modes) was not very successful with respect to the quality of the outcomes. The response rate did not increase substantially, and also the composition of the response did not improve much. The hope that introducing new modes would increase the response rate of low-response groups was not fulfilled.

The experiment showed that a mixed-mode survey can be very successful with respect to survey costs. The four-mode survey was substantially less expensive than the two-mode survey.

TABLE 5.9    **Ranges of the response rates (%) in the categories of auxiliary variables**

| Auxiliary variable | Categories | Range in SM4 (%) | Range in SM2 (%) |
|---|---|---|---|
| Household size | 2 | 52–64 | 54–67 |
| Gender | 2 | 60–62 | 63–65 |
| Age | 7 | 55–66 | 52–70 |
| Ethnicity | 4 | 41–64 | 44–67 |
| Degree of urbanization | 5 | 52–66 | 53–71 |
| Average | | 60 | 64 |

# 5.4 Summary

Survey researchers have control over many different aspects of the survey process. With the proper choice of a sampling frame, a sampling design, and an estimation procedure, they can obtain precise estimators of population characteristics. Unfortunately, not everything is under control. Survey researchers may be confronted with various phenomena that may have a negative impact on the quality and therefore the reliability of the survey outcomes. Some of these disturbances are almost impossible to prevent. After giving an overview of what can go wrong, the chapter concentrates on two types of problems: measurement errors and nonresponse errors.

Measurement errors occur when the answers given by respondents differ from the true answers. The reason can be that the respondent does not understand a question, does not know the true answer, or does not want to give the true answer. The nature and magnitude of measurement errors may be different for different modes of data collection. This chapter focuses on measurement errors in web surveys. An important aspect is the absence of interviewers in web surveys. This may have positive effects (for example, less socially desirable answers) but also negative effects (for example, more satisficing). This chapter stresses the importance of paying careful attention to the design of the web survey questionnaire, since small errors in the design may lead to large errors in the answers.

Another important issue is nonresponse. Particularly high nonresponse rates may cause large biases in estimates. It is important to distinguish various causes of nonresponse (noncontact, refusal, not-able), since they may have a different impact on estimates. The cause of nonresponse may be difficult to establish for web surveys.

It should always be attempted to correct for a possible nonresponse bias. Auxiliary variables are required for this. These variables must have been measured in the survey, and moreover their distribution in the population or complete sample must be available.

When considering mobile web surveys or only mobile survey, several challenges are existing, despite the progress made in understanding the mobile web phenomenon. Couper, Antoun, and Mavletova (2017) report an interesting literature overview of the open problems and experiments about different types of errors (coverage, measurement, break-off, and so on) and contests. A set of findings show that the proportion of sample persons completing a web survey on their mobile device is still low with respect to smartphone penetration. In general, mobile web surveys have lower response rates. Mobile surveys have higher break-off rates and longer completion times than PC web surveys. As regards measurement error, few differences between those who complete a web survey on a mobile device or on a PC are found.

Using mobile for completing a questionnaire, the technology aspects as well as the respondent behavior and the characteristics of the device may produce several effects on the participation and the data quality. It has to be advised that further research is needed to fully understand the components of the mobile survey errors. Thus, researchers should try to evaluate the role that different aspects are playing in their survey.

## KEY TERMS

**Acquiescence**: The tendency that respondents tend to agree with statements in questions, regardless of their content. They simply answer "yes."

**Data editing**: The activity of checking collected data for errors and, where possible, correcting these errors.

**Eligible**: A sample element is eligible for a survey if it belongs to the target population of the survey.

**Estimation error**: The deviation of the estimate from the true value caused by investigating only a sample instead of the complete population.

**Grid question**: See matrix question.

**Item nonresponse**: A type of nonresponse occurring when some questions have been answered but no answer is obtained for some other, possibly sensitive, questions. So, the questionnaire form has been partially completed.

**Matrix question**: A series of questions with the same set of answer options combined in a matrix. Each row of a matrix question represents a single question, and each column corresponds to an answer option.

**Measurement error**: An error occurring if the answer given by a respondent differs from the true answer. The reason can be that the respondent does not understand a question, does not know the true answer, or does not want to give the true answer.

**Memory error**: The error caused by respondents who forget to report events, particularly when they happened a long time ago or when they are not too salient.

**Missing at Random (MAR)**: Nonresponse depends on auxiliary variables only. Estimators will be biased, but a correction is possible if some technique is used that takes advantage of this auxiliary information.

**Missing Completely at Random (MCAR)**: Nonresponse happens completely independent of all survey variables. Estimators will not be biased.

**Non-differentiation**: A form of satisficing that typically occurs if respondents have to answer a series of questions each with the same set of response options. They tend to select the same answer for all these questions irrespective of the question content.

**Non-observation error**: A type of non-sampling error that occurs when intended measurements cannot be carried out. Two types of non-observation errors can be distinguished: under-coverage errors and nonresponse errors.

**Nonresponse**: The phenomenon that elements in the selected sample, and that are also eligible for the survey, do not provide the requested information, or that the provided information is not usable.

**Non-sampling error**: The error caused by problems that even can occur if the whole population is investigated (instead of a sample). Non-sampling errors can be divided in observation errors and non-observation errors.

**Not Missing at Random (NMAR)**: Nonresponse depends directly on the target variables of the survey. Estimators will not be biased, and correction techniques will not be successful.

**Observation error**: An error made during the process of obtaining answers from respondents and recording and further processing these answers. Three types of observation errors are distinguished here: over-coverage errors, measurement errors, and processing errors.

**Over-coverage error**: An error caused by elements that are included in the survey and that do not belong to the target population.

**Primacy effect**: The tendency of respondents to pick an answer early in the list of answer options of a closed question. Primacy effects typically occur in interviewer-assisted surveys.

**Processing error**: An error made during the phase of recording and processing the collected data.

**Random response model**: A model for nonresponse that assumes every element in the population to have an (unknown) response probability.

**Recency effect**: The tendency of respondents to pick an answer at or near the end of the list of answer options of a closed question. Recency effects typically occur in self-administered surveys.

**Response order effect**: The tendency that the answer selected by the respondent depends on its location in the list of answer options. Primacy and recency effects are special cases.

**Response rate**: The number of responding eligible elements in the sample divided by the number of eligible elements in the sample.

**Sampling error**: The error in the estimate introduced by the sampling design. It is caused by the fact that estimates are based on a sample from the population and not on a complete enumeration of the population.

**Satisficing**: The phenomenon that respondents do not do all they can to provide a correct answer. Instead they attempt to give a satisfactory answer with minimal effort.

**Selection effect**: The phenomenon that different subpopulations favor different data collection modes of a mixed-mode survey.

**Selection error**: An error occurring when the selection probabilities used for the computation of an estimate differ from the true selection probabilities.

**Socially desirable answer**: The tendency that respondents give answers that will be viewed as more favorable by others. This particularly happens for sensitive questions.

**Straight-lining**: The tendency that respondents give the same answer to all single questions in a matrix question. They simply check all answer options in the same column.

**Total error**: The combined effect of all phenomena that contribute to an estimated value that deviates from the true value of a population characteristic.

**Under-coverage**: The sampling frame does not cover completely the target population of the survey. There are persons in the population who do not appear in the sampling frame. They will never be selected in the sample.

**Unit nonresponse**: This type of nonresponse occurs when a selected element does not provide any information at all, i.e., the questionnaire form remains completely empty.

## EXERCISES

**Exercise 5.1**   Which of the sources of error below does not belong to the category of observation errors?

**a.** Measurement error.

**b.** Over-coverage.

**c.** Under-coverage.

**d.** Processing error.

**Exercise 5.2**   Memory effects occur if respondents forget to report certain events or when they make errors about the date of occurrence of events. To which source of errors do these memory effects belong?

**a.** Estimation error.

**b.** Under-coverage.

**c.** Measurement error.

**d.** Non-observation error.

**Exercise 5.3**   Which of the effects below may occur in a web survey?

**a.** Both primacy and recency effects.

**b.** Only primacy effects.

**c.** Only recency effects.

**d.** Non-observation error.

**Exercise 5.4** Which of the phenomena listed below is not a form of satisficing?

a. Response order effects.

b. Acquiescence.

c. Socially desirable answers.

d. Non-differentiation.

**Exercise 5.5** What is the best way to format a closed question?

a. With a set of checkboxes.

b. With a drop-down list.

c. With a set of radio buttons.

d. With a text input field.

**Exercise 5.6** In what situation are progress indicators must be effective?

a. In case of complex questionnaires with a lot of routing.

b. In case of long questionnaires without routing.

c. In case of short questionnaires without routing.

d. In case of questionnaires with a lot of checks.

**Exercise 5.7** Which phenomenon makes it difficult, if not impossible, to compute the response rate of a web survey?

a. Over-coverage.

b. Under-coverage.

c. Noncontact.

d. Refusal.

**Exercise 5.8** A survey is usually carried out to measure the state of a target population at a specific point in time (the reference date). The survey outcomes are supposed to describe the population at this date. Ideally, the fieldwork of the survey should take place at that date. This is not possible in practice. Interviewing usually takes place in a period of a number of days or weeks around the reference date.

Suppose a web survey is carried among employees of a company. Each employee has a company e-mail address. So there is a sampling frame. A sample of employees is selected from this sampling frame two weeks before the reference date. Selected employees are asked to complete the questionnaire in a period of four weeks: the two weeks between sample selection and reference date and the two weeks after the reference date.

Explain for each of the situations described below what kind of problem there is: nonresponse, under-coverage, over-coverage, or another sampling frame error:

- A selected employee died before the sample selection date.
- A selected employee died between the sample selection date and the reference date.
- A selected employee died between the reference date and the sample selection date.

**Exercise 5.9** A town council wants to do something about the traffic problems in the town center. There is a plan to turn it into a pedestrian area, so cars cannot go into the center any more. The town council wants to know what companies think of this plan. A simple random sample of 1,000 companies is selected. Each selected company is invited to participate in a web survey. They are asked whether they are in favor of the plan or not. Furthermore, the location of the company is recorded (town center or suburb). The results of the survey are summarized in the table below:

|              | Suburbs | Town center |
|--------------|---------|-------------|
| In favor     | 120     | 80          |
| Not in favor | 40      | 240         |

**a.** Determine the response percentage.

**b.** Determine the percentage of respondents in favor of the plan.

**Exercise 5.10** The local authorities of a town want to know how satisfied citizens are with public transport facilities in town. They conduct a web survey. The target population is defined as all citizens that used public transport at least once in the last year. A sample is selected from the population register of the town. Selected persons are sent a letter with the Internet address of the survey. The results of the survey are summarized in the table below:

| Result        | Frequency |
|---------------|-----------|
| Over-coverage | 320       |
| Refusal       | 240       |
| Noncontact    | 80        |
| Not able      | 40        |
| Response      | 440       |
| Total         | 1,120     |

Compute the response rate of the survey. Make clear how the response rate was computed and which assumptions were made.

# REFERENCES

Bethlehem, J. G. (2009), *Applied Survey Methods, a Statistical Perspective.* Wiley, Hoboken, NJ.

Bethlehem, J. G., Cobben, F., & Schouten, B. (2011), *Handbook of Nonresponse in Household Surveys.* Handbooks in Survey Methodology, Wiley, Hoboken, NJ.

Beukenhorst, D. & Giesen, D. (2010), *Internet Use for Data Collection at Statistics Netherlands.* Paper presented at the 2nd International Workshop on Internet Survey Methods, Daejoeon, South Korea.

Beukenhorst, D. & Wetzels, W. (2009), *A Comparison of Two Mixed-mode Designs of the Dutch Safety Monitor: Mode Effects, Costs, Logistics.* Technical paper DMH 206546, Statistics Netherlands, Methodology Department, Heerlen, the Netherlands.

Biemer, P. (2010), Total Survey Error. Design, Implementation, and Evaluation. *Public Opinion Quarterly,* 74, 5, pp. 817–848.

Bishop, G. F. (1987), Experiments with the Middle Response Alternative in Survey Questions. *Public Opinion Quarterly,* 51, pp. 220–232.

Bradburn, N. M., Sudman, S., & Wansink, B. (2004), *Asking Questions, The Definitive Guide to Questionnaire Design—For Market Research, Political Polls, and Social and Health Questionnaires.* Jossey-Bass, San Francisco, CA.

Buskirk, T. D. & Andrus, C. (2014), Making Mobile Browser Surveys Smarter: Results from a Randomized Experiment Comparing Online Surveys Completed via Computer or Smartphone. *Field Methods,* 26, pp. 322–342.

Callegaro, M. (2010), Do You Know Which Device Your Respondent Has Used to Take Your Online Survey? Using Paradata to Collect Information on Device Type. *Survey Practice,* 3, 6. https://doi.org/10.29115/SP-2010-0028.

Callegaro, M., Manfreda, K., & Vehovar, V. (2015), *Web Survey Methodology,* Sage Publications Ltd.

Christian, L. M., Dillman, D., & Smyth, J. (2007), Helping Respondents Get It Right the First Time: The Influence of Words, Symbols and Graphics in Web Surveys. *Public Opinion Quarterly,* 71, pp. 113–125.

Converse, P. E. (1964), The Nature of Belief Systems in Mass Public. In: Apter, D. (ed.), *Ideology and Discontent.* Free Press, New York, pp. 206–261.

Couper, M. P. (1999), Usability Evaluation of Computer Assisted Survey Instruments. *Proceedings of the Third ASC International Conference,* ASC, Edinburgh, pp. 1–14.

Couper, M. P. (2000), Web Surveys, A Review of Issues and Approaches. *Public Opinion Quarterly,* 64, pp. 464–494.

Couper, M. P. (2008), *Designing Effective Web Surveys.* Cambridge University Press, New York.

Couper, M. P., Antoun, C., & Mavletova, A. (2017). Mobile Web Surveys. A Total Survey Error Perspective. In: Biemer, P., de Leeuw, E., Eckman, S., Edwards, B., Krauter, F., Lyberg, L., Tucker, N., & West, B. (eds.), *Total Survey Error in Practice,* Wiley Series in Survey Methodology, Wiley, Hoboken, NJ.

Couper, M. P., Tourangeau, R., Conrad, F. G., & Crawford, S. D. (2004), What They See Is What They Get: Response Options for Web Surveys. *Social Science Computer Review,* 22, pp. 111–127.

Couper, M. P., Traugott, M., & Lamias, M. (2001), Web Survey Design and Administration. *Public Opinion Quarterly,* 65, pp. 230–253.

de Leeuw, E. D. (1992), *Data Quality in Mail, Telephone, and Face-to-Face Surveys.* TT-Publications, Amsterdam, the Netherlands.

de Leeuw, E. D. (2008), Choosing the Mode of Data Collection. In: de Leeuw, E. D., Hox, J. J., & Dillman, D. A. (eds.), *International Handbook of Survey Methodology*. Lawrence Erlbaum Associates, New York/London, U.K., pp. 113–135.

Derouvray, C. & Couper, M. P. (2002), Designing a Strategy for Reducing "No Opinion" Responses in Web-Based Surveys. *Social Science Computer Review*, 20, pp. 3–9.

Dillman, D. A. (2007), *Mail and Internet Surveys. The Tailored Design Method*. Wiley, Hoboken, NJ.

Dillman, D. A. & Bowker, D. (2001), The Web Questionnaire Challenge to Survey Methodologists. In: Reips, U. D. & Bosnjak, M. (eds.), *Dimensions of Internet Science*. Pabst Science Publishers, Lengerich, Germany, pp. 159–178.

Dillman, D. A., Smyth, J. D., & Christian, L. M. (2009), *Internet, Mail, and Mixed-mode Surveys, The Tailored Design Method*. Wiley, Hoboken, NJ.

Dillman, D. A., Smyth, J. D., & Christian, L. M. (2014), *Internet, Mail, and Mixed-mode Surveys, The Tailored Design Method*. Wiley, Hoboken, NJ.

Eckman, S. & Kreuter, F. (2017), Undercoverage-Nonresponse Tradeoff. In: Biemer, P., de Leeuw, E., Eckman, S., Edwards, B., Krauter, F., Lyberg, L., Tucker, N., & West, B. (eds.), *Total Survey Error in Practice*. Wiley, Hoboken, NJ, pp. 95–112.

Fricker, R. & Schonlau, M. (2002), Advantages and Disadvantages of Internet Research Surveys: Evidence from the Literature. *Field Method*, 15, pp. 347–367.

Fricker, S., Galesic, M., Tourangeau, R., & Yan, T. (2005), An Experimental Comparison of Web and Telephone Surveys. *Public Opinion Quarterly*, 69, pp. 370–392.

Galesic, M., Tourangeau, R., Couper, M. P., & Conrad, F. G. (2007), *Using Change to Improve Navigation in Grid Questions*. Paper presented at the General Online Research Conference (GOR '07), Leipzich, Germany.

Heerwegh, D. (2004), *Using Progress Indicators in Web Surveys*. Paper presented at the 59th AAPOR Conference, Phoenix, AZ.

Heerwegh, D. & Loosveldt, G. (2002), *An Evaluation of the Effect of Response Formats on Data Quality in Web Surveys*. Paper presented at the International Conference on Improving Surveys, Copenhagen, Denmark.

Heerwegh, D. & Loosveldt, G. (2008), Face-to-Face Versus Web Surveying in a High-Internet Coverage Population. Differences in Response Quality. *Public Opinion Quarterly*, 72, pp. 836–846.

Holbrook, A. L., Green, M. C., & Krosnick, J. A. (2003), Telephone Versus Face-to-Face Interviewing of National Probability Samples with Long Questionnaires. *Public Opinion Quarterly*, 67, pp. 79–125.

Holbrook, A. L. & Krosnick, J. A. (2010), Social Desirability Bias in Voter Turnout Reports. Tests Using the Item Count Technique. *Public Opinion Quarterly*, 74, pp. 37–67.

Holmberg, A. (2010), *Using the Internet in Individual and Household Surveys, Summarizing Some Experiences at Statistics Sweden*. Paper presented at the 2nd International Workshop on Internet Survey Methods, Daejoeon, South Korea.

Kaczmirek, L. (2010), Attention and Usability in Internet Surveys: Effects of Visual Feedback in Grid Questions. In: Das, M., Ester, P., & Kaczmirek, L. (eds.), *Social and Behavioral Research and the Internet, Advances in Applied Methods and Research Strategies*. Routledge, New York/London, U.K., pp. 191–214.

Kaczmirek, L., Neubarth, W., Bosnjak, M., & Bandilla, W. (2004). *Progress Indicators in Filter-based Surveys: Computing Methods and Their Impact on Drop Out*. Paper presented at the RC33, the 6th International Conference on Social Science Methodology, Amsterdam, the Netherlands.

Kalton, G., Roberts, J., & Holt, D. (1980), The Effect of Offering Middle Response Option with Opinion Questions. *Journal of the Royal Statistical Society, Series D,* 29, pp. 65–78.

Kalton, J. & Schuman, H. (1982), The Effect of the Question on Survey Responses: A Review. *Journal of the Royal Statistical Society. Series A (General),* 145, pp. 42–73.

Kish, L. (1967), *Survey Sampling.* Wiley, New York.

Kolbas, V. (2014), *The Measurement Effect in PC Smartphone and Tablet Surveys.* Paper presented at the ESRA Biannual Conference, Reykjavik, Iceland.

Kraan, T., Van den Brakel, J., Buelens, B., & Huys, H. (2010), *Social Desirability Bias, Response Order Effect and Selection Effects in the New Dutch Safety Monitor.* Discussion Paper 10004, Statistics Netherlands, The Hague/Heerlen, the Netherlands.

Kreuter, F., Presser, S., & Tourangeau, R. (2008), Social Desirability Bias in CATI, IVR and Web Surveys. *Public Opinion Quarterly,* 72, pp. 847–865.

Krosnick, J. A. (1991), Response Strategies for Coping with the Cognitive Demands of Attitude Measures in Surveys. *Applied Cognitive Psychology,* 5, pp. 213–236.

Krosnick, J. A. (1999), Survey Research. *Annual Review of Psychology,* 50, pp. 537–567.

Krosnick, J. A. & Alwin, D. F. (1987), An Evaluation of Cognitive Theory of Response-order Effects in Survey Measurement. *Public Opinion Quarterly,* 51, pp. 201–219.

Leenheer, J. & Scherpenzeel, A. (2013), Does It Pay Off to Include Non-internet Households in an Internet Panel? *International Journal of Internet Science,* 1, pp. 17–29.

Liu, M. & Cernat, A. (2018), Item-by-item Versus Matrix Questions: A Web Survey Experiment. *Social Science Computer Review,* 36, 6, pp. 690–706.

Lodge, M., Steenbergen, M. R., & Brau, S. (1995), The Responsive Voter: Campaign Information and the Dynamics of Candidate Evaluation. *American Political Science Review,* 89, pp. 309–326.

Lugtig, P. & Toepoel, V. (2016), The Use of PCs, Smartphones and Tablets in a Probability Based Panel Survey. Effects on Survey Measurement Error. *Social Science Computer Review,* 34, 1, pp. 78–94.

Mavletova, A. (2013), Data Quality in PC and Mobile Web Surveys. *Social Science Computer Review,* 7, pp. 191–205.

Millar, M. M. & Dillman, D. A. (2011), Improving Response to Web and Mixed-Mode Surveys. *Public Opinion Quarterly,* 75, 2, pp. 249–269.

Peterson, G., Griffin, J., LaFrance, J., & Jiao Li, J. (2017), Smartphone Participation in Web Surveys. In: Biemer, P., de Leeuw, E., Eckman, S., Edwards, B., Krauter, F., Lyberg, L., Tucker, N., & West, B. (eds.), *Total Survey Error in Practice,* Wiley Series in Survey Methodology, Wiley.

Saris, W. E. (1997), The Public Opinion About the EU Can Easily Be Swayed in Different Directions. *Acta Politica,* 32, pp. 406–436.

Scherpenzeel, A. (2009), *Start of the LISS Panel: Sample and Recruitment of a Probability-based Internet Panel.* Centerdata, Tilburg, the Netherlands.

Schuman, H. & Presser, S. (1981), *Questions and Answers in Attitude Surveys.* Academic Press, New York.

Schwarz, N., Hippler, H. J., & Noelle-Neumann, E. (1992), A Cognitive Model of Response Order Effects in Survey Measurement. In: Schwarz, N. & Sudman, S. (eds.), *Context Effects in Social and Psychological Research,* Springer-Verlag, New York, pp. 187–201.

Schwarz, N., Knäuper, B., Oyserman, D., & Stich, C. (2008), The Psychology of Asking Questions. In: de Leeuw, E. D., Hox, J. J., & Dillman, D. A. (eds.), *International Handbook of Survey Methodology.* Lawrence Erlbaum Associates, New York/London, U.K., pp. 18–34.

Sikkel, D. (1983), Geheugeneffecten bij het Rapporteren van Huisartsencontacten. *Statistisch Magazine*, 3, 4, pp. 61–64.

Smyth, J. D., Dillman, D. A., Christian, L. M., & Stern, M. J. (2006), Comparing Check-all and Forced Choice Formats in Web Surveys. *Public Opinion Quarterly*, 70, pp. 66–77.

Snijkers, G., Haraldsen, G., Jones, J., & Willimack, D. (2013), *Designing and Conducting Business Surveys*, Wiley, Hoboken, NJ.

Statistics Netherlands (2002), *Blaise Developer's Guide* Statistics Netherlands, Heerlen, the Netherlands.

Stern, M. J., Sterrett, D., Bilgen, I., Raker, E., Rugg, G., & Baek, J. (2015). *The Effects of Grids on Web Surveys Completed with Mobile Devices*. Paper presented at the AAPOR Annual Conference, Hollywood, FL.

Sudman, S. & Bradburn, N. (1982), *Asking Questions: A Practical Guide to Questionnaire Design*. Jossey-Bass, San Francisco, CA.

Sudman, S., Bradburn, N., & Schwarz, N. (1996), *Thinking about Answers: The Application of Cognitive Processes to Survey Methodology*. Jossey-Bass, San Francisco, CA.

Tiemeijer, W. L. (2008), *Wat 93,7 Procent van de Nederlanders Moet Weten over Opiniepeilingen*. Aksant, Amsterdam, the Netherlands.

Toepoel, V. & Lugtig, P. (2014), What Happens If You Offer a Mobile Option to Your Web Panel? Evidence from a Probability-Based Panel of Internet Users. *Social Survey Computer Review*, 32, pp. 544–560.

Tourangeau, R. (2017), Mixing Modes. Tradeoffs Among Coverage, Nonresponse, and Measurement Error. In: Biemer, P., de Leeuw, E., Eckman, S., Edwards, B., Krauter, F., Lyberg, L., Tucker, N., & West, B. (eds.), *Total Survey Error in Practice*. Wiley, Hoboken, NJ.

Tourangeau, R., Couper, M. P., & Conrad, F. (2004), Spacing, Position and Order, Interpretive Heuristics for Visual Features of Survey Questions. *Public Opinion Quarterly*, 68, pp. 368–393.

Tourangeau, R., Maitland, A., Rivero, G., Sun, H., Williams, D. T., & Yan, T. (2017), Web Surveys by Smartphone and Tablets: Effects on Survey Responses. *Public Opinion Quarterly*, 81, 4, pp. 896–929.

Villar, A., Callegaro, M., & Yang, Y. (2013), Where Am I? A Meta-Analysis on the Effects on the Experiments of Progress Indicators for Web Surveys. *Social Science Computer Review*, 31, 6, pp. 744–762.

Vis-Visschers, R., Arends-Tóth, J., Giesen, D., & Meertens, V. (2008), *Het Aanbieden van "Weet niet" en Toelichtingen in een Vragenlijst*. Report DMH-2008-02-21-RVCS, Statistics Netherlands, Heerlen, the Netherlands.

Wells, T., Bailey, J. T., & Link M. W. (2014), Comparison of Smartphone and Online Computer Survey Administration. *Social Science Computer Review*, 57, pp. 521–532.

Zaller, J. R. (1992), *The Nature and Origins of Mass Opinion*. Cambridge University Press, Cambridge, U.K.

# Chapter SIX

# Web Surveys and Other Modes of Data Collection

## 6.1 Introduction

### 6.1.1 MODES OF DATA COLLECTION

Survey data collection has evolved over the years. The period before the 1970s was the era of traditional interviewing using paper forms. There were three main ways to do this: face-to-face interviewing, telephone interviewing, and mail interviewing.

*Face-to-face interviewing* is a mode of data collection used already for the first censuses. Thus, it is not surprising that the first surveys were using it. Face-to-face interviewing means that developments of information technology since the 1970s have made it possible to use microcomputers for data collection. Thus, *computer-assisted interviewing* (CAI) emerged—a computer program containing the questions the interviewer had to ask. The computer took control of the interviewing process, and it checked answers to questions on the spot. CAI has three important advantages over traditional forms of interviewing:

- It relieves the interviewers of the task of choosing the correct route through the questionnaire. The interview software take care of this task. Therefore, interviewers can concentrate on asking questions and assisting respondents in getting the right answers.

*Handbook of Web Surveys*, Second Edition. Silvia Biffignandi and Jelke Bethlehem.
© 2021 John Wiley & Sons, Inc. Published 2021 by John Wiley & Sons, Inc.

- It can improve the quality of the collected data. Answers can be checked, and correction of the answers takes place during the interview. This is more effective than having to do it afterward in the survey agency.

- The data are written in the computer already during the interview. This results in a "clean" record. No more subsequent data entry and data editing are necessary. This considerably reduces time needed to process the survey data and thus improves the timeliness of the survey results.

CAI has different modes of data collection. They are the electronic analogues of the traditional modes of data collection.

*Computer-assisted personal interviewing* (CAPI) is a form of face-to-face interviewing where interviewers take their laptop computer to the homes of the respondents. There they start the interview program and enter the answers to the questions appearing on the screen. Interviewers send the collected data to the survey organization by means of the Internet. In return they may receive new names and addresses of persons to interview.

*Computer-assisted telephone interviewing* (CATI) is the electronic form of telephone interviewing. Interviewers call respondents from the call center of the survey organization. The interview software decides when and who to contact. If contact is established by telephone and the person agrees to participate in the survey, the interviewer starts the interview program. The first question appears on the screen. If this is answered and no error is detected, the software proceeds to the next question on the route through the questionnaire.

*Computer-assisted self-interviewing* (CASI), or sometimes also *computer-assisted self-administered questionnaires* (CASAQ), is the electronic analogue of mail interviewing. The electronic questionnaire is sent to the respondents. They run the software on their own computer, answer the questions, and send the answers back to the survey agency. Nowadays it is common practice to download the software from the Internet. The answers are returned electronically in the same fashion.

The rapid emergence and diffusion of the Internet in 1990s has led to a new mode of data collection: the *web survey*, sometimes also called *computer-assisted web interviewing* (CAWI). The respondents are invited to go to a specific page on the Internet. There they complete the questionnaire themselves. There are no interviewers involved. In fact, a web survey is a special type of CASI survey.

## 6.1.2 THE CHOICE OF THE MODES OF DATA COLLECTION

The choice of the mode of data collection is not always an easy one. There are two important, but often conflicting, factors: costs and quality.

Conducting a survey can be expensive and time consuming, particularly if the sample size is large and the questionnaire is long and complex. Many people may be involved in setting up and carrying out a survey. Depending on the survey, there may be researchers, questionnaire designers, interviewers, supervisors, data entry typists, analysts, etc. Staff costs may well be the largest component of the total survey costs. There may be other costs, like the costs of hardware and software. Equipping a large group of interviewers with laptops for a CAPI survey requires substantial investments. Setting up a smooth-running, large-scale web survey is not possible without installing several web servers. Hardware and software requirements are much less for mail surveys, but then there are printing and mailing costs.

To reduce the problems caused by nonresponse, one may decide to use incentives. Such incentives are most effective if delivering is at or before the time of the interview attempt and not promised as a reward after a completed interview. This means that costs of incentives are proportional to the sample size.

Large surveys cost more than small surveys. Unfortunately, the precision of the survey results relates to the sample size, too: the larger the sample, the more precise the estimates. Therefore, the higher the demands for precision are, the more expensive the survey will be. In many practical situations, the sample size will be a compromise between costs and precision.

Many different factors determine the costs of a survey. It will be clear that the mode of data collection is certainly one of them. Anyway, cost is only one aspect that plays a role in choosing the type of survey to conduct. Quality is another aspect. Quality can have many dimensions. For example, Eurostat, the statistical office of the European Union, distinguishes the following five dimensions for the quality of statistics in its European Statistics Code of Practice:

- *Relevance*: European statistics must meet the needs of users.

- *Accuracy and reliability*: European statistics has to be an accurate and reliable portrayal of reality.

- *Timeliness and punctuality*: European statistics must be disseminated in a timely and punctual manner.

- *Coherence and comparability*: European statistics should be consistent internally over time and comparable between regions and countries; it should be possible to combine and make joint use of related data from different sources.

- *Accessibility and clarity*: European statistics presentation has to be in a clear and understandable form, disseminated in a suitable and convenient manner, available, and accessible on an impartial basis with supporting metadata and guidance.

Some of these dimensions of quality may also be relevant for choosing the proper mode of data collection. This section focuses on the quality dimension *accuracy and reliability*.

A survey estimator is *reliable* if repeating the survey would result in approximately the same estimate. Reliability does not imply *validity*. An estimator is valid if it estimates what it intends to estimate. A reliable estimator can produce consistently wrong estimates, for example, if the estimator has a fixed bias. The reliability measure of an estimator is its variance or standard error. An estimator is *precise* if it has a small variance.

An estimator is *accurate* if repeatedly conducting the survey would result in estimates that are all close to the true value. See Example 6.1. The measure of the accuracy of an estimator is its mean square error. This quantity contains both a variance and a bias component.

---

### ▣ EXAMPLE 6.1  Accuracy and reliability of a bathroom scale

Suppose to conduct a survey about obesity. The interviewers using a bathroom scale determine the weight of all respondents. If someone who weighs 90 kg steps on the scale repeatedly and gets readings of 30, 150, 70, 120, 50, etc., the scale is not reliable. If all readings are close to 120, the scale is reliable but not accurate. If all readings are close to 90, the scale is reliable and accurate.

---

A number of quality aspects impact in a different way depending on the mode of data collection. Thus, they could play a role in deciding the mode of data collection. Hereunder, they describe some type of errors with respect to the mode:

- *Coverage.* Some data collection modes may suffer more from coverage problems. Problems occur if the mode of data collection requires use of a sampling frame that does not coincide with the target population of the survey. For example, a telephone survey may suffer from serious undercoverage if the sampling frame is a list of registered telephone numbers. Another example is a web survey, where those without Internet access will never be selected in the sample.

- *Nonresponse.* The occurrence of nonresponse may cause estimators of population characteristics to be biased. Nonresponse problems seem to be more severe if no interviewers are involved in the survey. An extensive overview of the relationship between nonresponse and the mode of data collection is in Bethlehem, Cobben, and Schouten (2011).

- *Measurement errors.* The impact of measurement errors on the answers to the survey questions was already described in Chapter 5. Measurement errors can have several causes. One main cause is *satisficing*. This is the phenomenon that respondents do not do all they can to provide a correct answer. Instead, they attempt to give a satisfactory answer with minimal effort. Satisficing comes in different forms:

- *Response order effects*: The tendency the answer selected by the respondent depends on its location in the list of answer options. Primacy effects occur in self-administered surveys (respondents typically pick an answer early in the list of answer options of a closed question), and recency effects occur in interviewer-assisted surveys (respondents typically pick an answer at or near the end of the list of answer options of a closed question).

- *Acquiescence*: This is the tendency that respondents tend to agree with statements in questions, regardless of their content. They simply answer "yes." There is less acquiescence in self-administered surveys than in interviewer-assisted surveys.

- *Status quo endorsement*: Surveys sometimes ask respondents to give their opinion about changes. A typical example is a question whether government should change its policy with respect to a specific issue. The easiest way to answer such a question without thinking is to select the option to keep everything the same. There seems to be less status quo endorsement in interviewer-assisted surveys.

- *Non-differentiation*: Form of satisficing that typically occurs if respondents have to answer a series of questions with the same set of response options. Satisficing respondents tend to select the same answer for all these questions irrespective of the question content. Literature suggests there is more non-differentiation in self-administered surveys.

- *Answering "don't know"*: This is a form of satisficing where respondents choose this answer to avoid having to think about a real answer. Not making it possible to answer "don't know" may also cause measurement errors as respondents not knowing the answer are forced to give one. Treating "don't know" in survey may depend on the mode of data collection. For example, it is generally advised to offer "don't know" in CAPI and CATI surveys only implicitly. Some research suggests not using "don't know" explicitly in a mobile web survey. If respondents skip the question because they do not know the answer, the question reappears, but now with "don't know" as one of the possible answers.

- *Arbitrary answer*: Respondents may decide an arbitrary answer in order to avoid having to think about a proper answer. They may also give an arbitrary answer if giving the proper answer is undesirable. The name of this behavior sometimes is "metal coin flipping." This phenomenon typically occurs in mobile web surveys for check-all-that-apply questions.

- *Socially desirable answers*: The tendency that respondents give answers viewed as more favorable by others. This particularly happens for sensitive questions. If a true answer would not make the respondents look good, they will refuse to answer or give a different answer. The literature shows that the effects of socially desirable answers are stronger in interviewer-assisted surveys. Truthful answers tend to be more in self-administered surveys.

- *Questionnaire design effects*: Depending on the mode of data collection, the questionnaire presentation is in a different way: questionnaire printed on paper or displayed on a device (computer, smartphone, tablet, etc.) screen. Electronic questionnaires can consist of a set of pages/screens where each page contains a separate question (the *question-based approach*) or it can consist of one form containing all questions (the *form-based approach*). Particularly for self-administered questionnaires, the design of the questionnaire is very important. See Dillman, Smyth, and Christian (2009, 2014) and Chapter 7 in this handbook for in-depth treatment of many of issues related to questionnaire design effect.

- *Checking*: Errors in the answers to the questions are checked by carrying out extensive checking during the interview. Most electronic questionnaires con tain *domain checks*. For example, for a question about the size of the household of the respondent, it is checked if the answer is a number within a specific range. Some CAI software systems also have the possibility to conduct *consistency checks*. For example, if the age of respondent is not older than 12 years, he/she cannot be married. If a problem arises, the interviewer or respondent is informed so that they can correct the error. Checks generally improve the quality of the collected survey data. Interactive checking is not possible for paper questionnaires.

- *Routing*: Electronic questionnaires may contain routing instructions. These instructions see to it that respondents only answer questions that are relevant to them and that they skip irrelevant questions. Usually routing forced for electronic questionnaires is used. For paper questionnaires, it is not possible to force the routing. Respondents may fail to follow instructions so that they answer the wrong questions or get lost in the questionnaire.

- *Timeliness*: The less time it requires to conduct a survey and to process the collected data, the more relevant the results will be. Computer-assisted modes of data collection will typically take less time than their paper analogues.

The next sections consider the various data collection modes in more detail. Criteria listed in this section are fully in line with the description of the modes.

## 6.2 Theory

### 6.2.1 FACE-TO-FACE SURVEYS

A face-to-face survey is a mode of data collection where interviewers visit the homes of the respondents or another location convenient for the respondent. Together, the interviewer and the respondent complete the questionnaire.

An advantage of face-to-face interviewing is that a visit of an interviewer can be longer than a telephone call. Usually, respondents allow interviewers to be in their homes for an hour or more. It is unlikely that a telephone call can take more than 30 minutes. Self-administered survey questionnaires should even be shorter. If more time is available for an interview, the questionnaire can ask more questions, but it becomes more complex.

Face-to-face surveys are expensive. The main reason is that interviewers do the data collection. They go from one address to the other. Only a part of their time is on interviewing. Travel also costs time. This limits the number of interviews they can do in one day. If there is no contact with persons at the selected address, callbacks are due. This will increase costs even more.

Paper questionnaires collection needs data to enter in the computer for further processing. Data entry is particularly costly and time consuming for large surveys. To clean the data, often a data editing procedure is applied. The collected data is checked, and detected errors are corrected. Data editing of large population surveys requires also substantial resources.

In case of a CAPI survey, interviewers are equipped with laptop computers. This is a major investment. Since data entering the computer is during the interview, no subsequent separate data entry is required. If, during the interview, also the entered data are checked, there is no need for a separate data editing procedure. On the one hand, use of and, on the other hand, integrating data entry and data editing in the interview decrease the costs.

For face-to-face surveys, the sampling frames usually consist of names and addresses. For example, Statistics Netherlands uses the population register for this purpose. This register covers the complete population, with the exception of illegal immigrants. The postal service in the Netherlands maintains a list of all addressed where a mail should be delivered. This list also covers the complete population. Thus, there are no coverage problems. In countries where no population register or address list is available, area sampling may be applied. This will not be the cause of serious coverage problems. Under-coverage may occur if incomplete sampling frames are used. An example is a telephone directory. Such a directory only contains registered telephone numbers. There will usually be a substantial amount of non-listed numbers. For example, only between 60% and 70% of the Dutch population is in the telephone directory.

An important aspect of face-to-face interviewing is that the interviewer and the selected person are together at the same location. This makes it easier for the interviewer to convince a person to participate. This is much more difficult to achieve by telephone or by sending written text. Consequently, response rates are higher for face-to-face surveys than for other types of surveys.

Goyder (1987) was one of the first to compare response rates of different modes of data collection. He analyzed several surveys conducted in the United States and Canada in the period from 1930 to 1980. The average response rate of face-to-face surveys was around 67%. The rate was lower for telephone surveys: 60%. The response was lowest for mail surveys: 58%. This result is similar to the one by Hox and de Leeuw (1994). They compared response rates of a number of surveys in Europe, the United States, and Canada. They found that face-to-face surveys had the highest response rates (on average 70%), followed by telephone surveys (67%) and mail surveys (61%).

The main difference between face-to-face and telephone surveys on the one hand and mail surveys on the other is the presence or absence of interviewers. The presence of interviewers can have advantages. They can provide additional information about the surveys and the questions. Thus, they can assist the respondents in answering the questions. This may help respondents in correctly interpreting questions. This reduces the risks of item nonresponse and therefore has a positive impact on the quality of the collected data.

Sending an advance letter has proved to increase response rates of face-to-face surveys. Such letters announce the visit of the interviewer. They also contain background information about the survey and explain why it is important to participate. They take away the surprise of an unexpected visit and provide legitimacy for the survey. Biemer and Lyberg (2003) note that an advance letter may also have a negative effect, as they give selected persons more time to think of an excuse not to participate. However, literature shows that the prevailing effect is that of higher response rates.

Physical impediments may lead to nonresponse. Doormen, gatekeepers, or locked gates may make it difficult or impossible for an interviewer to make contact with a person. Locked communal entrances or intercom systems prevent making face-to-face contact. As a result, the interviewer cannot show an identification card or a copy of the advance letter. It may even happen that the person is not living any more at that address and the interviewer is not informed.

Response order effects may occur in a face-to-face survey. The interviewer reads aloud the answer options for closed questions, making it difficult for respondents to remember all options. Since only the last few options are still in their short-term memory, they restrict their judgment to these options. As a result, there is a preference for options near the end of the list (recency effect). This effect reduces by using show cards. A *show card* contains the list of all possible answers to a question. It allows respondents to read the list at their own pace and select the answer reflecting their situation or opinion best.

*Acquiescence* is the tendency that respondents tend to agree with statements in questions, regardless of their content. They simply answer "yes." The reason is respondents only superficially think about the statement offered in the question. This will result in a confirmatory answer. Acquiescence is typically a problem for agree/disagree, true/false, or yes/no questions. The literature suggests that acquiescence is more common among respondents with a lower socioeconomic status. There is less acquiescence in self-administered surveys than in interviewer-assisted surveys. Respondents tend to agree more with statements made in questions if interviewers are present. Without interviewers, respondents may feel more anonymous and therefore will be more inclined to answer sensitive questions honestly. All this suggests that acquiescence may be a problem in face-to-face surveys.

Survey questions may ask respondents to express their opinion about changes. A typical example is a question that asks whether government should change its policy with respect to a specific issue. The easiest way to answer such a question without thinking is to select the option to keep everything the same. If the option of no change is not explicitly stated, not many respondents will insist on giving this answer. However, if this option is available, the number of people selecting it will increase substantially. *Endorsing the status quo* occurs more in self-administered surveys than in interviewer-assisted surveys. Therefore, there is expectation to be less of a problem in face-to-face surveys.

*Non-differentiation* may occur if respondents have to answer a series of questions with the same response options. The original idea was that this would make it easier for respondents to answer the questions. Changing the response options from question to question would increase the cognitive burden of respondents. However, keeping response options the same is also problematic because respondents tend to select the same answer for all these questions irrespective of the question content. Non-differentiation can be even more a problem if a series of questions with the same set of answer options is into a matrix question. See Figure 6.2 for an example. Respondents simply select all answers in the same column. This is often the column corresponding to the middle response option. This phenomenon is called *straight-lining*. Non-differentiation occurs more in self-administered surveys than in interviewer-assisted surveys. Therefore, it is not a serious problem in face-to-face surveys.

Respondents are sometimes not able to provide the required information, because they simply do not know the answer. It therefore seems reasonable to have "don't know" as one of the answer options. However, this may lead to a form of *satisficing* where respondents choose this option to avoid having to think about a real answer. It is common in CAPI/CATI software to have "don't know" implicitly. The screen shows only a list of substantive answer options. The interviewer reads to the respondents them. See Figure 6.4 for an example. If respondents insist they do not know the answer, the interviewer can record this by using a special key combination. Traditional face-to-face surveys might apply a more or less similar approach. Initially the respondents can choose from the list of substantive answers. If they insist they do not know the answer, the interviewer records the answer as "don't know" on the forms. An alternative approach is to start with a filter question. The first question for respondents asks whether they have an opinion about the specific issue. Only if they do, questions about it are administered. With these approaches, the treatment of "don't know" does not seem to be a problem in face-to-face surveys.

Respondents who want to avoid having to think about the proper answer may decide to pick just an *arbitrary answer*. This type of *satisficing* particularly occurs for a special type of question called the *check-all-that-apply question*. It is a closed question for which more than one answer can be selected. It is common practice for computer-assisted surveys to use square checkboxes for such questions. The respondents have to check all appropriate items in the list of answer options. This can be a lot of work. Instead of checking all relevant answers, they may just check some arbitrary answers and stop when they think they have checked enough answers. This problem typically occurs in mail and web surveys. It is not a problem for interviewer-assisted surveys where interviewers ask for each item in the list separately whether it applies to them. In fact, the check-all-that-apply question is replaced by a set of yes/no questions.

Interviewer-assisted surveys perform less well with respect to answering questions about sensitive topics. The presence of an interviewer may prevent a respondent to give an honest answer to a question about a potentially embarrassing topic.

Self-administered surveys (mail, web) perform better. Note that a sensitive question in CAPI may lead not only to item nonresponse but also to a *socially desirable answer*. See de Leeuw (2008) for more information.

The way the questionnaire is designed may have an effect on the way the questions are answered. This is particularly important for self-administered questionnaires. Respondents are usually not familiar with questionnaire forms. If it is not clear if and how questions must be answered, this may be a source of errors. Therefore, it is crucial to pay attention to questionnaire design. Design aspects are less important in interviewer-assisted surveys. The interviewers are in charge of navigating through the questionnaire, asking the questions, and recording the answers. They usually have received training to do their job. So, questionnaire design will not be much of an issue in face-to-face surveys.

It is not uncommon in computer-assisted surveys to have *consistency checks* in the questionnaire. These checks are carried out immediately after the relevant questions have been answered. If an inconsistency is detected, an error message is displayed on the screen. The interviewer discusses the problem with the respondent, which may result in correcting the answers to one or more questions. Research shows that these checks improve the quality of the collected data. Thus, it is good to have checks in the CAPI. It is not possible to implement extensive checking in paper questionnaire forms. This means that such a quality improvement cannot be obtained for face-to-face surveys with traditional paper questionnaire forms.

An important advantage of CAI is that *automatic routing* instructions can be included in the questionnaire. These instructions see to it that respondents only answer relevant questions, whereas irrelevant questions are skipped. This reduces the number of questions in the questionnaire. Moreover, it also avoids respondents to become irritated because they have to answer inapplicable questions. Automatic routing also reduces the workload of the interviewers. They are relieved of the task of making sure to follow the correct route through the questionnaire. In CAPI survey, automatic routing is implemented. It is impossible to do this in face-to-face surveys with paper questionnaires. Of course, printed routing instructions are available, but there are no guarantees that due to errors or confusion respondents end up in the wrong part of the questionnaire.

The fieldwork for face-to-face surveys is time consuming. Interviewers have to travel from one address to the others. This limits the number of interviews they can do on a day. There are substantial differences between computer-assisted data collection and data collection with paper forms in terms of time needed for subsequent processing. Traditional face-to-face data collection is in fact a sequential three-step process. First, there is data collection in the field, resulting in completed paper forms. Second, the data on the forms must be entered into the computer. Particularly for large surveys (in terms of number of questions and number of forms), this may take time. Third, the entered data has to be checked, and detected errors must be corrected. This also takes a considerable amount of time. As a result, the data are not yet ready for analysis straight after completion of the fieldwork. In contrast, a CAPI survey combines the three steps combined into one. The answers to the questions are immediately entered in the computer and checked during the interview. After completion of the fieldwork, almost no subsequent data processing is required. The data are ready for analysis. So timeliness is less of a problem for CAPI surveys.

Table 6.1 contains a summary of the effects of various phenomena in face-to-face surveys. A plus (+) indicates a positive effect or no negative effect and a minus (–) a negative effect.

**TABLE 6.1    Cost and quality aspects of face-to-face surveys**

| | | |
|---|---|---|
| Costs | – | Involving interviewers, who also have to travel, makes a face-to-face survey expensive. Use of laptops increases the costs even more |
| Coverage | + | Sampling frames for face-to-face interviewing generally do not exclude persons from the target population |
| Nonresponse | + | Face-to-face surveys have higher response rates than other modes of data collection |
| Response order effects | – | If the possible answers to a closed question are read out loud, there will be recency effects |
| Acquiescence | – | Respondents tend to agree with statements in questions, regardless of their content |
| Status quo endorsement | + | If interviewers are present, respondents are less inclined to take the easy way out (select no change) |
| Non-differentiation | + | If interviewers are present, respondents are less inclined to select the same answer for a set of questions |
| Answering "don't know" | + | Don't know is not explicitly offered, but it is accepted as an answer if respondents insist they do not know the answer |
| Arbitrary answer | + | Due to the presence of interviewers, respondents are less inclined to give an arbitrary answer |
| Socially desirable answers | – | Due to the presence of interviewers, respondents are more inclined to give socially desirable answers. This happens particularly for sensitive questions |
| Questionnaire design effects | + | Since the interviewers are in charge of completing the questionnaires, there will generally be no problems |
| Checking | +/– | A CAPI questionnaire with checks will improve the quality of the data. For face-to-face interviewing with paper questionnaires, the data may contain errors |
| Routing | +/– | A CAPI questionnaire with routing will improve the quality of data. There are no missing data due to answering the wrong questions. Routing errors may occur in paper questionnaires |
| Timeliness | – | The fieldwork of a face-to-face survey is time consuming. Data processing after the fieldwork is quicker for a CAPI survey than for a traditional survey |

## 6.2.2 TELEPHONE SURVEYS

A telephone survey is a mode of data collection where interviewers call selected persons by telephone. If contact is made with the proper person and this person agrees to cooperate, the interview is started and conducted over the telephone. Telephone interviewing has in common with face-to-face interviewing that there are interviewers asking the questions. This has advantages and disadvantages.

One disadvantage of telephone interviewing is that an interview cannot last as long as a face-to-face interview. To avoid problems, it is generally advised to limit the interview to at most 30 minutes.

CATI is the computerized form of telephone interviewing. It was one of the first modes of data collection to be computerized. From the point of view of the respondent, there is no difference between a traditional telephone survey and a CATI survey. They just answer questions by telephone without being aware of what is on the other end of the line: the interviewers can have a paper questionnaire form or a computer (or both).

An important component of a CATI survey is the call management system. Objective of such a system is to manage and schedule call attempts. It typically may involve the following tasks:

- Offering the interviewer the right telephone number at the right moment, taking into account possible appointments and the quota the interviewers can handle.
- Handling busy signals. Apparently, someone is at home. Therefore, the number is offered again after a short while (a few minutes).
- Handling no answers or answering machine calls. Typically, the number will be offered again at a different day and/or a different time of the day. For example, if there is no answer in the afternoon, the next call is in the evening.
- Managing appointments. If an interviewer makes an appointment with a respondent to call back at a specific date and time, the system must offer the call to this interviewer at the appropriate data and time.
- Producing summaries of the progress of the fieldwork.

Telephone surveys are expensive, but not as expensive as face-to-face surveys. The interviewers are a major cost component. Since they do not have to travel, but do their work in a call center, they can do more interviews per day than face-to-face survey interviewers. This implies that fewer interviewers are needed for a telephone survey. Moreover, there are also no travel costs.

Data collected by means of paper questionnaires have to be entered in the computer for further processing. This may require data entry personnel, particularly for large surveys. To clean the data, often a data editing procedure is carried out. The collected data are checked and detected errors are corrected. Data editing of large population surveys requires substantial resources.

There are two general approaches to sampling for telephone surveys. The first one is to use a list of telephone numbers. An example is a telephone directory. This sampling frame may suffer from serious under-coverage as many people have unlisted numbers. Particularly, numbers of mobile telephones may be missing. The rapidly increasing popularity of mobile phones makes the under-coverage problems even more substantial. For example, in the Netherlands, only about two-thirds of the telephone numbers can be found in the directory. Another problem is existence of *do-not-call registers*. Although interviewing over the telephone is not the same as selling products or services, many market research organizations avoid calling people in this register. For example, in the United Kingdom, the do-not-call registers contain more than 18 million households (in 2010), which is much more than half of approximately 26 million households. The conclusion can be the telephone directory becomes less and less useful as a sampling frame for a telephone survey.

Another approach to select a sample of telephone numbers is to apply *random digit dialing* (RDD). Taking into account the structure of telephone numbers, a computer algorithm generates random, but valid, telephone numbers. Such an algorithm will produce both listed and unlisted numbers. This guarantees complete coverage. However, also here telephone numbers in the do-not-call register should not be used.

RDD also has drawbacks. In some countries it is not clear what an unanswered number means. It can mean that the number is not in use, which is a case of over-coverage. Then no follow-up is needed. It can also mean that someone simply does not answer the phone, which is a case of nonresponse that has to be followed up.

Like in face-to-face surveys, telephone survey interviewers can use their skills to convince persons to participate. Another advantage is that interviewers can supply additional information about the survey and its questions. They can assist respondents in finding the proper answer to a question. The risk of item nonresponse is therefore reduced. Generally, this results in high response rates, although they may be somewhat lower than in face-to-face surveys. Interviewer assistance can also have a positive effect on the quality of the collected data.

All kinds of additional information can be used in the attempt to get contact and to obtain cooperation. Information like composition of the household or the ages of its members may help choose an optimal contact strategy. The amount of available information depends partly on the way in which the sample was selected. For example, Statistics Netherlands selects its samples for telephone surveys from the population register. Next, telephone numbers are added to the selected names/addresses by the telephone company. So all register information (gender, age, marital status, location) is available for everyone in the sample.

If the sampling frame contains address information, it is possible to send an advance letter. Such a letter announces the telephone call of the interviewer and therefore takes away the surprise of an unexpected call. The letter should provide information about the survey and explain why it is important to participate. To be effective, advance letters should also mention the research agency, and they must have an official letterhead. Such advance letters will help to increase response rates.

It should be noted that only listed telephone numbers can be linked to addresses and other sampling frame information. Unfortunately, not every telephone number is listed. This is a serious case of under-coverage. Note that if the sample is selected by means of RDD, no information at all is available about the selected addresses.

The fast rise of mobile telephones has not made it easier to conduct telephone surveys. Landline telephones are increasingly replaced by mobile telephones. Landline telephones are typically a means to obtain contact with households, whereas mobile telephones are linked to specific persons. Therefore, the probability of contacting a member of the household is higher for landline telephones. Moreover, if persons can only be contacted through their mobile telephones, it is often in situations that are not fit for interviewing. An additional problem is that sampling frames (telephone directories) in many countries do not contain mobile telephone numbers. A final complication to be mentioned here is that in countries, such as the Netherlands, Italy, and other European countries, people often switch from one telephone provider to another. However, it is sometimes possible, but it is not the rule, to keep the original number. For more information about the problems and possibilities of mobile telephones for interviewing, see Peterson et al. (2017) and Kuusela, Vehovar, and Callegaro (2006).

If RDD is used to select the sample, there is no information at all about the selected persons. This does not help the interviewers in their preparation of calls. It becomes more difficult to persuade reluctant persons. Lack of information from a sampling frame also makes nonresponse correction (for example, adjustment weighting) more difficult.

There may be physical impediments preventing interviewers making contact with respondents. One example is an answering machine. People may be at home, but still have the answering machine switched on. It is not clear whether it is good idea for the interviewers to leave a message. It may or may not help to get into contact at a next attempt. Groves and Couper (1998) note that sometimes answer machines give some relevant information about the people living at the address. This information may be useful for a next contact attempt.

If no contact is established with a selected person, and possibly also in case of an initial refusal, one or more subsequent attempts may be made to obtain participation. Fortunately, repeated call attempts are not so expensive (as compared with face-to-face surveys). So, it is relatively easy to do this in practice. It is not uncommon that survey agencies may make six call attempts or more before the case is closed as nonresponse due to no contact.

Some CATI systems (for example, Blaise) distinguish *call attempts* and *contact attempts*. A contact attempt consists of a series of call attempts within a short time interval. Several contact attempts, each with several call attempts, are made before a case is closed as nonresponse. For example, Statistics Netherlands makes at most three or four contact attempts each consisting of at most three call attempts. The time interval for contact attempts is one hour in case of no answer and five minutes in case of a busy number.

Many telephone companies have a *calling number identification* (CNID) service. It transmits the survey organization's number to the telephone of the selected person during the ringing signal. This information becomes visible on the telephone of this person. If persons do not recognize the number, or if the number is shown as an "unknown number," they may decide not to pick up to phone. Thus, CNID may be a cause of nonresponse.

Response rates of telephone surveys suffer from telemarketing activities. Typically people are called around dinner time (when the contact probability is high because people are at home) in attempts to sell products or services. This spoils the survey climate. This phenomenon has led to a more hostile attitude toward interviewers. It is therefore important that interviewers make clear at the very start of the interview that they are not selling anything. In some countries (for example, the United States, the United Kingdom, and the Netherlands), there are *do-not-call registers*. When people register, they will not be called any more by telemarketing companies. Such registers may help to improve the survey climate.

Incentives may help to increase response rates. Research has shown that incentives are most effective when they are given before the interview and not after it. To be able to give incentives, the addresses of the respondents must be available. This is typically not the case for RDD surveys. So the possibilities for giving incentives in RDD surveys are limited.

An effective call management system may reduce nonresponse due to noncontact. Models may even be developed to predict the optimal time to call. Of course, this requires auxiliary information about the selected persons.

Call management systems are able to display results of earlier attempts and other information on the computer screen of the interviewers. This information may help in persuading reluctant respondents. Thus the refusal rate may go down. See Wagner (2008) for details tuning CATI call management systems with the objective to reduce the nonresponse rate.

*Response order effects* can occur in telephone surveys. Since the interviewer reads out loud the answer options for a closed question, it is difficult for respondents to remember all options. Because it will be likely that only the last few options remain in their short-term memory, they restrict their judgment to these options. As a result, there is a preference for options near the end of the list (*recency effect*). In case of face-to-face surveys, this effect may be reduced by using show cards. This is not possible in telephone surveys.

Telephone survey interviews may suffer from *acquiescence* (the tendency that respondents tend to agree with statements in questions, regardless of their content). They only superficially think about the statement offered in the question. They simply give a confirmatory answer. Acquiescence is typically a problem for agree/disagree, true/false, or yes/no questions. Acquiescence seems to be more common among respondents with a lower socioeconomic status. Typically acquiescence occurs in interviewer-assisted surveys. Respondents tend to agree more with statements made in questions if interviewers are present. Without interviewers, respondents may feel more anonymous and therefore will be more inclined to answer sensitive questions honestly. All this suggests that acquiescence may be a problem in telephone surveys.

Survey questionnaires may contain questions asking for opinions about changes. For example, a question asks whether government should change its policy with respect to a specific issue. The easiest way to answer such a question without thinking is to say that everything should remain as it was. If the "no change" option is not explicitly offered, not many respondents will insist on giving this answer. However, if this option is available, the number of people selected will substantially increase. *Endorsing the status quo* occurs more in self-administered surveys than in interviewer-assisted surveys. Therefore it will be less of a problem in telephone surveys.

*Non-differentiation* may occur if respondents have to answer a series of questions with the same response options. This approach is problematic because respondents tend to select the same answer for all these questions irrespective of the question content. Non-differentiation can be even more a problem if a series of questions with the same set of answer options is combined into a matrix question. See Figure 6.2 for an example. Respondents simply select all answers in the same column. This is often the column corresponding to the middle response option. This phenomenon is called *straight-lining*. Non-differentiation occurs more in self-administered surveys than in interviewer-assisted surveys. Therefore it is not a serious problem in telephone surveys.

Respondents are sometimes not able to give an answer, because they simply do not know the answer. Therefore, it seems reasonable to have "don't know" as one of the answer options. However, this may lead to *satisficing*: respondents select "don't know" to avoid having to think about a real answer. It is common in CATI software to offer "don't know" implicitly. Only a list of substantive answer options is shown on the screen and read out to the respondents. See Figure 6.4 for an example. If respondents insist they do not know the answer, the interviewer can record this by using a special key combination. A more or less similar approach can be followed for traditional telephone surveys. First, a list of substantive answers is read out. If a respondent insists he does not know the answer, it is recorded as "don't know" on the form by the interviewer. An alternative approach is to use a filter question. First, respondents are asked whether they have an opinion about a specific issue. Only if they do, their opinion is asked about it. With these approaches, the treatment of "don't know" does not seem to be a problem in telephone surveys.

If respondents do not want to think about a proper answer, they may decide to just pick an arbitrary answer. This type of *satisficing* particularly occurs in *check-all-that-apply questions*. It is a closed question for which more than one answer can be selected. It is common practice for computer-assisted surveys to use square checkboxes for such questions. See Figure 7.25 in Chapter 7 for an example. The respondents are asked to check all appropriate items in the list of answer options. This can be a substantial cognitive task. Instead of checking all relevant answers, respondents may just check some arbitrary answers and stop when they think they have checked enough answers. This problem typically occurs in mail and web surveys. It is not a problem for telephone surveys where interviewers ask for each item in the list separately whether it applies to them. This comes down to splitting one check-all-that-apply question into a set of yes/no questions.

Interviewer-assisted surveys perform less well with respect to answering questions about sensitive topics. The presence of an interviewer may prevent a respondent to give an honest answer to a question about a potentially embarrassing topic. Self-administered surveys (mail, web) perform better. Note that a sensitive question may lead not only to item nonresponse but also to socially desirable answer. See de Leeuw (2008) for more information. So giving *socially desirable answers* may be a problem in telephone surveys.

Questionnaire design will not be an important issue in telephone surveys. Interviewers are in charge of navigating through the questionnaire, asking the questions, and recording the answers. They usually have received training to do their job.

Computer-assisted surveys may have *consistency checks* in the questionnaire. This helps to detect errors already during the interview. Problems can be repaired immediately. This improves the quality of the collected data. Therefore checks should be included in a CATI survey. It is not possible to implement extensive checking in paper questionnaire forms. This means that such a quality improvement cannot be obtained for telephone surveys with traditional paper questionnaire forms.

It is possible to make use of *automatic routing* in CAI questionnaires. Respondents will only have to answer relevant questions, whereas irrelevant questions are skipped. This reduces the number of questions that have to be answered. Moreover, it also avoids respondents to become irritated because they have to answer inapplicable questions. Automatic routing also reduces the workload of the interviewers. They are relieved of the task of making sure the correct route is followed through the questionnaire. Automatic routing can be implemented in CATI surveys. It is impossible to do this in paper questionnaires. Printed routing instructions can be included, but there is always a risk that respondents end up in the wrong part of the questionnaire.

The fieldwork of a telephone survey is not so time consuming. The length of the fieldwork period is determined by the number of cases that can be handled on a day. This depends on the size of the interviewer crew and the success rate of the call attempts. If paper forms are used, subsequent data processing will take time. This includes data entry and data editing. In case of a CATI survey, the answers to the questions are immediately entered in the computer and checked during the interview. After completion of the fieldwork, almost no subsequent data processing is required. So timeliness is even less of a problem for CATI surveys.

Table 6.2 contains a summary of the effects of various phenomena in telephone surveys. A plus (+) indicates a positive effect or no negative effect and a minus (–) a negative effect.

TABLE 6.2     **Cost and quality aspects of telephone surveys**

| | | |
|---|---|---|
| Costs | − | Involving interviewers makes a telephone survey expensive (but not as expensive as a face-to-face survey) |
| Coverage | − | Telephone surveys may suffer from severe under-coverage. This may be caused by unlisted telephone numbers and numbers in do-not-call registers |
| Nonresponse | + | Telephone surveys have higher response rates than self-administered modes of data collection |
| Response order effects | − | If the possible answers to a closed question are read out loud, there will be recency effects |
| Acquiescence | − | Respondents tend to agree with statements in questions, regardless of their content |
| Status quo endorsement | + | If interviewers are present, respondents are less inclined to take the easy way out (select no change) |
| Non-differentiation | + | If interviewers are present, respondents are less inclined to select the same answer for a set of questions |
| Answering "don't know" | + | Don't know is not explicitly offered, but it is accepted as an answer if respondents insist they do not know the answer |
| Arbitrary answer | + | Due to the presence of interviewers, respondents are less inclined to give an arbitrary answer |
| Socially desirable answers | − | Due to the presence of interviewers, respondents are more inclined to give a socially desirable answer. This happens particularly for sensitive questions |
| Questionnaire design effects | + | Since the interviewers are in charge of completing the questionnaires, there will generally be no problems |
| Checking | +/− | A CATI questionnaire with checks will improve the quality of the data. For telephone interviewing with a paper questionnaire, the data may contain errors |
| Routing | +/− | A CATI questionnaire with routing will improve the quality of data. There are no missing data due to answering the wrong questions. Routing errors may occur in paper questionnaires |
| Timeliness | + | The fieldwork of a telephone survey is not so time consuming. Data processing after the fieldwork is quicker for CATI than for traditional telephone surveys |

## 6.2.3  MAIL SURVEYS

Respondents complete the questionnaire themselves in mail surveys. There are no interviewers asking the questions and recording the answers. The respondents themselves read the questions and record the answers. Consequently, there are no interviewers attempting to persuade a reluctant person to fill in the form. There are no interviewers to explain unclear aspects and to assist in answering the questions. As a result, response rates and data quality will generally be lower in mail surveys than in face-to-face surveys and telephone surveys.

Mail surveys are much cheaper than interviewer-assisted surveys. The reason is the absence of interviewers. They are usually a major cost component. Mail surveys have other costs, such as printing costs and mail costs. These costs are much lower than interviewer costs.

Completed questionnaire forms will be returned to the survey organization. The next step is to enter the collected data into the computer for further processing. This may require data entry personnel, particularly for large surveys. Mail questionnaire forms will not be without errors. One of the advantages of CAI is that the answers can be checked during the interview. This is not possible for paper questionnaire forms. Therefore, often a data editing procedure is carried out. The collected data is checked and detected errors are corrected. Data editing of large population surveys requires substantial resources.

A list of addresses is required to be able to select a sample for a mail survey. Coverage problems depend on the extent to which this list covers the target population of the survey. For general population surveys, such a list may be obtained from the postal service organization. They often maintain a list of points where post can be delivered. Usually home addresses can be distinguished from business addresses. A disadvantage of these address lists is that they do not contain the names of the people living there. Therefore, their name cannot be included in the address on the envelope. Instead, a phrase like "The residents of ..." has to be used. According to Dillman, Smyth, and Christian (2009, 2014), such a less personalized approach may increase nonresponse. Usually the coverage of address lists is good.

In countries like the Netherlands and the Scandinavian countries, the sample can be selected from the population register. Such a register contains both names and addresses. This makes it possible to implement a personalized approach.

Several literature overviews indicate that response rates of mail surveys are generally lower than those of interviewer-assisted surveys. See, for example, Goyder (1987) and Hox and de Leeuw (1994). This is mainly caused by the absence of interviewers. Their efforts usually have a positive effect on response rates.

Nonresponse occurs in a mail survey if the questionnaire form is not returned to the survey agency. There is usually no information at all about the reasons for nonresponse. Here are some examples of what the reasons for nonresponse could be:

- The letter did not arrive at the indicated address (no contact).
- The people were not at home during the survey period (no contact).
- The letters were received but ignored (refusal).
- The people at the address did not understand the language the letter was written in (not able).

Bethlehem et al. (2011) stress the importance of distinguishing different types of nonresponse. Every type of nonresponse can have a different impact on survey outcomes and therefore should require its own treatment. Unfortunately, this is not possible to distinguish different types of nonresponse in mail surveys. This makes it more difficult to correct for the effects of nonresponse in such surveys.

To obtain a reasonable response rate, special efforts are required for contacting people and persuading them to participate in the survey. Furthermore, the design of the questionnaire is very important, as well as all other procedures, like advance letters, cover letters, reminders, and incentives. Dillman, Smyth, and Christian (2009, 2014) describe this in detail.

The mail to the respondents must contain not only the questionnaire form but also a cover letter. This letter should explain why participation is important and for what purpose the data will be used. The letter should preferably come from a high official in the organization. The letter should not look like it has been photocopied, but resemble an original letter (including colored letterheads and a signature). It will also help increase the response if the letter contains a clear statement that all collected data will be treated as confidential.

It should be as easy for the respondents to return the completed questionnaire. Therefore, the enclosing of a postage-paid return envelope in the letter to the sampled persons is recommended.

Reminders are important for increasing the response rate. If there is no response after two to four weeks, a follow-up letter is sent. It reminds persons that they have to send the completed questionnaire back. It is urgent they do it. This reminder letter should contain a replacement questionnaire (in case they lost the original copy).

An incentive can be included in the first letter to the sample persons. Incentives increase response rates. They work best when sent in advance. Examples of incentives are cash payment, lottery tickets, postage stamps, pens, or a donation to a charity organization in the respondent's name. Some research indicates that donations to charity organizations seem not to work as well as real monetary incentives.

Do not use long questionnaires. There is empirical evidence that long questionnaires reduce the response rate. For example, Dillman, Sinclair, and Clark (1993) show that shorter questionnaire forms increase response in census mail surveys.

A final practical advice is to avoid periods in which there is other heavy mail traffic. Examples are the period just before tax form submission deadline and the period before the Christmas holiday.

A special form of a mail survey is one that uses questionnaire drop-off and pickup. Personal delivery of a questionnaire by only slightly trained survey takers may increase the response. This approach also provides more information about nonrespondents.

*Response order effects* may occur for closed questions. Respondents have to pick the proper answer from a list of possible answer options, sometimes long. Instead of thinking carefully about the appropriate answer, they chose the first reasonable option. Mail survey questions may suffer from a special type of response order effect called *primacy effect*. This is the tendency to pick an answer early in the list of options. Reading to a list of possible options and considering each option require a considerable effort. Therefore, respondents may stop at the first reasonable option.

Interviewer-assisted surveys may suffer from *acquiescence*. This is the tendency to agree with statements in questions, regardless of their content. Without interviewers, respondents may feel more anonymous and therefore will be more inclined to answer sensitive questions honestly. This suggests that acquiescence is not a problem in mail surveys.

Surveys sometimes ask questions opinions about changes. The easiest way to answer is to say that everything should remain as it was. Not many respondents will insist on answering "no change" if this option is not offered. If this option is available, the number of people selected will substantially increase. *Endorsing the status quo* occurs more in self-administered surveys than in interviewer-assisted surveys. Therefore, it may be a problem in mail surveys.

*Non-differentiation* occurs if respondents have to answer a series of questions with the same response options. Respondents tend to select the same answer for all these questions irrespective of the question content. Non-differentiation can be even more a problem if a series of questions with the same set of answer options is into a matrix question (see Figure 6.2). Respondents simply select all answers in the same column (*straight-lining*). This is often the column corresponding to the middle (neutral) response option. Non-differentiation occurs more in self-administered surveys than in interviewer-assisted surveys. Therefore, it may be a problem in mail surveys.

The treatment of "don't know" in mail surveys requires careful consideration. The offer of "don't know" explicitly as one of the answer options has the advantage that respondents not knowing the answer can answer so. This approach accepts the existence of a group of persons that cannot answer the questions, and thus "don't know" value is like a substantive answer. This approach may suffer from *satisficing*. People not wanting to think about an answer or not willing to provide an answer have an escape in "don't know." Several authors have shown that explicitly offering "don't know" substantially increases the percentage of respondents choosing this option. See, for example, Sudman and Bradburn (1982).

To avoid satisficing, one may decide not to offer the option "don't know." Consequently, respondents always have to provide a substantive answer, even if they do not know the answer. According to Couper (2008), this violates the norm of voluntary participation and therefore may frustrate respondents, resulting in nonresponse. Dillman (2007) also strongly recommends not forcing respondents to answer. He warns for negative effects on respondent motivation, data quality, and response.

Satisficing reduction might be achieved by using a filter question. See Krosnick (1991) and Schuman and Presser (1981). This filter question asks whether respondents have an opinion about a specific issue. Only if they say they have, a next question asks what their opinion really is.

If respondents do not want to think about an answer, they may decide to show just an *arbitrary answer*. This type of satisficing particularly occurs in check-all-that-apply questions. Instead of checking all relevant answers, they may just check some arbitrary answer options and stop when they think they have checked enough. This problem typically occurs in self-administered surveys. Thus, it can be a problem in mail surveys.

A mail survey may perform better than an interviewer-assisted survey with respect to answering questions about sensitive topics. The absence of an interviewer may encourage a respondent to give an honest answer to a question about a potentially embarrassing topic. Therefore, there is a lower risk of *socially desirable answers*.

Questionnaire design is of crucial important in mail surveys. There are no interviewers assisting respondents in navigating through the questionnaire, asking the questions, and explaining and recording the answers. The respondents are completely on their own. Unclear questions and navigation may result in wrong answers or no answers at all.

A lot of attention to the design of the questionnaire form is crucial. It must look attractive to the respondents. The more personalized it is, the better it works, and the more likely it is that respondents will complete it. The clearer it is, the less likely it is they will get confused resulting in errors.

Navigation instructions (*routing instructions*, skip patterns) are a potential source of error. If respondents fail to follow the correct route through the questionnaire, wrong questions are answered, and right questions are left unanswered, which comes down to item nonresponse. Navigation instructions must be clear and unambiguous. Dillman, Smyth, and Christian (2009, 2014) advise to indicate jumps to other questions by arrows or other graphs with a similar meaning. See Figure 6.2 for an example.

Computer-assisted surveys may have *consistency checks* in the questionnaire. This helps to detect errors already during the interview and problems repaired immediately. This improves the quality of the collected data. It is not possible to implement extensive checking in paper questionnaire forms. This means that mail surveys do not benefit of such a quality improvement.

It is possible to make use of *automatic routing* in CAI questionnaires. Respondents will only have to answer relevant question, skipping irrelevant questions. This reduces the number of questions to answer. Moreover, it also avoids respondents to become irritated because they have to answer inapplicable questions. Automatic routing also reduces the workload of the interviewers. They are relieved of the task of making sure to follow the correct route through the questionnaire. It is impossible to do this in mail surveys. Of course, printed routing instructions can be included in the questionnaire form, but there is always a risk that respondents end up in the wrong part of the questionnaire.

The fieldwork of a mail survey may be time consuming. It takes time to send the empty questionnaire forms to the respondents. Respondents tend be slow in returning the completed forms. Often reminders have to be sent several times. It takes many weeks. Subsequent data processing will take also time. This includes data entry and data editing.

TABLE 6.3   **Cost and quality aspects of mail surveys**

| | | |
|---|---|---|
| Costs | + | Due to the absence of interviewers, costs are relatively low |
| Coverage | + | Available address lists have a good coverage of the population |
| Nonresponse | – | Mail surveys have lower response rates than interviewer-assisted modes of data collection |
| Response order effects | – | Answers to a closed question read by the respondent generate a primacy effects |
| Acquiescence | + | The tendency to agree, regardless of their content, with statements in questions is less for self-administered modes of data collection |
| Status quo endorsement | – | If no interviewers are present, respondents are more inclined to take the easy way out (select no change) |
| Non-differentiation | – | If no interviewers are present, respondents are more inclined to select the same answer for a set of questions |
| Answering "don't know" | – | It is generally advised to offer "don't know" as one of the answer options. Particularly for opinion questions, respondents may use this option as an escape for not giving substantial answer |
| Arbitrary answer | – | If no interviewers are present, respondents are more inclined to give an arbitrary answer |
| Socially desirable answers | + | If no interviewers are present, respondents are less inclined to give a socially desirable answer. This happens particularly for sensitive questions |
| Questionnaire design effects | – | Questionnaire design is critical in mail surveys. A suboptimal design may have severe consequences |
| Checking | – | It is not possible to include checks in paper questionnaire forms for mail surveys |
| Routing | – | Routing takes the form of printed instructions for the respondents. There is no guarantee the proper route will be followed |
| Timeliness | – | Collecting the completed questionnaire forms may take time. Reminders take even more time. Data processing is slower for mail surveys than for computer-assisted surveys |

Table 6.3 contains a summary of the effects of various phenomena in mail surveys. A plus (+) indicates a positive effect or no negative effect and a minus (–) a negative effect.

## 6.2.4 WEB SURVEYS

The basic feature of a web survey is that the design of the questionnaire is as a website accessed by respondents. Web surveys combine some of the aspects of self-administered surveys and computer-assisted surveys. On the one hand, a web survey looks like a mail survey, but with the questionnaire on the computer screen instead of on paper. On the other hand, a web survey may include facilities like error checking and routing.

Web surveys have become very popular in a short time. Not surprising, as web surveys seem to have some attractive advantages: (1) it is a simple means to get access to a large group of potential respondents, (2) questionnaire distribution is at very low costs, (3) surveys can be launched very quickly, and (4) they offer new, attractive possibilities, such as the use of multimedia (sound, animation, and movies). So web surveys appeal to be a fast, cheap, and attractive means of collecting large amounts of data. In some survey situations, a web survey is indeed an effective mode of data collection that produces high-quality data. However, conducting a web survey in a scientifically sound way is sometimes not so easy, particularly in case of large population surveys. Some of the advantages may be not so clear any more.

A web survey can be a cheap data collection instrument. There are no interviewers involved, and there are no mailing and printing costs. Of course, there are hardware and software costs, but a well-designed survey infrastructure can be used for many different surveys. The costs may rise for sample selection. If a list with e-mail addresses is available, sample selection is cheap and straightforward. The situation becomes more difficult for a general population survey where a sample selection is necessary. There is no sampling frame with e-mail addresses. Thus, a different mode to recruit persons for the survey is required. One way to realize this is to send a letter to the sample persons containing a link to the website and a unique identification code to start the questionnaire. If the persons do not respond, a telephone call is attempting to encourage them to complete the questionnaire. If this fails, visiting them at home is considered. It will be clear that such an approach implies a substantial rise of the costs of the web survey.

Depending on the target population of the survey, under-coverage may be a serious problem. In simple situations, like a survey among employees of a company or students at a university, there usually is a list of e-mail addresses. These lists cover the population completely. However, in many situations, the target population is usually wider than the Internet population. This typically applies to general population surveys. For example, the Internet coverage varies between 30% and 90% in European countries. Broadband Internet access is even lower. This prevents web surveys with advanced features (requiring broadband). See Chapter 10 for more details.

Unfortunately, population access to the Internet is unevenly distributed. A typical pattern found in many countries is that elderly, the low-educated, and ethnic minorities are severely underrepresented among those having access to the Internet.

If under-coverage in a web survey really is a problem, a possible solution could be to simply provide Internet access to those without Internet. An example of this approach is the LISS panel, described by Scherpenzeel (2008). This web panel has been constructed by selecting a random sample of households from the population register of the Netherlands. Selected households were recruited for this panel by means of CAPI or CATI. Cooperative households without Internet access were provided with equipment giving them access to the Internet.

Web surveys suffer from nonresponse. Since a web survey is a self-administered survey, they have a potential of high nonresponse rates. An additional source of nonresponse problems are technical problems of respondents having to interact with the Internet, see, e.g., Couper (2000), Dillman and Bowker (2001), Fricker and Schonlau (2002), and Heerwegh and Loosveldt (2002). Respondents need a browser to open and complete a web survey questionnaire. There are many different browsers available. Examples are Internet Explorer, Firefox, Safari, Opera, and Google Chrome. These browsers do not behave exactly in the same way. Therefore, a questionnaire may behave differently in different browsers. A feature may not even work in specific browser, preventing respondents to record their answers to the questions. This may result in nonresponse.

The Internet is in continuous development. Specific features may work in the most recent version of a browser, but not in earlier versions. Unfortunately, not all people have the latest version of their browser installed. Again, as a result, a specific feature in the questionnaire may or may not work. Some questionnaire features (for example, use of animation and video) require an Internet connection with substantial bandwidth, but not every Internet user has a broadband Internet connection. So, these features will not work properly on their computer.

Slow modem speeds, unreliable connections, high connection costs, low-end browsers, and incompatible browsers may frustrate respondents, often resulting in prematurely interrupting completion of the questionnaire. In order to keep the survey response up to an acceptable level, every measure must be taken to avoid these problems. This requires a careful design of web survey questionnaire instruments.

*Response order effects* may occur in web surveys. Instead of thinking carefully about the appropriate answer, the first reasonable option of a closed question is chosen. Web survey questions may suffer from a special type of response order effect called *primacy effect*. This is the tendency to pick an answer early in the list of options. Instead of reading the list of possible options and considering each option, respondents may stop at the first reasonable option.

Interviewer-assisted surveys may suffer from *acquiescence*. This is the tendency to agree with statements in questions, regardless of their content. Without interviewers, respondents may feel more anonymous and therefore will be more inclined to answer sensitive questions honestly. Therefore, acquiescence is not a problem in web surveys.

Survey questions sometimes ask opinions about changes. The easiest way to answer is to say that everything should remain as it was. Therefore, if the there is a "no change" option, many respondents will choose it. This phenomenon of *endorsing the status quo* typically occurs in self-administered surveys. Thus, it may be a problem in web surveys.

*Non-differentiation* occurs if respondents have to answer a series of questions with the same set of response options. Respondents tend to select (without thinking) the same answer for all these questions irrespective of the question content. Non-differentiation can be even more a problem if a series of questions with the same set of answer options is into a matrix question. Respondents simply select all answers in the same column (*straight-lining*). This is often the column corresponding to the middle (neutral) response option. Non-differentiation occurs more in self-administered surveys than in interviewer-assisted surveys. Therefore it may be a problem in web surveys.

There are various approaches possible to offer the option "don't know" as an answer to a question. On the one hand, it can be explicitly shown in one of the options. This will encourage *satisficing*. Many will see this option as an escape for having to think about the proper answer. On the other hand, one may decide to not offer the option "don't know." Then respondents always have to provide a substantive answer, even if they do not know the answer. Experts advise against this approach as it may lead to nonresponse.

Vis-Visschers et al. (2008) investigated an approach whereby respondents were first offered the question with only substantive answer options. If they did not select an option and attempted to skip the question, the question was offered again, but then "don't know" was included in the list. It turned out that some respondents did not understand this mechanism, as they complained that they could not answer "don't know." The other respondents less frequently selected "don't know."

Derouvray and Couper (2002) experimented with a similar approach. The answer option "don't know" was not offered for the question. If respondents attempted to skip the question without answering it, a new screen appeared offering two choices: (1) go back and answer the question, and (2) record the answer as "don't know" and proceed to the next question. This approach resulted in the lowest "don't know" rates.

If respondents do not want to think about an answer, they may decide to just pick an *arbitrary answer*. This type of *satisficing* particularly occurs in *check-all-that-apply questions*. Instead of checking all relevant answers, they may just check some arbitrary answer options and stop when they think they have checked enough. This problem typically occurs in self-administered surveys. So it can be a problem in web surveys.

A web survey may perform better than an interviewer-assisted survey with respect to answering questions about sensitive topics. The absence of an interviewer may encourage a respondent to give an honest answer to a question about a potentially embarrassing topic. Therefore, there may be less *socially desirable answers* in web surveys.

Designing a questionnaire for a web survey is to some extent similar to designing it for a mail survey. At first sight, one could say that a web survey questionnaire is nothing more than a paper form displayed on a computer screen. There are, however, also differences that may affect the answers to the questions. For example, a paper questionnaire page usually does not fit on the computer screen. This means the respondent has to scroll to see all parts of the page. Failing to do so may mean that some (not visible) questions are skipped, resulting in item nonresponse.

The designer of a web survey has the choice to display just one question per screen or to put more questions on the screen. One question per screen may be more appropriate if the questionnaire is to contain extensive routing instructions (skip patterns). However, this will increase the perceived length of the questionnaire, possibly resulting in (partial) nonresponse.

There are many more design issues that may affect response to a web survey. See Chapter 4. A detailed description is also given by Couper (2008).

CAPI and CATI survey questionnaires may contain extensive edit checks. These checks help to detect and correct inconsistencies in the answers of respondents. Such checks are not possible in mail surveys, and this may have a negative effect on the quality of the collected data. The designer of a web survey questionnaire can decide to include edit checks. This should increase data quality, but many (unfriendly) error messages may scare away respondents. So, reporting and treating errors should be implemented carefully.

Respondents are completely free in the way they complete a paper questionnaire. They can answer any question they like in any order. Web questionnaires can be designed such that routing is enforced like in CAPI or CATI surveys. This sees to it that only relevant questions are answered and irrelevant ones are skipped. Routing can only be implemented in a meaningful way if the questionnaire is processed on the basis of one question per screen (the *question-based* approach). It is difficult, if not impossible, to implement routing for a *form-based* approach.

Conducting a web survey is not time consuming. There are no interviewers involved. Questionnaires do not have to be sent by mail. There is no separate data entry phase. If the web survey questionnaire contains checks, there is no need for a subsequent data editing phase. In case of web panels, it is even possible to conduct a survey in one day. Carrying out a web survey may become more time consuming if there is no proper sampling frame. If respondents have to be recruited by means of a telephone or face-to-face survey, the survey will take more time.

Table 6.4 contains a summary of the effects of various phenomena in web surveys. A plus (+) indicates a positive effect or no negative effect and a minus (−) a negative effect.

## 6.2.5 MOBILE WEB SURVEYS

Mobile web survey is any survey taken on a mobile device like a smartphone or a tablet computer. A mobile web survey typically offers a mobile participant the opportunity to complete a web survey from a mobile device (tablet or smartphone), rather than from a PC.

Interest in mobile web surveys is growing and is related to the massive increasing diffusion of mobile devices in most countries. Mobile web surveys are a form of mixed mode naturally related to web surveys. The basic situation is that mobile devices connected to the Internet (especially smartphones) have a high penetration in the population. According to comScore, smartphones made up 67% of the mobile market. Smartphone users tend to be attached to their phones. Most smartphone users check their phones first thing in the morning and frequently throughout the day. Smartphones provide users with social connectivity, work productivity, and entertainment. Because they're never without their phones,

TABLE 6.4     Cost and quality aspects of web surveys

| | | |
|---|---|---|
| Costs | + | There are no interviewer, printing, and mailing costs. Recruitment can be expensive without a good sampling frame |
| Coverage | – | There is serious under-coverage in many situations |
| Nonresponse | – | Web surveys have lower response rates than interviewer-assisted modes of data collection |
| Response order effects | – | If the possible answers to a closed question are read by the respondent, there will be primacy effects |
| Acquiescence | + | The tendency to agree with statements in questions, regardless of their content, is less for self-administered modes of data collection |
| Status quo endorsement | – | If no interviewers are present, respondents are more inclined to take the easy way out (select no change) |
| Non-differentiation | – | If no interviewers are present, respondents are more inclined to select the same answer for a set of questions |
| Answering "don't know" | + | It is generally advised to offer "don't know" as one of the answer options. Particularly for opinion questions, respondents may use this option as an escape for not giving a substantial answer |
| Arbitrary answer | – | If no interviewers are present, respondents are more inclined to give an arbitrary answer |
| Socially desirable answers | + | If no interviewers are present, respondents are less inclined to give a socially desirable answer. This happens particularly for sensitive questions |
| Questionnaire design effects | – | Questionnaire design is critical in web surveys. A suboptimal design may have severe consequences |
| Checking | + | It is possible to include checks in web surveys. Presenting and treating errors is critical |
| Routing | + | Automatic routing can be implemented for the question-by-question implementation |
| Timeliness | + | A web survey can be conducted quickly. However, recruitment by means of mail/CAPI/CATI may take time |

smartphone users are quick to respond. This makes a survey fast and contributes to a higher response rate.

Therefore, when a person is checking his/her e-mail on the smartphone, he/she might receive the e-mail invitation for the survey on the smartphone whatever he/she is doing. He/she could decide to complete the survey soon or to do it later or directly refuse to participate. Independently from the decision of the researcher, it is up to the contacted sampled interviewee answering via smartphone or other mobile device.

The researcher needs therefore to take care that the web questionnaire is well adapted to the mobile device screens and other technical matters; otherwise the collected data could be affected from relevant errors. A mobile optimized survey is a survey easy to read and respond to using a small touchscreen device. The text is larger and the buttons are easier to click.

In some cases the researcher might decide to organize a pure mobile survey. That means he will collect data only on mobile. This is especially interesting for in-the-moment surveys, for instance, price information and consumption. The interview completion is just when the event is taking place; thus no memory effect could bias the result. In undertaking pure mobile surveys, many problems (often not possible to solve) arise to recruiting people according to probability-based sampling criteria.

In general, if the access to mobile devices is disallowed, the questionnaire can't be received and completed in a mobile device. This choice (disallowing access) is rarely used and useful only in specific situation like the abovementioned in-the-moment surveys. Standard situation is that a web survey is run and the questionnaire is allowed for mobile devices (for a full categorization and description of alternative choice for mobile devices participation is in Buskirk and Andrus (2012)). This situation is called mobile web survey; this handbook relies on it. Distinction among devices is not considered, i.e., the comparison of the impact of different devices on the results is not discussed in this handbook.

Like a web survey, the basic feature of a mobile web survey is that the questionnaire design is as a website accessed by respondents. Mobile web surveys are easy to administer by sending the target audience an e-mail with a link to a questionnaire or to download an app.

Mobile web surveys combine elements of traditional CAI systems, online data collection, and additional new elements. Smartphones, tablets, and phablets can support the administration of surveys in a number of ways—online/web surveys, application (or "app")-based surveys, voice interviews, or interviews via text messaging. Smartphones are true multimodal devices, and figuring out how to use them for surveys will be the big challenge for the next few years.

On the one hand, a mobile web survey looks like a web survey. On the other hand, mobile devices are designed to be used on the go; therefore, you may very likely be catching your audience on the way out the door, between meetings, or just before dinner. Thus, you need to make your survey short, no more than five minutes or no more than 10 questions. You will achieve a greater response rate, especially if you tell your audience up front exactly how little time the survey will take.

Mobile web surveys appear to be a fast, cheap, and attractive mean of collecting large amounts of data. In some survey situations, only mobile survey is an effective mode to collect data, for example, when hard-to-reach populations need to be reached or when in-moment data collection is useful to avoid memory effects. Mobile surveys are more versatile than online surveys in that respondents can send pictures, record their voice, or write notes/diaries all on their smartphone. This is especially helpful if the survey requires the respondent to complete specific tasks.

A mobile web survey is indeed an effective mode of data collection that produces high-quality data and contributes to improve the response rate. However, conducting a mobile survey or a mobile web survey in a scientifically sound way is sometimes not so easy, particularly in case of large population surveys. Some of the advantages may be not so clear any more.

A mobile survey can be a cheap data collection instrument. There are no interviewers involved, and there are no mailing and printing costs. Of course, there are hardware and software costs, but a well-designed survey infrastructure can be used for many different surveys. The costs may rise for sample selection. If an e-mail address list is available, sample selection is cheap and straightforward. The situation becomes more difficult if the selection of the sample is for a general population survey. There is no sampling frame with e-mail addresses. Thus, a different mode has to be used to recruit persons for the survey. One way is to send a letter to the sample persons containing a link to the website and a unique identification code to start the questionnaire. If the persons do not respond, they may be called by telephone in an attempt to encourage them to complete the questionnaire. And if this fails, it may be considered to visit them at home. It will be clear that such an approach implies a substantial rise of the costs. An alternative approach is to send an SMS. In such a case a complete list of the mobile phone numbers is necessary; this type of list is in general not available. This is a crucial problem to run a probability-based survey. Even if mobile phone users don't cover the whole general population, the diffusion of mobile devices, in particular smartphones, is largely increasing in European countries. In general, the Internet usage penetration has been growing, and this trend going on. In 2018 broadband Internet access (required for surveys with advanced features) is used by 86% of the households in the EU-28, 38 percentage points higher than the share recorded in 2008 (48%). A high increase is registered also to the use of Internet by individuals. See Chapter 10 for more details. Due to the lack of a list of mobile numbers and the still existing under-coverage of the general population, a probability-based sample of it is not possible using only mobile mode. Thus, mobile mode is usually part of a mixed-mode approach. There are however situations where a full list of e-mail addresses or of mobile phone numbers exists; for instance, employees of a company or students at a university are usually registered in the internal database with the description of basic variables, among them being e-mail address and mobile phone number. These registers cover the population completely.

Mobile web surveys as well as only mobile surveys suffer from nonresponse. They are self-administered surveys; they have a potential of high nonresponse rates. An additional source of nonresponse problems are technical problems of respondents. Respondent needs a browser to open and complete a web survey questionnaire. There are many different browsers available. These browsers do not behave exactly in the same way. Therefore, a questionnaire may behave differently in different browsers. A feature may not even work in specific browser, preventing respondents to record their answers to the questions. This may result in nonresponse.

Most importantly, the researcher should care about network speed and availability. People being located in large cities are more likely to have better network coverage than a person in remote areas. Data coverage is different from voice coverage. So while there may be an area with great voice coverage, that area may be lacking in data coverage. This could increase nonresponse and potentially skew sample toward respondents who live in areas with better network coverage.

Because of both time and screen size constraints, it's important to keep questions clear and concise. Open-ended questions and huge text boxes should be avoided. Every effort to limit the number of options in multiple-choice questions when possible is required. Scrolling through a long list of answer options will create fatigue and affect data quality. Question wording should be kept brief and clear.

The Internet is in continuous development. Specific features may work in the most recent version of a browser, but not in earlier versions. Unfortunately, not all people have the latest version of their browser installed. Again, as a result, a specific feature in the questionnaire may or may not work. Some questionnaire features (for example, use of animation and video) require an Internet connection with substantial bandwidth, but not every Internet user has a broadband Internet connection. So, these features will not work properly on their computer.

Slow modem speeds, unreliable connections, high connection costs, low-end browsers, and incompatible browsers may frustrate respondents, often resulting in prematurely interrupting completion of the questionnaire. In order to keep the survey response up to an acceptable level, every measure must be taken to avoid these problems. This requires a careful design of web survey questionnaire instruments.

*Response order effects* may occur in web surveys; mobile web and purely mobile survey suffer the same effect. Instead of thinking carefully about the appropriate answer, the first reasonable option of a closed question is chosen. One particular effect is the *primacy effect*, i.e., the tendency to pick an answer early in the list of options.

Interviewer-assisted surveys may suffer from *acquiescence*. This is the tendency to agree with statements in questions, regardless of their content. Without interviewers, respondents may feel more anonymous and therefore will be more inclined to answer sensitive questions honestly. Therefore, acquiescence is not a problem in mobile surveys as well as in mobile web surveys.

The phenomenon of *endorsing the status quo*, i.e., the tendency to choose "no change" when asked about changes, typically occurs in self-administered surveys. So it may be a problem in every type of mobile surveys.

*Non-differentiation* occurs if respondents have to answer a series of questions with the same set of response options. Respondents tend to select (without thinking) the same answer for all these questions irrespective of the question content. Non-differentiation can be even more a problem if a series of questions with the same set of answer options is combined into a matrix question. Respondents simply select all answers in the same column (*straight-lining*). This is often the column corresponding to the middle (neutral) response option. Non-differentiation occurs more in self-administered surveys than in interviewer-assisted surveys. Therefore, it may be a problem in every type of mobile surveys.

There are various approaches possible to offer the option "don't know" as an answer to a question. On the one hand, it can be explicitly shown in one of the options. This will encourage *satisficing*. Many will see this option as an escape for having to think about the proper answer. On the other hand, one may decide to not offer the option "don't know." Then respondents always have to provide a substantive answer, even if they do not know the answer. Experts advise against this approach as it may lead to nonresponse.

Mobile surveys combine elements of traditional CAI systems, online data collection, and additional new elements. Smartphones and tablets can support the administration of surveys in a number of ways: online/web surveys, application (or "app")-based surveys, voice interviews, or interviews via text messaging. Smartphones are true multimodal devices, and figuring out how to use them for surveys will be the big challenge for the next few years.

If respondents do not want to think about an answer, they may decide to just pick an *arbitrary answer*. This type of *satisficing* particularly occurs in *check-all-that-apply questions*. This problem typically occurs in self-administered surveys. So it can be a problem in mobile surveys.

Mobile surveys, like web surveys, may perform better than an interviewer-assisted survey with respect to answering questions about sensitive topics. The absence of an interviewer may encourage a respondent to give an honest answer to a question about a potentially embarrassing topic. Therefore, there may be less *socially desirable answers* in mobile surveys.

Designing a questionnaire for a mobile survey is to some extent similar to designing it for a web survey. At first sight, one could say that a mobile survey questionnaire is nothing more than a web questionnaire form displayed on a mobile screen. There are, however, many differences that may affect the answers to the questions. There are several technical differences; for example, the size and characteristics of the screen might affect the participation in different ways. If some questions are not visible questions, they are skipped, resulting in item nonresponse. Furthermore, the designer of a web survey has the choice to display just one question per screen or to put more questions on the screen. One question per screen may be more appropriate if the questionnaire is to contain extensive routing instructions (skip patterns). However, this will increase the perceived length of the questionnaire, possibly resulting in (partial) nonresponse. There are only a few examples of errors depending on the questionnaire design. There are many more design issues that may affect response to a mobile survey.

CAPI and CATI survey questionnaires may contain extensive edit checks. These checks help to detect and correct inconsistencies in the answers of respondents. Such checks are not possible in mail surveys, and this may have a negative effect on the quality of the collected data. The designer of a web survey questionnaire can decide to include edit checks. This should increase data quality, but many (unfriendly) error messages may scare away respondents. So, reporting and treating errors should be implemented carefully.

Respondents are completely free in the way they complete a paper questionnaire. They can answer any question they like in any order. Web questionnaires can be designed such that routing is enforced like in CAPI or CATI surveys. This sees to it that only relevant questions are answered and irrelevant ones are skipped. Routing can only be implemented in a meaningful way if the questionnaire is processed on the basis of one question per screen (the *question-based* approach). It is difficult, if not impossible, to implement routing for a *form-based* approach.

Conducting a mobile web survey is not time consuming. There are no interviewers involved. Questionnaires do not have to send by mail. There is no separate data entry phase. If the web survey questionnaire contains checks, there is no need for a subsequent data editing phase. In case of web panels, it is even possible to conduct a survey in one day. Carrying out a web survey may become more time consuming if there is no proper sampling frame. If respondents have to be recruited by means of a telephone or face-to-face survey, the survey will take more time.

Most importantly, researchers have to think about network speed and availability. People living in large cities are more likely to have better network coverage than a person living in some mountain regions or remote territories (islands, for example). Data coverage is different from voice coverage. So while there may be an area with great voice coverage, that area may be lacking in data coverage. This could potentially skew sample toward respondents who live in areas with better network coverage.

Speaking of skewed results, some target audiences may not even have a smartphone or a data plan due to economic constraints. In general population surveys, the results could definitely be skewed to more affluent and/or younger audiences.

Table 6.5 contains a summary of the effects of various phenomena in mobile web surveys. A plus (+) indicates a positive effect or no negative effect and a minus (–) a negative effect.

TABLE 6.5   **Cost and quality aspects of mobile surveys**

| | | |
|---|---|---|
| Costs | + | There are no interviewer, printing, and mailing costs |
| Coverage | – | There is serious under-coverage in many situations |
| Nonresponse | – | Mobile web surveys have lower response rates than interviewer-assisted modes of data collection and web surveys. Smartphone surveys have higher response rate than other devices, like tablets |
| Response order effects | – | If the possible answers to a closed question are read by the respondent, there will be primacy effects |
| Acquiescence | + | The tendency to agree with statements in questions, regardless of their content, is less for self-administered modes of data collection |
| Status quo endorsement | – | If no interviewers are present, respondents are more inclined to take the easy way out (select no change) |
| Non-differentiation | – | If no interviewers are present, respondents are more inclined to select the same answer for a set of questions |
| Answering "don't know" | + | It is generally advised to offer "don't know" as one of the answer options. Particularly for opinion questions, respondents may use this option as an escape for not giving a substantial answer |

*(Continued)*

**TABLE 6.5    Continued**

| | | |
|---|---|---|
| Arbitrary answer | − | If no interviewers are present, respondents are more inclined to give an arbitrary answer |
| Socially desirable answers | + | If no interviewers are present, respondents are less inclined to give a socially desirable answer. This happens particularly for sensitive questions |
| Questionnaire design effects | − | Questionnaire design is critical in mobile surveys and mobile web surveys. If the questionnaire is designed specifically for mobiles, more options than in traditional surveys are available, like videos, voice, etc. These possibilities make the surveys special. It the questionnaire is designed for a web survey, it should absolutely be adapted for mobile surveys; otherwise a lot of bias and errors will arise, and data quality will be very scarce |
| Checking | + | It is possible to include checks in mobile surveys. Presenting and treating errors is critical |
| Routing | + | Automatic routing can be implemented for the question-by-question implementation |
| Timeliness | + | A mobile survey can be conducted quickly, when an e-mail address of the selected people is available. Otherwise, recruitment by means of mail/CAPI/CATI may take time |

# 6.3 Application

The Blaise system is a software system that supports various modes of data collection. It was developed by Statistics Netherlands as a tool for making survey data collection faster and at the same time improving the quality of the data. Blaise can be used for large and complex questionnaires. It is used by many national statistical institutes in the world. Blaise is used in this section to illustrate the advantages and disadvantages of the different modes of data collection. More about the development of the Blaise system and its underlying philosophy can be found in Bethlehem and Hofman (2006).

A questionnaire is designed in Blaise using a scripting language. It defines the questions to be asked, the possible answers to the questions, the order in which the questions must be asked, and the conditions under which the questions have to be asked. Moreover, relationships can be defined that have to be checked. Once the Blaise questionnaire definition is ready, the system can generate the software tools for various modes of data collection: CADI (*computer-assisted data input*) for data entry of paper forms, CAPI, CATI, and web.

A small election survey questionnaire is used as an example. Figure 6.1 contains the specification of this questionnaire in the Blaise system. The first part of the questionnaire specification is the *Fields section*. It contains the definition of all questions. A question consists of an identifying name, the text of the question as

```
DATAMODEL ElectionSurvey "The 2010 Election Survey"
FIELDS
  Democracy
    "On the whole, are you very satisfied, fairly satisfied, not very satisfied
     or not satisfied at all with the way democracy works in the country?":
    (VerySat "Very satisfied",
     SomeSat "Somewhat satisfied",
     Neutral "Neither satisfied nor dissatisfied",
     SomeDis "Somewhat dissatisfied",
     VeryDis "Very dissatisfied")
  NewsTV
    "How much attention did you pay to the TV news about the election":
    (Lot "A lot", Fair "Fair amount", Some "Some", Little "Little", None "None")
  NewsRad
    "How much attention did you pay to the radio news about the election":
    (Lot "A lot", Fair "Fair amount", Some "Some", Little "Little", None "None")
  NewsPap
    "How much attention did you pay to the news in newspapers about the election":
    (Lot "A lot", Fair "Fair amount", Some "Some", Little "Little", None "None")
  NewsWeb
    "How much attention did you pay to the news on the Internet about the election":
    (Lot "A lot", Fair "Fair amount", Some "Some", Little "Little", None "None")
  Voted
    "Did you vote for the parliamentary election on June 2, 2110?":
    (Yes      "Yes, did vote",
     CouldNot "Could not vote",
     DidNot   "Could vote, but did not vote")
  Party
    "Which party did you vote for?":
    (Con   "Conservative Party",
     Soc   "Social Democratic Party",
     Lib   "Liberal Party",
     Green "Green Party",
     Oth   "Other party"), DONTKNOW
  SecParty
    "Which party was your second choice?":
    (Con   "Conservative Party",
     Soc   "Social Democratic Party",
     Lib   "Liberal Party",
     Green "Green Party",
     Oth   "Other party",
     None  "None"), DONTKNOW
  WhyNot
    "What is the reason you did not vote?":
    (NoTime  "No time, too busy",
     NotInt  "Not interested",
     PhysLim "Physical limitations",
     OthReas "Other reason")
  DatBirth
    "What is your date of birth?": DATETYPE
  MarStat
    "Are you presently married, living with a partner, divorced,
     separated, widowed, or have you never been married?":
    (Married "Married",
     Partner "Living with a partner",
     Divorce "Divorced",
     Separat "Separated",
     Widowed "Widowed",
     NevMarr "Never been married")

RULES
  Democracy
  NewsTV NewsRad NewsPap NewsWeb
  Voted
  IF Voted = Yes THEN
    Party SecParty
    Ord(Party) <> Ord(SecParty) "Your first choice was ^Party. Your second choice must
    be a different party."
  ELSEIF Voted = DidNot THEN
    WhyNot
  ENDIF
  DatBirth
  IF (Voted = Yes) OR (Voted = DidNot) THEN
    DatBirth <= (1992, 6, 2) "You are too young to vote!"
  ENDIF
  MarStat
ENDMODEL
```

FIGURE 6.1  **A questionnaire definition in Blaise**

presented to the respondents, and a specification of valid answers. For example, the question about voting behavior has the name *Voted*, the text of the question is "*Did you vote for the parliamentary election on June 2, 2110*?" and there are three answer options. Each option has a name (for example, *Yes*) and a text for the respondent (for example, "*Yes, did vote*").

Almost all questions in this sample questionnaire are closed questions. There is one exception: the question DatBirth asks for the date of birth. The answer must be a date.

By default the answer "don't know" is forbidden for all questions in Blaise. To allow this answer, the keyword *DONTKNOW* must be added to the question definition. This has been done for the questions *Party* and *SecParty* in Figure 6.1. It means that in a computer-assisted data collection mode, this answer option is implicitly available. It can be selected with the function key <*Ctrl-K*>.

The second part of the Blaise specification is the *Rules section*. Here, the order of the questions is specified, as well as the conditions under which they are asked. According to the rules section in Figure 6.1, every respondent must answer the questions *Democracy, NewsTV, NewsRad, NewsPap, NewsWeb*, and *Voted*. Only persons who voted (*Voted = Yes*) have to answer the questions *Party* and *SecParty*. Respondents who could vote but did not vote (*Voted = DidNot*) are asked why they did not vote (*WhyNot*). Finally, all respondents have to provide their date of birth (*DatBirth*) and marital status (*MarStat*).

The Rules section contains two checks. The first one checks that the second choice for a party is different from the first choice (*Ord(Party) <> Ord(SecParty)*). Note that a text label is attached to the check. This text appears in the error message on the screen. The second check produces an error message if a voter is younger than 18 years (*DatBirth <= (1992, 6, 2)*).

The election survey questionnaire has been formatted as a paper questionnaire in Figure 6.2. It is a disadvantage of a paper questionnaire that there is no software in charge of the proper route through the questionnaire. There are printed instructions like "Go to 7," but there is no guarantee that these instructions will be followed. There is always a risk that respondents end up in the wrong part of the questionnaire.

The four questions about paying attention to news about election campaign all have the same answer options. Therefore, they have been formatted as a matrix question. This saves space in the questionnaire, but there is also a risk of *straight-lining*: respondents may make it easy for themselves by selecting all answer options in the same column. This is not the only form of satisficing that may occur. Another is that respondents may not have an opinion or do not want to think about an opinion. As result, they choose the middle option for all four questions.

The question asking for the date of birth contains clear instructions as to how the date must be written down. These instructions are important. Without them an answer may be confusing. For example, if someone writes down 4-3-1951, it is unclear whether this means March 4 or April 3.

The two questions about party choice contain "don't know" as an explicit answer option. If there are respondents who really do not know the answer, this

---

1. **On the whole, how satisfied are you with the way the democracy works in the country?**

   ☐ Very satisfied
   ☐ Somewhat satisfied
   ☐ Neither satisfied nor dissatisfied
   ☐ Very satisfied
   ☐ Somewhat satisfied

2. **How much attention did you pay to news about the election campaign in each of the following media?**

   |  | A lot | Fair amount | Some | Little | None |
   |---|---|---|---|---|---|
   | Television | ☐ | ☐ | ☐ | ☐ | ☐ |
   | Radio | ☐ | ☐ | ☐ | ☐ | ☐ |
   | Newspapers | ☐ | ☐ | ☐ | ☐ | ☐ |
   | Internet | ☐ | ☐ | ☐ | ☐ | ☐ |

3. **Did you vote for the parliamentary election of June 2, 2010?**

   ☐ Yes        ➔ *Go to 5*
   ☐ Could not vote  ➔ *Go to 7*
   ☐ Could vote, but did not vote

4. **What is the reason you did not vote?**

   ☐ No time, too busy
   ☐ Not interested
   ☐ Physical limitations  ➔ *Go to 7*
   ☐ Other reason
   ☐ Don't know

5. **Which party did you vote for?**

   ☐ Conservative Party
   ☐ Social-democratic Party
   ☐ Liberal Party
   ☐ Green Party
   ☐ Other party
   ☐ Don't know

6. **Which party was your second choice?**

   ☐ Conservative Party
   ☐ Social-democratic Party
   ☐ Liberal Party
   ☐ Green Party
   ☐ Other party
   ☐ None
   ☐ Don't know

7. **What is your date of birth?**

   ☐☐ dd   ☐☐ mm   ☐☐☐☐ yyyy

8. **Are you presently married, living with a partner, divorced, separated, widowed, or have you never been married?**

   ☐ Married
   ☐ Living with a partner
   ☐ Divorced
   ☐ Separated
   ☐ Widowed
   ☐ Never been married

FIGURE 6.2 The paper version of the questionnaire

may be the best approach. Of course, this creates an escape route for those not wanting to give an answer.

After the completed paper forms have been sent back to the survey agency, the data must be entered into the computer. Some kind of data entry tool can be helpful to do this in an efficient way. Blaise implements an approach that is called *computer-assisted data input*. It is a combination of data entry and data editing. The CADI program can be automatically generated from the questionnaire definition.

Figure 6.3 shows an example of a screen of the CADI program for the election survey. The questions are indicated by their short identifying names. If necessary, the complete question text can be displayed by pressing a special function key.

Data entry typists enter the answers to the questions. They are completely free in the order in which they enter data. They are not constrained by routing instructions. The idea is to first copy all data from the form to the computer, after which the answers are checked and corrections can be made.

Special symbols in front of the input fields denote problems in the data. The answers to two questions *Party* and *SecParty* in Figure 6.3 are inconsistent. A diagnostic error message can be displayed by moving the cursor to the field and pressing a special function key ("*Your first choice was Conservative Party. Your second choice must be a different party*").

The two symbols for the questions *DatBirth* and *MarStat* indicate that these questions are on the route through the questionnaire and therefore have to be answered. These symbols will disappear once the answers have been entered in the respective fields.

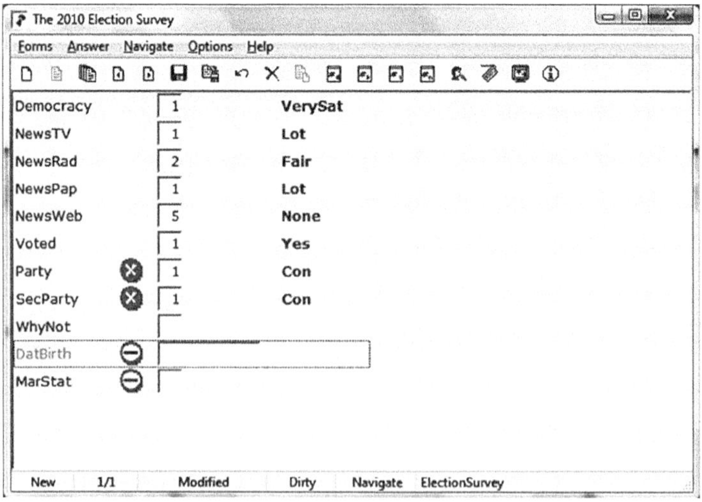

FIGURE 6.3  The data entry program

Note that because the paper questionnaire and the CADI program have been generated from the same source (the Blaise questionnaire specification), the form and the program are consistent with each other.

The Blaise system also supports CAPI and CATI. A program for CAI can be automatically generated from the questionnaire specification. Figure 6.4 shows an example of a screen for the election survey.

The screen is divided into two parts. The top half contains the current question to be answered. The bottom half is a condensed view of the questionnaire form. It shows which questions have already been answered and which questions still must be answered. This gives the interviewers an overview of where they are in the questionnaire.

Routing is forced in this questionnaire. It is only possible to move forward to the next question on the route if the current question has been properly answered. Changes can be made in the questionnaire by moving back to a previous question.

If an error is encountered in the answers, an error message will be displayed on the screen. Figure 6.5 contains an example. The message also shows the questions involved in the error. The interview can only go back to one of these questions and correct its answer. It is not possible to proceed to the next question on the route as long as the error has not been corrected.

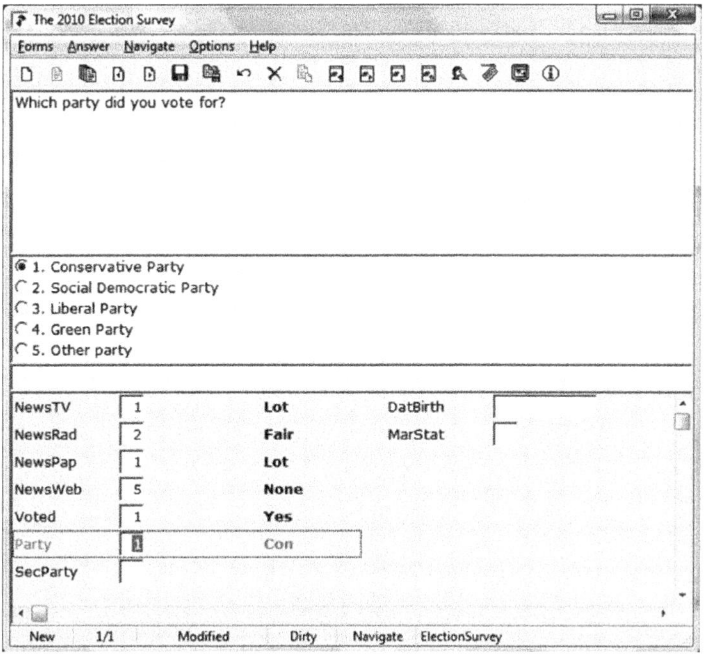

FIGURE 6.4 The computer-assisted interviewing program

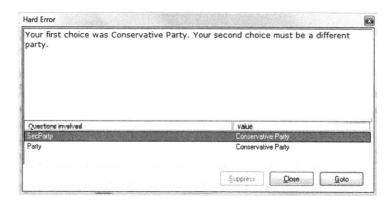

FIGURE 6.5 A detected consistency error

This approach of forced routing and forced error correction has shown to improve the quality of the collected data. The data contains fewer errors, and there is less item nonresponse.

The questions *Party* and *SecParty* do not have an option "Don't know." This option is implicitly available for these questions. If the respondents insist they really do not know the answer, the interviewer can press *<Ctrl-K>* to record the answer as "Don't know."

The Blaise system uses the same CAI program for CAPI and CATI. Some other tools are specific for either CAPI or CATI. For example, there is an extensive call management system for CATI. This system sees to it that the call of the right telephone number is at the right time. In case of a busy signal, there is a scheduling after a short while for a subsequent call attempt. If there is no answer, typically, there is a scheduling at a different day and/or time of the day for a new call attempt. Interviewers can also make appointments with respondents to call back at a specific date and time. This call management system helps to increase response rates.

It is also possible to carry out a web survey with Blaise. The system distinguishes two approaches:

- The *question-based approach*. There is only one question on the screen. After this question is completed, the answer is sent to the web server and checked. The system determines the next question on the route, and this question will appear on the screen. If the questionnaire contains routing instructions and checks, this approach is used. The question-based approach requires the respondent to be online while answering the questions.

- The *form-based approach*. The whole questionnaire is displayed on the screen as one form. If there are no routing instructions and checks, this approach is used. When all questions are completed, the answers are sent to the web server. The respondent does not need to be online while answering the questions.

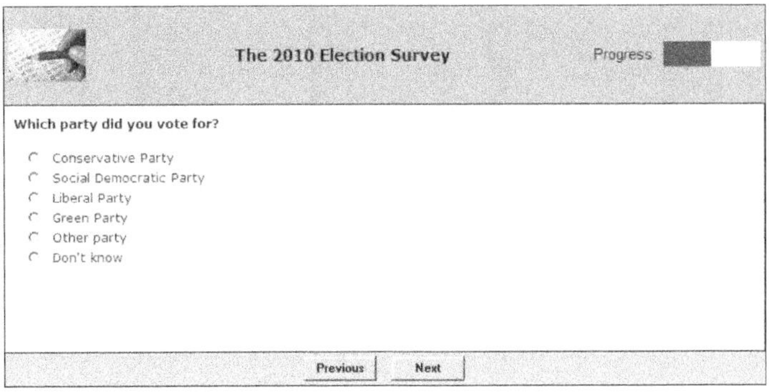

FIGURE 6.6  **Web survey question**

It is also possible to mix both approaches. The idea is to divide the questionnaire into a number of sub-questionnaires each without routing. There is a form for each sub-questionnaire. After a form has been completed, it is sent to the web server, where it is checked. Then a next form will be sent to the respondent.

Figure 6.6 contains an example of a screen for the election survey. Because the questionnaire contains a number of checks and route instructions, it has been implemented in the question-based approach.

The web questionnaire is somewhat more attractive by adding a logo. Note also that there is a progress bar in the top-right corner. This may help to keep respondents motivated, as long as the progress bar does not behave too wild due to large jumps in the questionnaire.

The problem of "don't know" is treated by including this answer as one of the answer options for the question. At the same time, respondents have to answer this question. They cannot proceed without answering it. This may help in preventing skipping questions too quickly. Nevertheless, there is always the risk that respondents answer "don't know" as the easy way out.

The routing is forced in this web questionnaire. Respondent can only move back to the previously answered question. They can only move forward to the next questionnaire on the route if they have properly answered the current question. The routing mechanism is the same as that of the CAPI/CATI program.

It is possible to include checks in the Blaise web questionnaire. This may help to increase data quality. However, note that many unfriendly error messages often frustrate respondents, resulting in (partial) nonresponse. A special recommendation is to pay careful attention to the way error message appears to respondents.

It is an attractive feature of the Blaise system, the generation from the same questionnaire specification of software tools for various modes of data collection. This forces these instruments to be consistent with each other. This makes the system suitable for mixed-mode data collection.

# 6.4 Summary

A researcher can choose from various modes of data collection. They all have their advantages and disadvantages with respect to cost, timeliness, and data quality.

Interviewer-assisted modes (CAPI, CATI) are expensive but produce good quality data. Other modes (mail, web) are less expensive, but a price may have to be paid in terms of data quality. With respect to response rates, the best option is to use an interviewer-assisted mode of data collection. This applies to face-to-face and telephone interviewing as well as to their computer-assisted analogues CAPI and CATI. The presence and efforts of interviewers often lead to higher response rates. Response rates are lower in web surveys.

Although mail and web surveys are not as good as the interviewer-assisted modes in terms of measurement errors, they perform better with respect to sensitive questions. Respondents tend to answer this type of questions better if there are no interviewers present.

Web surveys grow in popularity. Up until now their response rates are disappointingly low. This is partly due to the self-administered nature of this mode of data collection. Nonresponse may, however, also be caused by technical problems, like slow modems (no broadband connection) and old or incompatible browsers.

Web surveys may also suffer from under-coverage problems, particularly in case of large population surveys. This problem may solve itself in the future. In the meantime, mixed-mode surveys may help to get into contact with those without access to the Internet.

Another problem of web surveys is that often a proper sampling frame is lacking. By recruiting respondents by means of another mode of data collection (mail, CATI) this problem could be solved. A disadvantage of this approach is that it makes the web survey much more expensive and conducting it will also take more time.

Because respondents are on their own, when answering the questions, it is of crucial importance that the design of the questionnaire is such that it helps them to perform this task properly.

Due to the wide diffusion of mobile devices, especially smartphones, when administering a web survey, if access from mobile devices is not prevented, a mobile web survey is in practice sent out. However, there is no reason to prevent access from mobile devices except in special cases (long questionnaire, specific research issues, or target populations). Mobile web surveys are challenging and have advantages and disadvantages similar to web surveys. They may contribute to increase the response rate in some cases. There are quality risks if the researcher doesn't design the survey taking into account the mobile device designing requirements.

## KEY TERMS

**Accurate:** An estimator that always results in estimates close to the true value (if the survey is repeated).

**Acquiescence:** The phenomenon that respondents tend to agree more with statements in questions if there are interviewers present.

**Blaise**: A software package for computer-assisted interviewing and survey processing developed by Statistics Netherlands.

**Computer-assisted interviewing (CAI)**: A form of interviewing without a paper questionnaire. Computer program is presenting the questions.

**Computer-assisted personal interviewing (CAPI)**: A form of face-to-face interviewing in which interviewers use a laptop computer to ask the questions and to record the answers.

**Computer-assisted self-administered questionnaires (CASAQ)**: A form of data collection in which respondents complete the questionnaires on their own computer. See also CASI.

**Computer-assisted self-interviewing (CASI)**: A form of data collection in which respondents complete the questionnaires on their own computer. See also CASAQ.

**Computer-assisted telephone interviewing (CATI)**: A form of telephone interviewing in which interviewers use a telephone to ask the questions and to record the answers.

**Computer-assisted web interviewing (CAWI)**: A form of self-interviewing in which respondents complete the questionnaires on the Internet.

**Face-to-face survey:** A survey where interviewers visit the homes of the respondents (or another location convenient for the respondent). Together, the interviewer and the respondent complete the questionnaire.

**Mail survey**: A survey where respondents are receiving a paper questionnaire form for a survey. After completion, the questionnaire is sent back to the research organization.

**Mixed-mode survey:** A survey in which various modes of data collection are combined. Modes can be used concurrently (different groups are approached by different modes) or sequentially (nonrespondents of a mode are reapproached in a different mode).

**Non-differentiation**: A form of satisficing that typically occurs if respondents have to answer a series of questions each with the same set of response options. They tend to select the same answer for all these questions irrespective of the question content.

**Nonresponse**: The phenomenon that elements in the selected sample, and that are also eligible for the survey, do not provide the requested information, or that the provided information is not usable.

**Precise**: An estimator with a small variance. The precision is a quantification of the reliability.

**Primacy effect:** The phenomenon that respondents have a preference for the first option in the list of answers of a closed question. This typically happens in face-to-face and telephone surveys.

**Probability-based sampling**: A form of sampling where selection of elements is a random process. Each element must have a positive and known probability of selection.

**Random digit dialing (RDD)**: A form of sample selection for a telephone survey: some kind of computer algorithm is generating random telephone numbers.

**Recency effect:** The phenomenon that respondents have a preference for the last option in the list of answers of a closed question. This typically happens in mail and web surveys.

**Reliable:** An estimator that would result in (approximately) the same estimates if the survey is repeated.

**Response order effect:** The tendency that the answer selected by the respondents depends on its location in the list of answer options. Primacy and recency effects are special cases.

**Satisficing**: The phenomenon that respondents do not do all they can to provide a correct answer. They attempt to give a satisfactory answer with minimal effort instead.

**Socially desirable answer**: The phenomenon that respondents do not give the true answer but an answer that is more socially desirable.

**Status quo endorsement**: The tendency to answer that everything should be kept the same. This typically occurs in opinion questions about change.

**Straight-lining**: The tendency that respondents give the same answer to all single questions in a matrix question. They simply check all answer options in the same column.

**Telephone interviewing:** A survey where interviewers call selected persons by telephone. If contact is with the proper person and this person wants to cooperate, the interview is started and conducted over the telephone.

**Under-coverage**: The sampling frame does not cover completely the target population of the survey. There are persons in the population who do not appear in the sampling frame. They will never be selected in the sample.

**Valid**: An estimator that estimates what the estimator is intended to estimate.

**Web survey**: A survey where respondents complete the questionnaires on the Internet.

## EXERCISES

**Exercise 6.1**   Which mode of data collection is most expensive?

**a.** A face-to-face survey.

**b.** A telephone survey.

**c.** A mail survey.

**d.** A web survey.

**Exercise 6.2**   Should error message be included in a web survey?

**a.** Yes, because they always improve data quality.

**b.** Yes, but careful attention should be paid to their design.

**c.** No, because they increase item nonresponse.

**d.** No, because they lead to socially desirable answers.

**Exercise 6.3**   Why is random digit dialing (RDD) sometimes preferred for a telephone survey instead of random sampling from a telephone directory?

**a.** RDD sampling provides more auxiliary information about nonrespondents.

**b.** Response rates are lower in an RDD survey.

**c.** RDD guarantees full coverage of the population.

**d.** An RDD sample is less expensive than a sample form directory.

**Exercise 6.4**   What does acquiescence mean?

**a.** This is the tendency to not answer sensitive questions.

**b.** This is the tendency to give more extreme answers.

**c.** This is the tendency to disagree with what the interviewers say.

**d.** This is the tendency to agree with what the interviewers say.

**Exercise 6.5**   How, in web surveys, to reduce the primacy effect in closed questions?

**a.** By randomizing the order of the answer options.

**b.** By putting the answer options in the reverse order.

**c.** By reducing the number of answer options.

**d.** By increasing the number of answer options.

**Exercise 6.6**   Which of the following options is not an advantage of computer-assisted interviewing (CAI) as compared with traditional modes of data collection?

**a.** Data quality is higher due to included checks.

**b.** The software is in charge of routing through the questionnaire.

**c.** CAI leads to higher response rates.

**d.** Data can be processed quicker.

**Exercise 6.7**   What is the effect of the mode of data collection on an opinion question with possible answers: strongly disagree, disagree, neutral, agree, and strongly agree?

**a.** More people will select the neutral category in mail and web surveys.

**b.** More people will select the neutral category in CAPI and CATI surveys.

**c.** More respondents will select the extreme options strongly disagree and strongly agree.

**d.** The mode of data collection does not influence the answer patterns.

**Exercise 6.8**   In which situation a web survey is expected to use the question-based approach (and not the form-based approach)?

**a.** If the questionnaire contains many questions.

**b.** If the respondent must be able to complete the questionnaire offline.

**c.** If the questionnaire contains matrix questions.

**d.** If there are checks and routing instructions in the questionnaire.

**Exercise 6.9**   How, in a web survey questionnaire, to treat "don't know"?

**a.** Do not present it as one of the answer options, and force respondents to answer.

**b.** Do not present it as one of the answer options and do not force respondents to answer. If the question is skipped, record it as "don't know."

**c.** Do not present it as one of the answer options and do not force respondents to answer. If the question is skipped, give the respondent two options: (1) answer the question, or (2) record the answer as "don't know."

**d.** Present it as an answer option, but less obvious elsewhere on the screen.

**Exercise 6.10**   What kind of problem may check-all-that-apply questions cause in web surveys?

**a.** Acquiescence.

**b.** Selecting an arbitrary answer.

**c.** Straight-lining.

**d.** Socially desirable answer.

## REFERENCES

Bethlehem, J. G., Cobben, F., & Schouten, B. (2011), *Handbook of Nonresponse in Household Surveys*. Wiley, Hoboken, NJ.

Bethlehem, J. G. & Hofman, L. P. M. B. (2006), Blaise—Alive and Kicking for 20 Years. *Proceedings of the 10th Blaise, Users Meeting*. Statistics Netherlands, Voorburg/Heerlen, the Netherlands, pp. 61–88.

Biemer, P. P. & Lyberg, L. E. (2003), *Introduction to Survey Quality*. Wiley, Hoboken, NJ.

Buskirk, T. D. & Andrus, C. (2012), Smart Surveys for Smart Phones: Exploring Various Approaches for Conducting Online Mobile Surveys via Smartphones. *Survey Practice*, 5, 1. https://doi.org/10.29115/SP-2012-0001.

Couper, M. P. (2000), Web Surveys: A Review of Issues and Approaches. *Public Opinion Quarterly*, 64, pp. 464–494.

Couper, M. P. (2008), *Designing Effective Web Surveys*. Cambridge University Press, Cambridge, UK.

de Leeuw, E. D. (2008), Choosing the Method of Data Collection. In: de Leeuw, E. D., Hox, J. J., & Dillman, D. A. (eds.), *International Handbook of Survey Methodology*. Lawrence Erlbaum Associates, New York, NY, pp. 113–135.

Derouvray, C. & Couper, M. P. (2002), Designing a Strategy for Reducing "No Opinion" Responses in Web-Base Surveys. *Social Science Computer Review*, 20, pp. 3–9.

Dillman, D. A. (2007), *Mail and Internet Surveys: The Tailored Design Method*. Wiley, Hoboken, NJ.

Dillman, D. A. & Bowker, D. (2001), The Web Questionnaire Challenge to Survey Methodologists. In: Reips, U. D. & Bosnjak, M. (eds.), *Dimensions of Internet Science*. Pabst Science Publishers, Lengerich, Germany.

Dillman, D. A., Sinclair, M. D., & Clark, J. R. (1993), Effects of Questionnaire Length, Respondent-friendly Design and a Difficult Question on Response Rates for Occupant-addressed Census Mail Surveys. *Public Opinion Quarterly*, 57, pp. 289–304.

Dillman, D. A., Smyth, J. D., & Christian, L. M. (2009), *Internet, Phone, Mail and Mixed-mode Surveys: The Tailored Design Method*. Wiley, Hoboken, NJ.

Dillman, D. A., Smyth, J. D., & Christian, L. M. (2014), *Internet, Phone, Mail and Mixed-mode Surveys: The Tailored Design Method*. Wiley, Hoboken, NJ.

Fricker, R. & Schonlau, M. (2002), Advantages and Disadvantages of Internet Research Surveys: Evidence from the Literature. *Field Methods*, 15, pp. 347–367.

Goyder, J. (1987), *The Silent Minority: Nonrespondents on Sample Surveys*. Westview Press, Boulder, CO.

Groves, R. M. & Couper, M. P. (1998), *Nonresponse in Household Interview Surveys*. Wiley, New York, NY.

Heerwegh, D. & Loosveldt, G. (2002), *An Evaluation of the Effect of Response Formats on Data Quality in Web Surveys*. Paper presented at the International Conference on Improving Surveys, Copenhagen.

Hox, J. J. & de Leeuw, E. D. (1994), A Comparison of Nonresponse in Mail, Telephone and Face-to-face Surveys. Applying Multilevel Modeling to Meta-analysis. *Quality & Quantity*, 28, pp. 329–344.

Krosnick, J. A. (1991), Response Strategies for Coping with the Cognitive Demands of Attitude Measures in Surveys. *Applied Cognitive Psychology*, 5, pp. 213–236.

Kuusela, V., Vehovar, V., & Callegaro, M. (2006), *Mobile Phones—Influence on Telephone Surveys*. Paper presented at the Second International Conference on Telephone Survey Methodology, Florida, USA.

Peterson, G., Griffin, J., LaFrance, J., & Li, J. (2017), Smartphone Participation in Web Surveys, Choosing Between the Potential or Coverage, Nonresponse, and Measurement Error. In: Biemer, P., de Leeuw, E., Eckman, S., Edwards, B., Kreuter, F., Lyberg, L., Tucker, N., West, B. (eds.), *Total Survey Error in Practice* (Wiley Series in Survey Methodology). Wiley, Hoboken, NJ.

Scherpenzeel, A. (2008), An Online Panel as a Platform for Multi-disciplinary Research. In: Stoop, I. & Wittenberg, M. (eds.), *Access Panels and Online Research, Panacea or Pitfall?* Aksant, Amsterdam, pp. 101–106.

Schuman, H. & Presser, S. (1981), *Questions and Answers in Attitude Surveys*. Academic Press, New York, NY.

Sudman, S. & Bradburn, N. (1982), *Asking Questions: A Practical Guide to Questionnaire Design*. Jossey-Bass, San Francisco, CA.

Vis-Visschers, R., Arends-Tóth, J., Giesen, D., & Meertens, V. (2008), *Het Aanbieden van 'Weet niet' en Toelichtingen in een Vragenlijst*. Report DMH-2008-02-21-RVCS, Statistics Netherlands, Heerlen, the Netherlands.

Wagner, J. (2008), *Adaptive Survey Design to Reduce Nonresponse Bias*. Doctoral Dissertation, University of Michigan, Ann Arbor, USA.

# CHAPTER SEVEN

# Designing a Web Survey Questionnaire

## 7.1 Introduction

A web survey is a survey in which data is collected using the World Wide Web. As for all surveys, the aim of a web survey is to investigate a well-defined population. Such populations consist of concrete elements, such as individuals, households, or companies. It is typical for a survey that information is collected by means of asking questions of the representatives of the elements in the population. To ask questions in a uniform and consistent way, a questionnaire is used. There are various ways in which a questionnaire can be offered on the Internet:

- Pop-ups can be used to direct respondents to the questionnaire while they are visiting another site. This approach is particularly useful when the objectives of the survey relate to the website being visited, such as evaluating the website.

- E-mails are sent to people in a panel, a mailing list of customers, or other people who might qualify for the survey. The e-mail contains a link that directs them to the web survey questionnaire.

- Respondents can be directed to the website following a recruitment interview, either by telephone or face-to-face. Postal contact can be used to push participation to web, too.

*Handbook of Web Surveys*, Second Edition. Silvia Biffignandi and Jelke Bethlehem.
© 2021 John Wiley & Sons, Inc. Published 2021 by John Wiley & Sons, Inc.

Pop-up surveys are not considered in this chapter. Such surveys are usually conducted for simple evaluation purposes and consist of only a few straightforward questions. Instead focus is on quite complex surveys, where the web survey questionnaire may be directed to individuals, households, or businesses.

To make advances in the social sciences and to take informed decisions for public policy and businesses, a high-quality data collection process is essential for capturing representative information. The design of the questionnaire is a crucial factor in the survey process, as this determines the quality of the collected data. In all data collection modes, the questionnaire is important, yet self-administered web or e-mail surveys rely even more heavily on the quality of the questionnaire. The recent diffusion of mobile web surveys brings to the attention new problems for the questionnaire quality.

Designing a web survey is not the same as designing a website. In other words, web surveys are not the same as the websites. Their goal is different, and their structure should be different as well. The web design aspects of the questionnaire do not solve all problems related to web surveys. However, a well-designed web-based survey questionnaire can help in reducing non-sampling errors, such as measurement errors and nonresponse. A badly designed web questionnaire causes errors occurring at various steps of survey process described in Chapter 3 (Figure 3.1). For example, at the data collection step, participation in the survey could be reduced as well as item nonresponse or full questionnaire completion. Nonresponse affects the estimation step by modifying estimation errors. Thus, errors such as measurement errors and nonresponse can increase. Therefore, data quality can decrease.

Web surveys are self-administered. Because of this they are similar to mail or fax. In terms of data collection, the major differences between web surveys and other forms of data collection are the same as between self-completion (mail or fax) and interviewer-assisted surveys (face-to-face and telephone). It is, however, interesting to note that some literature has found that differences in the mode of data collection do not always imply differences in survey results. For instance, Cobanoglu, Warde, and Moreno (2001) have shown that mean scores for data collection via a web-based questionnaire are the same as for other self-completion methods, such as mail and fax. More recently, Biffignandi and Manzoni (2011), in an experimental design comparing paper and web surveys, found no difference in evaluation scores relative to the data collection mode. It is advised that switching from one mode to another mode could improve measurement error, but could worse response rate. Thus, since literature presents contrasting results on measurement error (especially with reference to answers to scales), it is suggested to consider total survey error (TSE) perspective referring on other types of errors, rather than focusing only on measurement error.

Even if the completion mode (self-administered versus interviewer assisted) potentially has no effect on survey results, many issues should be taken in account in designing web survey questionnaires able of capturing information correctly. Only in this way can neutrality and objectivity be preserved.

A point to be taken in account in choosing the mode of data collection is that self-administered questionnaires have particular advantages. One is that a longer list of answer options can be offered if the questionnaire is printed on paper or displayed on the computer screen. This is particularly true for web surveys and mobile web surveys. A high level of detail is difficult to obtain for other modes of data collection. Furthermore, a self-administered survey is more effective for addressing sensitive issues (such as medical matters or drug use). Answers to web survey questions suffer less from social desirability bias as respondents answer more truthfully. This means that web survey data on "threatening" issues, where respondents may feel a need to appear socially acceptable, are likely to represent much better how the survey population really feels.

Some surveys are carried out regularly. Examples are periodical surveys on prices, production, or international trade conducted by national statistical institutes. The transition of such surveys from a traditional paper mode to the web mode should take in account the fact that when the visual layout of the questions is not consistent with past experience and expectations, respondents may perceive (or even effectively undergo) a greater response burden and confusion. In transition situations (and even when a mixed-mode approach is adopted), the key factor is to determine whether a questionnaire design should be preferred that fully takes advantage of the possibilities offered by the Internet or that a more "plain" questionnaire should be used that is similar to previous questionnaires.

When constructing a web questionnaire, many design principles of paper questionnaires can be applied. Actually, the basic principles for paper questionnaire design can be copied to the web. Examples are principles like using short questions, avoiding double questions, and avoiding (double negative) questions. Nevertheless, data collection with online questionnaires is still a relatively new methodology. Research has been carried out and is still in progress to retest the principles of traditional paper questionnaires in the context of Internet research and to identify (and solve) new problems. However, the design of a web survey questionnaire is more complex than the design of a paper questionnaire. Web surveys allow for a wide range of textual options, format, and sophisticated graphics, none of which are usually attainable with e-mail surveys. Tools and procedures are available that allow for improving quality and simplifying compilation. For example, web surveys provide additional formats and response control such as preventing multiple answers when only one is called for and links that provide the respondents with direct reference to definitions or examples at multiple points in the survey. Moreover, the questionnaire may include route instructions that see to it that respondents only answer relevant questions, whereas irrelevant questions are skipped. There are various ways to visualize questions on the screen (both computer screens and mobile device screens), for instance, scrolling or paging, using images or not, or use of progress indicators or not. According to Mavletova and Couper (2014), a scrolling design appears to make the process of completing the survey easier and more engaging for mobile web respondents. Specifically, it significantly decreases the completion time, produces fewer reported technical problems, and increases the reported level of satisfaction among respondents.

Another advantage of use of the Internet for data collection is that it provides a lot of extra information (so-called paradata) about the questionnaire completion process, such as completion time, the number of accesses to the questionnaire website, the number of clicks, and completion patterns. Moreover, data about the respondents can already be imported in the questionnaire before they start answering the questions. Examples are the values of stratification variables (type and size of the company) and variables that are included in the sampling frame (address, household composition).

An important matter in designing a web questionnaire is the recent high diffusion of Internet access through mobile devices (smartphones, tablets, etc.). If a survey is not prevented to mobile devices, the questionnaire might be accessed (and even completed) through a mobile device. Since more and more people have only smartphones and not PCs or laptops, preventing the use of them may lead to serious bias. Thus the recommended solution is the researcher not to prevent the access of the questionnaire to mobile devices, except if a special related to the topic of the survey exists. Rather he should check that the structure of the questionnaire in the mobile devices is clear. Some rules to formatting questions in a more suitable way for small screens have to be considered. Programs to optimize the questionnaire for mobile device adopted. Advanced software let multiple targets made operations enabling to give a number of shaping for the same questionnaire. For example, Net-Survey automatically detects the browser, and questionnaire has adapted versions to different display screens are set up.

Provided that the questionnaire is optimized for mobile surveys, measurement equivalence has been confirmed for web and mobile surveys through a series of randomized controlled experiments in which respondents were kept similar across devices. Consistent results have been found in the literature (Buskirk and Andrus, 2012; Mavletova and Couper, 2013).

Summing up, web questionnaire design provides new insights in certain traditional basic principles of questionnaire design, and it draws the attention to new methodological issues, too. In this chapter, focus is on design issues that are specific to web survey questionnaires (with some advice about mobile devices). It gives an overview of the various possible question types and shows how these question types should be formatted. Special attention is paid to handling "don't know" in web surveys. The effects it can have on the answers to the questions have to be considered.

# 7.2 Theory

## 7.2.1 THE ROAD MAP TOWARD A WEB QUESTIONNAIRE

Adequate *questionnaire construction* is critical for the success of a survey. Inappropriately formatted questions, incorrect ordering of questions, incorrect answer scales, or a bad questionnaire layout can make the survey results meaningless, as they may not accurately reflect the views and opinions of the participants. Before entering into the detailed problems related to questionnaire design, some important general guidelines are mentioned. These guidelines help to obtain a good questionnaire design and to successfully complete the web survey:

- *Conduct a pretest:* Before going into the field, test the questionnaire on a small group of respondents from the target population. This helps to check whether the survey accurately captures the intended information. This is a primitive way of testing. A lot more can be found about questionnaire testing in Converse and Presser (1986). Questionnaires of web surveys are now increasingly evaluated by qualitative testing methods and redesigned to reduce the burden for respondents and to increase data quality, especially in official statistics. However some testing innovation has been recently adopted by well famous survey/panel research societies, like Ipsos and Gallup. For example, Ipsos is undertaking the study of facial emotions to measure them and as possible tool to check questionnaire design (Ipsos, 2016). Pretesting web questionnaires shall improve their usability, functionality, and comprehensibility. Innovative qualitative testing includes also innovative approaches like emotional reactions (for example, facial movements and eye movements) and cognitive interviewing. Eye tracking can be a very useful method in pretesting questionnaires because it delivers insights into the unconscious behavior of probands. For example, the perception and understanding of navigation, error messages, and instructions is well captured from the eye-tracking experiment. For example, at the Swiss Federal Statistical Office (FSO), a three-step approach is applied. At first eye movements and facial expressions (in real time) are observed while respondents deal with the questionnaire. Afterward, it is run a cognitive interviews, in which the video of eye movements during the questionnaire completion is shown to the probands. Third step consists in analyzing the eye-tracking data (fixation length, number of fixations, etc.). Mixing the two methods of evaluation weaknesses and strengths of each method is balanced to obtain a high-quality questionnaire. Pretest should include also the check of the layout in mobile devices.
- *Pay attention to the way the survey is presented.* The way the potential respondents are approached and the survey is presented to them are vital to obtaining their participation. Failure to carry out a proper contact approach may easily result in refusal to respond. It has to be borne in mind that when you are conducting a web survey, you are conducting a mobile survey, if it has not been decided to prevent mobile access to the survey. Therefore, the contact presentation should be simple, and the way to access to the questionnaire should clearly offer the possibility to complete the survey using a different device. SMS invitation is an effective way for mobile surveys.
- *Include instructions.* Wherever needed, question-specific instructions should be incorporated into the survey instrument. Avoid offering instructions on separate sheets, in booklets, or on web pages. If instructions are included in a question in the paper version of the questionnaire, they should also appear in the electronic version of the questionnaire. Help facilities can be more

powerful in web surveys than in interviewer-assisted surveys. A first advantage is that the help information is always presented in the same, consistent way. It does not depend on the interviewers. A second advantage is that web survey respondents need not ask a person for help. They can just click on a button or link to open help window. Respondents in an interviewer-assisted survey may be reluctant to ask the interviewer for clarification, because this would mean admitting ignorance. Moreover, they also may feel embarrassed by asking for clarification about everyday concepts, although the terms may be used in atypical or ambiguous ways. It seems unlikely that web survey respondents will be as embarrassed to ask a computer more information. Nevertheless, the literature on social presence suggests computers do trigger similar self-presentation concerns and so could potentially prevent requests for clarification. See Tourangeau, Couper, and Steiger (2003) for a discussion of social presence in web surveys.

- **Pay attention to technical aspects.** On the one hand, the limitations of the available hardware and software may restrict what can be done with a web survey, both at the design stage and at the data collection stage. On the other hand, powerful hardware and software may open new possibilities.

- *Design stage aspects.*
  There are nowadays many different technological environments available for the survey researcher. There is difference in users, browsers, and settings. In addition, there is a variety of devices providing Internet access (laptops, desktop computer, smartphones, tablets, etc.). They vary not only on browsers but also in screen size and resolution and in many customizations. Some technical solutions are now available to minimize differences across devices and the browsers; however attention has to be devoted to optimize the questionnaire usability for alternative device characteristics. See Dillman, Smyth, and Christian (2014) for more details. If the questionnaire becomes inaccessible, or difficult to complete, for some groups of respondents, response rates will drop. Moreover, high difference in downloading the questionnaire and in the interaction might lead to bias and drop-off for specific groups of respondents. Incorrect visualization of questions will result in measurement errors and thus in biased results. This is especially relevant in mobile web surveys.
  Another important aspect is that the current stage of the technology makes it possible to control access to the questionnaire with a unique identification code for each sample element. This approach is recommended since it guarantees that only selected elements can obtain access to the questionnaire. The unique code can also help to prevent someone completing a questionnaire more than once. The unique identification code should be included in the e-mail or letter sent to the selected survey elements. Such codes are often part of the link to the survey website. Using code access to the questionnaire usually has no effect on survey participation.

EXAMPLE 7.1 **The library study**

The library of the Faculty of Economics of Bergamo University conducted a customer satisfaction survey in 2007. The target population of the survey consisted of all students of the faculty. Since all students get an e-mail address when they register for the university, it was decided the use a web survey.

Objective of the survey was the evaluation of the library services (opening times, books availability, interlibrary delivery, etc.), room space, and equipment (computers). The questionnaire contained also some questions about the use of the Internet, particularly about the use of e-mail.

Students were contacted by e-mail. The e-mail message contained a link to the questionnaire and a unique identification code. The students needed to enter the code in order to get access to the questionnaire. Figure 7.1 contains the access screen.

**Università di Bergamo**
**Servizi bibliotecari**

Username

Password

Invia

FIGURE 7.1 **Using a unique code to get access to the questionnaire. Source: University of Bergamo**

A total of 1,273 students completed the questionnaire. A simple constant progress indicator was used to inform students about their progress in answering the questions. The page number and the total number of pages were reported in every page. It was a simple, non-graphic message showing the question number and the total number of questions. Since the questionnaire was simple, providing information about the progress of the completion process is an adequate approach. Figure 7.2 shows the screen presenting the question about the frequency of the checking of the e-mail address where they have received the survey link access.

The results of the question in Figure 7.2 give some insight in the e-mail checking behavior of the students. Table 7.1 summarizes these results.

FIGURE 7.2 **A question of the survey and the page number. Source: University of Bergamo**

TABLE 7.1    Frequency of checking e-mail

| Checking | Number | Percentage |
|---|---|---|
| Once or more every a day | 533 | 41.9 |
| Once or more every week | 552 | 43.4 |
| Once or more every month | 145 | 11.4 |
| Less than once a month | 43 | 3.4 |
| Total | 1,273 | 100.0 |

It is clear that many survey respondents check often their e-mail. It should be noted that students who never check their e-mail, or only will do so now and then, will not be well represented in the survey.

It is important to keep respondents motivated while they are in the process of completing a self-administered questionnaire. A feature that may help to accomplish this in a web survey is to include a *progress indicator*. This is a textual or a graphical device that gives feedback about the progress of the respondent in the questionnaire.

▣  EXAMPLE 7.2 **A progress indicator**

Figure 7.3 shows an example of a progress indicator that was used in a radio listening survey. This indicator took the form of a black bar that increases in length as the respondent progresses through the questionnaire.

The questionnaire of the radio listening survey was fairly straightforward, with a limited number of questions and without complex routing. This makes a progress indicator ideal as a tool for motivating respondents to complete the questionnaire.

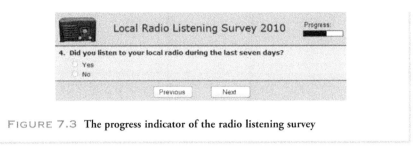

FIGURE 7.3 **The progress indicator of the radio listening survey**

Progress indicators are widely used in web surveys. However, their usefulness depends on the characteristics of the survey and the context. The prevailing view among survey designers in early time of web surveys seemed to be that this information was appreciated by respondents. The feeling was that they will be more likely to continue answering questions if they have some sense of where they are in the process. However, a negative effect is also possible: if progress is slow, respondents may become discouraged, resulting in breaking off their task. Progress indicators are useful in surveys related to individuals or households and even in quite simple surveys. In complex surveys and in business surveys (especially in surveys requesting business data, such as financial and financial data), progress indicators are less useful since data must be collected before completing the questionnaire.

Recently many studies have been carried out to test hypotheses concerning the effect of progress indicators on survey participation. Prevailing conclusion is that constant progress indicator don't significantly increase participation. The results suggest that when progress seems to surpass the expectations of the respondent, feedback can significantly improve completion rates. When progress seems to lag behind what they expect, feedback reduces engagement and completion rates. Furthermore, eventual effectiveness of the progress indicator varies depending on the speed of the indicator itself (Villar, Callegaro, and Yang, 2013).

- *Data collection stage aspects.*
  Progress in the development of Internet technology has greatly improved the quality and possibilities of online access of households and businesses participating in surveys. However, the survey designer has to be aware that one of the major problems of web surveys is that it is never clear how the questionnaire will appear on the computer of the respondents. Their browsers may not be able to display certain pictures or symbols. A low screen resolution may require respondent to scroll to make different parts of the questionnaire visible. A low bandwidth may slow down navigation through the questionnaire. There are many different configurations, including the platform or operating system. There are differences in hardware (desktop, laptop, mobile phone, smartphone, tablet, etc.), screen size, Internet browser, versions of the same Internet browser, processor speed, available memory, and so on. All these differences may cause problems for

large, complex, sophisticated web survey questionnaires. However, if the questionnaire is kept simple, one may expect that respondents will not encounter technical problems, whatever the environment they use. Nevertheless, survey designers should be aware of potential problems and therefore should test their survey instruments in many different environments. Mobile web surveys may have more complex technical problems. Form factors like screen size, display width, display height, and pixel per inch vary in smartphones. They probably will change in the near future. Thus, an important point is to understand the least common denominators of these factors at the survey design stage. Understanding differences in operating systems and the types of scripting allowed (e.g., Adobe Flash, Java scripting, etc.) is also important for determining whether certain question types can be displayed or implemented in the mobile format. If a survey is completed from smartphones, users can browse the Internet via native mobile browsers that are running on the operating system of the device. If a mobile device accesses a website without a mobile version, the user will still be able to navigate the page. However, differences in screen size will usually require the user to perform gestures or scrolling in order to browse the entire content. Therefore, if a smartphone accesses a nonmobile version of a web survey, then it is likely that the respondent will see only a portion of the content of question. The smartphone user would likely have to scroll horizontally and/or vertically as well as to zoom in to select a response or press the next button. Nevertheless, allowing smartphone users to access such web surveys requires neither content modification nor development of a mobile version of the survey. Buskirk and Andrus (2012) call such online surveys completed via smartphone Passive-Mobile Browser Surveys (P-MBS). These surveys not require additional formatting of the questionnaire, programming cost are similar to those of web surveys, and the number of the interviewees that can be contacted and that can participate can be larger. On the other side, there are some disadvantages since the questionnaire may not be viewable without additional effort of scrolling or zooming. For example, a multiple-choice question may require zooming in order to select desired response. Answering questions becomes boring and longer. As a consequence, higher survey drop-off is registered, even for short surveys. Longer surveys may be difficult to process due to long time in loading, in transferring data, and in the adapting the format.

A preferable approach consists of reformatting and deploying the web survey as a mobile survey. Buskirk and Andrus (2012) call these surveys Active-Mobile Browser Surveys (A-MBS). This type of survey has many advantages, like limited horizontal scrolling, navigational tools clearly visible, graphical format adequate to mobile screen, and other user-friendly format aspects. However, there are some limitations and disadvantages, like GPS and other data capture don't work, implementation costs are a bit higher, constant Internet connection is required, and the speed of the connection can be somewhere long or difficult. Therefore, longer questionnaire could not be friendly managed by the interviewed individual. Sometimes format could be not adequate, i.e., some scrolling is required, or some markup languages are incorrectly presented.

Other approaches to mobile web surveys are based on the use of apps. However, the use of apps is not without disadvantages and limitations in the current cultural and technical environment. For example, people using apps do not cover the whole population, neither the Internet user population; therefore, coverage problems are increased. Changes are to be expected in the future of the mobile surveys due to constant technological change and digital culture in the interviewee. New generation is more familiar in handling apps as well as in writing fast and without problems on small screens. As far as new generation is getting older, the mobile users will generally be more skilled. The advice is to stay compatible with the current Internet access trends.

Another important issue arises when a mixed-mode approach is applied, i.e., if in addition to mobile web the respondent could complete the questionnaire using one or more alternative modes, like paper or telephone. For instance, in some cases, respondents are allowed to choose the mode they prefer for completing the questionnaire. Smyth, Olson, and Kasabian (2014) evaluate whether changes in question format have different effects on data quality for those responding in their preferred mode than for those responding in a non-preferred mode for three question types (multiple answer, open ended, and grid). Their conclusion is that on the multiple answer and open-ended items, those who answered in a non-preferred mode seemed to take advantage of opportunities to satisfice when the question format allowed or encouraged it (e.g., selecting fewer items in the check-all than the forced-choice format and being more likely to skip the open-ended item when it had a larger answer box), while those who answered in a preferred mode did not. There was no difference on a grid-formatted item across those who did and did not respond by their preferred mode. However, it was found that a fully labeled grid reduced item missing rates compared with a grid with only column heading labels. The effect of tailoring to mode preference on commonly used questionnaire design features is especially relevant for those who no choice of the preferred mode is allowed.

- *Make it attractive to complete the questionnaire.* Web surveys are self-administered surveys. There are no interviewers to motivate respondents to complete the questionnaire. Respondents will be motivated if the questionnaire is attractive. The decision to participate is to a large extent determined by respondent motivation, which depends mainly on the topic of the survey, and satisfaction. The design of the web questionnaire should provide the respondents with as much pleasure and satisfaction as possible in order to increase their commitment in completing the questionnaire. This helps in convincing respondents to answer all survey questions and to complete the questionnaire. Item nonresponse and partial nonresponse are minimized. The so-called interactional information system approach (a questionnaire not only collects data but also provides respondents with relevant information) may be useful in generating and maintaining interest in the survey, thereby increasing response rates. This approach is in line with experimental studies that highlight the importance of placing interest-related questions early in the questionnaire as this would be attrition due to lack of interest. See Shropshire, Hawdon, and Witte (2009).

- *Apply questionnaire design principles.* As previously stated, web questionnaire design can partly be based on the principles for paper questionnaire construction. These criteria are a valuable starting point, but also two specific aspects of web surveys should be considered. First, it should be borne in mind that even small differences in question wording or of a stimulus embedded in the question display may greatly affect the answers given. Second, it should be noted that checks can be included in a web survey questionnaire. This feature is also available in computer-assisted surveys modes like CAPI and CATI, but not in a paper-based approach. Every change in web survey questionnaire may make it easier or harder to complete it or may convey different messages about what kind of information is expected. Summing up, this may have a serious impact on the collected data.

The recipe for designing a web survey questionnaire in a scientifically sound way involves the following specific ingredients:

**a.** The proper technology (hardware, platform, software, and so on).

**b.** Well-organized survey management, both in the design phase and in the data collection phase.

**c.** Availability of skills for defining question and response format, being aware of the possible impact of the offered response choices and of the possible ways in which the questionnaire can be visualized and customized using the available Internet technology.

**d.** Anticipating the collection of other data (paradata, auxiliary information) during the survey that may be linked to the interview. This aspect is typical of web surveys and—if used appropriately—may improve participation and the quality of the survey process.

Technical aspects will not be discussed in the subsequent chapters. Attention will also not be paid to organizational issues, as these issues depend very much on the context in which the survey is being conducted and the objectives of the survey.

The focus is on various aspect of questionnaire design (point c), with special attention to the issues that are typical for web questionnaires. In addition, Section 7.2.5 introduces the concept of paradata. An attempt is made to answer the question how auxiliary information and data from other linked sources can be used during the survey process (point d).

Web questionnaires started as simple electronic analogues of paper questionnaires. Over time, web questionnaires have evolved. There is ample research about applying new technological innovations. Clear criteria have been formulated, but some issues are still under investigation. Criteria will be discussed that are related to the following themes:

- Formatting the text of the questions;
- Formatting the answers of the questions (closed questions, open questions);
- Paradata collection.

## 7.2.2 THE LANGUAGE OF QUESTIONS

The design of a question requires decisions to be made with respect to its format and layout. The format of the question includes aspects like the wording of the question and the type of answer (the answer format) that must be given. The layout of the question includes aspects like font type, font size, and use of colors. All these aspects are called the language of questions.

Two very basic types of survey questions can be distinguished: open questions and closed questions. A *closed question* offers a (not very long) list of answer options. The respondent has to select one option. A closed question may be used if there are a limited number of possible answers and the researcher wants to avoid the respondent to overlook an answer. Such a question may also be used if the researcher wants to define his own classification of possible answers. A closed question is a tool to measure a categorical (qualitative) variable. Figure 7.4 shows an example of a closed question (from a radio listening survey).

A special case of a closed question is a *check-all-that-apply* question. This is an open question for which more than one answer option may be selected. Another special case is a closed question with ordered categories. An example is an opinion question with the answer options *Strongly agree, Agree, Neutral, Disagree,* and *Strongly disagree.*

```
1. In the last seven days, what type of music did you listen to most?
   ⊙  Chart / Top 40
   ⊙  Dance
   ⊙  Rock
   ⊙  R & B
   ⊙  Hip-hop
   ⊙  Country
   ⊙  Folk
   ⊙  Easy listening
   ⊙  Jazz
   ⊙  Classical
   ⊙  New age
   ⊙  Other music
```

FIGURE 7.4 **Closed question**

The other basic question type is the open question. For such a question, the respondents may enter any answer they like. There are no restrictions other than the length of the answer. Use open questions where there are a very large number of possible different answer options, where the researcher does not know all possible answer options, or where one requires the respondents to give the answer in their own words. Space provided for the open answer should be of moderate length, but not very short. The length of a given answer depends on the amount of space provided. So, if the researcher wants long texts, he should offer a lot of space. Figure 7.5 shows an example of an open question.

If the list of answer options is very long, if the complete list of answer options is unknown, or if there may be unanticipated answers, one may decide to use a question type that is mix of a closed and an open question (hybrid question). The main options are listed, and all possible answers can be dealt with by selection the option "Other, please specify" and entering the answer if it were an open question. See Figure 7.6 for an example.

The question language of a survey includes the wording of the text of the questions, instructions for answering them, and visual aspects such as font size, font type, color, layout, symbols, images, animation, and other graphics. Couper (2000, 2008) includes visual language in the question language since it is intended to supplement the written language. In fact, the way the questionnaire is visualized can have a great impact on the way the questions are answered. Abundant nonfunctional use of graphical effects may draw away attention from the text, alter the meaning of words, and complicate easy and straightforward understanding of the questions. In summary, the various visualization aspects together affect the perception of the survey questionnaire and the response burden. A more

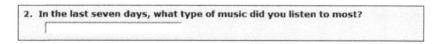

**2. In the last seven days, what type of music did you listen to most?**

FIGURE 7.5  **Open question**

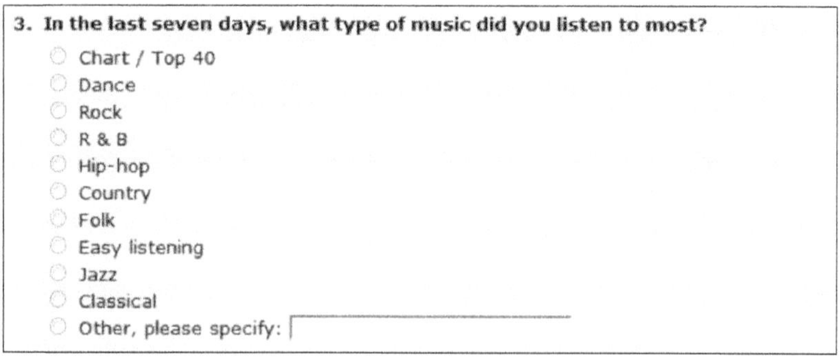

**3. In the last seven days, what type of music did you listen to most?**
- ○ Chart / Top 40
- ○ Dance
- ○ Rock
- ○ R & B
- ○ Hip-hop
- ○ Country
- ○ Folk
- ○ Easy listening
- ○ Jazz
- ○ Classical
- ○ Other, please specify:

FIGURE 7.6  **A hybrid question**

detailed description of the *visual language*, as proposed by Redline and Dillman (1999), includes three different types of languages:

- *Graphic language*. This language consists of fonts, font sizes and font enhancements (bold, italics, underline), borders, and tables. When used in a functional way, it helps respondents to move their eyes across the page and to comprehend the questionnaire.
- *Symbolic language*. This language is used mainly in questionnaires for navigation purposed. Symbols like arrows help leading respondents through the survey questions in the proper order and answering the proper questions.
- *Numeric language*. This language is used in questionnaires for numbering questions and for numbering answer options, too.

Visual language is an auxiliary language. It may help to make the questionnaire more attractive. However, the questionnaire designer should be aware that this may affect the way in which respondents may interpret questions. For instance, if graphics or pictures are used, bear in mind that respondents tend to interpret questions in the context of those graphics and pictures. For example, choosing to present a female or a male could let the question perceived in a different perspective; moreover, how the person is dressed or apparently healthy versus ill affects the perception of the questions and therefore the answer. Example 7.3 discusses an experiment in using pictures in a questionnaire.

---

📑  EXAMPLE 7.3  **Use of pictures in questions**

Couper, Tourangeau, and Kenyon (2004) describe an experiment in which questions were offered in three different ways: (1) without a picture, (2) with a picture indicating low-frequency behavior, and (3) with a picture indicating high-frequency behavior. One of these questions asked how many times one went out to eat in the last month. The low-frequency behavior picture was one of an intimate, exclusive restaurant, and the high-frequency behavior picture was one of a fast food restaurant.

When the low-frequency picture was included in the questionnaire, the frequency of going out to eat was 9.9%. When no picture was included, the percentage was higher: 12.0%. And when the high-frequency picture was included, the percentage was even 13.6%. Apparently, use of pictures partly determines what people mean by "going out to eat."

---

Many experiments with traditional surveys have shown that minor changes in question wording may substantially change responses. This also applies to web surveys. However, not only the wording of questions is important in web surveys but also the presentation of the questions. Minor changes in presentation may seriously affect the answers given.

In mobile web surveys, every type of language should be carefully evaluated. Symbolic languages are especially important, since used for navigational purposes. Therefore, symbols should be evident and appear in the right position also on the mobile device screen.

## 7.2.3 BASIC CONCEPTS OF VISUALIZATION

**7.2.3.1 Answer Spaces.** The *answer space* is the area on the computer or mobile device screen where respondents type their answers to the (open) questions. Answer space requires careful design. They should be easy to locate by standing out visually, and it should also be clear to respondents what is expected of them. In specific situations it may be helpful to include extra instructions.

Research results show that increasing the size of the answer box has little effect on early respondents of the survey but substantially improves the quality of the answers of late respondents. Including instructions and explanations improves data quality for both early and late respondents. Early respondents are receiving information they need and the quality of their answer is better, whereas late respondents take their time for understanding the question, if no instructions are given.

A consistent questionnaire design is important. If the same type of question is asked, the same type of answer space must be used. This reduces the cognitive task of the respondents. Research shows that respondents use all available information to help them to formulate an answer. That is, in addition to the questions themselves, they use the information provided by the response categories and the answer space. See Sudman, Bradburn, and Schwarz (1996; Sudman and Bradburn, 1973). Inconsistencies in the web questionnaire will confuse them and may lead to lower data quality or nonresponse.

For questionnaires in general, it is advised to surround answer spaces by a frame in a contrasting color. This clearly separates the answer spaces from the rest of the questionnaire page. This is particularly important in the following situations:

- The questionnaire has a light background color, so there is not enough contrast to distinguish white answer spaces from the rest of the page.
- The questionnaire is subject to key-from-image (KFI) processing. KFI involves separating the questionnaire forms into single pages, scanning the pages on high-speed scanners, and storing a digital image of each page in a central repository. Data entry operators then key the information from the images into the census system.
- The route through the questionnaire depends on the answer to a previous question. It is particularly important that such a filter question is answered correctly. See also Example 7.4

■ EXAMPLE 7.4 **Adding a frame to an answer space**

Question 25 of the R&D survey of the Italian Statistical Institute (ISTAT) asks whether the company has any branch or subsidiary abroad performing research and development. In addition, firms having such a branch or subsidiary have to provide information about R&D expenditures and personnel. Thus, a different answer scheme is provided related to the answer to the main question. So, the filter question in Figure 7.7 (possible answers yes or no) determines whether the subsequent matrix question has to be answered.

FIGURE 7.7 **Stressing the importance of a filter question with a frame.** Source: www.istat.it

Some research suggests framed answer spaces should be preferred because they decrease the cost of keying forms or increase accuracy if questionnaires are optically scanned and verified. It is sometimes also believed that framed answer spaces often require less interpretation on the part of the data entry typist or data editor. There is, however, no experimental evidence that this is the case. Recently, Singer and

Couper (2017) have suggested that adding a few open questions in the web questionnaire may yield important insights and is useful in motivating respondents. The analysis of textual answers is at the time being facilitated due to the diffusion of programs processing techniques like sentiment analysis and text mining. Thus, the use of open question could be encouraged in researches where more information from the respondent is desired and they should be better motivated in survey participation.

Cognitive testing of questionnaire instruments has revealed that respondents do not have a strong preference for open answer spaces or framed spaces, as long as the answer spaces are sized appropriately for the information being requested. In deciding in favor of or against framed answer spaces, it is better to rely on testing on respondents, data entry typists, and data editors.

As a general rule, it is desirable to use the same type and physical dimensions of answer spaces when requesting similar information. For example, if percentages or euro amounts are asked for in different parts of the questionnaire, it will help respondents if the same type of answer space is used and also if the same additional information and instructions is included (Christian, Dillman, and Smyth, 2007; Couper, Traugott, and Lamias, 2001).

If respondents are asked to enter values or amounts, they should do so in proper unit of measurement. They should be helped with that. Errors caused by entering amounts in the wrong unit of measurement are not uncommon. For example, a company enters its turnover in dollars instead of in thousands of dollars. To avoid or reduce these problems, question instructions should make clear what is expected of respondents. This can be accomplished by adding words or symbols near the answer space. For example, for some survey questionnaires, "000" is printed next to answer space to indicate that respondents should report in thousands of euros (or dollars). Other survey questionnaires have ".00" next to the answer space to make clear that responses are to be rounded to the nearest euro/dollar. There is no empirical evidence what works best for respondents. The main point here is that answer spaces should be consistent within a questionnaire.

---

**EXAMPLE 7.5** **Asking values in R&D survey**

The R&D Survey of ISTAT, the Italian Statistical Institute, is a compulsory business survey collecting data on research activities and expenses. This survey asks responding companies to report values in thousands of euros. To make this clear to the respondents, the column heading contains the instruction ("Report in thousands of euros"). Moreover, the zeroes necessary to form the complete number are already next to the answer space. See Figure 7.8 for an example.

FIGURE 7.8 **Helping the respondent to record values in the proper unit of measurement. Source: www.istat.it**

If respondents are asked to enter percentages, this should be clearly mentioned. The percent symbol should be placed adjacent to the input field. If relevant, it should be emphasized that percentages have to sum up to 100%. This can be done by placing "100%" at the bottom of the column. It is also possible to automatically compute the sum of the percentages and check whether this total is equal to 100%.

**7.2.3.2 Use of Color.** When respondents are presented with visual information in the questionnaire, they quickly decide which elements to focus on (Lidwell, Holden, and Butler, 2003; Ware, 2004). Regular and simple visual features are easier to perceive and remember. This is the so-called Gestalt principle of simplicity. Moreover, respondents are more likely to perceive answer spaces or response categories as being related to one another if they have the same color. *This is the Gestalt principle of similarity.*

To facilitate the comprehension process, white answer spaces should be displayed against a light-colored or shaded background questionnaire screen. As a result the small answer space tends to "rise" above the colored background. It is an identifiable graphic area that becomes an object of interest. Therefore, it is seen as more prominent. White answer boxes against colored backgrounds are especially important for use in optical imaging and scanning systems. In this case, there is a contrast between the white answer space and the colored background. Therefore, it is not necessary any more to put a frame around the answer frame.

Dividing lines tend to focus visual attention on the area around answer spaces, rather than the answer spaces themselves. Therefore, it is recommended not to use them unless necessary.

### 7.2.3.3　Use of Images.
Research literature has shown that small changes in the visual presentation of questions can lead to substantial variations in the response distribution and in the amount of time taken to completing questions in self-administered surveys. If the burden of answering the questions becomes too high, there is a serious risk: respondents may interrupt the completion of the questionnaire, resulting in (partial) nonresponse. Therefore, it is important to know why respondents break off questionnaire completion and which question formats require most time to complete.

One of the issues is the inclusion of images in the questionnaire, as they definitely affect the answers of the respondent. Thus, the decision to use images, and the choice of images, has consequences in terms of measurement error.

For example, if a picture of a person is included, several aspects of this person (male or female, ill or healthy, sportsman or dull civil servant, and so on) might greatly affect how the question is perceived and therefore the answer that is given. If a house picture is presented, the type of building can influence the answer letting the interviewee to refer to a house in the countryside rather than in the city center, to a villa rather than to a flat, and so on.

Unformatted images in a mobile web page will account for a disproportionate amount of overall display (Peytchev and Hill, 2010). The images will be too large with respect to the size of the text of the questionnaire.

**▣　EXAMPLE 7.6　Survey about mobile phones**

Including images can stimulate participation in the survey and clarify the characteristics of products about which questions are asked. These images should draw attention and stimulate interest of respondents in a neutral manner.

The example in Figure 7.9 is taken from a study by Arruda Filho and Biffignandi (2011). In a first study, the focus of the survey was on the characteristics of the mobile phone, not on the brand. Objective was to gain insight into new product development and to decide which features are important for consumers. The image in the question was kept simple and without displaying a brand name. Respondents were asked to give their opinion using a Likert scale.

In Study 2, respondents were asked to compare brands. Therefore, mobile phone images were related to brand names. The comparison is between a new brand (called *Mandarina*) and a well-established brand (*Nokia*). Figure 7.10 contains the two questions for the Mandarina mobile phone.

To create a perception of a simple, unknown brand, the Mandarina image was kept simple, while the well-established brand was represented by a more appealing image. Figure 7.11 contains the two questions for the Nokia mobile phone.

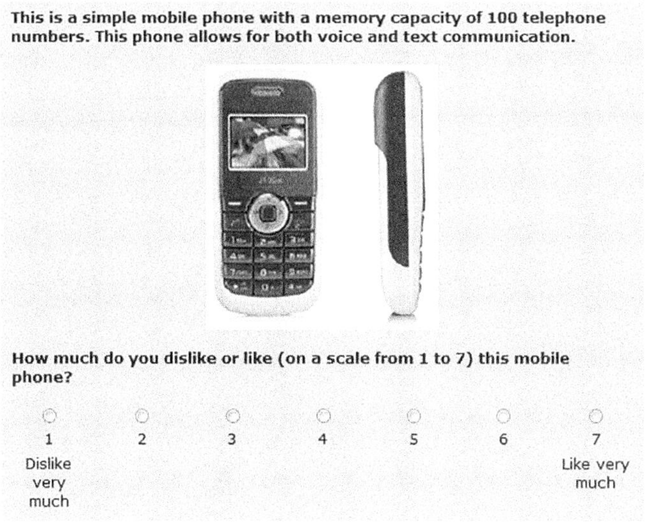

This is a simple mobile phone with a memory capacity of 100 telephone numbers. This phone allows for both voice and text communication.

How much do you dislike or like (on a scale from 1 to 7) this mobile phone?

| ○ | ○ | ○ | ○ | ○ | ○ | ○ |
|---|---|---|---|---|---|---|
| 1 | 2 | 3 | 4 | 5 | 6 | 7 |

Dislike very much        Like very much

FIGURE 7.9  A mobile phone survey question (study 1)

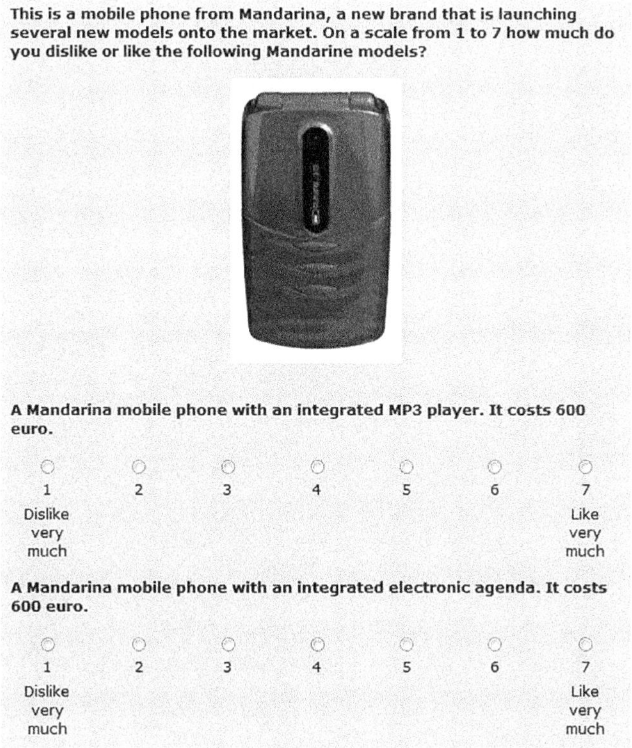

This is a mobile phone from Mandarina, a new brand that is launching several new models onto the market. On a scale from 1 to 7 how much do you dislike or like the following Mandarine models?

A Mandarina mobile phone with an integrated MP3 player. It costs 600 euro.

| ○ | ○ | ○ | ○ | ○ | ○ | ○ |
|---|---|---|---|---|---|---|
| 1 | 2 | 3 | 4 | 5 | 6 | 7 |

Dislike very much        Like very much

A Mandarina mobile phone with an integrated electronic agenda. It costs 600 euro.

| ○ | ○ | ○ | ○ | ○ | ○ | ○ |
|---|---|---|---|---|---|---|
| 1 | 2 | 3 | 4 | 5 | 6 | 7 |

Dislike very much        Like very much

FIGURE 7.10  Questions for the Mandarina mobile phone (study 2)

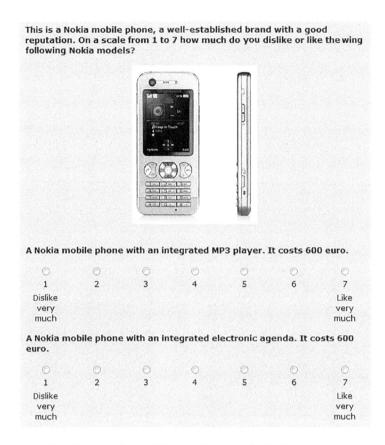

This is a Nokia mobile phone, a well-established brand with a good reputation. On a scale from 1 to 7 how much do you dislike or like the wing following Nokia models?

A Nokia mobile phone with an integrated MP3 player. It costs 600 euro.

| ○ | ○ | ○ | ○ | ○ | ○ | ○ |
|---|---|---|---|---|---|---|
| 1 | 2 | 3 | 4 | 5 | 6 | 7 |
| Dislike<br>very<br>much | | | | | | Like<br>very<br>much |

A Nokia mobile phone with an integrated electronic agenda. It costs 600 euro.

| ○ | ○ | ○ | ○ | ○ | ○ | ○ |
|---|---|---|---|---|---|---|
| 1 | 2 | 3 | 4 | 5 | 6 | 7 |
| Dislike<br>very<br>much | | | | | | Like<br>very<br>much |

FIGURE 7.11  Questions for the Nokia mobile phone (study 2)

## 7.2.4 ANSWERS TYPES (RESPONSE FORMAT)

The two basic question types were already mentioned: open questions (people answer the question in their own words) and closed questions (the answer is selected from a list of possible answer options). The answer options can take the form of an unordered list, like in Figure 7.4. The answer options can also take the form of an ordered list like a rating scale (*Strongly agree, Agree, Neutral, Disagree,* and *Strongly disagree*). Another frequently occurring example is a yes/no question where the respondent has to select either *Yes* or *No*. It is also possible to allow the respondent to select more than one answer option. Then the closed question takes the form of a *check-all-that-apply question*.

There are various ways to visualize an answer on the computer screen. Every comment on the response format in web surveys holds for mobile web surveys, if not differently described.

It is important that the visual format corresponds to what is expected of the respondents. The HTML language that is used to define web pages allows for the following constructs:

- Radio buttons;
- Checkboxes;
- Drop-down boxes and list boxes;
- Text boxes and text areas.

Radio buttons, checkboxes, and drop boxes are alternative solutions for closed questions. Text boxes and text areas are used for open questions. In the following sections, different answer types are described and discussed. It is remarked again that the visual aspects of questionnaire design are very important. There is more to web questionnaire design than just mimicking the paper questionnaire on the computer screen. Small changes in the design and layout may have large consequences for the answers to the questions.

**7.2.4.1 Radio Buttons.** *Radio buttons* should be used for closed questions where respondents can select only one answer option. The answer options must be mutually exclusive, and together they must cover all possible answers.

Initially, no option is selected. This forces the respondents to select an option. They cannot continue to the next question without thinking about the current question. An answer is selected by clicking the radio button corresponding to the answer option. If the respondent clicks another answer option, the previously selected option is deselected. Not more than one option can be selected.

To avoid confusion among respondents, radio buttons are used to implement closed questions with one answer. It is technically possible to use radio buttons also for check-all-that-apply questions, but this is not recommended.

One limitation of the use of radio buttons is that their size is not related to the size of the font attached to it. Radio buttons keep the same size irrespective of the font size. It is advised to use a font of approximately the same size as the radio button size.

The answer options should be presented in a logical order. Alphabetical ordering is not recommended because it creates problems for multilingual questionnaires. This would mean a different order for a different language. Data processing would become more complex due to different order of the items in various languages.

A point of discussion is always whether to include answer options "Don't know" or "Does not apply." On the one hand, the danger of including "Don't know" is that people may select it to avoid having to think about an answer. On the other hand, if people really do not know the answer, they must have the possibility to answer so. This is generally more a point of discussion of opinion questions than for factual questions.

▓ EXAMPLE 7.7 **Answering a closed question**

A simple example shows what happens when a closed question is answered if answer options are denoted by radio buttons. The question is, "What kind of product were you thinking about while filling in this questionnaire?" There are five possible answers: "Electronics," "Clothes/ accessories," "Watches," "Cars," and "Other product." The respondents can only give one answer because they can only think of one product. Therefore, radio buttons are the appropriate form to define this question. The possible answers are mutually exclusive. The option "Other product" guarantees a respondent that always an option is appropriate.

Initially, the question looks like in Figure 7.12. No answer has yet been selected. This is done to prevent respondents from skipping the questionnaire (because it already has been answered). This would lead to bias toward the default answer option. Figure 7.13 shows the situation after the respondent clicked the radio button for the option "Cars."

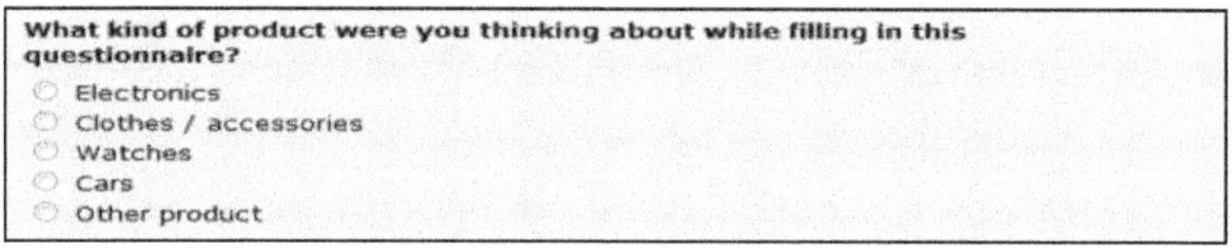

**What kind of product were you thinking about while filling in this questionnaire?**
○ Electronics
○ Clothes / accessories
○ Watches
○ Cars
○ Other product

FIGURE 7.12 **A closed question in its initial state**

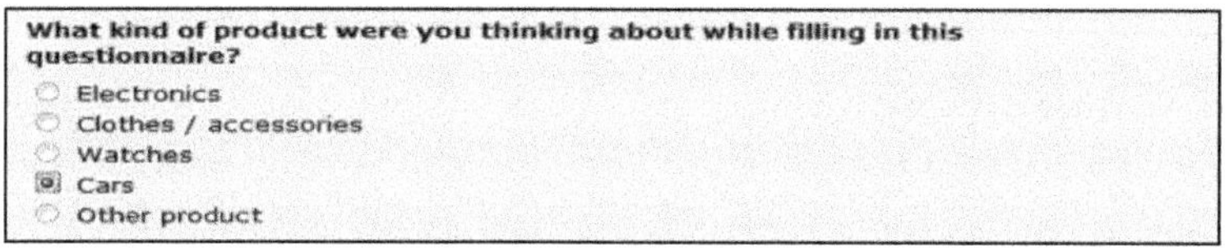

**What kind of product were you thinking about while filling in this questionnaire?**
○ Electronics
○ Clothes / accessories
○ Watches
◉ Cars
○ Other product

FIGURE 7.13 **A closed question after selecting an answer**

If the respondent decides that he made a mistake and that "Electronics" is the proper answer, he can just click the radio button corresponding to this option. The result is shown in Figure 7.14. The answer "Cars" has been deselected automatically.

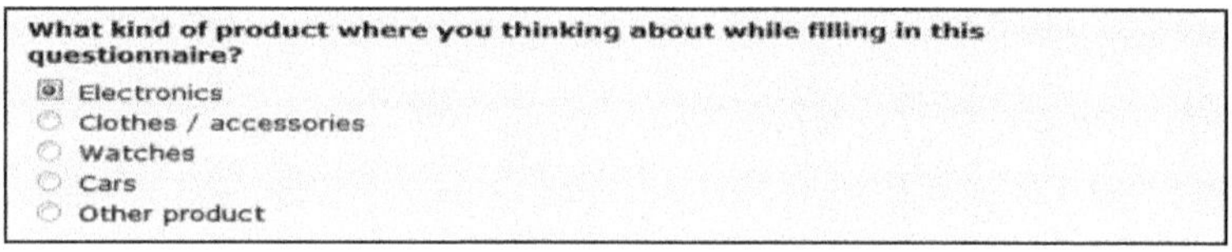

**What kind of product where you thinking about while filling in this questionnaire?**
◉ Electronics
○ Clothes / accessories
○ Watches
○ Cars
○ Other product

FIGURE 7.14 **A closed question after selecting another answer**

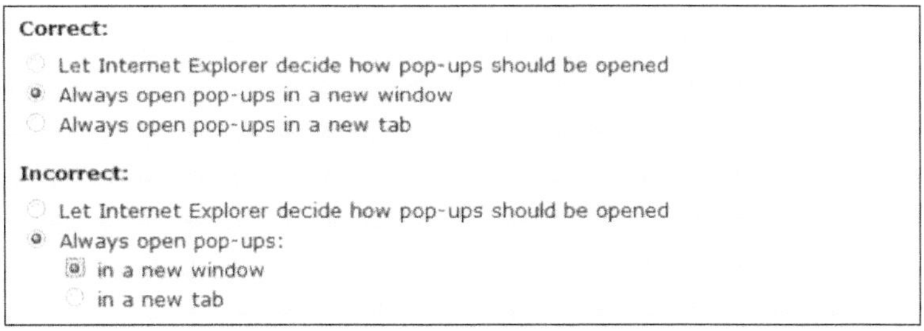

FIGURE 7.15  A closed question with nested answers

In the examples up until now, there was just a list of possible answer options. It is also possible to offer more complex structures. Figure 7.15 contains an example of nested options. Such a hierarchical structure should be avoided as it may cause confusion among respondents. It is preferred to keep all options at the same level.

A closed question can be used to measure a value on a scale. This very often takes the form of a *Likert scale*. When responding to a Likert scale question, respondents specify their level of agreement to a statement. The scale is named after its inventor, the psychologist Rensis Likert. Often five ordered response options are used, although some researchers advocate using seven or nine levels. A typical Likert questions has the response options *Strongly agree, Agree, Neither agree nor disagree, Disagree*, and *Strongly disagree*. Figure 7.16 shows an example of a Likert scale question.

A Likert scale question measures an ordinal variable, i.e., the answer options have a natural ordering. It is even possible to assign numerical values to the response options (1 = strongly agree, 2 = agree, etc.) so that in fact a numerical variable is measured and for examples mean scores can be computed.

A Likert scale question has the advantage over a simple yes/no question that it allows respondents to give a more differentiated answer than just yes or no. It even allows respondents to have no opinion at all. The other side of the coin is that respondents can select the neutral middle option to avoid having to give an opinion.

Use of Likert scales is not without problems. The description of the response categories must be accurate as possible. All respondents must interpret the descriptions in the same way. This is not always easy to realize. What does "Strongly

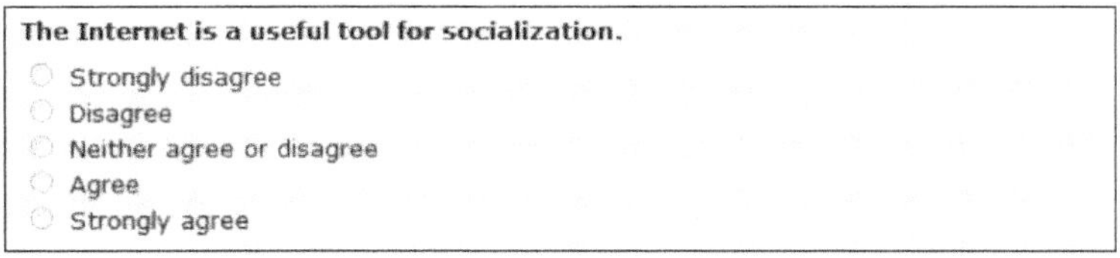

FIGURE 7.16  A Likert scale question

agree" mean? What is the difference with "Agree"? Likewise, respondents find it hard to distinguish "good" from "very good" and "very good" from "excellent." Sometimes closed questions ask for the frequency with which activities are carried out. Words like "often" or "sometimes" should be avoided as it is unclear what they mean. These words are sometimes called "vague quantifiers." They could be understood differently by different respondents. Problems could increase even more in multilingual questionnaires if these words have to be translated in different languages. For question asking about frequencies, a solution is to relate the activities to concrete time periods, like "every day," "at least once a week," etc.

Figure 7.15 shows just one way to display a Likert scale question on the screen. Figure 7.17 shows another way to do it. The response categories are now placed horizontally. This may correspond better to how respondents visualize a scale.

It is also possible to replace the response category labels by numbers. For example, the radio buttons in Figure 7.18 could be numbered 1 to 5. If this is done, the question text must explain that 1 means "Strongly disagree" and that 5 corresponds to "Strongly agree." Experience has shown that numbered scales are difficult to handle for people. For example, scales that are marked "1 to 5," with 5 being the highest, require more cognitive efforts than scales with labels such as "poor" or "excellent." Some studies (see, for instance, Christian, Parsons, and Dillman, 2009) show that response times are longer for scales with numeric labels, but there are no differences in response patterns. Numbered response categories may help if they are combined with labels. Figure 7.18 contains two examples. In the first example, all categories are labeled, whereas in the second example only the endpoints are labeled.

Respondents interpret the endpoint label for the low end of the scale as more negative or extreme when negative numbers are used. Research also indicates that

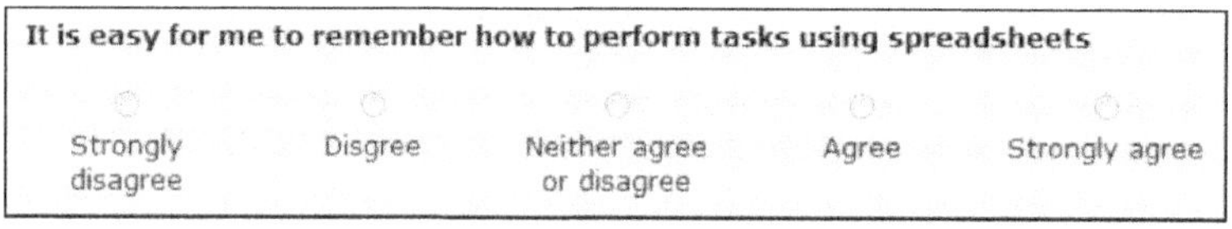

FIGURE 7.17  A Likert scale question with horizontal categories

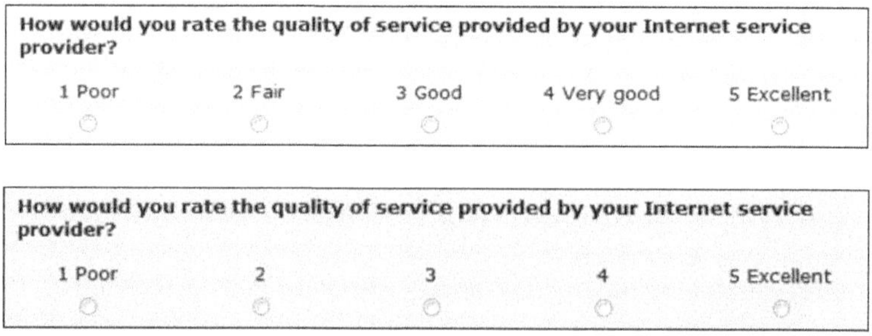

FIGURE 7.18  Likert scale questions with numbered and labeled categories

scales with numeric labels produce results that are similar to scales without numeric labels. This suggests a hierarchy of features that respondents pay attention to, with text labels taking precedence over numerical labels and numerical labels taking precedence over purely visual cues, such as color. See also Tourangeau, Couper, and Conrad (2007).

Some researchers prefer five-point scales (or seven-point scales) because they offer respondents a "neutral" middle point. Other researchers prefer an even number of response options (for example, a four-point scale), because they "force" people to select a negative or positive answer in order to avoid people going into the neutral answer. Forcing people is limiting the respondent in showing its effective position; thus the approach is trying to modify the real situation. It is to be critical in applying odd scales.

---

■ EXAMPLE 7.8 **Asking about the use of mobile phones**

Sometimes an extra response option is added to the scale indicating that the question does not apply. This is illustrated in Figure 7.19. A five-point Likert scale has been used. Note that all response options have been numbered. Only the extreme options and the middle option have a text label.

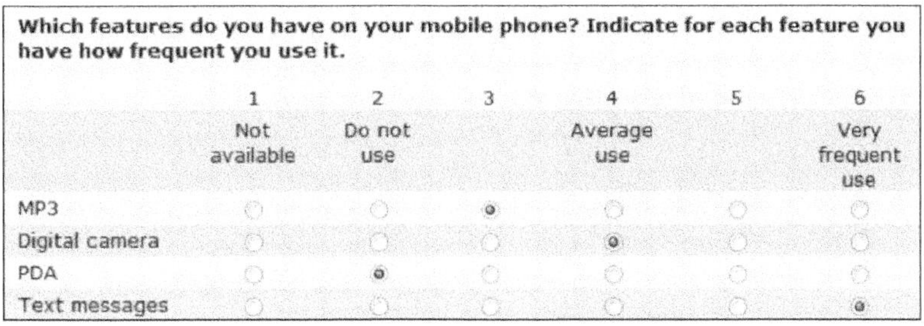

| Which features do you have on your mobile phone? Indicate for each feature you have how frequent you use it. | | | | | | |
|---|---|---|---|---|---|---|
| | 1 | 2 | 3 | 4 | 5 | 6 |
| | Not available | Do not use | | Average use | | Very frequent use |
| MP3 | ○ | ○ | ◉ | ○ | ○ | ○ |
| Digital camera | ○ | ○ | ○ | ◉ | ○ | ○ |
| PDA | ○ | ◉ | ○ | ○ | ○ | ○ |
| Text messages | ○ | ○ | ○ | ○ | ○ | ◉ |

FIGURE 7.19 **Adding an extra option to a Likert scale question**

Note that the first response option is not part of the Likert scale. It has been added for respondents that do not have the specific options in their mobile phone.

In fact, this is a matrix question containing four separate questions. To make it easier for the respondents to find the proper answer for each question, the rows have alternate background colors.

---

The treatment of "Don't know" is always a point of discussion in surveys. The general rule is applied here to make a distinction between factual questions and opinion/attitudinal questions. In case of a factual question, the respondents should know the answer. Therefore "Don't know" is not offered as one of the response options. The situation is different for opinion and attitudinal questions. If respondents are asked about their opinion about a specific issue, it is possible that they do not have an opinion. Therefore "Don't know" should be offered as a possible response option. There is a risk, however, that respondents select this option to avoid having to express their opinion. This is called satisficing.

If "Don't know" and possibly also "No opinion," are included in the list of response options, the question arises at which position to put in the list. Tourangeau, Couper, and Conrad (2004) have conducted experiments where these non-substantive options were at the bottom of the list and visually separated from the other options by means of a dividing line (see Figure 7.20). The result was that more respondents selected "Don't know" because their attention was drawn to it. Without the dividing line they observed an upward shift in the answers, as many respondents tended to select an answer in the visual middle of the list and they considered "Don't know" and "No opinion" part of the rating scale.

Derouvray and Couper (2002) experimented questions where the "Don't know" option was displayed in a smaller and lighter font so that its visual prominence was reduced. This did not affect the number of respondents selecting this option.

Not offering "Don't know" in opinion questions is not an option. This would imply that respondents always have to provide a substantive answer, even if they do not know the answer. According to Couper (2008), this violates the norm of voluntary participation. Respondents should have the possibility to not answer a question. Forcing respondents to answer may frustrate respondents resulting in break-off. Also, Dillman, Smyth and Christian (2014) strongly recommends not forcing respondents to answer. He warns for detrimental effects on respondent motivation, on data quality, and on the risk of break-off.

Matrix questions (or grid questions) include a series of survey items with the same answering categories on a relatively similar topic. They are presented as a whole, and answering categories are shown only once. They form a block of items that demand more screen size or a different type of navigation. Matrix questions seem to have some advantages. A matrix question takes less space on the questionnaire form than a set of single questions, and it provides respondents with more

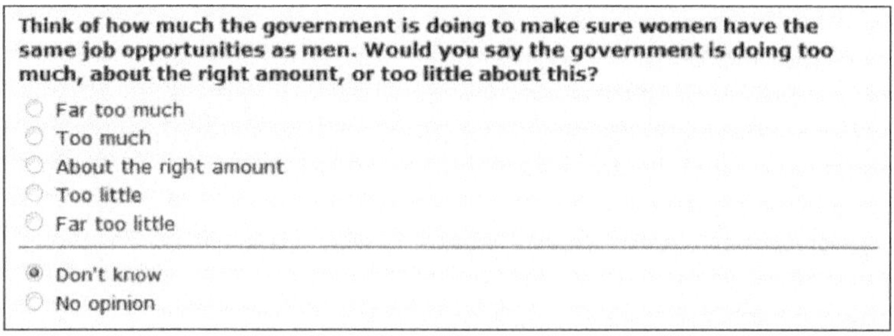
**Think of how much the government is doing to make sure women have the same job opportunities as men. Would you say the government is doing too much, about the right amount, or too little about this?**

- Far too much
- Too much
- About the right amount
- Too little
- Far too little

---

- Don't know
- No opinion

FIGURE 7.20 **Including non-substantive options**

oversight. Therefore, it can reduce the time it takes to answer questions. Couper, Traugott, and Lamias (2001) indeed found that a matrix question takes less time to answer than a set of single questions.

However, matrix question has some disadvantages. The format is not recommended for mobile screens, due to the large size with respect to the size screen. On smartphones grid questions require scrolling, or navigation is changed through accordion, carrousel, or other modifications of the navigation.

Moreover, according to Dillman, Smyth and Christian (2014), answering a matrix question is a complex cognitive task. It is not always easy for respondents to link a single question in a row to the proper answer in the column. Moreover, respondents can navigate through the matrix in several ways: row-wise, column-wise, or a mix. This increases the risk of missing answers to questions, resulting in a higher item nonresponse. Shading the rows of the matrix, like in Figure 7.21, may help to reduce this problem.

Dillman, Smyth and Christian (2014) advise to limit the use of matrix questions as much as possible. If they are used, they should not be too wide to too long. Preferably, the whole matrix should fit on a single screen. This is not so easy to realize as different respondents may have different screen resolutions on their computer screens. If respondents have to scroll, either horizontally or vertically, they may easily get confused, leading to wrong or missed answers.

Several authors, for example, Krosnick (1991) and Tourangeau, Couper, and Conrad (2004), express concern about a phenomenon that is sometimes called *straight-lining*. Respondents give the same answer to all questions in the matrix. They simply check all radio buttons in the same column. Often this is the column corresponding to the middle response option. For example, respondents could make it easy for themselves by selecting all radio buttons in column 4 (average use) for the question in Figure 7.21.

If the response options of a closed question form an ordinal scale, this scale must be presented in a logical order. This also applies to matrix questions where the scale is displayed in a horizontal fashion. The leftmost option must either be the most positive one or the most negative one. Tourangeau, Couper, and Conrad (2004) found that response times were considerably longer if the response options were not ordered or if the midpoint (e.g., no opinion) was the last option. Respondents tend to use the visual midpoint in a response scale as an anchor or reference point for judging their own position. They get confused when the visual midpoint does not coincide with the midpoint of the Likert scale. Thus, it is advised to avoid matrix answers.

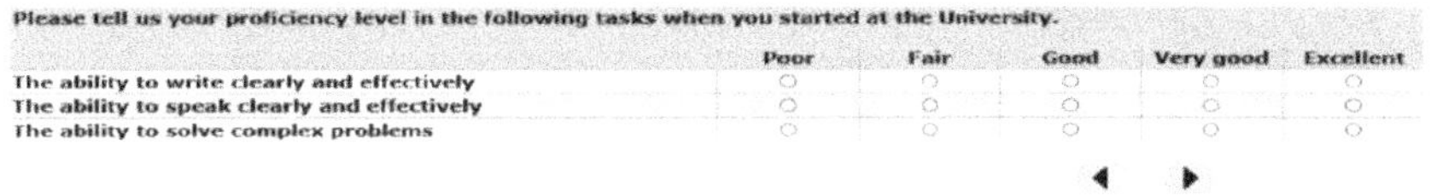

Please tell us your proficiency level in the following tasks when you started at the University.

| | Poor | Fair | Good | Very good | Excellent |
|---|---|---|---|---|---|
| The ability to write clearly and effectively | ○ | ○ | ○ | ○ | ○ |
| The ability to speak clearly and effectively | ○ | ○ | ○ | ○ | ○ |
| The ability to solve complex problems | ○ | ○ | ○ | ○ | ○ |

◀  ▶

FIGURE 7.21 A three-item scale question on a tablet or computer browser

Researchers are still evaluating if criteria for web are fitting to mobile devices. Input style, like sliding scale, is of relevance for touch-centric input devices (smartphones or tablets). The use of slider asks for the critical decision of the choice of optimal starting position need. One more point is if respondents, using less touch capable devices, equally prefer touch-centric input styles. While an outside starting position comes as a good choice for slider questions, completed via computer, the experimental study by Buskirk, Saunders, and Michaud (2015) concludes no evidence is found. However, their results are not to be considered a valid solution across different devices. How the use of sliders works across devices requires more studies.

---

### ▦ EXAMPLE 7.9 Question structure

A three-item scale question is an example question structure that shows the challenge of developing a survey for both a browser and a device. Figure 7.21 shows how the question would look on a tablet or computer browser. The question uses a 5-point, fully labeled scale *Poor, Fair, Good, Very good,* and *Excellent.* The respondents are expected to touch or click on 1 choice for each item. If they are on a computer browser, they click the right-arrow button. If they are on a tablet, they might swipe to the next page.

In Figure 7.21 the respondent sees the entire question and immediately knows that there are three items to answer. It is obvious that the lead question applies to all rows.

For a smartphone, there is more limited space, and thus the question is displayed across three screens (Figure 7.22). The respondent does not know how many items are there (unless this is stated in the text). A convention is

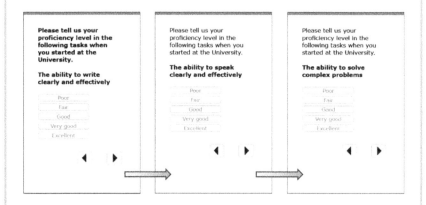

FIGURE 7.22 A three-item scale question on a smartphone

used here to repeat the header text on all three screens. This is unbolded for the second and third screens. In some questions, when a response button is touched, the screen automatically advances. This is signified by the arrows in the figure. The navigation buttons indicate to the respondent that he/she can go back to the previous item or forward to the next item.

The question structure above is a specific case of a multi-item scale structure. Two parameters that would guide the specification writer and the generation of the instrument include (1) the number of items and (2) the scale to be used, including whether the scale should be (a) fully labeled or (b) ascending or descending.

### 7.2.4.2 Drop-Down Boxes.

A different way to select one option from a list of answer options is using a *drop-down box*. This solution may be considered if the list of answer options is very long. Radio buttons are less effective for such lists. They would require a lot of space, and the respondents lack oversight. Figure 7.23 shows an example of a drop-down box in its initial state. The question asks for the country of birth, and the list of countries is invisible.

To open the list, the respondent has to click on "Select a country." If this list is very long, it only becomes partially visible. See Figure 7.24 for an example. It depends on the browser used how long this list is. If the list is longer, scroll bars are provided to make other items visible. The respondent selects an answer by clicking on it in the list.

Drop-down boxes have a number of disadvantages. In the first place, it requires more actions to select an answer (as compared with radio buttons). Three actions have to be performed: clicking the box, scrolling to the right answer, and clicking this answer. In the second place, there can be serious primacy effects if only part of the list is displayed: respondents tend to select an option in the visible part of the list. In the third place, it is unclear how much space the question requires on the screen.

It is possible to modify the behavior of the drop-down box so that it always shows a fixed number of items in the list. This has the advantage that it is clear how much space the question requires. However, also here there may be serious primacy effects, see Couper et al. (2004).

Note that in fact "Select a country" is the first item in the list. It could be removed from the list, but then the first country (Afghanistan) would be visible in the initial state. This could be the cause of a primacy effect. The text "Select

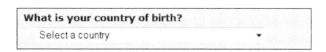

FIGURE 7.23  **A drop-down box in its initial state**

FIGURE 7.24  **A drop-down box after opening the list**

a country" sees to it that all items in the list are treated equally and it also provides a clue as to what the respondents should do. When the list is long, drop-down boxes are not user friendly to be answered. Multiple choice, i.e., selection of more than one item, is possible using drop-down boxes, but it is very complicated to make a correct choice.

Drop-down boxes are difficult especially for mobile screens; due to small size of the screen if the list of item is long, reading them to get the suitable choice becomes uncomfortable and boring.

The general advice is that, where possible, radio buttons should be preferred. The advantages and disadvantages of various answer formats of closed questions are also discussed by Couper (1999), Heerwegh and Loosveldt (2002), and Dillman, Smyth and Christian (2014).

**7.2.4.3  Checkboxes.**  *Checkboxes* are used for answering closed questions for which more than one answer is allowed. Such a question is also called a *check-all-that-apply question*. Figure 7.25 shows an example of such a question. A checkbox is shown on the screen as a square box. It can have two states: an empty white box indicates the option is not selected, and a tick mark indicates the option is elected. The state of the checkbox can be changed by clicking on it with the mouse.

Checkboxes permit the user to select multiple answers from the set of answer options. For example, two options have been selected in the checkbox in Figure 7.25. All answer options must be mutually exclusive and together cover

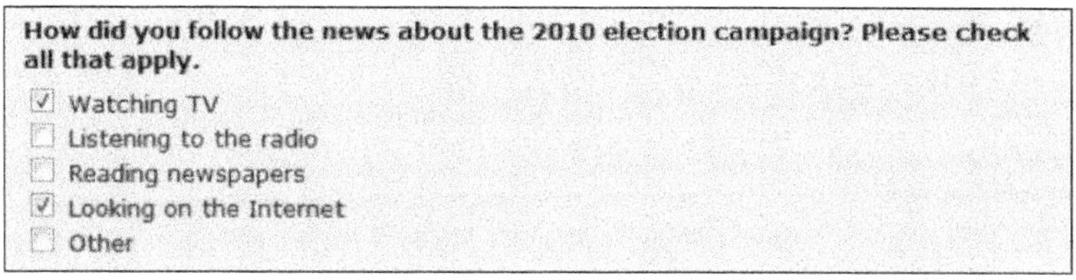

FIGURE 7.25 **A check-all-that-apply question**

all possible answers. Note that the option "Other" in Figure 7.25 guarantees that always an answer can be selected.

For specific questions it may be possible that none of the answer options apply to the respondent. That can be dealt with by checking no option. However, this makes it easy for a satisficing respondent to just skip the question without attempting to answer it. This can be avoided by including the option "None of the above" and forcing the respondent to at least check one answer. Of course, the questionnaire software must prevent selecting "None of the above" in combination with another answer.

If the list of answer options of a check-all-that-apply question is long, selecting the proper answer may mean a lot of work for respondents. Instead of checking all relevant answers, they may just check some arbitrary answers and stop when they think they have checked enough answers. Moreover, satisficing respondents tend to read only the first part of the list, and not the complete list. This causes a bias toward answers in the first part of the list. A solution to this problem may be replacing checkboxes by radio buttons, like in Figure 7.26. Smyth et al. (2006) have shown that the format in Figure 7.20 leads to more selected options and also respondents take more time to answer the questions. This is an indication that the format as in Figure 7.20 may cause satisficing.

If there is a series of check-all-that-apply questions all with the same set of possible answers, one may consider combining them in a matrix question (grid question). It was already mentioned that a matrix question has some advantages. It takes less space on the questionnaire form than a set of single questions, and it provides respondents with more oversight. Therefore, it can reduce the time it takes to answer questions. However, answering a matrix question is a complex cognitive task. Respondents can navigate through the matrix in several ways:

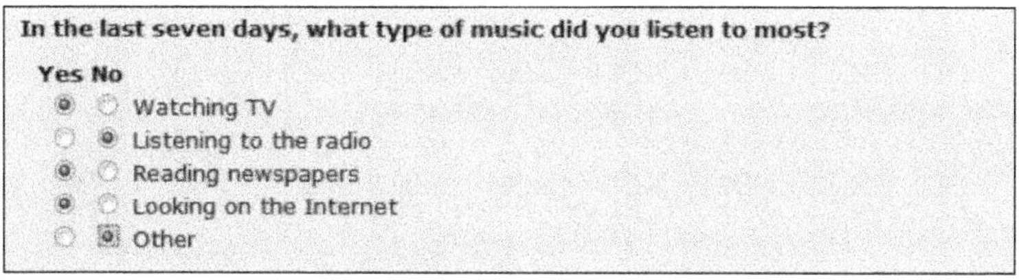

FIGURE 7.26 **A check-all-that-apply question with radio buttons**

row-wise, column-wise, or a mix. This increases the risk of missing answers to questions, resulting in a higher item nonresponse. Dillman, Smyth and Christian (2014) advise to limit the use of matrix questions as much as possible. If they are used, they should not be too wide nor to too long. Preferably, the whole matrix should fit on a single screen. This is not so easy to realize as different respondents may have set different screen resolutions on their computer screens. If respondents have to scroll, either horizontally or vertically, they may easily get confused, leading to wrong or missed answers.

### 7.2.4.4 Text Boxes and Text Areas.

Text boxes and text areas are used in web survey questionnaires to record answers to open questions. Open questions have the advantage that the respondents can reply to a question completely in their own words. No answers are suggested. There are also several disadvantages. It takes more time to answer such questions. Processing the answers is much more complex. The answers have to be analyzed, coded, and put into manageable categories. This is time consuming and error prone.

Since there is not much guidance as to what is expected of the respondents, they may easily forget things in their answers, put the focus differently, or have difficulty putting their thoughts into words. This may be the cause of measurement errors. If the question lacks focus, the answer will also lack focus. For example, the question "When did you move to this town?" could elicit responses like "When I got married," "Last year," "When I started my study at the university," "When I bought a house," and "Last year."

Open questions must be treated with considerable caution. Nevertheless, in some specific research contexts, they may offer a number of advantages. No other solution may be available to retrieve such information. Respondents answer open questions in their own words and therefore are not influenced by any specific alternatives suggested by the interviewer. If there are no clear ideas as to which issues may be the most important to the respondents, open questions are required. They may reveal findings that were not originally anticipated.

An open question can be implemented by means of a *text box*. An example is the first question in Figure 7.24. It provides space for just one line of text. The length of the text box can be specified by the questionnaire designer. This implementation of an open question should typically be used in situations where short answers are expected.

The second question in Figure 7.27 is an implementation of an open question by means of a *text area*. This provides for an answer space consisting of several lines of text. The width and height of the area can be specified by the questionnaire designer. The scroll bars even suggest that the text can be longer than the size of the box. Research shows that indeed the second format leads to longer answers than the first format.

If the researcher has good prior knowledge of the question topic and is able to generate a set of likely (but not exhaustive) response options, an alternative approach could be a hybrid question. This is a combination of a closed and an open question (Figure 7.28). All known options are listed, and a different answer can be taken care of by the option "Other, please specify."

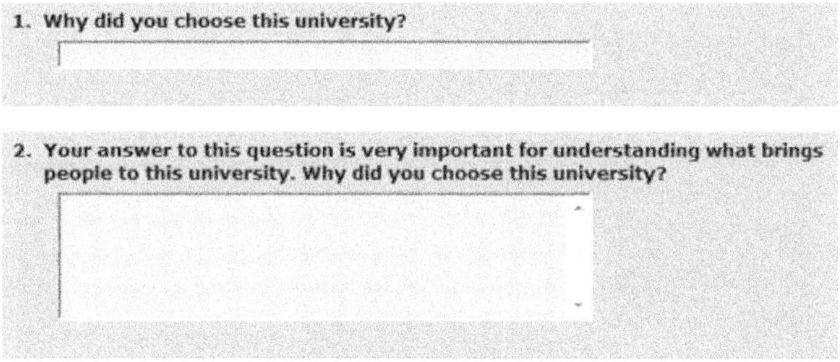

FIGURE 7.27 Formatting an open question

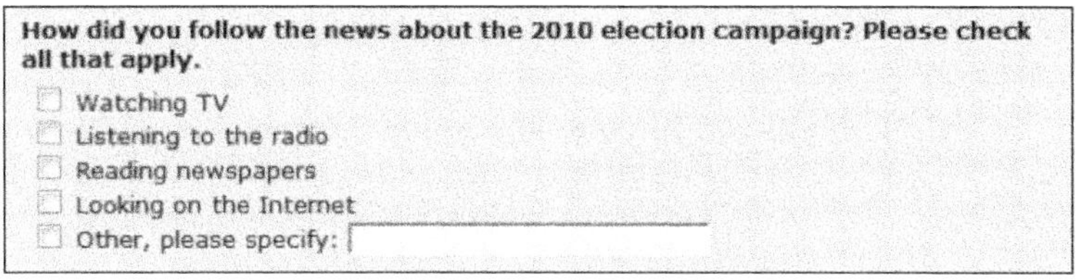

FIGURE 7.28 Combining an open and a closed question

## 7.2.5 WEB QUESTIONNAIRES AND PARADATA

**7.2.5.1 Definition of Paradata.** The concept of paradata emerged in the era of computer-assisted interviewing (CAI). *Paradata* are defined as data that are generated during the fieldwork of the survey. Early examples of paradata are related to keystroke files and audit trails that were automatically generated by many CAI systems. They were a by-product of the CAI system and used for technical purposes, such as error diagnosis and recovery from failure. Already early in their existence, it was realized that paradata can also provide insight in the process of asking and answering questions and how respondents interact with computer systems. Therefore, they may help to improve data collection. Early applications of paradata can be found in Couper, Hansen, and Sadosky (1997) and Couper, Horm, and Schlegel (1997). Couper (1998) was the first to coin the term paradata. Heerwegh (2003) describes the use of paradata in web surveys.

In Chapter 3 the steps of the web survey process are presented and discussed. Each survey step has some risks of error. In the TSE perspective, decisions at one step may produce errors at different steps of the process, and sometimes there are relationships between different risks of errors. Paradata can be useful to understand errors that might occur at different steps and could provide information to model the errors, even alongside the data collection process. For example, an adaptive approach could contribute to optimize the sample, while the field period is undergoing. See Chapter 8 for adaptive design. Challenges of the paradata for several types of errors, like nonresponse, measurement, and coverage, and for the improvement in the survey production process are presented and discussed in various chapters of Kreuter's (2013) edited book.

Paradata can be particularly useful in web surveys and in mobile web surveys. There are no interviewers, so other means have to be found for obtaining insight in the activities of respondents while they are completing the questionnaire form. Examples of paradata are the keys pressed by respondents, the movement of the mouse, changes they have made, time it takes to answer a question or the complete questionnaire, use of help functions, etc.

Two types of paradata are collected during web questionnaire completion: server-side paradata and client-side paradata. *Server-side paradata* are collected by software tools running at the server where the survey resides. These are usually data measured at the level of the questionnaire. Examples of server-side paradata include the download time of the survey questionnaire, the number of times the survey web page was accessed, the time spent in each visit, identifiers of respondents, the type of browser used, and the operating system on the computer of the respondent. Server-side paradata is usually stored in *logfiles* for later analysis.

▓ **EXAMPLE 7.10** Audit trials

In 2006, a pilot study was carried out by Statistics Netherlands to examine the feasibility of introducing an electronic questionnaire for the Structural Business Survey. Snijkers and Morren (2010) describe the details of this study. Figure 7.29 shows a sample screen (in Dutch) of the downloadable questionnaire form.

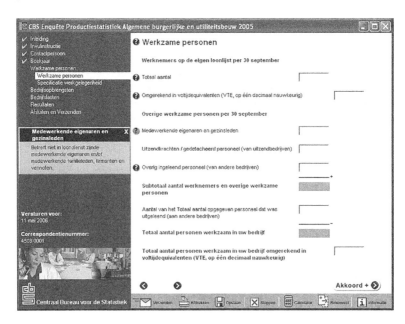

FIGURE 7.29 **A screen of the Structural Business Surveys**

Audit trails were used to examine the response behavior in this survey. Audit trails are a specific kind of paradata. They contain information on all events occurring on the computer while the respondent completes the questionnaires. For example, they record all key presses, all mouse clicks, and the time at which they occurred. This information can be used to answer questions like:

- Do respondents complete the questionnaire in one session or in several sessions?
- How long does it take to complete a questionnaire as a whole?
- How do respondents navigate through the questionnaire?
- Do respondents use the print function? How often and at what moments in the completion process do they print the questionnaire?
- Do respondents first browse through the questionnaire before they start filling it in (to get an overview of the questionnaire), or do they start filling it in right away question by question?

The survey was conducted with the Blaise system. This survey software can automatically generate audit trail. The audit trail data was imported in the statistical analysis package SPSS. Figure 7.30 shows a small part of the data for one respondent.

Each line represents data on one event. The respondent is identified (id), the data and time, a code for the specific action (for example, entering an answer field, exit a field, clicking on help, click the save button), the question (field) for which the event occurred, and the answer entered by the respondent.

| | id | date | time | action | field | value |
|---|---|---|---|---|---|---|
| 1 | 12345 | 22-MAY-2007 | 10:43:17 | 0 | 0 | |
| 2 | 12345 | 22-MAY-2007 | 10:43:58 | 1 | 40 | |
| 3 | 12345 | 22-MAY-2007 | 10:44:19 | 2 | 40 | 01-04-2005 |
| 4 | 12345 | 22-MAY-2007 | 10:44:19 | 14 | 40 | |
| 5 | 12345 | 22-MAY-2007 | 10:44:29 | 14 | 27 | |
| 6 | 12345 | 22-MAY-2007 | 10:44:48 | 14 | 21 | |
| 7 | 12345 | 22-MAY-2007 | 10:44:51 | 1 | 48 | |
| 8 | 12345 | 22-MAY-2007 | 10:45:15 | 3 | 48 | |
| 9 | 12345 | 22-MAY-2007 | 10:50:20 | 8 | 48 | |
| 10 | 12345 | 22-MAY-2007 | 10:51:05 | 10 | 48 | |
| 11 | 12345 | 22-MAY-2007 | 10:51:05 | 99 | 0 | |
| 12 | 12345 | 20-JUN-2007 | 8:52:28 | 0 | 0 | |
| 13 | 12345 | 20-JUN-2007 | 8:53:09 | 8 | 15 | |
| 14 | 12345 | 20-JUN-2007 | 8:54:50 | 13 | 15 | |
| 15 | 12345 | 20-JUN-2007 | 8:54:53 | 11 | 15 | |
| 16 | 12345 | 20-JUN-2007 | 8:55:02 | 9 | 15 | |
| 17 | 12345 | 20-JUN-2007 | 9:35:58 | 99 | 0 | |
| 18 | 12345 | 29-JUN-2007 | 15:10:38 | 0 | 0 | |

FIGURE 7.30 **Blaise audit trail data imported in SPSS**

*Client-side paradata* are collected by software tools running on the computers of the respondents. These data describe how the respondents answer the questions in the questionnaire. They give insight in which questions are answered, in what order, and whether all relevant questions have been answered. Many activities can be recorded, like keys that have been pressed and mouse movement across the screen. As time stamps can be associated with each activity, it is, for example, possible to measure how long it takes to answer a question. All these data may provide insight on how easy or difficult it is for the respondents to complete the questionnaire. Recording paradata is often implemented by embedding JavaScript code in the survey questionnaire.

Callegaro (2013) classifies paradata in web surveys in device-type paradata and questionnaire navigation paradata. Device-type paradata include information about the browser used, the browser window size and other technical browse information, the operating system and its language, the screen resolution, the IP address of the device used, and the GPS coordinates. Device-type paradata are collected at session level; when analyzing this data, the researcher has to take into account that the respondent could complete the questionnaire in more than one session. Questionnaire navigation paradata describe the whole process of the questionnaire completion, for example, time spent in answering the question, mouse clicks, change of answers, scrolling, and last question answered before dropping out the survey. Questionnaire navigation paradata are useful to understand the quality of the questionnaire as well as the behavior of the respondents. Extended application of this type of paradata is a promising way to improve the survey quality facing the challenges of the continuous changing survey environment.

Paradata are also useful to take under control different sources of error at different stages of the survey process. Technological advancements and trends are continuously changing the characteristics of some available paradata. Researchers are recommended to try to incorporate new paradata based on technological aspects and study design characteristics. It is possible to classify paradata according to the stage of the survey process they are available. For example, in a longitudinal panel perspective, at the starting phase, contacts from previous waves, item missing data, response speed indices, previous mobile device use can be considered. At recruitment phase indicators such as e-mail opening, times/dates of contacts, time to open contact could provide interesting information. At access phase, paradata like time from open contact to access and device characteristics, login attempts are signals to be used to understand the participation behaviour. Considering response phase additional information becomes available such as response time and, item missing data. It is also possible to follow the navigation process and response changes, mouse movements, touch events, and device characteristics used for questionnaire completion. McClain et al. (2019) present this type of paradata classification and an extended discussion.

### 7.2.5.2 Use of Paradata. Initially, researchers collected paradata to study the response behavior of different groups. Analysis of these paradata provides insight in this response behavior and therefore may help to direct resources and efforts aimed at improving response rates.

Analysis of paradata is now increasingly used to understand or even control respondent behavior. An interesting new area of research is the application of methods and concepts from cognitive psychology to the development of web questionnaires and of new computerized methods of data collection. This type of research is referred to here as the "behavioral approach." In this approach, the focus turns toward understanding why someone responds in a web surveys while others do not and whether and in what ways these two groups may differ on key variables of interest.

Using client-side paradata it is possible to better understand how respondents construct their answers, including data about the time it takes to answer a question and also about possible changes in their answers. Results suggest that the visual layout of survey questions affects not only the number but also the types of changes respondents make. The theory does not concern itself with the impact of nonresponse on estimates although it is not unlikely that different phenomena may cause a different type of nonresponse bias.

Objective of the behavioral research approach is to answer questions such as who is more likely to respond, why does nonresponse occur, who is likely to be hard to reach, and how does interest in the survey topic affects the willingness to participate. As an ultimate goal, behavioral studies aim to acquire an understanding of how respondents construct their answers in their natural setting, which in turn facilitates web questionnaire design and tailoring.

From the methodological point of view, behavioral analyses are related to the Cognitive Aspects of Survey Methodology (CASM) movement. In many empirical studies the *theory of planned behavior* (TPB) is applied. This theory was proposed by Ajzen (1985) and is described in Ajzen (1991). The main objective of TPB is to obtain a more comprehensible picture of how intentions are formed. This theory is an extension of the theory of reasoned action (Ajzen and Fishbein, 1980). The TPB specifies the nature of relationships between beliefs and attitudes. According to these models, people's evaluations or attitudes toward behavior are determined by their accessible beliefs about such behavior, where a belief is defined as the subjective probability that the behavior will produce a certain outcome. Specifically, the evaluation of each outcome contributes to the attitude in direct proportion to the person's subjective belief in the possibility that the behavior produces the outcome in question.

The central factor of this theory is the individual intention to perform a given behavior. The first postulate is that intention is the result of three conceptual determinants. Human behavior is guided by the following three kinds of considerations:

- "Behavioral beliefs." They produce a favorable or unfavorable *attitude toward behavior*. It is the degree to which a person has a favorable or unfavorable evaluation or appraisal of the behavior in question (Ajzen, 1991). When new issues arise requiring an evaluative response, people can draw on relevant information (beliefs) stored in memory. Because each of these beliefs carries evaluative implications, attitudes are automatically formed.

- "Normative beliefs." They result in *subjective norms* (SN). This refers to individuals' perceptions of others' opinions of their behavior. SN has been shown to be a predictor of behavior (Bagozzi, Davis, & Warshaw, 1992; Fishbein & Ajzen, 1975, 2010; Mathieson, 1991). In the context of web surveys, SN would be the amount of influence a person's superiors (i.e., employers, parents, or spouse) would have in influencing a choice to participate in the survey.

- "Control beliefs." They give rise to *perceived behavioral control (PBC)*. It is the perceived social pressures to perform, or not, a certain behavior (Ajzen, 1991), i.e., the subject's perception of other people's opinions of the proposed behavior. This pressure can have, or have not, an influential role. For example, in France, the failure of a company is negatively perceived, whereas in the United States, a person can undergo several failures and yet often undertake new attempts. PBC is presumed to not only affect actual behavior directly but also affect it indirectly through behavioral intention. PBC refers to the perception of individual of whether or not they can perform a particular behavior. In the context of a web survey, PCB would be defined as whether or not an individual could use the web tools to successfully participate and engage in the survey. Therefore, PBC would be similar to computer self-efficacy (CSE), see Bagozzi, Davis, and Warshaw (1992). CSE is defined as the judgment of one's capability to use a certain aspect of information technology (Agarwal, Sambamurthy, and Stair, 2000; Compeau and Higgins, 1995; Gist, 1989; Gist, Schwoerer, and Rosen, 1989). Attitudinal models (such as the abovementioned TPB) use multiple constructs to predict and explain behavior. The concept of perceived ease or difficulty of performing a behavior (Ajzen, 1991) was introduced into the TPB to accommodate the non-volitional elements inherent, at least potentially, in all behaviors (Ajzen, 2002a).

This theory assumes that human social behavior is reasoned, controlled, or planned in the sense that it considers the likely consequences of the studied behavior (Ajzen and Fishbein, 2000). This model has been applied for the prediction of many types of human behavior, such as electoral choice and intention to stop smoking.

Combining attitude toward behavior, the SN, and PBC leads to *behavioral intention* (Ajzen, 2002b). As a general rule, the more favorable the attitude toward behavior and SN is, and the greater the PBC, the stronger the person's intention to perform the behavior in question should be. Finally, given a sufficient degree of actual control over the behavior, people are expected to carry out their intentions when the opportunity arises (Ajzen, 2002b).

In its simplest form, the TPB can be expressed as the following statistical function:

(7.1)     $BI = (W1)AB[(b) + (e)] + (W2)SN[(n) + (m)] + (W3)PBC[(c) + (p)],$

where

$BI$ = behavioral intention

$AB$ = attitude toward behavior

$(b)$ = the strength of each belief

$(e)$ = the evaluation of the outcome or attribute

$SN$ = social norm

$(n)$ = the strength of each normative belief

$(m)$ = the motivation to comply with the referent

$PBC$ = perceived behavioral control

$(c)$ = the strength of each control belief

$(p)$ = the perceived power of the control factor

$W1, W2, W3$ = empirically derived weights/coefficients

For instance, Tourangeau (2003) proposes an extended TPB model to explain the intentions of potential respondents in participating in web surveys. His model is shown graphically in Figure 7.31.

As a proxy of the burden on the respondent, often the time spent to complete the web questionnaire is used, i.e., the actual number of minutes spent working on the questionnaire. While this is not a perfect measure, it is one that indicates the

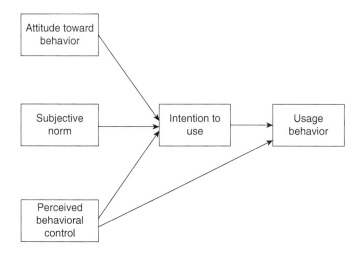

FIGURE 7.31 **Theory of planned behavior**

intensity and duration of the time spent performing the task of participating in the survey.

## 7.2.6 TRENDS IN WEB QUESTIONNAIRE DESIGN AND VISUALIZATION

### 7.2.6.1 The Cognitive Approach to Web Questionnaire Design.
The World Wide Web has features that are not available for paper questionnaires. These features enable a new way of thinking about the design and self-administration of survey questions. In particular, web survey interaction can be conceived of as a dialogue consisting of turns of interaction between a user (respondent) and the system (the interviewing agent). From this viewpoint, inspired by the collaborative comprehension of elements of cognitive psychology and psycholinguistics, each system action (presenting a question, prompting for an answer) and each respondent action (clicking to proceed, reading a question, typing a number as an answer) corresponds to some physical or mental action in a face-to-face interview. Conceiving interaction this way not only highlights and clarifies the function of each action but also opens up possibilities that survey designers can implement in web surveys. The task of the so-called cognitive approach is to assess how respondents go about answering questions.

There is widespread agreement about the cognitive processes involved in answering questions optimally. See, for example, Cannell, Miller, and Oksenberg (1981), Tourangeau and Rasinski (1988), and Willis (2005). Specifically, respondents are presumed to execute each of four steps:

1. Interpreting the question and deducing its intent;
2. Searching their memories for relevant information and then integrating whatever information comes to mind into a single judgment;
3. Translating this judgment into a response by selecting one of the alternatives offered by the question.

Each of these steps can be quite complex, involving considerable cognitive work. A wide variety of motives may encourage respondents to perform these activities, including the desire for self-expression, interpersonal response, intellectual challenge, self-understanding, altruism, or emotional catharsis. Their efforts can also be motivated by the desire to assist the survey sponsor, e.g., to help employers in improving working conditions, to help businesses in designing better products, or to help governments in formulating better-informed policies. To the extent that such motives inspire a respondent to perform the necessary cognitive tasks in a thorough and unbiased manner, the respondent may be said to be *optimizing*. The web questionnaire designer hopes all respondents will optimize throughout a questionnaire. This is often an unrealistic expectation. Some people may agree to complete a questionnaire as result of a relatively automatic compliance process or because they are required to do so. Thus, they may agree merely to provide answers, with no intrinsic motivation to produce high-quality answers. Other respondents may satisfy whatever desires motivated them to participate after answering a first set of questions and become fatigued, disinterested, or distracted as a questionnaire progresses further.

Rather than make the effort necessary to provide optimal answers, respondents may take subtle or dramatic shortcuts. In the former case, respondents may simply be less thorough in comprehension, retrieval, judgment, and response selection. They may be less thoughtful about a question's meaning, search their memories less comprehensively, integrate retrieved information less carefully, or select a response choice less precisely. All four steps are executed, but less diligently as in the case of optimization. Instead of attempting the most accurate answers, respondents settle for merely satisfactory answers. The first answer a respondent considers that seems acceptable is the one offered. This response behavior is termed *weak satisficing* (Krosnick, 1991; borrowing the term from Simon [1957]).

In other cases, respondents skip the retrieval and judgment steps altogether. That is, respondents may interpret each question superficially and select what they believe will appear to be a reasonable answer. The answer is selected without reference to any internal psychological cues specifically relevant to the attitude, belief, or event of interest. Instead, the respondent may look easily defended if necessary. If no such cue is present, the respondent may select an answer completely arbitrarily. This process is termed *strong satisficing*.

It is useful to see optimizing and strong satisficing as the two ends of a continuum indicating the degrees of thoroughness with which the response steps are performed. The optimizing end of the continuum involves complete and effortful execution of all four steps. The strong satisficing end involves little effort in the interpretation and answer reporting steps and no retrieval or integration at all. In between are intermediate levels, like weak satisficing.

The likelihood of satisficing is thought to be determined by three major factors: task difficulty, respondent ability, and respondent motivation (Krosnick, 1991). *Task difficulty* is a function of both question-specific attributes (e.g., the difficulty of interpreting a question and of retrieving and manipulating the requested information) and attributes of the questionnaire's administration (e.g., the pace at which an interviewer reads the questions and the presence of distracting events). *Ability* is shaped by the extent to which respondents can perform complex mental operations, practiced at thinking about the topic of a particular question, and equipped with preformulated judgments on the issue in question. *Motivation* is influenced by need for cognition (Cacioppo et al., 1996), the degree to which the topic of a question is personally important, beliefs about whether the survey will have useful consequences, respondent fatigue, and aspects of questionnaire administration (such as interviewer behavior) that either encourage optimizing or suggest that careful reporting is not necessary. Efforts to minimize task difficulty and maximize respondent motivation are likely to pay off by minimizing satisficing and maximizing the accuracy of self-reports.

Web surveys can incorporate audio and video to more closely approximate an interviewer-assisted survey. A so-called cognitive interview provides basic information for understanding the impact of audio, video, and even interviewers' faces. All this information allows for further improvements in a web questionnaire. Web survey researchers are increasingly conducting experiments where they introduce pictures of faces of interviewers. This kind of cognitive interview is labor intensive. Current results suggest that bringing features of human dialogue into web surveys can exploit the advantages of both the interviewer-assisted and self-administered interviews. Nevertheless, it should be noted that introducing interviewers may also introduce biases.

🖳  EXAMPLE 7.11  **Cognitive interviewing**

A simple example shows how cognitive interviewing can be used in web surveys. The Samplonion Survey Research Institute (SSRI) wants to improve the design of its web survey questionnaires. The hope is to increase response rates. An experiment is set up to compare various survey designs.

A lab experiment is organized involving a company that provides this kind of services. The lab is equipped with audio and video recording equipment using multiple cameras and two-way communication with an observation room. Each room contains a computer on which a web survey questionnaire can be completed.

A group of 100 respondents takes part in this experiment. They are divided in five subgroups:

1. 20 respondents are invited to complete the web questionnaire without any special external stimulus, just with the standard survey design. This is the control group.

2. 20 respondents are invited to complete the questionnaire with audio support: a male voice introduces the survey and asks the respondent to participate in the survey.

3. 20 respondents are invited to complete the questionnaire with video and audio support: an animated male face appears on the screen and asks the respondent to participate in the survey.

4. 20 respondents are invited to complete the questionnaire with audio support: a female voice introduces the survey and asks the respondent to participate in the survey.

5. 20 respondents are invited to complete the questionnaire with video and audio support: an animated male face appears on the screen and asks the respondent to participate in the survey.

The behavior of the respondents can be viewed and recorded. By studying this and other information, it may become clear in which circumstances and at which points in the questionnaires respondents face problems, make mistakes, or do not know what to do. By comparing the groups, insight may be obtained in the effects of audio and video support.

# 7.3 Application

This section describes a web survey on the values of consumers with respect to purchases of trendy and vogue products. This survey is described in detail by Biffignandi and Manzoni (2011).

Students were asked to complete a questionnaire made up of two different parts. In the first part, the respondents were asked to indicate which consequences a certain product attribute is linked to. In the second part, they had to indicate which consumers' values are linked to certain consumers' consequences. Both product attributes and consequences were identified in a previous study using a different sample.

The survey was administered to three experimental groups determined by different data collection modes (paper questionnaire versus computer-based questionnaire) and different measurement scales used in the questionnaire. There was a 5-point Likert numerical scale and a 2-point Likert scale (yes/no). The following hypotheses were tested:

$H_1$: The mode of data collection (paper versus computer-based) influences the respondents' responses.

$H_2$: The type of Likert scale (5-point versus 2-point) influences the respondents' responses.

$H_3$ 2-point Likert scale is equal to the 5-point Likert scale if the latter is recoded such that the first four items are assigned to value 0 and the last item is assigned to value 1.

A field experiment was designed using various combinations of the mode of data collection and different scales. The experimental design is summarized in Table 7.2.

For the analysis of the survey results, the questionnaire for group 3 (computer based, with a 5-point Likert scale) was recoded using four different dichotomization criteria: 1 versus $2 + 3 + 4 + 5$, $1 + 2$ versus $3 + 4 + 5$, $1 + 2 + 3$ versus $4 + 5$, and $1 + 2 + 3 + 4$ versus 5.

The results for the different groups (including the recoded versions) were compared by adopting the computer-based dichotomy questionnaire as a benchmark. With respect to comparing the web-based and paper-based questionnaires,

TABLE 7.2 **The design of the consumers' values experiment**

| Group | Mode of data collection | Type of Likert scale |
|---|---|---|
| 1 | Paper | Dichotomy scale |
| 2 | Computer based | Dichotomy scale |
| 3 | Computer based | Five-level Likert numerical scale |

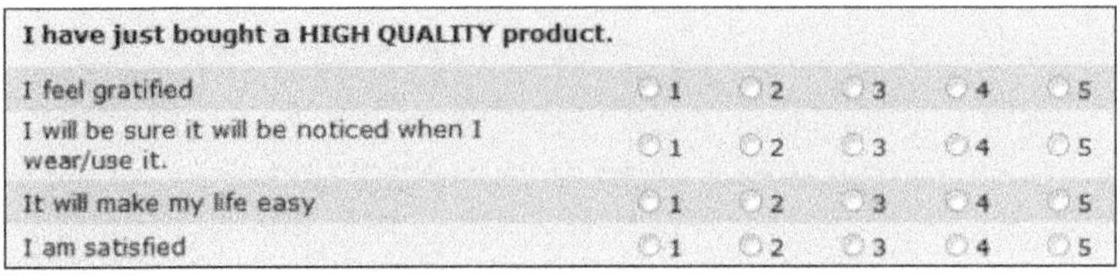

FIGURE 7.32 A matrix question in the Consumers' Values Survey

results showed that the mode of data collection had no influence on the responses of the respondents.

With respect to comparing the 2-point Likert scale with the 5-point Likert scale, analysis showed that the recode $1 + 2 + 3 + 4$ versus 5 provided the same results as the 2-point scale. This confirms the theory that respondents are unsatisfied until completely satisfied. In other words, respondents are aware of not being unsatisfied only after realizing they are satisfied. As an example, Figure 7.32 contains an example of one of the survey questions. The meaning of the scale values was explained in the survey questionnaire.

# 7.4 Summary

A web survey is a tool that can collect information in a cheap and timely manner, producing good quality data with a reduced response burden. To achieve this, the questionnaire designer should apply not only traditional survey research principles but also more recent research results that are more specific for online data collection. Together, they provide a general framework for designing web surveys.

A well-designed web survey requires research that pays attention to several different aspects. One aspect is to define the research tasks and the environment required for this. Another aspect is to select the appropriate hardware and software. There should be a balance between technological sophistication and user accessibility. Adaption of the questionnaire for using smartphone in survey participation is crucial.

With respect to the survey questionnaire, careful attention must be paid to labeling the answers for closed questions and all-that-apply questions. Navigation through the questionnaire must be easy and intuitive, such that only relevant questions are answered and irrelevant ones are skipped. Smartphone and mobile device wide diffusion allows for receiving a questionnaire on these devices. If access to the questionnaire is not blocked at the survey design stage, attention has to be paid to optimize the questionnaire for mobile devices. A long and complex questionnaire is difficult to be managed in a mobile device.

Another important aspect is the way the questionnaire is visualized on the screen. Proper use of symbols, colors, and images may help in this respect. Offering answer spaces that are sized appropriately for the information being requested improves the likelihood that respondents will provide answers and good quality information.

A final aspect to pay attention is to collect paradata. This information may help to detect problems in the survey instrument. Analysis of these data and, more generally, research on cognitive interviewing can help in improving web survey design in the future. Paradata together eventually with the use of administrative data may be useful for applying adaptive design at data collection stage.

Summing up, the web questionnaire design process is challenging work requiring knowledge of different fields, such as (1) the subject-matter topic of the survey, (2) computer technology, (3) textual language (labeling, wording, graphics, and so on), (4) use of paradata, and (5) cognitive methods. The most promising direction in the field of web survey design research is user-centered design methods. This approach seems to be worthwhile in terms of conducting effective web survey and collecting good quality data. When working on cross-cultural international surveys, a critical aspect to be considered is the questionnaire translation. Aspects of the wording and the language in different culture should be considered. An important aspect in web questionnaire design is testing the questionnaire. If a mixed-mode approach is adopted, decision has to be made on the format of the questionnaire across various modes.

## KEY TERMS

**Check-all-that-apply question**: A question allowing respondents to select more than one answer from a list of answer options. Checkboxes are used to select answer options.

**Checkbox:** A graphical user interface element. It is a small rectangular box allowing respondents to select an associated option. An option is selected or deselected by clicking on it. A selected option is denoted by a tick mark.

**Closed question:** A question allowing respondents to select exactly one answer from a list of answer options. Radio buttons or drop-down boxes can be used to select answer options. Radio buttons are preferred.

**Drop-down box:** A list of answer options from which one answer can be selected. Initially, the list is invisible. The list can be opened by clicking on its title. Longer lists will only be partly visible. Other parts of the list can be shown by using a scroll bar.

**Likert scale**: A type of closed question where the answers constitute a scale. It allows respondents to indicate how closely their feelings match the question or statement on a rating scale. Likert scales are useful measuring the degree of respondents' feelings or attitudes concerning some issue.

**Open question:** A question that does not have a set of possible answers associated with it. The verbatim response of the respondent is recorded. Such a question is used if respondents are required to respond completely and freely.

**Ordinal question**: A type of closed question that allows respondents to rank their answer to a question. Ordinal questions are used for determining priorities or preferences of respondents.

**Radio button:** A graphical user interface element. It is a small circle box allowing respondents to select the associated option. An option is selected by clicking on it. A selected option is denoted by a dot in the circle. An option is deselected by selecting another option.

**Theory of planned behavior (TPB)**: A psychological theory that links attitudes to behavior.

## EXERCISES

**Exercise 7.1**  Closed questions must have:

**a.** Only numerical answers.

**b.** Text areas for each option.

**c.** Mutually exclusive categories.

**d.** Checkboxes.

**Exercise 7.2**  Suppose a customer satisfaction survey on motorbike preferences is conducted. What is best way to obtain an answer on the question "Indicate which motorbike model your prefer"?

**a.** A set of checkboxes.

**b.** An open question.

**c.** A set of radio buttons.

**d.** A drop-down box.

**Exercise 7.3**  A web survey contains the following question:

Do you agree or disagree with the following statement: School is a place where I usually feel great?
- ☐ Strongly agree
- ☐ Agree
- ☐ Neither agree or disagree
- ☐ Disagree
- ☐ Strongly disagree

Which of the following statements is correct?

**a.** The box format is a correct.

**b.** The answers form a Likert scale.

**c.** The wording is not correct.

**d.** This is a check-all-that-apply question.

**Exercise 7.4**   Which of the following statements about a web questionnaire is correct?

**a.** It is an interviewer-assisted survey.

**b.** It allows for collecting server-side paradata.

**c.** The use of open questions is prohibited.

**d.** As much as possible matrix questions should be used.

**Exercise 7.5**   Which of the following statements are advantages of web questionnaires over paper questionnaires?

**a.** They allow for customization.

**b.** Radio buttons and checkboxes can enforce different treatment.

**c.** They do not allow for partial completion.

**d.** "Don't know" answers are not possible.

**Exercise 7.6**   For what kind of answer formats are radio buttons used in web surveys?

**a.** For selecting one answer.

**b.** For selecting more than one answer.

**c.** For recording any text.

**d.** For recording amounts and values.

**Exercise 7.7**   How should "don't know" be treated in web surveys?

**a.** It should never be offered as an answer option.

**b.** It should always be offered as an answer option.

**c.** It should only be offered as an answer option in opinion questions.

**d.** It should only be offered as an answer option in factual questions.

**Exercise 7.8**   What can be said about the use of pictures in web survey questions?

**a.** They should never be used as they may suggest a wrong reference framework for the respondent.

**b.** They should always be used as they suggest the right reference framework for the respondent.

**c.** They should always be used as they make the questionnaire more attractive.

**d.** They should only be used to help the respondents to understand the question correctly.

## REFERENCES

Agarwal, R., Sambamurthy, V., & Stair, R. (2000), Research Report: The Evolving Relationship Between General and Specific Computer Self-efficacy—An Empirical Assessment. *Information Systems Research*, 11, pp. 418–430.

Ajzen, I. (1985), From Intentions to Actions: A Theory of Planned Behavior. In: Kuhl, J. & Beckman, J. (eds.), *Action-control: From Cognition to Behavior*. Springer Verlag, Heidelberg, Germany, pp. 11–39.

Ajzen, I. (1991), The Theory of Planned Behavior. *Organizational Behavior and Human Decision Processes*, 50, pp. 179–211.

Ajzen, I. (2002a), Perceived Behavioral Control, Self-efficacy, Locus of Control, and the Theory of Planned Behavior. *Journal of Applied Social Psychology*, 32, pp. 665–683.

Ajzen, I. (2002b), Residual Effects of Past on Later Behavior: Habituation and Reasoned Action Perspectives. *Personality and Social Psychology Review*, 6, pp. 107–122.

Ajzen, I. & Fishbein, M. (1980), *Understanding Attitudes and Predicting Social Behavior*. Prentice Hall, Englewood Cliffs, NJ.

Ajzen, I. & Fishbein, M. (2000), Attitudes and the Attitude-behavior Relation: Reasoned and Automatic Processes. *European Review of Social Psychology*, 11, pp. 1–33.

Arruda Filho, E. J. M. & Biffignandi, S. (2011), *Utilitarian Guilt Pushed by Hedonic Preference: Perceived Values to a Society Choice's Justification*. Research Report, CASI/DMSIA, Bergamo, Italy.

Bagozzi, R. P., Davis, F. D., & Warshaw, P. R. (1992), Development and Test of a Theory of Technological Learning and Usage. *Human Relations*, 45, pp. 659–687.

Biffignandi, S. & Manzoni, A. (2011), *The Mode and Scale Measurement Impact in a Consumer Behaviour Survey*. Research Report, CASI/DMSIA, Bergamo, Italy.

Buskirk, T. D. & Andrus, C. (2012), Smart Surveys for Smart Phones: Exploring Various Approaches for Conducting Online Mobile Surveys via Smartphones. *Survey Practice*, 5, 1, pp. 1–11.

Buskirk, T. D., Saunders, T., & Michaud, J. (2015), Are Sliders Too Slick for Surveys? An Experiment Comparing Slider and Radio Button Scales for Smartphone, Tablet and Computer Based Surveys. *Methods, Data and Analyses*, 9, 2, pp. 229–260.

Cacioppo, J. T., Petty, R. E., Feinstein, J. A., & Jarvis, W. B. G. (1996), Dispositional Differences in Cognitive Motivation: The Life and Times of Individuals Varying in Need for Cognition. *Psychological Bulletin*, 119, pp. 197–253.

Callegaro, M. (2013), Paradata in Web Surveys. In: Kreuter, E. (ed.), Improving Surveys with Paradata, Analytic Uses of Process Information. Wiley, Hoboken, NJ, pp. 259–279.

Cannell, C. F., Miller, P. V., & Oksenberg, L. (1981), Research on Interviewing Techniques. In: Leinhardt, S. (ed.), *Sociological Methodology*. Jossey-Bas, San Francisco, CA.

Christian, L. M., Dillman, D. A., & Smyth, J. D. (2007), Helping the Respondents Get It Right the First Time: The Influence of Words, Symbols. *Public Opinion Quarterly*, 71, pp. 113–125.

Christian, L. M., Parsons, N. L., & Dillman, D. A. (2009), Measurement in Web Surveys: Understanding the Consequences of Visual Design and Layout. *Sociological Methods & Research*, 37, pp. 393–425.

Cobanoglu, C., Warde, B., & Moreno, P. J. (2001), A Comparison of Mail, Fax and Web-based Survey Methods. *The Market Research Society*, 43, pp. 441–452.

Compeau, D. R. & Higgins, C. A. (1995), Computer Self-efficacy: Development of a Measure and Initial Test. *MIS Quarterly*, 19, pp. 189–211.

Converse, J. M. & Presser, S. (1986), *Survey Questions, Handcrafting Standardized Questionnaire*, 63 Sage University Paper Series on Quantitative Applications in the Social Sciences. 07-063. Sage Publications, Beverly Hills, CA.

Couper, M. P. (1998), Measuring Survey Quality in a CASIC Environment. *Proceedings of the Section on Survey Research Methods of the American Statistical Association*, pp. 41–49.

Couper, M. P. (1999), Usability Evaluation of Computer Assisted Survey Instruments. *Proceedings of the Third ASC International Conference*, Edinburgh, pp. 1–14.

Couper M. P. (2000), Web Surveys: A Review of Issues and Approaches. *Public Opinion Quarterly*, 64, pp. 464–494.

Couper, M. P. (2008), *Designing Effective Web Surveys*. Cambridge University Press, New York.

Couper, M. P., Hansen, S. E., & Sadosky, S. A. (1997), Evaluating Interviewer Use of CAPI Technology. In: Lyberg, L., Biemer, P., Collins, M., de Leeuw, E., Dippo, C., Schwarz, N., & Trewin, D. (eds.), *Survey Measurement and Process Quality*. Wiley, New York, pp. 267–287.

Couper, M. P., Horm, J., & Schlegel, J. (1997), Using Trace Files to Evaluate the National Health Interview Survey CAPI Instrument. In: *Proceedings of the Section on Survey Research Methods*. American Statistical Association, Alexandria, pp. 825–829.

Couper, M. P., Tourangeau, R., Conrad, F. G., & Crawford, S. D. (2004), What They See Is What They Get: Response Options for Web Surveys. *Social Science Computer Review*, 22, pp. 111–127.

Couper, M. P., Tourangeau, R., & Kenyon, K. (2004), Picture This! Exploring Visual Effects in Web Surveys. *Public Opinion Quarterly*, 68, pp. 255–267.

Couper, M. P., Traugott, M. W., & Lamias, M. J. (2001), Web Survey Design and Administration. *Public Opinion Quarterly*, 65, pp. 230–253.

Derouvray, C. & Couper, M. P. (2002), Designing a Strategy for Reducing "No Opinion" Responses in Web-Based Surveys. *Social Science Computer Review*, 20, pp. 3–9.

Dillman, D. A., Smyth, J. D., & Christian, L. M. (2014), *Internet, Phone, Mail, and Mixed Mode Surveys: The Tailored Design Method*. 4th edition. Wiley, Hoboken, NJ.

Fishbein, M. & Ajzen, I. (1975), *Belief, Attitude, Intention and Behavior: An Introduction to Theory and Research*. Addison-Wesley, Reading, MA.

Fishbein, M. & Ajzen, I. (2010), *Predicting and Changing Behavior. The Reasoned Action Approach*. Routledge, London, UK.

Gist, M. (1989), The Influence of Training Method on Self-efficacy and Idea Generation Among Managers. *Personnel Psychology*, 42, pp. 787–805.

Gist, M. E., Schwoerer, C. E., & Rosen, B. (1989), Effects of Alternative Training Methods on Self-efficacy and Performance in Computer Software Training. *Journal of Applied Psychology*, 74, pp. 884–891.

Heerwegh, D. (2003), Explaining Response Latency and Changing Answers Using Client-side Paradata from a Web Survey. *Social Science Computer Review*, 21, pp. 360–373.

Heerwegh, D. & Loosveldt, G. (2002), *An Evaluation of the Effect of Response Formats on Data Quality in Web Surveys*. Paper presented at the International Conference on Improving Surveys, Copenhagen, Denmark.

Ipsos (2016), White paper, Emotion Measurement. realesit.com.

Kreuter, E. (ed.) (2013), Improving Surveys with Paradata, Analytic Uses of Process Information. Wiley, Hoboken, NJ.

Krosnick, J. A. (1991), Response Strategies for Coping with the Cognitive Demands of Attitude Measures in Surveys. *Applied Cognitive Psychology*, 5, pp. 213–237.

Lidwell, W., Holden, K., & Butler, J. (2003), *Universal Principles of Design*. Rockport, Gloucester, MA.

Mathieson, K. (1991), Predicting User Intentions: Comparing the Technology Acceptance Model With the Theory of Planned Behavior. *Information Systems Research*, 2, pp. 173–191.

Mavletova, A., & Couper, M. P. (2013), Sensitive Topics in PC Web and Mobile Web Surveys: Is There a Difference?. *Survey Research Methods*, 7(3), pp. 191–205.

Mavletova, A. & Couper, M. P. (2014), Mobile Web Survey Design: Scrolling Versus Paging, SMS Versus E-mail Invitations. *Journal of Survey Statistics and Methodology*, 2, pp. 498–518.

McClain, C., Couper, M., Hupp, A., Keutsch, F., & Peterson, G. (2019), A Typology of Web Survey Paradata for Assessing Total Survey Error. *Social Science Computer Review*, 37, 2, pp. 196–215.

Peytchev & Hill. (2010), Experiments in Mobile Web Survey Design—Similarities to Other Modes and Unique Considerations. *Social Science Computer Review*, 28, 3, pp. 319–335.

Redline, C. D. & Dillman, D. A. (1999), *The Influence of Symbolic, Numeric and Verbal Languages on Navigational Compliance in Self-Administered Questionnaires*. Paper Presented at the International Conference on Survey Nonresponse, Portland, OR, USA.

Shropshire, K. O., Hawdon, J. E., & Witte, J. C. (2009), Topical Interest Web Survey Design: Balancing Measurement, Response, and Topical Interest. *Sociological Methods & Research*, 37, pp. 344–370.

Simon, H. A. (1957), *Models of Man: Social and Rational*. Wiley, New York.

Singer, E. & Couper, M. P. (2017), Some Methodological Uses of Responses to Open Questions and Other Verbatim Comments in Quantitative Surveys. *Methods, Data and Analyses*, 11, 2, pp. 115–134.

Smyth, J. D., Dillman, D. A., Christian, L. M., & Stern, M. J. (2006), Comparing Check-all and Forced-choice Question Formats in Web surveys. *Public Opinion Quarterly*, 70, pp. 66–77.

Smyth, J. D., Olson, K., & Kasabian, A. (2014), The Effect of Answering in a Preferred versus Non-preferred Survey Mode on Measurement. *Survey Research Methods*, 8, 3, pp. 137–152.

Snijkers, G. & Morren, M. (2010), *Improving Web and Electronic Questionnaires: The Case of Audit Trails*. Paper presented at the European Conference on Quality in Official Statistics (Q2010), Helsinki, Finland.

Sudman, S. & Bradburn, N. M. (1973), *Asking Questions*. Jossey-Bass, San Francisco, CA.

Sudman, S., Bradburn, N. M., & Schwarz, N. (1996), *Thinking About Answers*. Jossey-Bass, San Francisco, CA.

Tourangeau, R. (2003), Cognitive Aspects of Survey Measurement and Mismeasurement. *International Journal of Public Opinion Research*, 15, pp. 3–7.

Tourangeau, R., Couper, M. P., & Conrad, F. (2004), Spacing, Position and Order, Interpretative Heuristics for Visual Features of Survey Questions. *Public Opinion Quarterly*, 68, pp. 368–393.

Tourangeau, R., Couper, M. P., & Conrad, F. (2007), Color, Labels, and Interpretative Heuristics for Response Scales. *Public Opinion Quarterly*, 71, pp. 91–112.

Tourangeau, R., Couper, M. P., & Steiger, D. M. (2003), Humanizing Self—Administered Surveys: Experiments on Social Presence in Web and IVR Surveys. *Computers in Human Behavior*, 19, pp. 1–24.

Tourangeau, R. & Rasinski, K. A. (1988), Cognitive Processes Underlying Context Effects in Attitude Measurement. *Psychological Bulletin*, 103, pp. 299–314.

Villar, A., Callegaro, M., & Yang, Y. (2013), Where Am I? A Meta-Analysis of Experiments on the Effects of Progress Indicators for Web Surveys. *Social Science Computer Review*, 31, 6, pp. 744–762.

Ware, C. (2004), *Information Visualization: Perception for Design*. Morgan Kaufmann, San Francisco, CA.

Willis, G. B. (2005), *Cognitive Interviewing: A Tool for Improving Questionnaire Design*. Sage, Thousand Oaks, CA.

# CHAPTER EIGHT

# Adaptive and Responsive Design

## 8.1 Introduction

So far in this handbook, it has been assumed that the same design features are applied to all sample elements, i.e., that data collection design is uniform or non-adaptive. The mixed-mode example in Section 7.2.1 provides a glance on designs that do vary over different sample elements. This chapter explains how survey design features can be adapted to specific characteristics of sample elements in order to achieve a trade-off between costs and quality. These techniques, where different design features are assigned to different sampled elements explicitly, are referred to as *adaptive or responsive survey designs*. Particularly, in this chapter, the focus is on adaptive and responsive survey designs (ASD/RSD) that include web as a design feature to be varied among elements. ASD and RSD are a natural choice for surveys that employ web data collection; response rates are relatively low, and extra effort may be allocated to subpopulations with the lowest response rates.

The origin of ASD and RSD is to be found in the medical statistics literature, where treatments are varied beforehand over patients and/or depend on the responses of patients (see, e.g., Heyd and Carlin, 1999; Murphy, 2003). The motivation for using such designs is the uncertainty about the optimal design when planning clinical trials and the costs and risks associated with the treatments.

*Handbook of Web Surveys*, Second Edition. Silvia Biffignandi and Jelke Bethlehem.
© 2021 John Wiley & Sons, Inc. Published 2021 by John Wiley & Sons, Inc.

ASD and RSD present similarities to other techniques in the survey sampling literature, like balanced sampling and quota sampling. It is important to clarify their distinctive features at the start. Balanced sampling (Deville and Tillé, 2004) attempts to select samples that are balanced to the population. On the other hand, ASD and RSD attempt to balance response to a given sample by optimizing the allocation of design features given the sample, but not the inclusion in the sample. Also, ASD and RSD should be distinguished from quota sampling. Like quota samples, they often attempt to achieve balance within the responding sample. However, the balance of response is assessed against a probability sample and not against the population.

The basic idea underlying ASD and RSD is to assign different design features to different elements in the sample, according to their characteristics. The objective is to provide efficient trade-offs between survey quality and survey costs. The rationale is that different elements may prefer or react differently to different treatments. ASD (Wagner, 2008; Schouten, Calinescu, and Luiten, 2013) and RSD (Groves and Heeringa, 2006) have subtle differences. The most striking is that RSD anticipates on design decisions during data collection, whereas ASD prespecifies adaptations and interventions. In RSD, the data collection is divided into phases. A new phase employs the outcomes from previous phases to distinguish effective treatments and identify costs associated with treatments. The allocation of design features is done in a way that each phase reaches its "phase capacity," which is the optimal trade-off between quality and costs. What characterizes RSDs is the fact that experiments are done as part of the main study, not as part of an earlier pretest or pilot study, and the data from all these cases are included in the final data sets and used to produce final estimates. RSDs are motivated by surveys having a long data collection period and where little is known about the sample before data collection starts and/or little information about the effectiveness of treatments is available from historic data. On the other hand, ASDs refer to short data collection periods with limited possibility of intervention and strong prior knowledge. In a sense, any RSD phase can be adaptive. In this chapter, the focus is primarily on ASDs. It is generally assumed that historic data are available and that effective design features have been identified before data collection starts.

Adaptation to the elements can be done before data collection starts on the basis of previous waves of the same survey or similar surveys or during data collection on the basis of observations made on sample elements. The first case is called static, and the second case is called dynamic, analogous to medical statistics literature. The richer the sampling frame and/or linked data from administrative records or previous surveys, the more ASD will tend to be static. Dynamic designs are more demanding in terms of processing survey data and implementation.

As a stand-alone survey mode, web does offer design features that can be adapted. Features are the content of the advance letter, the type of invitation (mailed letter, phone text message, e-mail), the number and type of reminders, the use of incentives, the design and layout of the survey web portal, a push to mobile devices (tablets, smart phones), and the content of the questionnaire.

However, due to the absence of interviewers, web offers relatively little paradata (process data observed during data collection) when a sample person never logs in to the survey website, so most adaptation must be based on auxiliary data available at the outset of the survey, so-called baseline covariates. Furthermore, because web response rates are relatively low, web is often combined with other survey modes, e.g., mailed questionnaires, telephone, and face-to-face (F2F). ASDs for web surveys, therefore, seem most interesting in multimode surveys.

A natural question that arises in this context is whether such adjustments by design on a specified set of variables are useful even when the same variables are employed for nonresponse adjustment methods. Schouten et al. (2016) provide some theoretical considerations and empirical evidence based on 14 survey data sets that, on average, a design with a more representative response has smaller nonresponse biases, even after adjustments on the characteristics for which representativeness was evaluated. Särndal and Lundquist (2014) found gains in balancing the respondents set, over and beyond those obtained by calibrating the sample. Further, it is worth noting that a more balanced sample leads to less variability in adjustment weights, which of course is a desirable property as large variation in adjustment weights may inflate standard errors. This does not mean that one should not apply adjustment at the estimation stage nor that one should blindly assume that more balanced response will always lead to bias reduction, even after adjustment. Indeed, a completely balanced response is not feasible in practice even after applying these techniques, and nonresponse adjustment weighting will always be necessary. Further, ASD and RSD need to be applied properly with careful selection of the main ingredients, as explained in Section 8.2.

Note that changing protocol will, in general, change the makeup of the response and not just bring in "more of the same." Peytchev, Baxter, and Carley-Baxter (2009) demonstrated that a major change in protocol produced changes in the study estimates.

Tourangeau et al. (2017) provide an overview of efforts to evaluate these strategies experimentally or via simulations. The overall conclusion is that although the field seems to have embraced these new tools, most of the evaluation studies suggest they produce marginal reductions in cost and nonresponse bias. Multiple reasons are provided as possible explanations. Above all, ASD and RSD have been developed in response to an increasingly difficult survey climate, where it is extremely difficult to make dramatic improvements. Other issues limiting the effectiveness of these designs include weakly predictive auxiliary variables, ineffective interventions, and slippage in the implementation of interventions in the field. These problems are not, however, unique to responsive or adaptive designs.

The literature presents many different versions of ASD and RSD. We will not cover all the possibilities. The aim of this chapter is mainly to provide a basic introduction and give some guidance in the implementation where web is possibly a design feature. Section 8.2 describes the theory and introduces the main concepts underlying these techniques. Section 8.3 presents a realistic application. Section 8.4 summarizes the chapter.

# 8.2 Theory

In this section, the theoretical framework for ASD and RSD is presented. We set out the basic terminology and ingredients needed for the implementation of these techniques. In Section 8.2.1, the concepts of stratification of the population and strategy are introduced. Section 8.2.2 is devoted to quality and cost functions, and Section 8.2.3 discusses methods for strategy allocation and the optimization problem.

## 8.2.1 TERMINOLOGY

In this section, we introduce two notions that are extremely important for the implementation of ASDs, namely, that of stratification and strategy. Both are part of step 3 of the web survey process (Chapter 3). Notice that the concept of stratification of the population was already introduced earlier in the book (e.g., Chapter 4), with reference to sampling and estimation. In ASD the same concept is used to identify groups to which different design features will be assigned. This is a crucial aspect for the efficacy of ASD. Sample elements are clustered into subgroups on the basis of a number of variables. These auxiliary variables are available from the sampling frame, administrative data, and paradata. Depending on the sampling frame used for drawing the sample, a number of variables may be available from the frame, such as gender and age. Further, in many countries, rich registers can be linked to the sample, e.g., from tax or housing records. In such cases, several variables may become available from administrative data, such as income, employment, type and number of vehicles owned by the household, and type and value of the dwelling in which the household lives. The third source of auxiliary variables is paradata. Paradata are process data observed during data collection. Examples of paradata are observations of the interviewers about the state of the dwelling or the neighborhood and the interviewers' assessment of the propensity to respond or the propensity to be contacted. With web mode, due to the absence of an interviewer, relatively little paradata are available when a sample element never logs into the survey website. Examples of paradata with web are the device used, the number of break-offs, and the number of accesses to the questionnaire website.

A particularly rich framework with respect to the availability of auxiliary variables that could be used for the implementation of ASDs is that of longitudinal panel surveys, where a wealth of information on panel members is available. These variables can have been collected during the profile survey and over time or can also be paradata type variables (Lynn, 2015).

Notice that the above auxiliary variables may have different levels of availability. Frame and register data are available at the start of the survey. They are also called *baseline covariates*. In this case, strata are formed before data collection starts. On the other hand, paradata generally become available during data collection. They may not be available during all phases of data collection, so in this case strata are formed during data collection.

A crucial point is the choice of the variables to be used to form subgroups. The effectiveness of ASD depends strongly on the relevance of these variables. These auxiliary variables need to be related to both the response behavior and target variables. Each subgroup should be relatively homogeneous both in terms of response propensities and in terms of key survey variables. These requirements are analogous to those for successful weighting nonresponse adjustments. As in this case, ASDs that are based on variables with weak relation to target variables may be counterproductive and may even increase imprecision (Little and Vartivarian, 2005). The utility of ASDs, regardless of adjustment in the estimation phase, is proportional to the strength of the association between auxiliary variables and the target variables. The survey designer should be warned that the absence of relevant auxiliary variables should warrant against use of ASDs. So it is important that the survey researcher should think about auxiliary variables already in the survey design phase.

In general, a straightforward way to choose auxiliary variables is to start from considering those that are used for weighting adjustments. A possible alternative is based on the use of so-called partial $R$-indicators. To identify population subgroups that should be targeted, these indicators have been introduced by Schouten, Shlomo, and Skinner (2011). Schouten and Shlomo (2017) show how to employ partial $R$-indicators for these purposes.

In what follows, we denote the homogeneous subgroups by $G = \{1, \ldots, G\}$ and their relative sizes by $W_g$, with $\sum_g W_g = 1$. For simplicity, we assume that a simple random sample of size $n$ has been drawn from the population. However, in general, the sampling design can be included in the optimization of the ASD.

Another key concept in ASDs is that of strategy. A survey strategy is a list of survey design features. The literature presents many design features (Groves et al., 2002), and these can be varied for different groups in the ASD. One of the most prominent features is the survey mode (web, mail, telephone, or F2F). Particularly, web has a strong quality-cost differential. It is cheap but has low response rates and has different measurement properties than interviewer-administered modes. Special care need to be taken when including web as a possible design feature in an ASD. There are also other design features that may be varied. Most common are length of the period of data collection, invitation or advance letters, number and type of remainders, level and type of incentives, length of the questionnaire, sample size, and sampling design. Other features are available in the longitudinal setting, including post-wave thank you letters, between wave motivation mailings of findings or other information, keep-in-touch mailings or phone calls, birthday cards, change-of-address cards, and so on (see, e.g., Lynn, 2016; Bianchi and Biffignandi, 2017). Further, longitudinal surveys are increasingly developing participant websites where sample members can log in, read information, and take part in various activities. For example, the U.K. Household Longitudinal Study (UKHLS) has a web page for participants, where information on the relevance of the survey, the interview, confidentiality, and the use of the data is provided, together with request of information in case of possible change of address and missed mailings. Further, there are also commercial web panels where members can login and see results of other surveys. The nature of these web-based interactions could also be targeted (Lynn, 2015).

We denote strategies by $s$ and the set of $M$ candidate strategies by $S$. So $S = \{\emptyset, s_1, \ldots, s_M\}$, where $\emptyset$ denotes the empty strategy. The empty strategy means that no action is taken. Usually, $\emptyset$ is included in the survey strategy set. Examples of strategies are

$$s_1 = \{\text{Invitation letter 1, web questionnaire, incentive}\}$$

$$s_2 = \{\text{Invitation letter 1, web questionnaire, no incentive}\}$$

$$s_3 = \{\text{Invitation letter 2, face} - \text{to} - \text{face interview, incentive}\}$$

$$s_4 = \{\text{Invitation letter 2, face} - \text{to} - \text{face interview, no incentive}\}.$$

Strategies can also be defined considering different data collection phases. For example, in a survey with two phases (main data collection and follow-up), one may define the survey strategy set as

$$S = \{ \emptyset, \text{Web}\} \times \{ \emptyset, \text{F2F}\}.$$

This means that four strategies are considered: $s_1 = (\emptyset, \emptyset)$, $s_2 = (\emptyset, \text{F2F follow-up})$, $s_3 = (\text{Web}, \emptyset)$, and $s_4 = (\text{Web}, \text{F2F follow-up})$. Notice that in general some strategy may be discarded based on deductive arguments. For example, strategy $s_1$ where no action is applied in any of the two phases is generally excluded.

■ EXAMPLE 8.1 **Adaptive survey design**

Suppose that a mixed-mode survey of size $n = 2{,}500$ is conducted. The available modes are web, telephone, and face-to-face (F2F). We assume that contact details in the three modes are available for everyone in the sample. From a previous wave of the survey, it is clear that different age groups of the population prefer to respond in different modes. It may therefore be useful to set up an ASD. The three strategies in the design are

$$s_1 = (\text{web}), s_2 = (\text{phone}), \text{and } s_3 = (\text{F2F}).$$

The variable *age* is the stratification variable. This variable is available from the sampling frame. Hence, it is a baseline covariate and it is available before data collection starts. Three strata are considered: young $g_1$ (<25), middle-aged $g_2$ (25–55), and elderly $g_3$ (>55). Their relative sizes in the sample are

$$W_1 = 21\%, W_2 = 63\%, \text{and } W_3 = 16\%.$$

A possible ASD could assign strategy $s_1$ to young persons, strategy $s_2$ to middle-aged persons, and strategy $s_3$ to the elderly. In this ASD, only one design feature is varied among elements, namely, the survey mode. It is assumed that all other features have standard implementations, such as advance letters, refusal conversion techniques, etc.

**EXAMPLE 8.2 Adaptive survey design**

In the same setting as Example 8.1, a different mixed-mode survey is conducted. Overall, data collection lasts five weeks. Given that web is the cheapest method of data collection, web is offered first to all elements in the sample. After two weeks of data collection, elements are classified as respondents or nonrespondents.

For the remaining three weeks, three survey modes are considered for follow-up: web, phone, and F2F. Hence the ASD has three candidate strategies:

$$s_1 = (\text{web}, \varnothing), s_2 = (\text{web}, \text{web}), s_3 = (\text{web}, \text{phone}), \text{and } s_4 = (\text{web}, \text{F2F}).$$

The set of candidate strategies can also be written as

$$S = \{\text{web}\} \times \{\varnothing, \text{web}, \text{phone}, \text{F2F}\}.$$

Again, age is chosen as stratification variable, leading to three subgroups: young, middle-aged, and elderly.

Stratification is done using age, which is available at the start of data collection:

$$G_1 = \{\text{young}, \text{middle} - \text{aged}, \text{elderly}\}.$$

A possible ASD in this case is to assign strategy $s_1$ to middle-aged persons, strategy $s_2$ to young, strategy, and strategy $s_4$ to elderly.

Assume now we can observe the device used for accessing the web survey (not recorded, smartphone or tablet, pc). We have nine strata now, namely,

$$G_2 = \{\text{young}, \text{middle} - \text{aged}, \text{elderly}\} \times \{\text{not recorded}, \text{smartphone or tablet}, \text{pc}\}.$$

A possible ASD could assign strategy $s_2$ to young and middle-aged who have accessed the web survey using smartphone, tablet, or pc; strategy $s_3$ to young and middle-aged whose device has not been recorded; and strategy $s_4$ to elderly (any device).

## 8.2.2 QUALITY AND COST FUNCTIONS

The choice of the quality function depends on which errors one wants to account for, i.e., step 1 of the web survey process (Chapter 3). So far, literature on ASD has focused mainly on bounding the impact of nonresponse error. However, in contexts where web is a design feature, possibly in combination with other survey modes, measurement error is expected to play a role and needs to be accounted for. Further, also other aspects may be taken into consideration. In a recent paper, Beaumont, Haziza, and Bocci (2014) remarked that ASD and RSD should also take variance into account. Below we review some indicators used as indirect measures of nonresponse error and how they are used as quality functions in ASD. Given the introductory level of this chapter, we do not introduce indicators that also consider measurement error in detail, but only explain how nonresponse indicators can be extended to this case.

For the sake of simplicity, we assume that elements in the same group $g$ are assigned the same strategy $s_g$. The basic ingredient for building quality functions is that of response propensities. Let $R_i$ denote an indicator variable taking value 1 if element $i$ in the population responds and value 0 otherwise. Response propensities are defined as the conditional probability that an element responds, $R_i = 1$, given that it belongs to a certain subgroup $g$ and that it receives a certain strategy:

$$(8.1) \qquad \rho_i = P(R_i = 1 \mid g, s).$$

For simplicity of exposition, we assume that response propensities are based on the same subgroups as the subgroups used for strategy allocation. Notice however that, in general, the choice of subgroups may be different. Since $\rho_i$'s are constant within a group $g$ given a certain strategy $s$, we denote them by $\rho_g(s)$.

The most used quality function is the *response rate*. In ASD, the response rate is represented as the mean of response propensities. This is justified by the fact that the expected value of the response propensities is equal to the (expected) response rate, $P(R_i = 1)$. So, the expected response rate for a design that assigns strategy $s_g$ to group $g$ is given by

$$(8.2) \qquad \bar{\rho} = \sum_{g \in G} W_g \rho_g (s_g).$$

Another popular quality indicator is the *R-indicator* (Schouten, Cobben, and Bethlehem, 2009), which is defined as a transformation of the sample variance of the response propensities:

$$(8.3) \qquad R = 1 - 2S,$$

where $S$ is the standard deviation of the response propensities. In ASD, $S^2$ is expressed as

$$(8.4) \qquad S^2 = \sum_{g \in G} W_g \left( \rho_g (s_g) - \bar{\rho} \right)^2.$$

The $R$-indicator varies between 0 and 1, where a value close to one is optimal. The rationale for this indicator is that an absence of variation in the response propensities ($S$ close to 0, thus $R$ close to 1) means that response is a random subsample of the original sample (with respect to the stratification variables). A closely related quality indicator is the *coefficient of variation* of the response propensities:

$$(8.5) \qquad CV = \frac{S}{\bar{\rho}}.$$

The coefficient of variation sets an upper bound to the absolute bias of response means (Schouten, Cobben, and Bethlehem 2009). The use of the coefficient of variation is recommended when a survey mainly estimates population means or totals.

Notice that the response rate, $R$-indicator, and coefficient of variation are subject to imprecision, which can be rather large. Shlomo, Skinner, and Schouten (2012) provide formulas for standard error estimation for $R$-indicators. Codes in SAS and $R$ for its computation can be downloaded from the RISQ website (http://www.risq-project.eu) together with a manual and test data set.

A number of other quality indicators have been proposed in the literature. Among them, we would like to mention the *balance indicator* (Särndal, 2011), which is very similar to the $R$-indicator; the *coefficient of variation of the nonresponse adjustment weights* (Särndal and Lundström, 2010), which is similar to the coefficient of variation; and the *fraction of missing information* (Wagner, 2010; Andridge and Little, 2011).

The quality indicators presented so far do not take measurement error into consideration. They assume that measurement error is affected in the same way by all design features. While this assumption is acceptable in some cases, it is certainly not tenable for design features like the survey mode. As extensively explained in Chapter 3, the survey mode choice step in a survey process is strictly related to the types of errors that may arise at later stages of the survey. Particularly, survey mode is known to affect nonresponse and measurement error simultaneously. For example, the web mode is known to provide low response rates and to be prone to measurement error. Further, the survey mode may be related to target variables. Indeed, different kinds of persons are more likely to respond in different modes.

Calinescu and Schouten (2016) introduced quality functions that take both nonresponse and measurement errors into account. The basic idea relies on modeling the underlying causes for differences in measurement error between design features. This is done by first introducing the definition of so-called response quality indicators, which are 0–1 indicators summarizing the quality of the response over multiple survey items. These indicators are included in the ASD by estimating their propensities to occur for different population subgroups. These propensities are called "response quality propensities" and have an analogous interpretation to the response propensities (that correspond to the occurrence of an element response).

Whether or not to use quality functions that take measurement error into consideration depends on the design features that are varied in the design. One should ask which errors are affected by each design feature. For example, the number of calls in a telephone survey is clearly related to the contact rate and, hence, the response rate. It is also related to nonresponse bias, as harder-to-contact persons are known to have different values on some key variables (e.g., employment-related variables). However, there is no clear link to measurement error.

The analytic introduction of this topic is rather complex and goes beyond the scope of this chapter. The interested reader is referred to Calinescu and Schouten (2016).

Let us now move to the cost function. Cost functions may include both overhead cost components and other variable cost components. In general, variable costs depend on the sample size, while overhead costs do not. Overhead costs may come from data collection staff, sampling, and processing of samples. Variable costs may be associated with training and instruction of interviewers, incentives, and processing of paper questionnaires. In case web is a survey design feature, overhead costs may come from initial setup costs in computers, servers, and software; web questionnaire design and implementation; help centers for respondents; backup devices; extra power supply; etc. Once the procedure is set up, variable costs are the costs of mailed advance letters and reminders and possibly incentives.

When computing the cost function in an ASD framework, it is however important to restrict to costs relative to the design features that are varied in the design. For example, when it is the incentive that is varied between subgroups, then costs need not include components related to advance letters. On the other hand, when mode is the tailored design feature and one possible mode is F2F, then traveling costs play an important role in the cost evaluation.

If we denote by $c_g(s_g)$ the cost incurred by allocating strategy $s_g$ to subgroup $g$, the cost function may be written as

$$(8.6) \qquad C = n \sum_{g \in G} W_g c_g \left( s_g \right).$$

The reader is referred to Calinescu and Schouten (2016) for a detailed example on the definition of $c_g(s)$. Again, for the sake of simplicity, we assume costs per sample element based on the same subgroups used for strategy allocation. This could be further generalized.

Finally, it has to be noticed that in the definition of quality and cost functions, we have assumed that elements in the same group are assigned the same strategy. These functions can be further extended to the case where elements in the same group receive different strategies. This can be done by allocating strategies to elements in a group with properly defined probabilities. See Calinescu and Schouten (2016) for more details.

### 8.2.3 STRATEGY ALLOCATION AND OPTIMIZATION

In this subsection, we discuss methods for assigning strategies to subgroups. In general, potential gains are envisaged for ASD when $c_g(s)$ and $\rho_g(s)$ vary substantially over groups in $G$ so that benefits could be achieved by assigning different strategies to different subgroups.

For the practical implementation of such methods, some basic ingredients need to be estimated, namely, the response propensities $\rho_g(s)$ and the costs per sampled element $c_g(s)$. In case one takes into account measurement error, it is necessary to estimate the response quality propensities as well.

The estimation of the response propensities is typically based on a logistic regression model that includes the auxiliary variables and the design features as covariates. In order to be able to run such regression models, membership and strategy allocation need to be known, for both respondents and nonrespondents in the data set. In case different design features are applied to different phases of data collection, response propensities can be estimated for each phase separately. The cumulative propensities can then be derived from phase-specific propensities. An example is presented in Section 8.3. We denote the estimated response propensities by $\hat{\rho}_g(s)$. Associated to each response propensity is a standard error that reflects the uncertainty of the estimate.

Next, one needs to estimate costs per sampled element. We denote such estimated costs by $\hat{c}_g(s)$. They also are accompanied by standard errors, reflecting the imprecision due to variation in interview length, number of visits/calls, etc. Usually, standard errors for costs per sampled element are estimated from cost sheets that are applied to all individual sample records in historic survey data.

For the estimation of $\rho_g(s)$ and $c_g(s)$, data from past surveys are used, preferably, from the same survey or similar surveys. Alternatively, as proposed in Groves and Heeringa (2006), one may use earlier phases of the data collection, where experiments done during an initial phase of data collection are used to inform key design decisions for later phases. Again, longitudinal surveys may be very informative for this purpose, as an experiment mounted at one wave can inform the design at the next wave (Bianchi and Biffignandi, 2018).

It is important that response propensities are estimated from randomized experiments, where the strategies under consideration have been assigned randomly. So, the data set used for estimation should include proper information related to this aspect. Further, randomization should have been performed over subgroups or over sample elements with different values on the auxiliary variables defining subgroups.

---

### ■ EXAMPLE 8.3 Estimating response propensities

Consider the first ASD in Example 8.1. There were three subgroups based on age categories (young, middle-aged, elderly) and three possible strategies, namely, $s_1$ = (Web), $s_2$ = (Phone), and $s_3$ = (F2F). If all three strategies are applicable to all three strata, in total nine response probabilities need to be estimated:

$$\rho_g(s), s \in S = \{s_1, s_2, s_3\}, g \in G = \{\text{young, middle} - \text{aged, elderly}\}.$$

These response propensities can be estimated if the three strategies have been randomly assigned to each age subgroup in a previous wave of the same survey or similar survey or pilot study.

Assume that data are available from a previous wave of the same survey. Response propensities are computed using logistic regression, where the dependent variable is the 0–1 response indicator and covariates are *Age* in three categories (young, middle-aged, elderly) and mode in three categories (web, phone, F2F). Results are presented in Table 8.1.

So, for example,

$$\hat{\rho}_1(s_1) = \hat{P}(R_i = 1 \mid \text{young, web}) = 0.55.$$

TABLE 8.1    **Estimated response propensities for the three strata and three strategies**

|  | Stratum | | |
|---|---|---|---|
| Strategy | Young | Middle-aged | Elderly |
| Web | 0.55 (0.09) | 0.45 (0.06) | 0.20 (0.08) |
| Phone | 0.30 (0.10) | 0.50 (0.05) | 0.45 (0.07) |
| F2F | 0.40 (0.11) | 0.40 (0.04) | 0.60 (0.06) |

Standard errors of response propensities are given in brackets.

## ▇ EXAMPLE 8.4 Quality functions

Given the estimated response propensities computed in Example 8.3, the quality functions for the ASD presented in Example 8.1 can be derived, namely, the response rate, the R-indicator, and the coefficient of variation.
Response rate:

$$\hat{\bar{\rho}} = \sum_{g=1}^{3} W_g \hat{\rho}_g\left(s_g\right) = 0.21 \times 0.55 + 0.63 \times 0.50 + 0.16 \times 0.60 = 0.53.$$

Before computing the $R$-indicator, we compute the variance of the response propensities

$$\hat{S}^2 = \sum_{g=1}^{3} W_g \left(\hat{\rho}_g\left(s_g\right) - \hat{\bar{\rho}}\right)^2 = 0.21 \times \left(0.55 - 0.53\right)^2$$

$$+ \, 0.63 \times \left(0.50 - 0.53\right)^2 + 0.16 \times \left(0.60 - 0.53\right)^2 = 0.0014.$$

The $R$-indicator then is

$$\hat{R} = 1 - 2\hat{S} = 1 - 2\sqrt{0.0014} = 0.925,$$

and the coefficient of variation is

$$C\hat{V} = \frac{\hat{S}}{\hat{\bar{\rho}}} = \frac{0.037}{0.53} = 0.070.$$

Notice that the computed value for the $R$-indicator is very high (the maximum value that it can take is 1). This is due to the fact that the estimated response propensities for the ASD under consideration do not show large variability. The value observed for the coefficient of variation is quite close to zero, meaning that the relative variability of the propensities, taking into account their average value, is small.

Of course, since response propensities are estimated with uncertainty, also response rate, $R$-indicator, and coefficient of variation are subject to imprecision, which can be rather large.

**▓ EXAMPLE 8.5 Estimating costs**

In Example 8.1, assume that the following information is available from a previous wave of the same survey:

- The cost of an invitation letter with reference to a website where the web questionnaire can be found is 3 euros.
- The interviewer's hourly rate is 30 euros.
- An average of 3.5 call attempts is needed for young, 2.5 for middle-aged, and 2.2 for elderly.
- An interview lasts on average 22 minutes and nonrespondents take 2.5 minutes of the interviewer time on average.
- Average traveling costs is 100 euros per interview.

From this information, it is possible to estimate costs per sample element for each combination of stratum and strategy:

$$\hat{c}_1(s_1 = \text{web}) = 3$$

$$\hat{c}_1(s_2 = \text{CATI}) = 30 \times [3.5 \times 22 \times \rho_1(s_2) + 3.5 \times 2.5 \times (1 - \rho_1(s_2))]/60 =$$
$$= 30 \times [3.5 \times 22 \times 0.30 + 3.5 \times 2.5 \times 0.70]/60 = 14.6$$

$$\hat{c}_1(s_3 = \text{F2F}) = 100 + 30 \times [22 \times 0.40 + 2.5 \times 0.60]/60 = 105.2$$

Complete results are contained in Table 8.2.

The expected cost for the ASD that assigns strategy $s_1$ (web) to young, $s_2$ (phone) to middle-aged, and $s_3$ (F2F) to elderly is therefore

$$\hat{C} = n \sum_{g=1}^{3} W_g \hat{c}_g(s_g) = 2,500 \times (0.21 \times 3 + 0.63 \times 15.3 + 0.16 \times 107.1)$$
$$= 68,512.5.$$

**TABLE 8.2** Estimated costs per sample element for the three strata and three strategies

| | | Stratum | |
|---|---|---|---|
| Strategy | Young | Middle-aged | Elderly |
| Web | 3.0 (0.0) | 3.0 (0.0) | 3.0 (0.0) |
| Phone | 14.6 (0.4) | 15.3 (0.5) | 12.4 (0.4) |
| F2F | 105.2 (3.1) | 105.6 (3.0) | 107.1 (3.6) |

Standard errors are given in brackets.

Notice that to derive the cost function for a specific ASD, we do not need to compute costs per sample element for each combination of stratum and strategy. It is enough to compute it for combinations considered in the specific ASD.

As for methods used to assign strategies to subgroups, literature proposes several possibilities. The approaches vary with respect to several aspects: explicit focus on mathematical optimization, certainty to be effective, ability to be linked to candidate strategies, reproducibility, and reliance on the accuracy of response propensity estimates. First, Schouten and Shlomo (2017) propose a structured trial-and-error approach, where a proxy measure of nonresponse is monitored and, when it is above a certain threshold, nonrespondent profiles are derived using partial $R$-indicators and additional effort is made for subgroups that turn out to be underrepresented. This method is feasible in practice and pragmatic: it has some mathematical rigor and allows a quality-cost trade-off. However, there is no guarantee that the adaptation leads to optimal allocation and better accuracy. A second possible approach is prioritization of sample elements in data collection based on estimated response propensities (Peytchev et al., 2010; Wagner, 2013). Response propensities are estimated during data collection. The lowest propensity cases have higher priority and receive more effort. This approach is not linked to a specific quality objective function, but it aims at equalizing response propensities. The goal is generally to target low propensity cases in an effort to reduce the overall variation in response propensities. With this method, there is some guarantee that quality is improved. However, this approach is sensitive to the accuracy of response propensity estimates, and it makes it harder to link effective data collection strategies as the sorted cases do not have an easy translation into characteristics. Examples 8.6 and 8.7 report on two experiments that use this method to assign strategies to subgroup. A third approach consists of setting different stopping rules for different subgroups to decide whether continued efforts should be made to subgroups or not (Lundquist and Särndal, 2013). See also Bianchi and Biffignandi (2014). This method does not make use of explicit optimization, but stopping rules are set on the basis of quality functions. So, this approach has some guarantee that quality is improved. Further, it allows for participation of expert knowledge to decide the most effective strategy with each subgroups, and it is mildly sensitive to the accuracy of response propensities. The fourth method is the one that we present in more detail, and it is an optimization method. If all input parameters are estimated accurately, then this approach leads to optimal improvement of quality. However, the approach is sensitive to inaccurate input parameters. Furthermore, the optimization problem can be rather complex and, in some cases, computationally intractable.

⊞ EXAMPLE 8.6 **Experiment in the Dutch Survey of Consumer Sentiments**

> Luiten and Schouten (2013) report an experiment in the Dutch Survey of Consumer Sentiments, where mode (including web) is a design feature that is varied among elements. On the basis of earlier rounds of the same survey, contact and cooperation propensities were first estimated. Next, the data collection was designed to encompass two phases. In the first phase, cases with low estimated cooperation propensities were sent a mail questionnaire; those with the highest estimated propensities were invited to complete a web survey; and those with intermediate estimated propensities were given a choice between mail and web. In the second phase, nonrespondents were followed up by telephone, and cooperation was manipulated by assigning specific interviewers to specific sample elements and contact was manipulated by timing, spacing, and prioritizing calls. The control group for the experiment was the regular Survey of Consumer Sentiments, which is a computer-assisted telephone interviewing only survey. Overall, the tailored strategy was successful in significantly increasing representativeness while maintaining the level of response and costs.

⊞ EXAMPLE 8.7 **Experiment in the High School Longitudinal Study**

> Rosen et al. (2014) report on a study where low propensity cases were targeted to reduce nonresponse bias. The study refers to the High School Longitudinal Study in 2009, which involved the collection of data on students and their parents. The experiment refers to the collection of data on parents. First, response propensities for parents' response were estimated. After an initial period of data collection by telephone and the web, the remaining parents in the lowest propensity quartile were followed up by face-to-face. As for parents in the other three quartiles, they were followed up by telephone and the web. This strategy increased the response rate in the lowest propensity quartile and reduced the average relative bias.

In a general optimization design, one wants to maximize the response rate in Equation (8.2), given a number of constraints on cost and quality. Typical constraints are:

**1.** Cost constraint, where one specifies a total available budget $B$:

$$(8.7) \qquad \hat{C} = n \sum_{g \in G} W_g \hat{c}_g \left( s_g \right) \leq B;$$

**2.** Representativeness constraint, requiring a minimum level of representativeness $\alpha$ on the auxiliary variables included in the response propensity model:

$$(8.8) \qquad\qquad \hat{R} = 1 - 2\hat{S} \geq \alpha,$$

where $\hat{S}^2 = \sum_{g \in G} W_g \left( \hat{\rho}_g(s_g) - \hat{\bar{\rho}} \right)^2$; and

**3.** Precision constraint, requiring a precision for certain subgroups, which is usually operationalized as a lower bound $n_g$ to the expected number of respondents in each subgroup $g$:

$$(8.9) \qquad\qquad nW_g\hat{\rho}_g(s_g) \geq n_g, \quad \forall g \in G.$$

Notice that subgroups for precision constraints may be taken differently from the subgroups that are used for assigning strategies. Alternatively, the precision constraint may also be simplified to an overall constraint on the expected number of respondents.

If the number of strata and strategies is relatively small (like in the application presented in Section 8.3), one could look directly for the solution through an extensive grid search. The objective function and the constraints are evaluated for each admissible stratum and strategy, and the solution with highest value for the objective function that satisfies the constraints is selected. In general, the number of decision parameters is too large to apply a direct search, and one needs to set up a proper optimization problem. In such situations, the objective function and the constraints need to be rewritten using proper decision variables, and the solution of the problem requires numerical optimization for nonlinear problems, which can be done using standard statistical computer software like SAS or R. Packages like *nloptr* in R can be used. The recommendation is to use various starting values in the optimization. One such starting value should be the current nonadaptive design so that any optimum, even if local, would represent an improvement. So, for example, one should use as one of the possible starting values in the optimization problem the case where F2F is assigned to every element, if one wants to set up an ASD where mode (web, telephone, F2F) is the only survey design feature that is varied among elements and where the same survey has been carried out F2F so far. The optimization problem outlined above is one of the most common in practice. However, this is not the only possibility. One may also minimize other quality functions than the response rate or specify different constraints or even minimize costs subject to constraints on quality. It should be clear that depending on the choice of the constraints and indicators, different ASDs may be obtained as a solution to the optimization problem.

Finally, it is important to remark that the input parameters to the ASD (the response propensities and costs per element) are estimated and, therefore, subject to imprecision and bias. Estimation is based on historical survey data, which has sampling error and may be outdated or incomplete. The imprecision and bias of the input parameters may affect the performance of the ASD. It is therefore important to assess the sensitivity and robustness of ASDs to inaccuracy in the estimated response propensities and cost parameters. See Section 8.3 for an example on how to carry out sensitivity analyses in ASD.

▣ EXAMPLE 8.8 **Optimization problem**

Assume there is only one possible alternative ASD to the one described in Examples 8.1, 8.3, 8.4, and 8.5, that is, assign strategy $s_1$ (web) to young and middle-aged and strategy $s_3$ (F2F) to elderly. We would like to choose the ASD that maximizes the response rate and with R-indicator greater than 0.90 while keeping costs below 70,000.

This is a simple optimization problem as only two ASDs are admissible. In order to choose the best design, we should compute the response rate, $R$-indicator, and cost function for the alternative design:

$$\hat{\bar{\rho}} = \sum_{g=1}^{3} W_g \hat{\rho}_g(s_g) = 0.21 \times 0.55 + 0.63 \times 0.45 + 0.16 \times 0.60 = 0.50$$

$$\hat{S}^2 = \sum_{g=1}^{3} W_g \left( \hat{\rho}_g(s_g) - \hat{\bar{\rho}} \right)^2 = 0.21 \times (0.55 - 0.50)^2$$

$$+ 0.63 \times (0.45 - 0.50)^2 + 0.16 \times (0.60 - 0.50)^2 = 0.0037$$

$$\hat{R} = 1 - 2\hat{S} = 1 - 2\sqrt{0.0037} = 0.878$$

$$n \sum_{g=1}^{3} W_g \hat{c}_g(s_g) = 2,500 \times (0.21 \times 3 + 0.63 \times 3 + 0.16 \times 107.1) = 49,140$$

The second ASD has an estimated $\hat{R}$ that does not satisfy the constraint on representativity. However, given the moderate sample size, this may be due to sampling variation. The $R$-indicator is only an estimate of $R$ with a margin of error. So there is no guarantee that the true $R$ is smaller than 0.90. For this reason, we have computed the 95% confidence interval for $R$ (0.861, 0.895), which does not contain 0.90. (Formulas for standard error estimation for $R$-indicators are provided in SAS and R codes that can be downloaded from the RISQ website.) Given that the second ASD does not satisfy the constraint on representativity, we choose the first design.

# 8.3 Application

In this section, we will demonstrate the various elements and choices in ASD using a realistic, but somewhat simplified, application.

As argued in the Introduction to this chapter, web is often combined with other survey modes, e.g., mailed questionnaires, telephone, and F2F. For this reason, in the application, we consider a design with multiple modes including web. The adaptation comes from the optional use of incentives in web and an interviewer follow-up.

The first step is the identification of relevant population groups from available auxiliary data. Suppose three covariates can be linked for the survey. First, there is age of the sample person, $X_1$, as a baseline covariate. Second, in case the web survey mode is applied, it can be observed whether a person logs in to the survey website but breaks off before completing the survey. This binary indicator is labeled $X_2$. The third covariate, $X_3$, is also binary and becomes available when F2F interviewers report whether there are signs of life at the sample addresses. Age is, generally, relevant and links to key survey variables in almost all surveys. Also younger, middle-aged and older age groups tend to have different web response rates. In F2F, response rates are also different but not as distinct as in web. However, the two paradata covariates, break-off and neighborhood status, may only be relevant in some surveys. In conjunction with features of the survey questionnaire and survey items, break-off may be a sign of low motivation and/or low ability to do a survey. Such persons may be relevant for some surveys, e.g., a labor force survey (LFS) or a survey on income and living conditions. Persons who break off have much higher response rates in interviewer follow-ups. Addresses at which weak signs of life are reported may be very interesting for a survey, e.g., a housing survey or a health survey, but response rates for persons registered at such addresses tend to be low in interviewer modes. In the application, we consider the setting of a LFS and use covariates $X_1$ and $X_2$ to construct six strata:

$$G = \{ < 25, 25 - 45, 45 + \} \times \{\text{no break} - \text{off}, \text{break} - \text{off}\},$$

where the three age classes are based on the publication cells of the LFS. In the following, we simply call them young (<25), middle-aged (25–45), and elder (>45). Obviously, at the outset of the survey, only the stratification into age groups can be made; the break-off stratification applies only to sample persons that have received web as part of their data collection strategy.

For simplicity, we assume a simple random sample without replacement, although in the optimization of the ASD also the sampling design can be included. The three age groups are of approximately equal size in the population, but break-off rates are different for the three age groups. They are set at 20% for the young, 10% for the middle-aged, and 5% for the elder. These break-off rates are slightly higher than is observed in practice, for illustration purposes, but the age pattern is realistic. The population distribution is

$$(Y \& NB, Y \& B, M \& NB, M \& B, E \& NB, E \& B) = (26\%, 7\%, 30\%, 3\%, 31\%, 2\%).$$

The second step is to determine the quality and cost functions in the ASD optimization. In the application, we first focus on nonresponse and ignore measurement error differences between strata and strategies. LFS users require a specified precision for each of the age groups, and minimal numbers of respondents are derived from these precision requirements. For each age stratum, 2,000 respondents are needed. Furthermore, the total budget is limited and is set at 360k euros, i.e., an average of 60 euros per respondent. In addition, we choose to maximize the response rate given constraints on the variance of response propensities through the R-indicator.

The third step is the selection of possible data collection strategies that allow us to moderate response propensities for the age and break-off groups. Suppose that web is viewed as the default mode for the LFS, but due to low response rates, especially for the young and elderly, additional effort is needed. A survey design with three data collection phases is considered, where web is optional in the first phase and the two subsequent phases use F2F visits (Figure 8.1). In the first phase, there are three possible actions: no action, a web invitation plus unconditional incentives, or a web invitation without incentives. In the second phase, there are two possible actions: no action, or a standard F2F strategy with a limited number of visits. Finally, in the third phase, there are again two possible actions: no action or a follow-up with F2F.

We can write the strategy set as

$$S = \{\emptyset, \text{Web} + , \text{Web} - \} \times \{\emptyset, \text{F2F}\} \times \{\emptyset, \text{F2F} + \},$$

where $\emptyset$ represents "no action." Since we like to apply at least some action, the strategy in which no action is applied in any of the phases will be excluded from the strategy allocation. Furthermore, the extended F2F action in phase 3 can only be applied to phase 2 F2F nonrespondents, i.e., the strategies $s = (\emptyset, \emptyset, \text{F2F+})$, $s = (\text{Web} - , \emptyset, \text{F2F+})$, and $s = (\text{Web} + , \emptyset, \text{F2F+})$ are not admissible. The number of possible strategies is eight.

The fourth step is the estimation of survey design parameters. In the application, these are response propensities per strategy and stratum and costs per sample element per strategy and stratum.

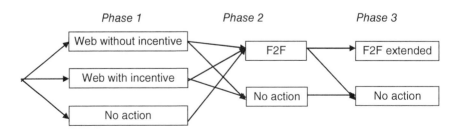

FIGURE 8.1  **Data collection phases**

Table 8.3 contains estimated response propensities for phase 1 and estimated unconditional and conditional response propensities for phases 2 and 3. Consider, for example, a young person that would break off in phase 1 web (Y&B). Without any action, the (unconditional) response propensity in phase 2 is estimated as 35%. The estimated value is the same as for a young person that would not break off, since actual break-off is not observed. The (conditional) response propensity in phase 3, given nonresponse in phase 2, is 20%. The (unconditional) response propensities in phase 1 under web– and web+ are obviously equal to 0%. The phase 2 and 3 conditional response propensities are equal to 60% and 25%, regardless of the use of an incentive. Cumulative propensities, i.e., after phase 2 or phase 3, for the eight strategies can be derived from Table 8.3. They are lowest for $s =$ (Web $-$, $\emptyset$, $\emptyset$). The two strategies $s =$ ($\emptyset$, F2F, F2F+) and $s =$ (Web $+$, F2F, F2F+) have comparable response propensities that are higher than all other strategies. For example, the cumulative propensities for young persons that break off for phases 2 and 3 are, respectively, $0\% + 100\% \times 60\% = 60\%$ and $0\% + 100\% \times 60\% + 100\% \times 40\% \times 25\% = 70\%$.

The various strategies have very different costs per sample element. Table 8.4 presents estimated marginal costs per phase for each stratum. Cumulative costs can be derived by combining marginal costs with the response propensities of Table 8.3. Costs are lowest for $s =$ (Web $-$, $\emptyset$, $\emptyset$) and highest for $s =$ ($\emptyset$, F2F, F2F+). For example, the cumulative costs of strategy $s =$ (Web $+$, F2F, F2F+)

TABLE 8.3    **Estimated response propensities for the six strata in phases 1, 2, and 3**

| Phase 1 action | Phase | Stratum | | | | | |
|---|---|---|---|---|---|---|---|
| | | Y&NB | Y&B | M&NB | M&B | E&NB | E&B |
| No | 1 | — | | — | | — | |
| | 2 | 0.35 (0.01) | | 0.45 (0.01) | | 0.55 (0.01) | |
| | 3 | 0.20 (0.02) | | 0.25 (0.02) | | 0.10 (0.02) | |
| Web– | 1 | 0.15 (0.02) | 0 | 0.25 (0.02) | 0 | 0.30 (0.02) | 0 |
| | 2 | 0.30 (0.02) | 0.60 (0.03) | 0.35 (0.02) | 0.70 (0.03) | 0.30 (0.02) | 0.80 (0.03) |
| | 3 | 0.15 (0.02) | 0.25 (0.03) | 0.10 (0.02) | 0.30 (0.03) | 0.05 (0.02) | 0.25 (0.03) |
| Web+ | 1 | 0.20 (0.02) | 0 | 0.35 (0.02) | 0 | 0.40 (0.02) | 0 |
| | 2 | 0.30 (0.02) | 0.60 (0.03) | 0.35 (0.02) | 0.70 (0.03) | 0.30 (0.02) | 0.80 (0.03) |
| | 3 | 0.15 (0.02) | 0.25 (0.03) | 0.10 (0.02) | 0.30 (0.03) | 0.05 (0.02) | 0.25 (0.03) |

Standard errors are given within brackets. For phase 2 with no action in phase 1, unconditional propensities are given. All other propensities are conditional.

TABLE 8.4   **Estimated marginal costs per sample element for the six strata in phases 1, 2, and 3**

| Phase 1 action | Phase | Stratum | | | | | |
|---|---|---|---|---|---|---|---|
| | | Y&NB | Y&B | M&NB | M&B | E&NB | E&B |
| No | 1 | — | | — | | — | |
| | 2 | 40 (1) | | 35 (1) | | 40 (1) | |
| | 3 | 50 (1) | | 50 (1) | | 50 (1) | |
| Web− | 1 | 2.5 | 2.5 | 2.5 | 2.5 | 2.5 | 2.5 |
| | | (0.1) | (0.1) | (0.1) | (0.1) | (0.1) | (0.1) |
| | 2 | 35 (2) | 40 (2) | 30 (2) | 50 (2) | 35 (2) | 55 (2) |
| | 3 | 40 (2) | 55 (2) | 45 (2) | 55 (2) | 35 (2) | 50 (2) |
| Web+ | 1 | 7.5 | 7.5 | 7.5 | 7.5 | 7.5 | 7.5 |
| | | (0.1) | (0.1) | (0.1) | (0.1) | (0.1) | (0.1) |
| | 2 | 35 (2) | 40 (2) | 30 (2) | 50 (2) | 35 (2) | 55 (2) |
| | 3 | 40 (2) | 55 (2) | 45 (2) | 55 (2) | 35 (2) | 50 (2) |

Standard errors are given within brackets.

for young persons that break off after phases 2 and 3 are, respectively, $7.5 + 1.0 \times 40 = 47.5$ and $7.5 + 1.0 \times 40 + 1.0 \times 0.6 \times 55 = 80.5$.

Now, with the identification of strata and strategies, the determination of quality and cost functions, and the estimation of survey design parameters, we have all ingredients to perform the ASD optimization. We have, however, not yet been specific about the model for the $R$-indicator. Given that break-off is not measured for sample elements that receive no action in phase 1, we evaluate the variance of response propensities against age groups. Based on responses to a range of other surveys, we set a lower threshold of 0.85.

In the optimization, we assume that each stratum receives one strategy, i.e., strategy allocation probabilities are either 0 or 1. The number of possible designs equals $(2 + 3^2 + 3^2)^3 = 20^3 = 8,000$. For persons with no action in phase 1, it will be unknown throughout data collection whether they would have broken off. As a consequence the phase 1 action must be the same for all persons in an age group, and, if no action is applied in phase 1, also phases 2 and 3 must be the same. The number of possible designs is small enough for a direct search or brute force optimization, i.e., we evaluate the response rate, costs, and $R$-indicator for each possible design and select the design with the highest response rate that satisfies the constraints.

Figure 8.2 shows box plots for the response rate, costs per response, and $R$-indicator over all possible designs. Around 37% of the candidate designs has an expected $R$-indicator value above 0.85. If we would raise the threshold to 0.90, then 20% out of the designs still are sufficiently balanced with respect to age. Thirty-three percent out of the designs has an expected cost per response below

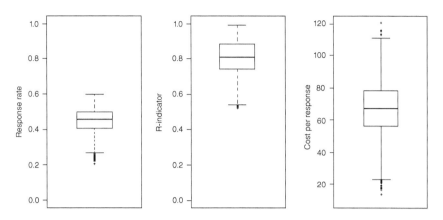

FIGURE 8.2 Boxplots of the response rate, R-indicator, and costs per respondent for all possible LFS designs

the 60 euros threshold. This percentage is 5% and 93% for thresholds of 40 and 90 euros, respectively. Only 7% out of all designs satisfies both the 0.85 *R*-indicator constraint and the 60 euros cost constraint.

The design that has the highest response rate and that satisfies the constraints allocates the six strata to (web− → F2F, web− → F2F, web− → F2F, web− → F2F, web+ → F2F, web+). Except for the eldest age group, all strata are assigned to web without incentives. Except the eldest age group with break-off, web non-respondents in all strata receive a standard F2F approach. This design has a response rate of 50.2%. The costs per response are 59.7 euros, which are just below the 60 euros threshold. The *R*-indicator is 0.913, which is well above the 0.85 threshold.

Suppose we like to evaluate the impact of the cost constraint on the performance. Figure 8.3 shows the optimal response rates when the cost constraint ranges from 40 to 90 euros. The optimal response rate increases steadily from a value close 40% to a value close to 60%. For cost thresholds between 70 and 90 euros, the increase in response rate is very small. The *R*-indicator values are not optimized but only constrained to be higher than 0.85. As a result, they do not show a clear pattern, although for higher cost levels also the *R*-indicator values are somewhat higher. While varying the cost constraint, the structure of the strategy allocation to the six age × break-off strata changes drastically. For the 40 euros threshold, the allocation is (web− → F2F, web− → F2F, web− → F2F, web− → F2F, web+, web+), i.e., a F2F follow-up is applied only to the youngest and middle-aged groups. For the 90 euros threshold, the allocation is (web+ → F2F → F2F+, web+ → F2F → F2F+, web+ → F2F → F2F+, web+ → F2F → F2F+, web+ → F2F, web+ → F2F → F2F+). In that setting, also the extended F2F approach is applied except for the eldest age group without break-off.

Implementation of the optimal designs requires specialized logistics and infrastructure and flexible monitoring tools. The optimal designs under all cost

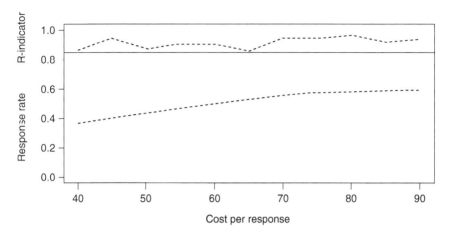

F I G U R E  8 . 3  **Optimal response rates (lower line) and R-indicator (upper line) values for cost constraints ranging from 40 to 90 euros per respondent. The R-indicator threshold (0.85) is depicted as a horizontal line**

constraint apply web to all strata, but incentives and F2F follow-up are varied. This implies that these designs are more challenging to conduct and monitor than a uniform design.

So far, we have ignored that the survey design parameters, i.e., the response propensities and costs per sample element, are estimated from historic survey data. As a consequence, they are subject to imprecision. Tables 8.3 and 8.4 contain estimated standard errors that reflect the imprecision. The standard errors for costs are a result of variation in the number of reminders that needs to be sent, in the number of F2F visits that needs to be made to establish contact, and in the F2F interview duration. The uncertainty about the true values of the survey design parameters may affect the performance of the ASD. Variability may be so large that the actual response rate, costs, and R-indicator may be very different and may not pass the thresholds. Obviously, the ASD may also be pessimistic, and performance ASD is above expectations. In order to assess the impact of imprecise survey design parameters, we perform a sensitivity analysis as follows: to generate response propensities, we need a probability distribution with support [0, 1]. For simplicity and convenience, we assume a beta distribution for the response propensities with shape parameters that match the expectation and variance in Table 8.3. We assume normal distributions for the costs per sample element with an expectation and variance as in Table 8.4. From the distributions, we repeatedly (and independently) draw response propensities and costs per sample element. These draws are used to recompute the expected response rate, costs, and R-indicator of the optimal ASD. Figure 8.4 shows the variation in response rate, costs, and R-indicator that results from drawing 1,000 of such survey design parameters.

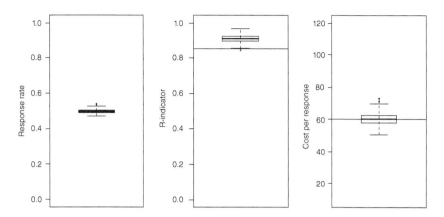

FIGURE 8.4 **Boxplots of the response rate, costs per respondent, and R-indicator for 1,000 random draws of survey design parameters**

Figure 8.4 shows that, in this example, there is some variation in the response rate, little variation in the $R$-indicator, and relatively a lot of variation in the costs per response. Note that the variation in costs depends both on the variation in the number of respondents and on the variation in the individual costs. The $R$-indicator is robust and always satisfies the 0.85 threshold. However, the costs per response are larger than 60 euros in 49% of the samples. If we take a budget overrun of 5% as acceptable, then still in 19% of samples the required budget is too large. In order to reduce the probability that realized costs are too large, the cost threshold may be lowered to, say, 55 euros. This choice will, obviously, change the structure of the optimal design as well as the optimal response rate.

In the application, we fully ignored measurement error differences between web and F2F responses. When survey mode is one of the design features, then such a simplification is too naïve. Calinescu and Schouten (2016) propose to include measurement error in ASD by considering response quality propensities. A response quality propensity is the propensity that a specified (undesirable) answering behavior, such as underreporting, straight-lining, or item nonresponse, occurs for a respondent from a particular stratum. The behaviors that are considered depend on the context and the items in the questionnaire. For instance, for the LFS questionnaire, designers may consider do-not-know answers or refused answers to be a potential source of differences between the modes. There are various ways to include the propensities to refuse an answer or to provide a do-not-know answer. They could be added to the optimization by adding a constraint on the maximal overall rate of occurrence. Alternatively, they could be incorporated in the R-indicator constraint by multiplying response propensities and quality response propensities.

## 8.4 Summary

This chapter described ASDs where different sample elements are treated differently in terms of data collection strategies. Sample subgroups are identified using either baseline covariates (frame data or register data) or paradata. Survey strategies are defined as different specifications of survey design features. The distinctive feature of ASDs is that adaptation to the elements is done before data collection starts. Survey designs where an initial design phase is used to learn how effective and costly different design features are called RSDs.

The chapter introduces the main ingredients that are needed in order to implement such designs, namely, stratification of the population, survey strategies, response propensities and costs, quality functions, and cost functions. Often the aim of ASDs is to optimize response rates by assigning different strategies to different sample elements while imposing constraints on the level of costs and quality. However, other quality functions may also be optimized, e.g., the coefficient of variation of the response propensities, or the dual problem may be considered where costs are minimized subject to quality constraints only.

### KEY TERMS

**Adaptive survey design**: A survey design where sample elements with different characteristics are approached using different data collection strategies. The assignment of strategies to elements is done before data collection starts.

**Baseline covariate**: Sampling frame data and register data. Baseline covariates are available before data collection starts.

**Coefficient of variation**: A quality indicator, defined as the $R$-indicator divided by the response rate.

**Cost function**: It is a function used to measure the overall cost of a certain ASD. In order to compare different ASDs, it is essential to restrict to costs relative to the design features that are varied in the designs.

**Design phase**: A separate part of data collection in responsive survey designs where the design features are varied. The initial design phase is different as it is used to identify and evaluate the effectiveness and costs associated with different design features and for elements with different characteristics.

**Paradata**: Process data observed during data collection.

**Quality function**: It is a function that measures the quality of the collected data. Quality functions are optimized in the ASD but are also used to set constraints. Examples of quality functions are the response rate, the $R$-indicator, and the coefficient of variation.

**Response propensity**: It is the probability of response of a population element, given the values of a set of covariates.

**Response quality propensity**: It is the probability that response from a population element is of "good quality" (in a properly specified sense), given the values of a set of covariates.

**Response rate**: The number of responding eligible elements in the sample divided by the number of eligible elements in the sample. In the ASD context, the response rate is estimated before data collection starts as the average estimated response propensities.

**Responsive survey design**: A survey design where sample elements with different characteristics are approached using different data collection strategies. In such designs, the data collection is divided into phases. A new phase employs the outcomes from previous phases to distinguish effective treatments and identify costs associated with treatments. The allocation of design features is done in a way that each phase reaches its "phase capacity," which is the optimal trade-off between quality and costs.

**$R$-indicator**: A quality indicator measuring the extent to which a response deviates from the representative response in term of a specified set of covariates.

**Survey strategy**: A sequence of survey design features.

**Survey strategy set**: The set of all possible survey strategies.

**Survey design feature**: A single aspect of a survey design such as contact mode, survey mode, or level of incentives.

## EXERCISES

**Exercise 8.1**    This exercise employs the GPS data set that is available at the handbook website [http://www.web-survey-handbook.com/]. They require a statistical software package, such as R, SAS, SPSS, or STATA, that supports generalized linear models.

The GPS survey design consists of two data collection phases that both take one month. After one month, phase 1 is ended, and remaining nonrespondents may receive a follow-up in phase 2. After the second month, data collection is stopped. The phase 2 follow-up is expensive and is a candidate design feature for adaptation through an ASD.

Table 2 of the GPS documentation at the handbook website gives the variables available in the GPS survey data set. The variables *Resp1* is a binary indicator for response in phase 1. The variable *Response* is a binary indicator for response when a case also receives a follow-up in phase 2. *Region, Urban, Phone, HouseVal, Ethnic*, and *HHType* are auxiliary variables that are used to form ASD strata.

In the construction of the GPS adaptive survey design, the focus is on the representativeness of response. For this purpose, $R$-indicators (R) and coefficients of variation (CV) need to be estimated and evaluated:

**a.** Estimate response propensities for a main effect regression model with *Region, Urban, Phone, HouseVal, Ethnic,* and *HHType* for the response after two months. Because the response indicator is a binary variable (either 0 or 1), use a logistic or probit regression model. The response indicator (*Response*) is the dependent variable, and the six auxiliary variables are entered in the model as explanatory variables as main effects, i.e., without interactions. Save predicted values for the response propensities. What are the values of *R* and *CV*?

**b.** Repeat the same steps for the 0–1 response indicator after the first month of data collection, *Respons1*. Have the *R* and *CV* improved after the second month?

**c.** Make box plots of the estimated response propensities after phases 1 and 2. Split the box plots over the classes of each of the selected auxiliary variables. What auxiliary variables are the most interesting for adaptation?

**Exercise 8.2** With reference to Exercise 8.1, build an adaptive survey design for the GPS by optimizing the use of the phase 2 follow-up in month 2 of the survey. Apply the follow-up only to particular nonrespondents in phase 1 in such a way that the variation in response propensities after phase 1 is reduced. What population groups will you target? What would you do if budget is limited so that only 25% of phase 1 nonrespondents can be assigned to phase 2?

## REFERENCES

Andridge, R. R. & Little, R. J. A. (2011), Proxy Pattern-Mixture Analysis for Survey Nonresponse. *Journal of Official Statistics*, 27, pp. 153–180.

Beaumont, J.-F., Haziza, D., & Bocci, C. (2014), An Adaptive Data Collection Procedure for Call Prioritization. *Journal of Official Statistics*, 30, pp. 607–621.

Bianchi, A. & Biffignandi, S. (2014), Responsive Design for Economic Data in Mixed-Mode Panels. In: Mecatti, F., Conti P. L., & Ranalli, M. G. (eds.), *Contributions to Sampling Statistics*. Springer, Cham, Switzerland, pp. 85–102.

Bianchi, A. & Biffignandi, S. (2017), Targeted Letters: Effects on Sample Composition and Item Non-response. *Statistical Journal of the International Association for Official Statistics*, 33, pp. 459–467.

Bianchi, A. & Biffignandi, S. (2018), Survey Experiments on Interactions: A Case Study of Incentives and Modes. In: Lavrakas, P. J., de Leeuw, E., Holbrook, A., Kennedy, C., Traugott, M. W., & West, B. T. (eds.), *Experimental Methods in Survey Research: Techniques that Combine Random Sampling with Random Assignment*. Wiley, Hoboken, NJ.

Calinescu, M. & Schouten, B. (2016), Adaptive Survey Designs for Nonresponse and Measurement Error in Multi-Purpose Surveys. *Survey Research Methods*, 10, pp. 35–47.

Deville, J. C. & Tillé, Y. (2004), Efficient Balanced Sampling: The Cube Method. *Biometrika*, 91, pp. 893–912.

Groves, R. M., Dillman, D., Eltinge, J., & Little, R. (2002), *Survey Nonresponse*. Wiley, New York.

Groves, R. M. & Heeringa, S. (2006), Responsive Design for Household Surveys: Tools for Actively Controlling Survey Errors and Costs. *Journal of the Royal Statistical Society Series A: Statistics in Society*, 169 (Part 3), pp. 439–457.

Heyd, J. M. & Carlin, B. P. (1999), Adaptive Design Improvements in the Continual Reassessment Method for Phase I Studies. *Statistics in Medicine*, 18, pp. 1307–1321.

Little, R. & Vartivarian, S. (2005). Does Weighting for Nonresponse Increase the Variance of Survey Means? *Survey Methodology*, 31, pp. 161–168.

Luiten, A. & Schouten, B. (2013), Tailored Fieldwork Design to Increase Representative Household Survey Response: An Experiment in the Survey of Consumer Satisfaction. *Journal of the Royal Statistical Society: Series A (Statistics in Society)*, 176, pp. 169–189.

Lundquist, P. & Särndal, C. E. (2013), Aspects of Responsive Design for the Swedish Living Conditions Survey. *Journal of Official Statistics*, 29, pp. 557–582.

Lynn, P. (2015), Targeted Response Inducement Strategies on Longitudinal Surveys. In: Engel, U., Jann, B., Lynn, P., Scherpenzeel, A., & Sturgis, P. (eds.), *Improving Survey Methods: Lessons from Recent Research*. Routledge, New York, pp. 322–338.

Lynn, P. (2016), Targeted Appeals for Participation in Letters to Panel Survey Members. *Public Opinion Quarterly*, 80, pp. 771–782.

Murphy, S. A. (2003), Optimal Dynamic Treatment Regimes. *Journal of the Royal Statistical Society, Series B*, 65, pp. 331–355.

Peytchev, A., Baxter, R. K., & Carley-Baxter, L. R. (2009), Not All Survey Effort Is Equal: Reduction of Nonresponse Bias and Nonresponse Error. *Public Opinion Quarterly*, 73, pp. 785–806.

Peytchev, A., Riley, S., Rosen, J., Murphy, J., & Lindblad, M. (2010), Reduction of Nonresponse Bias in Surveys Through Case Prioritization. *Survey Research Methods*, 4, pp. 21–29.

Rosen, J. A., Murphy, J., Peytchev, A., Holder, T., Dever, J., Herget, D., & Pratt, D. (2014), Prioritizing Low Propensity Sample Members in a Survey: Implications for Nonresponse Bias. *Survey Practice*, 7. https://doi.org/10.29115/SP-2014-0001.

Särndal, C. E. (2011), The 2010 Morris Hansen Lecture: Dealing with Survey Nonresponse in Data Collection, in Estimation. *Journal of Official Statistics*, 27, pp. 1–21.

Särndal, C. E. & Lundquist, P. (2014), Accuracy in Estimation with Nonresponse: A Function of Degree of Imbalance and Degree of Explanation. *Journal of Survey Statistics and Methodology*, 2, pp. 361–387.

Särndal, C. E. & Lundström, S. (2010), Design for Estimation: Identifying Auxiliary Vectors to Reduce Nonresponse Bias. *Survey Methodology*, 36, pp. 131–144.

Schouten, B., Calinescu, M., & Luiten, A. (2013), Optimizing Quality of Response Through Adaptive Survey Designs. *Survey Methodology*, 39, pp. 29–58.

Schouten, B., Cobben, F., & Bethlehem, J. (2009), Indicators for the Representativeness of Survey Response. *Survey Methodology*, 35, pp. 101–114.

Schouten, B., Cobben, F., Lundquist, P., & Wagner, J. (2016), Does Balancing Survey Response Reduce Nonresponse Bias?. *Journal of the Royal Statistical Society, Series A*, 179, pp. 727–748.

Schouten, B. & Shlomo, N. (2017), Selecting Adaptive Survey Design Strata with Partial R-Indicators. *International Statistical Review*, 85, pp. 143–163.

Schouten, B., Shlomo, N., & Skinner, C. (2011), Indicators for Monitoring and Improving Representativeness of Response. *Journal of Official Statistics*, 27, pp. 231–253.

Shlomo, N., Skinner, C., & Schouten, B. (2012), Estimation of an Indicator of the Representativeness of Survey Response. *Journal of Statistical Planning and Inference*, 142, pp. 201–211.

Tourangeau, R., Brick, M., Lohr, S., & Li, J. (2017), Adaptive and Responsive Survey Designs: A Review and Assessment. *Journal of the Royal Statistical Society A*, 180 (1), pp. 203–223.

Wagner, J. (2008), Adaptive Survey Design to Reduce Nonresponse Bias, PhD thesis. University of Michigan, Ann Arbor, USA.

Wagner, J. (2010), The Fraction of Missing Information as a Tool for Monitoring the Quality of Survey Data. *Public Opinion Quarterly*, 74, pp. 223–243.

Wagner, J. (2013), Adaptive Contact Strategies in Telephone and Face-to-Face Surveys. *Survey Research Methods*, 7, pp. 45–55.

# CHAPTER NINE

# Mixed-mode Surveys

## 9.1 Introduction

A mixed-mode survey is a survey in which various data collection modes are combined. Examples of data collection modes are interviewer-assisted data collection (face-to-face and by telephone) and self-administered data collection (by mail or by the Internet).

There have been mixed-mode surveys in the era before web surveys appeared. There are early applications where telephone interviewing was used as an alternative to self-completion of mail questionnaires. There also have been examples of large-scale mixed-mode surveys with mail, telephone interviewing, and face-to-face interviewing as modes of data collection. In the last 20 years, mixed-mode surveys have been using the Internet both for web only survey and as one of the data collection modes. See, for example, Biemer and Lyberg (2003), Christian, Dillman, and Smyth (2005) and de Leeuw (2005), Scherpenzeel and Das (2011), and Dillman, Smyth, and Christian (2014). An example of mixed-mode survey administered in the early 2000s when Internet use was not yet the usual communication tool inside businesses and institutions is described in Example 9.1.

Nowadays, the mixed-mode can be considered as the standard approach to collect data in social surveys as well as in economic surveys. In large population surveys, web only surveys suffer from coverage problems. Thus, there is the need to mix the mode with other types of surveys as it only results in nonresponse-bias. Mobile devices are a rather recent survey tool; it is possible to receive and deliver

*Handbook of Web Surveys*, Second Edition. Silvia Biffignandi and Jelke Bethlehem.
© 2021 John Wiley & Sons, Inc. Published 2021 by John Wiley & Sons, Inc.

the web questionnaire on these devices. Mobile devices diffusion has been growing, hence when a web survey is administered in practice a mobile web mode is
adopted. In principle, it is in itself a kind of mixed-mode. Moreover, mobile
web mode is becoming a common component of a mixed-mode survey. This
means that the participant to the survey chooses whether to use mobile devices
or PC. When mobile is chosen, the type of device has to be selected (for example
smartphone or tablet). Thus, mixed-mode survey design with mobile web component has to face additional challenges.

Chapter 3 (Figure 3.2) shows a flowchart with the steps to choose the mode
for a web survey, i.e. web only/mobile web or mixed-mode with web component.

Using the Internet as one of the data collection modes in a mixed-mode survey
offers new opportunities, but also creates challenges. Recent literature has discussed various aspects of this type of surveys. Both opportunities and challenges
are discussed in this chapter. The focus will be especially on challenges and critical
problems.

Mixed-mode surveys are becoming more and more popular. The two most
important reasons for considering such a survey approach are reducing survey costs
and keeping response at an acceptable level. Another reason that may play a role is
that a sampling frame for web only surveys or mobile web only surveys is often not
available. Thus, other mode of data collection (mixed-mode) is used to overpass
the abovementioned problem. Hereunder a short description of some of these
aspects; they are discussed in some more detail in the following sections.

 EXAMPLE 9.1 A mixed-mode survey on customer satisfaction

The management of the library of the University of Bergamo carried out a
survey on customer satisfaction. To recruit people, a simple random sample
was selected from the administrative database of the students of the university. Students were called by telephone and invited to participate in the
survey. If they agreed, they were offered the choice to answer the survey
questions in a telephone interview or to provide their e-mail address so that
they could fill in the questionnaire on the Internet.

A link to the questionnaire on the Internet and a unique access code
were sent to these e-mail addresses.

In this survey, telephone interviewing was chosen in the recruitment
phase and two concurrent modes (telephone and web) in the response
phase. This is a simple example of a mixed-mode survey with web component. Such an approach is useful if a list of e-mail addresses is not available.

Note that every student had a university provided e-mail address, but
at the time the survey took place (2006) students did not use the university
e-mail address very often. Consequently, many e-mails turned out to be
undeliverable or remained unread. For this reason, recruitment by telephone was chosen and the mixed-mode approach adopted.

The inclusion of web into a mixed-mode design has the potential to reduce costs, increase response rates, and improve quality and sample composition (Groves and Lyberg, 2010; Couper, 2011).

a. *Reduce costs of the survey*: Traditionally, large household surveys are interviewer-assisted surveys. Data are collected by means of telephone or face-to-face interviewing. This makes these surveys expensive. Interviewer costs are often a major cost component. There are no interviewers in a web survey. This makes web data collection an attractive alternative mode in a mixed-mode survey.

b. *Increase response rates*: Survey organizations in many countries are faced with reducing response rates. Daikeler, Bošnjak, and Lozar Manfreda (2020), considering 114 experimental comparisons between web and other survey modes, found that web surveys yielded lower response rates than other modes. Initially, mixed-mode surveys have been suggested as a way to increase response rates. See, for example, Groves et al. (2004). One possibility is offering potential respondents different modes of data collection (e.g. mail or web) from which they can choose one. The initial idea was that if people had the possibility to select their preferred mode, the response rate might increase. Another possibility is to determine, beforehand, what the most effective mode will be for people, based on their personal characteristics. For example, one could opt face-to-face interviewing for the elderly and the Internet for young people. There is, however, to bear in mind that young people in several studies have showed higher break-off rates and lower response rate. To date, the impression is that mixed-mode surveys can approximate response rates of traditional face-to-face surveys but not increase them unless face-to-face/CAPI is the first mode. Higher response rates are achieved mostly when an additional mode is included, for example because part of the population does not have access to the mode. This will usually increase costs. It should be noted, however, that by now, it is known that web response rates are lowest of all modes. Also, changing from a single mode survey to a mixed-mode survey may require substantial efforts, testing, and resources.

c. *Improve coverage*: Using web and mobile web as a survey tool brings to the need of mixed mode to overpass coverage problems. A frame of web user and/or of mobile user does not exist in large population; moreover, not the whole population is on the Internet. Using mixed-mode, it becomes possible to reach the whole target population, thus also non-Internet users. For instance, recruitment mode (telephone or paper) can help in getting the contact with the selected units, afterwards it is possible to select a specific participation mode (for example, the one most preferred by the interviewed).

Mixed-mode surveys are also attractive for International Surveys (cross-national and/or cross-cultural surveys), a type of survey which is highly relevant at the time being. International Surveys has to consider different survey traditions in different countries and different coverage patterns; to overcome these problems, mixing modes allows to adapt the survey methodology to the single country context.

Summing up, mixed-mode surveys have several advantages, but there are also many methodological problems and challenges.

There are different approaches in implementing a mixed-mode survey. The first approach is using different modes concurrently (parallel, concurrent). The second is using modes in sequence (sequential approach). There are different possible ways to set in sequence the different modes: for instance, web first, telephone as a second mode and finally face to face. There is another approach, which is called hybrid (also partially concurrent partially sequential), e.g. concurrent and sequential are combined in an effective way. Each mode has its advantages and its drawbacks. Mode selection effect produces differences in sample composition but also in measurement. Details about how to implement the mixed-mode approach, recommendations and some literature evidence of the challenges of different approaches are given in the following sections.

Mode-specific errors in terms of coverage, nonresponse, and measurement biases need to be considered. Their net effect is called a mode effect. Recent literature has focused on methods to reduce such effects. Then the question is about the survey representativeness and accuracy of the estimates.

The effectiveness of mixed-mode surveys depends on their design (concurrent, sequential, or a combination) and the data collection modes used. The main question is whether a mixed-mode survey allows for accurate, unbiased estimation of the population characteristics. Next sections are devoted to this.

Mode effects are of two type: (desired) wanted differential selection effects and unwanted effects.

Wanted mode effect is due to nonobservation errors (for instance, coverage, and nonresponse) across different modes. The surveyor should take advantage of this type of mode effect to improve survey quality by overpassing noncoverage problems and reducing nonresponse. The wanted mode effect is managed mainly at design stage. Concurrent approach is useful to reduce coverage and nonresponse.

Unwanted effect is due to observation errors and represents the differential measurement across different modes. Mixing modes might increase the measurement error. The surveyor can estimate the unwanted mode measurement effects while controlling for the wanted selection effects. Adjustment for unwanted differential measurement error is also necessary in some cases. Several recent studies focus on the methodology for unwanted differential measurement error and on experimental analyses, as discussed in the following sections.

In general, there are four options to reduce mode method effects: prevent through questionnaire design, avoid through data collection design (potentially adaptive), adjust through estimation design, or stabilize through mode calibration.

When mixing modes additional information is required to manage wanted mode effects and to control unwanted effects. Some auxiliary variables, mainly demographic variables, are the most used information. Moreover, other available data could be considered, for example, existing survey data, eventually longitudinal data, administrative data, or a reference survey. In this last case, additional data should be collected, for instance, the surveyor could run a single mode survey on a subset of the sample and ask new questions and again questions already collected in the main survey.

In several situations, coverage of the whole target population for the web data collection is mined from the absence of a list of e-mail contacts of the sampled units. Thus, to get the contact e-mail address to send the online questionnaire a sampling frame of the target population (containing postal mail of each population unit) is used to draw a probability-based sample. A letter asking for participation in the survey and the preferred mode is sent to the sampled units. Those choosing web should also provide the e-mail address. The advance letter is also useful to motivate people in responding to the survey.

Summing up, improving coverage and response is possible by using multiple modes of contact, (e.g., advance letters, push-to the-web methods). One strategy to improve coverage and increase response is to use mixed-mode approach with web component to provide those without Internet with Internet whether they prefer to respond using the online questionnaire. On the contrary, those who prefer to respond off-line with paper mail surveys are provided with a paper questionnaire (Bosnjak et al., 2018, e.g., GESIS panel).

This chapter gives an overview of mixed-mode surveys and the related concerns. The focus is on what mixed-mode surveys are, why they should be used, how they should be used for good data quality and how future trend looks like. One of the serious concerns of mixed-mode surveys is the occurrence of mode effects. This is the phenomenon that the same question is answered differently when asked in a different mode. In this chapter, mode effects are discussed and also how to deal with them.

There is growing interest in studying and experimenting methods to manage and analyze mixed-mode surveys. Several papers have been published in the recent years. Eurostat has been very active on mixed-mode surveys, attempting to stimulate exchange of experiences among NSIs (National Statistical Institutes) and with other researchers, too. Between 2011 and 2014 there was an ESSnet on Data Collection for Social Surveys using multiple modes (abbreviated to DCSS). The European project "Mixed Mode Design in Social Surveys" (MIMOD), lead by ISTAT has been working in the period 2017–2019. Nevertheless, new experiments and results are continuously appearing and Eurostat is very active on this front. ESSnet on Multi-Mode Data Collection (abbreviated to MMDC, led by ISTAT) is at follow-up stage.

Changing from a single mode survey to a mixed-mode survey may require substantial efforts and resources. There is continuous experimental research to provide documentation, suggestions, and recommendations to implement high-quality mixed-mode surveys.

This chapter provides an overview and discussion about the mixed-mode survey design and management.

## 9.2 The Theory

### 9.2.1 WHAT IS MIXED-MODE?

In this book, the expanded definition of mixed-mode is adopted. It includes also surveys using the mode of the recruitment different from the data collection. This is a definition used by many authors, like Schouten et al. (2013). Some authors exclude the mode in the recruitment survey (Tourangeau, 2017).

Before explaining what a mixed-mode survey is, the concept of mode is introduced. *Mode* refers either to the approach used to contact potential respondents or to the way the data are collected. So, if a sample of individuals is sent a letter in which they are requested to complete a questionnaire on the Internet, the mode for recruitment is mail and the mode for data collection is the web. There are many different modes of data collection. Major contemporary modes are face to face, telephone, mail, and the Internet, i.e. web surveys or mobile web surveys (surveys through PC or various mobile devices). Modes may use the same basic technology but differ in how it is used. The data collection instrument refers to the technology used to actually record the answers to the questions. The instrument may be a paper questionnaire, a laptop with a computer-assisted interviewing program, or a web questionnaire. The same instrument may be used for different modes. For example, the same paper questionnaire could be used for both a mail survey and a face-to-face survey. And, it is not unlikely that the same computer program is used for both web survey and telephone survey. Sometimes the same questionnaire is used for web surveys and mobile web surveys. In some case the web questionnaire is set equal to the paper questionnaire. Chapter 6 describes major currently used modes, whereas Chapter 7 is about questionnaire construction for web mode with some comments on the adaption to mobile devices.

Mixed-mode surveys are defined as surveys that use multiple modes of data collection.

There are different approaches to implement a mixed-mode survey. A first approach is using different modes concurrently (parallel). The sample is divided into groups and each group is approached with a different mode. *Concurrent mixed-mode* data collection is illustrated in Figure 9.1.

Sometimes, literature has reported concurrent mixed-mode that offers more than one mode to sample members at a time (e.g., both mail and web) letting the

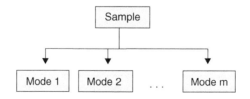

FIGURE 9.1   Concurrent mixed-mode data collection

respondent to choose the mode to respond. There have been some studies about the response rates in mixed-mode surveys. The findings do not seem to be consistent. Previous research has shown that offering respondents their preferred mode can increase response rates. Quigley et al. (2000) describe some studies showing increased response rates in mixed-mode surveys. However, some studies have found that offering respondents a choice of modes decreases the response rate. Dillman, Clark, and West (1995) found that providing alternative response modes does not necessarily improve response rates. Griffin, Fischer, and Morgan (2001) reported a lower response rate when changing from mail to mixed-mode. Medway and Fulton (2012) reported that offering a web alternative to a mail survey lowered response.

To date, the impression is that mixed-mode surveys can approximate response rates of traditional face-to-face surveys but not increase them unless face to face/CAPI is the first mode. It is also a common finding that offering choice makes respondents inactive and leads to lower response rates. Higher response rates are achieved mostly when an additional mode is included, for example because part of the population does not have access to the mode. This will usually increase costs. The strategy is commonly used in longitudinal surveys. For example, the longitudinal panel.

At first, conclusion appears that allowing respondents to choose how they complete the survey makes it more likely that they will not complete the survey at all. This is called "paradox of mode choice." A hypothesis about this is that the interviewed could perceive the choice action as an additional response burden and as a result not participate in the survey.

However, if mobile web approach is adopted, recent studies showed that letting respondent to choose the device to be used (this consequently implies letting the mode to be chosen, too) is more effective. People are more actively committed to the survey participation if the decision of the tool is up to them and they may use what they find more comfortable.

Taking this into account, providing to the respondent the possibility of choosing the mode to complete the questionnaire could be not recommended if the target population of the survey is expected not to be on the web; otherwise if a web push strategy is preferred and the target population is expected to be web oriented (at least in a good portion), extending the choice to the respondent could be a valid alternative approach,

This is a new interesting perspective in selecting the way the mode is assigned (i.e. voluntary choice or preassigned mode).

Concurrent approach is useful when increasing response rate is not a priority task. It is useful when coverage problems are to be faced.

A second mixed-mode approach is the sequential approach. All individuals in the sample are approached using one mode. The nonrespondents are then followed up by a mode different from the one used in the first approach. This process is repeated for a number of modes. The sequential mixed-mode approach is illustrated in Figure 9.2.

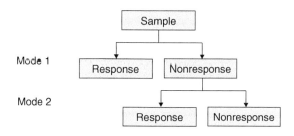

FIGURE 9.2 Sequential mixed-mode data collection

Nowadays, sequential mode is preferred. The mode sequence should be decided according to the prevailing task of the researcher: are the costs to be reduced? Should a high response rate be obtained?

It is suggested to start with a relatively inexpensive mode of data collection (typically, mail, or web) and then switch to more expensive modes, such as telephone or face-to-face interviewing. However, there is some exception, if the task is a high response rate, a sequence of modes beginning with one that typically produces lower response rates can depress the final response rate and perhaps affect representativeness as well.

If the main objective is to keep survey costs as low as possible, a sequential mixed-mode survey could start with a mail questionnaire or a questionnaire on the web. Nonrespondents can, for example, be followed up by CATI. Nonrespondents after CATI could be followed up by CAPI. So, the survey starts with the cheapest mode and ends with the most expensive one. For instance, the American Community Survey (American Community Survey, 2014) is an example of a cross-sectional survey in which multiple modes are ordered in sequence, starting with the cheaper (mail/online) modes and following-up nonrespondents with interviews.

Sometime tailoring mode approach is adopted or the more expensive modes may be used to follow up with nonresponding household. For instance, since attrition rates (given that a sample member has responded in the first wave of a panel survey) are often quite low compared to nonresponse to the initial wave of a panel study, it may make sense to use modes of data collection that are more costly but likely to achieve higher response rates in the first wave of the survey and to use less expensive modes in later waves. In the USA, the Current Population Survey (CPS) follows this strategy. The first wave of data collection is face-to-face and the next three waves are done mainly by telephone. The same pattern is followed in the final four waves of data collection, with the fifth wave done face-to-face and the final three done (as much as possible) by telephone. Other major federal panel surveys in the USA (such as the National Crime Victimization Survey) follow similar strategies. Thus, depending on the context, it is possible that the mixed-mode approach starts with expensive modes of data collection for the first wave (to maximize coverage and minimize nonresponse) but switches to less expensive methods in latter waves (to minimize overall cost).

Alternative mode combinations could be used; for instance, choice to either complete a paper questionnaire or to complete it on the web. Alternative modes are used sometime to recruit people participation to a survey and/or to collect e-mail addresses for the survey on the Internet.

At recruitment step, using multiple modes of contact is often a necessity and current best practice in the case of probability-based online panels (Blom et al., 2015). In establishing online panels in Europe (e.g. GIP, ELIPSS, and GESIS Panel) and in the USA (e.g., Knowledge network/GFK), the use of web is an important mode. Examples of the differences in mixing modes are found in Blom et al. (2015) and Bosnjak et al. (2018).

The Understanding Society Innovation Panels also provides a good example in a longitudinal setting of both a sequential CATI-CAPI mixed-mode experiment (Cernat, 2015) and a sequential Web-CAPI experiment (Bianchi, Biffignandi, and Lynn, 2017). Example 9.2 describes some studies on the Understanding Society longitudinal panel as regards allocation, nonresponse, and measurement.

Another application of a sequential mixed-mode design is a survey in which only one data collection mode is used, but for which, recruitment takes place with a different mode. A typical example is a web survey for which the sample is selected from an address register. Selected individuals are sent a letter with a link to the questionnaire on the web. This is combination of mail recruitment and web data collection.

---

**▦ EXAMPLE 9.2 Understanding Society longitudinal panel**

Understanding Society is a UK Household Longitudinal Study to collect information on changes to the household and individual circumstances on subjects such as health, work, education, income, family, and social life. Interviews are carried out face to face in respondents' homes by trained interviewers or through a self-completion online. Online interviewing was used for the first time on the main Understanding Society survey at Wave 7. At Wave 8 (2016–2018) 40% of households was issued "CAWI first."

Each household was assigned to a participation mode. The allocation to mode was decided by a model-based approach. This model identified the households with the lowest propensity to respond by web, which were then allocated to CAPI first. The remaining households were randomly allocated to CAWI first or CAPI first so that, overall, 40% of households were issued CAWI first. Some experimental studies, based on previous waves of Understanding Society panel and on Innovation Panel (a sample of 1500 UK households used by researchers as a test-bed for innovative ways of collecting data), have tried to evaluate the impact of the use of a mixed-mode approach with web component. Bianchi, Biffignandi, and Lynn (2017) evaluate the effect of a mixed-mode on several aspects related to data quality in a longitudinal survey. The design was design sequential web first, followed by face-to-face follow-up of nonrespondents to the web phase versus simple face-to-face design.

Findings show that this design is a cost-saving alternative to single-mode face-to-face.

As regards response rates, for individual participation no difference between mode treatments was detected. (Both in cumulative response rate and in response rate in each wave.) Also, no differences were found in either the original sample or the refreshment sample as a whole.

Contrasting results are found along waves for household participation. All in all, the results provide a signal of a general improvement in Wave 7 with respect to the previous waves.

Little evidence of mode treatment is affecting sample composition.

Cost savings related to the use of the web in the mixed-mode design could be substantial. The field costs regard only two-thirds of mixed-mode households that need to be issued to a face-to-face interview. Furthermore, the mean number of interviewer visits to a sample household was actually lower in the mixed-mode group than in the primarily face-to-face group. Both the situation allows for expecting interesting cost savings.

The sequential strategy is commonly used with cross-sectional surveys (in which respondents are asked to provide data only once.

There are some advantages in the longitudinal surveys (i.e. panel surveys in which recruited sample members are interviewed more than once). Approaching panel members in the most cost-efficient mode possible by considering the contact information for sample members collected at the first wave. This permit gains and cost savings. Previous waves provide detailed characteristics of respondents in different modes to study mode effect. Thus, it was possible to compute the probability across different modes sample members had to respond. This allows targeting of particular mode strategies at specific subgroups, in the framework of adaptive survey design (Calinescu and Schouten, 2016). Adaptive and responsive designs represent new approaches to mixed-mode data collection. These designs differentiate effort in data collection to different sample units based on available frame data, administrative data and/or paradata recorded during the survey. For example, only some nonrespondents to a web survey are targeted for follow-up in CAPI, based on age or on web break-off. This differentiation may be linked to survey modes, but also to other design features, such as timing and number of calls and visits, use of incentives, etc. For details and examples about adaptive survey design see Chapter 8 of this handbook.

MIMOD survey (e.g. a survey organized in the context of the project) has investigated the use of adaptive design in the interviewed NSIs. It came out that only two NSIs (Portugal and The Netherlands) use this approach. Other eight NSIs declared they were not so sure whether their data collection designs could be "adaptive" or "responsive." This last result confirms how difficult this topic is and that this type of design needs to be further explored.

When mixed-mode design is used with longitudinal surveys the mode composition of the mix may vary. In surveys repeated over time, attrition rate (i.e. given that a sample member has responded in the first wave of a panel survey he might not answer in the subsequent wave: this is called attrition) tends to lower response rate in the subsequent waves. It is therefore important to attain higher response rates in the first wave (Lynn, 2017). This is because nonresponding sample members cannot be replaced by new sample members. Thus, response rates and cumulative response rates are more important in the longitudinal framework than in cross-sectional surveys. In this case, it may make sense to use—at least in the first wave—modes of data collection that are more costly but likely to achieve higher response rate. Less expensive modes could be introduced in successive waves. When a panel is moving from face to face to mixed-mode including web or mobile web, sample members have prior experience of the interview in another mode and prior knowledge of the survey content. The chances of response in web mode (even in the absence of an interviewer) are greater as the respondent is already introduced to the survey and to the questionnaire. This situation could affect also the measurement error.

It has been mentioned that a mixed-mode survey can take many different forms. It can be a concurrent or sequential design, different modes can be used, the order of the modes in a sequential design has to be defined. It also makes a difference whether interviewer-assisted modes are used or self-administered modes. All these choices have an impact on the cost of the survey and the quality of the results. There are other factors that may play a role in designing a mixed-mode survey, like

- the complexity of the information being collected
- the time it takes to complete the interview (time burden)
- the sensitivity of the topics covered in the questionnaire
- surveys often have a long tradition and will not change modes from one year to the other in order to maintain comparability
- budget is a strong motive; face to face is often simply too expensive.

With respect to data collection, there is a substantial difference between interviewer-assisted modes of data collection (e.g. CAPI and CATI), on the one hand, and self-administered data collection (web, mail) on the other. Interviewers carry out the fieldwork in a CAPI or CATI mode. There are no interviewers in a self-administered mode. Therefore, quality of collected data may be lower due to higher nonresponse rates and more errors in the answers to the questions.

De Leeuw (2008), Dillman, Smyth, and Christian (2014), and Dillman and Edwards (2016) discuss differences between various data collection modes. They observe that a positive effect of the presence of interviewers is that they are in control of the interview. They lead the respondent through the interview. They see to it that the right question is asked at the right moment. If necessary, they can explain the meaning of a question. They can assist respondents in getting the right answers to the question. Interviewers can motivate respondents, answer questions for clarification, provide additional information and remove causes for misunderstanding. All this will increase the quality of the collected data.

The presence of interviewers can also have a negative effect. It will lead to more socially desirable answers for questions about potentially sensitive topics. Giving socially desirable answers is the tendency that respondents give answers that will be viewed as more favorable by others. This particularly happens for sensitive questions about topics like sexual behavior and use of drugs. If a true answer would not make the respondents look good, they will refuse to answer or give a different answer. This phenomenon is described by Tourangeau and Yan (2007); they also show that the detected mode effects seem to be consistent with earlier literature on social desirability bias. A meta-analysis by de Leeuw (1992) shows that the effects of socially desirable answers are stronger in interviewer-assisted surveys. Respondents tend to give more truthful answers in self-administered surveys. Kreuter, Presser, and Tourangeau (2008) showed that sensitive questions were answered more truthfully in web surveys as compared to CATI surveys. IVR had an intermediate position in this. For nonsensitive questions the differences between modes were much smaller. This confirms that mode effects are particularly important for sensitive questions. The presence of interviewers leads to more socially desirable answers to sensitive questions. Another effect is that there is less item nonresponse in web surveys.

More recently, Klausch, Hox, and Schouten (2013) have shown systematic differences between interviewer modes (CAPI and CATI) and self-administrative modes (mail and web) in a large experimental survey in the Netherlands. Similarly, Cernat et al. (2016) have shown that CATI and CAPI have higher levels of social desirability compared with web when measuring depression. Kappelhof (2017) describes how in special surveys of non-Western immigrants in the Netherlands, after correcting for mode selection effects, there is less social desirability online than in interviews.

Chapter 6 compares web surveys with other types of surveys. Many more aspects are described there for which surveys can differ.

One alternative mode of data collection in mixed-mode is based on the voice response or interaction. This mode has not been extensively used. A few studies have considered the effect in mixing voice response with other modes. However, with the diffusion of mobile devices, especially smartphones, interaction, and voice response may become, in the near future, an interesting alternative, whose effects should be investigated.

Knapp and Kirk (2003) have compared three types of self-administered data collection instruments: a web questionnaire, a paper questionnaire and Interactive Voice Response (IVR). In case of IVR, a computer system calls the respondents and asks the questions. The respondents answer like in a normal telephone call, after which their answer is processed by means of a voice recognition system, or they use the keypad of their telephone to type an answer. Knapp and Kirk (2003) found no differences between the three modes.

Summing up the findings on the differences between web, IVR and CATI, one conclusion is that there is no best mode. Each mode has its advantages and disadvantages, with respect to unit nonresponse, item nonresponse and measurement errors. Since the findings for IVR always seem to take a middle position between web and CATI, only the two extremes are compared:

- With respect to unit nonresponse, CATI surveys have higher unit response rates than web survey. So CATI is better.
- With respect to item nonresponse, CATI surveys have lower item response rates than web surveys. So, web surveys are better.
- With respect to social desirability bias, CATI surveys are more affected than web surveys. So, web surveys are better.

The rise of voice input on smartphones opens the possibility to use voice input to answer open questions in mobile web surveys. Revilla, Couper, and Ochoa (2018) carried out a survey asking respondents about their current use of voice input in everyday life and about their willingness to use voice input in a survey to answer open questions. All respondents had Internet access on smartphones. The use of voice input on smartphone is a future interesting approach. Research is needed to explore the likely use of voice input in surveys, but also to explore how the use of voice input could affect data participation, data quality, and respondent satisfaction with the survey.

An experiment about the effect of mode choice on i-Phone users (1260) contacted on their i-Phones by either a human or an automated interviewer via voice or SMS text. Four modes: human voice, human text, automated voice, and automated text were administered in the experiment (Conrad et al., 2017a). This study is a good example of new perspectives and challenges arising with the use of mobile devices.

Interesting recommendation is that leaving the respondent free in choosing the mode and or device provides better precise answers and a greater number of completed interviews. In mobile devices switching mode is quick and simple; the mode selected by the interviewed makes responding convenient for him and helps to attend to the tasks. Self-determination of the device creates satisfaction and commitment in the respondent, contributing to better mobile web survey quality.

A word of warning is at place when comparing research results in the literature and when different factors affecting the mixing mode are considered in the studies. All this makes it not very easy to design a mixed-mode survey. The following general observations can be made:

- Each mode of data collection has its own specific limitations and sources of errors;
- There is no unique, best mixed-mode design;
- The choice of modes may depend on the source of error that is most important for a specific survey;

- There are substantial differences between self-administered modes of data collection and interviewer-assisted modes of data collection with respect to response rates and measurement errors;
- Government household surveys dealing with nonsensitive topics such as work, education, and expenditure may be relatively immune to changes in the mode of data collection;
- Government business surveys that deal with topic like innovation, finance, employment may also be relatively immune to changes in the mode of data collection;
- Government surveys need to adapt their surveys to the new survey modes (mobile web); complex surveys they are managing should be critically reviewed and a heavy transition phase toward new mixed-mode design has to be faced;
- The answers to sensitive topics are substantially affected by the survey mode. Chapter 8 discusses the choice of mode targeted to preferences of subpopulations most targeted to preferences of subpopulations most.

## 9.2.2  WHY MIXED-MODE?

It is widely recognized that survey response rates have been declining almost in any country and area of research. Also, nonresponse is decreasing. See, for example, the international comparison by de Leeuw and De Heer (2002), de Leeuw, Hox, and Luiten (2018), and Luiten, Hox, and de Leeuw (2020). The decreasing trend of response rate is still a problem for the surveys. It is mainly due to:

- Increasing noncontact rates. Demographic changes (there are more single person households), socio-economic changes (there are more couples both of whom are employed) and new technological developments (mobile phones, Skype, and answering machines) make it increasingly difficult to contact persons that have been selected for the survey;
- Increasing refusal rates. Refusals are growing, since there is an increasing reluctance of the general public to take part in surveys. This may be caused by:

  ○ *An increase in the number of requests to participate in surveys.* For example, aside from official statistics surveys, many research institutes and commercial marketing research companies are conducting surveys. This is a problem for both household surveys and businesses surveys;
  ○ *A perceived heavy response burden.* Companies complain about the administrative burden imposed by government. They think they have to fill in too many forms. Although the share of statistical surveys in this administrative burden is usually very small, companies often have the perception of a high response burden;

○ *A decreasing trust in surveys.* There is already much information available in registers and other administrative sources. There are also many surveys that collect even more data. There are good surveys, but there are also many bad surveys. It becomes more and more difficult for people to distinguish the good from the bad. All this reduces trust in surveys.

**9.2.2.1 Response Rates.** In many surveys it is assumed that offering the right mode of data collection to the selected individuals will increase the response rates. Letting respondents choose their own mode of participation to the survey looked in the past as an appealing approach to improve response rates. Evidence in several surveys has been contradicting the usefulness of this approach in increasing response rate. Another way to try and achieve higher response rates is by predicting before the start of the fieldwork which mode of data collection is the most fit for the specific individuals or businesses.

Response rates are not the same for each mode of data collection. The highest response rates are obtained for face-to-face interviewing, closely followed by telephone interviewing. Response tends to be lower for self-administered modes. The lower the response is, the higher the risk of substantial nonresponse bias. First web survey experiences showed that response rates of web surveys based on probability-based sampling are not higher than around 40%.

Decreasing response rates often have a negative impact on survey results. There are many examples where nonresponse leads to biased estimates of population characteristics. The lower the response, the larger the bias will be. Therefore, it is important to reduce nonresponse as much as possible. This requires extra efforts during data collection. One can think extra training for interviewers, more and better supervisors and offering incentives for participation in the survey.

Implementing a survey with a mixed-mode design to obtain a high response rate is not recommended. Firstly, literature has shown—as already mentioned above—that in many cases response rates are not improved. Moreover, nonresponse research has shown that response of specific groups may depend on the mode of data collection. For example, young people may prefer the web for completing a questionnaire, whereas the elderly appreciate being visited by interviewers. This could imply a web survey for the young and a CAPI survey for the elderly. Of course, this approach requires the age of each person in the sample to be known in advance. This is not always possible. As was already mentioned, different strategies can be applied to assign modes to the individuals in the sample. One is to let the respondents choose their own favorite mode, and the other is to preassign modes to individuals based on their characteristics. The latter requires these characteristics to be available before data collection starts. One other important reason is that search of high response rates has not to be stressed. High response rate in the past have been considered as main indicator of survey quality. It is now clear that increasing response rate might decrease quality, if we consider other indicators like bias, measurement error.

To increase response rate, a frequently used approach, mainly in panel sur-veys, is providing incentives. Incentives in mixed-mode surveys are more fre-quently offered when self-administered modes are adopted e.g. mixed-mode surveys with web component. The absence of interviewers who generally try to obtain the participation of the sampled units creates a critical situation as regards response rate; therefore, strategies to increase it are needed.

Whatever the form (monetary or nonmonetary), incentives are a good way not only to increase response rates, but also to thank respondents for their time. The advance letter may contain an incentive (unconditional incentive) or the notice that this will be received after questionnaire completion (conditional incen-tive). The information about the incentives may also be sent also along with an information brochure, where there are detailed information on the survey and some interesting results from previous surveys. Examples of monetary uncondi-tional incentive are €5, sometime followed by a promised incentive of €30 in addi-tion to the unconditional one. Examples of nonmonetary incentives are calculators or books of stamps offered unconditionally.

The use of incentives in mixed-mode web surveys is an important issue with relevant implications both for survey quality in the broad perspective of total sur-vey error and for costs.

Many factors, like timing, amount, type of incentive, could be considered with reference to survey incentives to evaluate the impact on response rate. Inter-esting is also the use of different strategies for specific subgroups. Furthermore, in mixed-mode surveys, a combination of prepaid and promised incentives can be effective in pushing respondents to a new mode (AAPOR, 2019). For instance, the proportion of respondents who complete via a web instrument in a web survey with mail follow-up can be increased when a small prepaid incen-tive is followed by a larger promised incentive paid to those who respond by web (Biemer et al. 2018).

Experimental designs allow for identification of the factor effects. Experi-ments should be carefully planned, and their tasks clearly defined. Several studies are about incentive effects. Characteristics that could contribute to a targeted field-work (where the targeted feature is incentive and/or mode) and response rate and costs are considered key factors (Groves and Heringa, 2006; Luiten and Schouten, 2013). Example 9.3 presents a case study about the impact of incentives on survey participation.

 EXAMPLE 9.3 **Mixed-mode and incentives**

Bianchi and Biffignandi (2020) focus on the comparison of a mixed-mode design with a web component and a CAPI-only design, and on the use of incentives. Outcome measures include response rate and other quality indi-cators. A single wave (Wave 5) of a longitudinal panel, the Understanding Society Innovation Panel (IP5), is considered.

In Wave 5, a mode and incentives experiment has been adopted. The experiment considered two parallel modes: a sequential mixed-mode and single mode face-to-face (CAPI) interview. Sequential mixed-mode was: on line first (first step), face to face in the follow-up (second step). Single mode (face to face) was fielded parallel to the mixed-mode second step. On line option remained opened all along the field activities, thus it was always possible to switch to on line interview. Allocation of each sampled unit to one of two mode treatments was random. The research data are the database including the results of an experimental design based on three different unconditional incentives £5, £10, £20, or £30 and one more conditional incentive (equal to £10), if all household members completed the web survey within an established deadline. The experiment, the results and the original contribution of the study are summarized hereunder.

The experiment on incentives was nested within mode treatments. So, mode treatments and incentive treatments were fully crossed. It aimed to test ways of increasing web response rates.

Two types of experiments took place. Main experiment was on unconditional prepaid incentive, offered in an advance/prenotification letter. The incentive was a High Street gift voucher and the value of the voucher was manipulated experimentally. Panel members were randomly assigned to receive £5, £10, £20, or £30.

The second experiment was based on a conditional incentive, mentioned in the advance letters to all household members in this treatment group. Half of the households were offered an additional conditional incentive (equal to £10): if all eligible household members completed the web survey within two weeks.

Bianchi and Biffignandi (2020) propose an approach for the analysis of experimental data on survey incentives impact. Adequate statistical techniques are identified and used for the study. Thus, ANOVA, followed by multiple comparisons, considering multiplicity of comparisons by Tukey's HSD post hoc adjustments is proposed. Attrition bias effects are tested using logit regression models including treatment, a number of covariates, and their interaction with treatment. Adjustment for the sample design has been applied. Two issues are considered: the effectiveness of different level of incentives on survey participation and the characteristics that could contribute to a targeted fieldwork, where the targeted feature is incentive and/or mode.

Findings are that higher unconditional incentives (£30) are significant in increasing participation in both types of surveys (CAPI and mixed-mode). In mixed-mode (with web component) incentive levels differ significantly for females, for people living in urban centers, and age classes ranging from 41 to 70 years.

A regards quality, an indicator of item nonresponse is computed and results are compared with reference to the mode and incentives. Incentive levels are significantly different for mixed-mode, whereas no evidence on item nonresponse is found.

Costs evaluation is defined in terms of reduction of interviewers' visits to households. The number of interviewer visits to households and the percentage of households fully responding by web in the mixed-mode group are considered. An additional analysis focuses on the administration of conditional incentives (additional incentive for households fully responding via web). Whereas unconditional incentive did not show CAPI savings, the conditional incentives have achieved a greater number of households fully respondent by web.

As regards targeted incentive/mode strategies, the usefulness of conditional incentives, high level of unconditional incentives and usefulness of mixed-mode with respect to participation and quality is confirmed. Some evidence of impact of incentives on different subgroups is found, too. More strategies with respect to targeted subgroup might be evaluated considering a variety of specific survey features and studies (Bianchi and Biffignandi, 2013; Bianchi and Biffignandi, 2017).

It is difficult to give a simple practical recommendation for the application of incentives because the effects are strongly dependent on the survey mode, on other measures to increase the rate of return of questionnaires, on the target population, and on the topic of the survey. In large-scale face-to-face surveys, a conditional monetary incentive of around €10 is currently offered. In low-budget surveys, low-value incentives are definitely to be recommended because meta-analyses have shown that even incentives that are small in monetary terms increase response rate.

Large online panels, probability and nonprobability based, are usually offering incentives for participation; examples are ELLIPS (i.e. Étude Longitudinale par Internet Pour les Sciences Sociales) and GESIS Panel.

In official statistics the use of incentives in mixed-mode surveys is still limited. Their efficacy in enhancing response rates is not given for granted by all NSIs. Moreover, incentives are increasing survey costs and this does not match with budget constraints.

The countries offering incentives, according to MIMOD project survey (year 2018) nearly half of the European countries considered the study (14 out of 31) use incentives. NSIs are giving an incentive only for some specific surveys. As regards the incentives communication strategies, there are differences among NSIs: half of them are adapting communication of each survey, while the other half is keeping the communication equal for all the social surveys. It is interesting to notice that incentives are given for Household Budget Survey (HBS) in 8 countries out of 14. Incentives are provided for the Adult Education Survey (AES) in 7 out of 13 countries, whereas for the European Union Statistics on Income and Living Conditions EU-SILC wave 2 in 7 countries out of 15. For

Harmonized European Time Use Surveys (HETUS/TUS) all three NSIs running this survey with mixed-mode offer an incentive to respondents. A detailed description of the management of incentives by the NSIs is reported in the MIMOD project survey results Annex 3 (MIMOD PROJECT-2017-2019). Example 9.4 shortly describes the incentive strategy in a German survey.

**EXAMPLE 9.4** **European Health Interview Survey (EHIS), in Germany**

> For European Health Interview Survey (EHIS), in Germany, younger participants were offered a 10€-voucher after completing the questionnaire. This was done because participation rates in younger age groups are remarkably lower than in older age groups. Older age groups were offered to participate in a lottery (50€-voucher), once the questionnaire was filled out. However, it is a common practice to offer an incentive only to those households completing the questionnaire. For taking part in the survey, respondents are intended to get a small present, such as a shopping bag or a gift card for stores. Sometimes they are proposed to participate to lotteries (of i-Pads or other prizes).

**9.2.2.2 Costs.** If the focus of the survey design is on reducing the costs, a mixed-mode approach may be a way to realize this. Particularly, an expensive interviewer-assisted survey can be replaced by a mixed-mode one where one or more of the modes is a self-administered one.

A mixed-mode design aimed at keeping costs at a low level, is a sequential design that starts with the cheapest mode, for example mail or web. Nonrespondents are followed-up with a less cheap mode (CATI). The final mode could be the most expensive one. Beukenhorst and Wetzels showed the direct data collection costs of the Dutch Safety Monitor could be reduced by 40% in this way.

Several authors, for example Hochstim (1967) and Voogt and Saris (2005), argue that a sequential mixed-mode design offers advantages with respect to both response rates and costs. Indeed, de Leeuw (2005) shows that conducting a follow-up by telephone after an initial mail questionnaire mode improves response rates. The situation is, however, not clear if the web is included as one of the modes. Beukenhorst and Wetzels showed that sequential mixed-mode while reducing costs did not increase the response rate of the Safety Monitor.

**9.2.2.3 Data Quality.** Chapter 6 of this handbook compares different modes of data collection. It is shown there that each data collection mode has its own advantages and disadvantages with respect to data quality. Therefore, the combined effect on the quality of the survey data, depends on the actual mix of modes. Some effects are summarized here:

- *Response order effects.* Respondents in interviewer-assisted surveys show a preference for the last options in the list of answer options of closed questions (recency effect). In self-administered modes of data collection there is a preference for the first options in the list (primacy effect);
- *Acquiescence*: Respondents tend to agree with statements in questions, regardless of their content. They simply answer "yes." There is less acquiescence in self-administered surveys than in interviewer-assisted surveys;
- *Status quo endorsement.* If respondents are asked to give their opinion about changes, they tend to select the option to keep everything the same. There seems to be less status quo endorsement in interviewer-assisted surveys;
- *Non-differentiation.* This occurs when respondents have to answer a series of questions with the same set of response options. Respondents tend to select the same answer for all these questions irrespective of the question content. This is a form of satisficing. There is more non-differentiation in self-administered surveys;
- *Answering "don't know."* This is a form of satisficing where respondents choose this answer to avoid having to think about a real answer. Not making it possible to answer "don't know" may also cause measurement errors as respondents not knowing the answer are forced to give one. The way "don't know" is treated in survey may depend on the mode of data collection;
- *Arbitrary answer.* Respondents may decide to just pick an arbitrary answer in order to avoid having to think about a proper answer. They may also give an arbitrary answer if giving the proper answer is considered undesirable. This behavior is sometimes also called "metal coin flipping." This phenomenon typically occurs in web surveys for check-all-that-apply questions;
- *Socially desirable answers.* This is the tendency that respondents give answers that will be viewed as more favorable by others. This particularly happens for sensitive questions. If a true answer would not make the respondents look good, they will refuse to answer or give a different answer. The literature shows that the effects of socially desirable answers are stronger in interviewer-assisted surveys. Respondents tend to give more truthful answers in self-administered surveys.

These phenomena have different effects in different data collection modes. As a consequence, the same questions may be answered differently in different modes. In the context of mixed-mode surveys they are called *mode effects*.

It should be noted that changing a survey design, for example from a single-mode face-to-face survey to a mixed-mode survey (including the web as one of the modes), will lead to changes in mode effects. This will hinder comparing statistics over time. Observed changes in figures may be due to real changes in the phenomena measured, but these changes may also be caused by changes in mode effects.

Even if recent literature on data quality in mixed-mode surveys has been rather extensive, it remains unclear whether introducing a second mode of data collection actually improves (or degrades) data quality with respect to both non-response and measurement error bias and how better disentangle selection effect and measurement effect. Moreover, it is unclear whether the common practice of starting with a self-administered mode (e.g., mail) followed by an interviewer-administered mode (e.g., CATI) is preferable from a total bias perspective. The benefits of employing a mixed-mode sequence in terms of reducing one or more bias sources should be weighed against the potential for increasing the total bias in survey estimates, particularly when the estimates are affected by counteracting biases.

Implementing a sequential mixed-mode design is expected to produce the bias-reducing benefits. Doing so however, there is a potential for increasing the total bias in certain survey estimates by negating the offsetting effects of counteracting biases. Thus, as with any bias reducing intervention, it is useful to consider the expected direction and relative magnitude of different biases and how the intervention is likely to impact the different biases both individually and jointly.

### 9.2.2.4 Coverage Problems.

Coverage errors occur if the target population of the survey does not coincide with the sampling frame used. Under-coverage is a type of coverage problem that may have serious consequences. *Under-coverage* denotes the phenomenon that elements of the target population are not represented in the sampling frame. Therefore, these elements cannot be selected in the sample for the survey. If elements in the sampling frame differ from those not in the sampling frame, the survey may produce incorrect figures.

Under-coverage occurs in telephone survey if the sample is selected from a telephone directory. People with unlisted numbers are excluded from the survey and often also people with only a mobile phone. Under-coverage may also occur if the Internet is used to select persons for a survey and the target population is wider than just those with access to Internet. See Chapter 10 for an extensive description of under-coverage in web surveys.

A mixed-mode survey may help to reduce under-coverage problems. One approach is to divide the population into subpopulations and to assign to each subpopulation the data collection mode that is most appropriate for that group. For example, if a sample of individuals is selected from a population register, the age of all selected individuals is known. Therefore, it can be decided to approach young people using the web, and the elderly by means of a visit of an interviewer.

It should be noted that the extent and effects of under-coverage may change if the survey design is modified. There can be groups of people in the new survey that were not included in the previous survey. So, changes over time can be real changes in figures, but they can also be an error caused by a change in coverage. In fact, the surveyed population is not the same as before and therefore the statistics can be different.

**9.2.2.5 Selection Errors.** In some mixed-mode surveys it is left to the respondents to choose the mode of data collection. For example, respondents receive an invitation letter in which they are invited to participate in the survey. If they agree, they can choose to complete the questionnaire on the Internet or they can fill in the paper questionnaire that is included in the letter. Another example is a sequential mixed-mode survey where the selected persons are first asked to complete the questionnaire on the Internet. Next, those not responding are called by telephone.

The effect of these approaches is that specific groups choose specific modes. When the collected date are compared across modes, causes in their differences are not clear. On the one hand, differences can be caused by mode effects, and on the other, differences may be due to real differences between the two groups. Recent literature, is proposing methods and experiments to disentangle selection error from other errors, like measurement error. This issue is discussed in the next part of this chapter.

**9.2.2.6 Cognitive Efforts.** The data collection mode is a feature that affects not only the design and content of the questionnaire but also the respondent's motivation and cognitive process.

Tourangeau (1984) introduced his cognitive response model to describe the process of answering survey questions. This model helps in explaining mode effects. More details can be found in Roberts (2007), Bowling (2005), Jäckle, Roberts, and Lynn (2010), and Ariel et al. (2008). Only a short overview of this model and its relation to mode effects is given here.

The cognitive response model consists of four steps:

*Comprehension.* Respondents attempt to understand the meaning of the question. Comprehension is influenced by the presentation of the questions. Aural presentation by an interviewer may lead to recency effects and visual presentation in a self-administered mode to primacy effects. The presence or absence of interviewer can also have an effect. For example, an interviewer can always help by explaining the meaning of a question. If the respondent has a paper questionnaire form, he can page through it and look to other questions. This may help understanding what the survey is all about.

*Retrieval.* To be able to answer the question, respondents must collect relevant information. Their long-term memory is an important, and sometimes the only, source for this. The process of retrieving can differ substantially across modes. In case of self-administered modes of data collection, respondents can take as long as they want or need to perform this task. In case of interviewer-assisted data collection, respondents will feel pressure to answer as quickly as possible.

*Judgment.* The respondents assess whether the retrieved information is adequate for answering the question. They do this by comparing the available information to the meaning of the question. The presence of interviewers can have a positive effect. They may help if the respondents are unable to reach a positive judgment. They do this by making suggestions or probing.

*Response.* The respondents report or record the answers to the question. To do this they have to put their answer in the proper format, for example by selecting the right answer option of a closed question. In case of interviewer-assisted data collection, respondents may decide to change their initial answer if they consider it socially undesirable.

# 9.3 Methodological Issues

There are many unanswered questions with respect to the reliability of the outcomes of mixed-mode surveys. Differences (or similarities) between the outcomes of modes can be caused by differences between the respondents or by differences in measurement.

Mixed-mode surveys can be defined in so many ways, and there are so many phenomena that may affect the outcomes, that it is impossible to provide simple, general answers. There have been many studies, but their conclusions almost always apply to specific situations, i.e. specific target populations or specific survey designs. The number of studies related to mixed-mode surveys with the web is growing. Several interesting findings are explained in the recent literature. However, still several issues need new experiments and studies.

Using web as a component of a mixed-mode survey and the possibility to complete the questionnaire using mobile devices (mobile web survey) encompasses further methodological questions.

In this case, there are two issues to be investigated: the impact of web on mixed-mode measurement error and the impact of different mobile devices.

The impact on mixed-mode measurement error of the use of mobile devices and also of different types of devices is not yet deeply investigated. Furthermore, due to the technological innovation and the cultural change on digital knowledge these issues are subject continuously to new results.

Mobile web surveys have been described in Chapter 6. Literature has been initially focusing on the comparison between smartphones and PCs in terms of data quality. For example, Antoun, Couper, and Conrad (2017) found little impact on data quality when smartphone is used. Using smartphones produces high quality data even if the environment is more distracting. People using smartphones can provide high-quality response as long as they are presented with question formats that are easy to use on small touchscreens. Several other studies conclude that mobile use has no negative effect. However, in several cases, questionnaire completed on mobile phones have lower response rates, higher break-off rates and longer completion times than do web surveys on the computer. For a review see Couper, Antoun, and Mavletova (2017) and Wells (2015). Antoun and Cernat (2020) compare factors affecting completion time when Pc or smartphone is used for participation. They found that respondents took about 1.4 times longer when using smartphones than PCs. Respondents who had relatively low levels of familiarity and experience using smartphones presented larger differences.

Various mobile survey experiments show that mobile survey responses are sensitive to the presentation of frequency scales and the size of open-ended text boxes, as are responses in other survey modes (Wells, Bailey, and Link, 2014). Several studies were focusing on the impact on the questionnaire characteristics and the answers on mobile device versus PC and mainly no evidence for differences were found (see Chapter 7). However, there are suggestions for better accommodation of the questionnaire for mobile devices: for instance, about grids, sliders, do not know use and open ended questions. Lugtig and Toepoel (2016) considering within person behavior in the longitudinal setting of two consecutive waves of the LISS panel, found no change in primacy, in missing, in switch from PC to mobile. Growing interest is now on understanding the impact of different mobile devices in answers of mobile web survey as a component of mixed-mode surveys.

Mobile web surveys are becoming mixed-device surveys (de Leeuw and Toepoel, 2018; de Leeuw, 2018; Toepoel and Lugtig, 2015). Considering both PC and mobile survey completion as subtypes of the web mode, the collected data could be questionable both from the survey methodology and data quality results perspective. Mobile devices differ from laptop in several characteristics (screen size, environment where usually it is located, etc.); they also may be used at any time and in any location where the individual is, both in a fixed location and in movement (such as travelling). Moreover, there are differences between mobile devices. Smartphones and tablets differ in screen size and several aspects related to technology. In addition, the task of smartphone is mainly for communication and for "continuous connection" with the social environment both through telephone calls and other web modes (whatsapp, social network, e-mail, etc.); it is essentially a personal communication tool. Tablet is more likely to be shared with other members of the household; it has larger screen size; in some sense, it is more similar to the laptop as the way it is used. Taking all this into account, research on the impact of the alternative mobile devices is an important issue to design mixed-device (and also mixed-device mixed-mode surveys) and to process survey results and related errors. Some recent literature has been published on experiments involving different mobile devices. Penetration of various mobile devices has been growing during the last years. Due to the fast and continuous change in the digital skills of the participants to the survey and to the modified technological environment the results obtained in the literature become rather quickly obsolete.

Conrad et al. (2017b, special issue) considering smartphones concluded that respondents who chose their mode of interview were more likely to complete interviews they had started than respondents assigned to their mode of interviewing.

There was no evidence that mode choice led to a demographically different respondent sample than did assigning respondents to a mode. Some research has been considering apps effect together with smartphones (Callegaro, 2013).

As regards the impact of different mobile devices, there is a still limited, but fast-growing literature. Existing research has found differences between smartphone and tablet responses in surveys. Longer response time and lower response rates could be expected, whereas no relevant differences with respect to PC survey completion are likely (de Bruijne and Wijnant, 2013). It has been suggested that responses to surveys using tablets are more similar to PC responses than smartphone responses (Struminskaya et al., 2015). All in all, considering a range of data quality measures (completion times, rates of missing data, straightlining, and the reliability and validity of scale responses) a few effects of the type of mobile is found (Tourangeau et al., 2017; Tourangeau et al., 2018).

Lugtig and Toepoel (2016) evaluate which devices (tablet computer, smartphone, and laptop) respondents use over time (two consecutive waves in a panel). Considering measurement error associated with each device, it is shown that measurement errors are larger on tablets and smartphone than on PCs; moreover, at individual level no change in measurement error is found when switching the device. The message arising from this study is that self-selection in the choice of the device is affecting greatly measurement error in tablets and smartphones data. As regards the choice of the mobile device, Read (2019), analyzing the Understanding Society Innovation Panel Wave 9, proposes a new perspective based on previous studies on interviewer effect. The effect of device model is analyzed also. Results suggest that further research should be undertaken to support the conclusion of the effect of the device model.

Considering demographic variables and behavioral and context variables, Haan, Lugtig, and Toepoel (2019) evidence that there are subgroups of respondents more likely to use a mobile device for survey participation (young, females, foreign citizenship people, and lower-income households). As regards factors affecting the selection of the device, ease of use is the strongest predictor. The other variables, ownership, and usefulness, did not predict device use.

Some other studies are handling the demographic variables and sociodemographic background of device use, for example, Bosnjak et al. (2018) and Maslovskaya, (2019). This last paper provides a comparative analysis across five United Kingdom panels of the characteristics of the respondents with respect to the used devices. The UK surveys are (Understanding Society Innovation Panel Waves (wave 7 and 8), European Social Survey (ESSMM) experiment carried out in 2012, 1958 National Child Development Study (NCDS), Community Life Survey (CLS), and Wave 4 of the Second Longitudinal Study of Young People in England (LSYPE2). All these surveys are treated as cross-sectional in this study. Several socio-demographic variables are analyzed: age, gender, marital status, employment, religion, household size, children in household, household income, number of cars, and frequency of Internet use. Logistic regression modeling ad bivariate analyses show these variables are significantly associated with device used across surveys. Age, gender, employment status, household size, and education as factors related to device choice is supported by experimental results found in other countries. General findings are that the age of the respondents (younger people more likely to answer using mobile devices), gender (female more likely to answer using mobile devices), employment (employed respondents more likely to use smartphones) are especially significant.

The study of changes in measurement error over time, associated with a switch of devices over two consecutive waves of a panel suggests that the higher measurement error in tablets and smartphones is associated with self-selection of the sample into using a particular device (Lugtig and Toepoel, 2016).

Very few studies consider device choice and characteristics of respondents choosing different devices along time.

Recommendations are to accommodate the questionnaire for web surveys with attention to optimization of the survey instrument for small screens. Further experiments are needed to investigate the factors affecting mobile devices choice and errors.

However, a general recommendation when using web or mobile web is to design shorter mobile questionnaires, for example by avoiding multi-item pages, text inputs, or by using a split-ballots design. When mobile devices are also allowed for questionnaire completion a good solution is to let people choose the mode and the device.

There is no general theoretical model (yet) for mixed-mode surveys that can help to take a decision about the best survey design. Therefore, still a lot of experiments are needed to investigate various open questions like questionnaire design, survey design, and inference. The subsections of this part will be addressed to attempt to answer the following basic methodological questions with respect to mixed-mode surveys:

- *How to design a questionnaire for a mixed-mode survey?* Should the same questionnaire be used in each mode or should there be versions that are optimal for each specific mode?
- *How to mix modes?* Is it better to have a sequential or a concurrent design? And which modes should be used in a specific survey context?
- *How to compute response rates?* How can the performance of various modes (in terms of response rates) be compared? And how can a mixed-mode survey be compared with a single-mode survey, also?
- *How to compute measurement error?*
- *How to make statistical inference for a mixed-mode survey?* How can estimation procedures account for mode effects? Is it possible to disentangle mode effects from selection effects and other effects?

## 9.3.1 PREVENTING MODE EFFECTS THROUGH QUESTIONNAIRE DESIGN

Each mode of data collection has its advantages and disadvantages. The effects of different phenomena may vary across modes. This means that the same question may be answered differently. Consequently, observed differences in figures may be not be "true" differences but deviations caused by measurement problems. These mode effects should be avoided as much as possible. First of all, mode effect should be prevented through the questionnaire design.

The web questionnaire toolbox opens many opportunities for quality improvements. The technology introduces new sources of error as well. Thus, questionnaire design should be carefully studied.

The survey researcher is faced with the question whether to use the same questionnaire across modes. If he does, there are mode effects. He could also use a different questionnaire for each mode. These questionnaires must be designed such that, although they are different, they measure the same concepts. The questionnaires must be cognitively equivalent. This is not so easy to realize. Three approaches are discussed here to deal with this problem.

The first approach is the *unimode approach* as proposed by Dillman (2007). The idea is to use the same questionnaire in each mode, but define the questions in such a way that mode effects are minimized. Some examples of his guidelines are:

- The text of a question must be the same across modes;
- The number of answer options of a closed question must be kept as small as possible;
- The text of the possible answers to a closed question must be the same across modes;
- The order of the answer options to a closed question must be randomized;
- Include all answer options also in the text of a closed question;
- Develop equivalent instructions for skip patterns.

It may not be easy to develop a questionnaire which completely satisfies all unimode guidelines. Particularly for attitudinal questions, it may turn out to be necessary to still define mode-dependent versions of questions.

A properly designed unimode questionnaire should remove mode effects related to question interpretation. However, this approach cannot take advantage of the specific features that every mode offers. For example, a researcher may have to abandon the idea of randomizing the order of the answer options of a closed question, because it is not possible to do this in the paper questionnaire mode. Another example refers to displaying instructions about how to answer questions: in mixing paper and web mode, the web questionnaire cannot use pop-up windows since this is not possible in a paper questionnaire.

Another potential problem is the mode effects caused by the presence or absence of interviewers. The unimode approach may not be able to remove these effects completely. Differences may particularly remain for sensitive questions.

According to Dillman (2017) if unified mode construction is to be used on smartphones, the needs of such devices are likely to be the major determinant of how items are presented across all modes.

The message is again that transition to mixed-mode with mobile web component implies a complete revision of the structure of already existing questionnaire and a redesign of the questionnaire. An omnimode approach, which consists of rebuilding questionnaires for mixed-mode from scratch instead of adapting existing questionnaires to the web mode should be investigated, as suggested in the MIMOD project report (2017–2019).

The second approach to diminish mode effects is designing mode specific questionnaires. Each questionnaire should be optimal for its corresponding mode. Optimal means that the questions are defined such that the answers given are as close as possible to the "true" value.

As an example, consider answering a factual question in a CAPI survey and a web survey. In case of a CAPI, there is always an interviewer who can assist the respondent in understanding and answering the question. In case of a web survey or mobile web there is no interviewer assistance. The respondents are on their own. Nevertheless, it is possible to develop some kind of interactive help system for web surveys. There could be help-buttons on the screen giving access to additional information about the question. It is even possible that an animated interviewer appears.

Summing up, mode specific should provide perception of the same stimulus. Each questionnaire is optimized for corresponding mode, functional equivalence should be assured, measuring the same concept is the ultimate task. It has to be advised that it is difficult to ensure equivalence.

When more than one mode is offered for survey participation, answers to the same questions could differ across mode. Thus, in designing the questionnaire in a mixed-mode survey, the researcher should evaluate the cross-mode exchangeability of responses or comparability of results obtained in alternative response modes. The measurement of the mode effect therefore has to consider if the selection of a response category has a probability attributable to the survey administration mode. For example, Mariano and Elliot (2017) provide evidence of across ordinal response categories. The authors apply *item response theory* (IRT) and propose a method for adjusting cross-mode results. The method to adjust for the survey administration mode is a Bayesian hierarchical IRT model. Cluster-level parameter estimates of the latent trait are used for observed groupings of respondents of interest.

It may not be easy to design optimal mode specific questionnaires. It will require a fair amount of experimentation with different formats to obtain the best one. Moreover, the best question for one survey context may not be the best question in another survey context.

The questionnaire length is a major critical problem when completing questionnaire on mobile devices. Toepoel and Lugtig (2018) proposes and experimented a modularization approach. The way to understand how to optimize the questionnaire construction across modes is still an open and challenging question. The problem is crucial in official statistics where complex questionnaires are usually used. Recent research advocates for a mobile device first questionnaire design, or, at the least, of a rigorous account of the mobile device option in questionnaire design. Two main reasons stand for this (see MIMOD report). The first reason is that smartphones have become a dominant communication channel and cannot be ignored in design. The second reason is that issues with usability and comprehension on smartphones reveal the measurement error prone questions and question blocks. Such a viewpoint, however, has implications for ESS model questionnaires and ESS survey guidelines. Multi-device surveys introduce additional challenges for the questionnaire design.

This is an obstacle that is often put forward to introduction of new devices is questionnaire length. As most ESS surveys are long and, consequently, demanding when filling in on a smartphone, it is imperative to prevent "speed" and "stimulate" a relaxed manner leading to better quality and less measurement errors. A responsive design should facilitate the respondent filling in on a smartphone screen, however, most questions and answer texts are cognitively demanding due to their specific content or response task, as, for example, long reference periods. When reflecting on the fitness criteria and the experiences from the test interviews at CBS and SSB it is a harsh job to find a good modus in redesigning, i.e., responsive design and collecting valid and reliable data comparable over devices and modes (also interviewer based). This leads to a question for future discussion: is it feasible to find possibilities to shorten/redesigning an ESS model questionnaire making it user friendly to fill in on a smartphone?

If a survey is repeated at regular intervals, like often happens in national statistical institutes, there is also a maintenance challenge. Changes in the survey must now be implemented and tested in several questionnaire forms instead of in one.

A third approach to dealing with mode effects is to identify one primary mode of data collection. This is seen as the most important mode of data collection. This is the benchmark for all other modes of data collection. The questionnaire is optimized to get the best answers in the primary mode. The questionnaires for other modes must be designed such that there will be no mode differences with the primary mode.

Pierzchala (2006) determined the main factors responsible for most mode effects. He identified three dimensions. These are:

1 Presentation—aural versus visual presentation;
2 Administration—self-administered versus interviewer-assisted;
3 Behavior—dynamic versus passive questionnaires.

Questionnaires for computer assisted interviewing are usually dynamic. They have forced routing and perform consistency checks. Paper questionnaires are passive.

The first and third dimension relate to the way in which information is transmitted. The second is related to the medium used for this. Based on these three dimensions, Pierzchala (2006) introduces *disparate modes* as modes that differ on at least one of these dimensions. Furthermore, the larger the degree of disparity is, the higher the risk of mode effects will be. See Table 9.1.

CAPI and CATI are similar in presentation, administration and behavior of the questionnaire. The degree of disparity of the other modes of data collection is

TABLE 9.1    Degree of disparity between data collection modes

| Mode combination | CAPI/CATI | Mail | Web |
|---|---|---|---|
| CAPI/CATI | | 3 | 2 |
| Mail | 3 | | 1 |
| Web | 2 | 1 | |

indicated with respect to both CAPI/CATI. Web surveys have a dynamic questionnaire. Presentation is visual instead of aural and web surveys are self-administered. Therefore, the degree of disparity between web surveys and CAPI/CATI surveys is 2. Mail surveys share none of the aspects with CAPI/CATI which results in a degree of disparity equal to 3. Mail surveys and web surveys are similar in self-administration and visual presentation, but mail surveys have a passive questionnaire. Therefore, their degree of disparity is 1.

It can be concluded from Table 9.1 that the largest mode differences can be expected for a mix of mail with CAPI or CATI, followed by a mix of web surveys with and CAPI or CATI. A combination of mail and web surveys has a reduced risk of mode effects. Combining CAPI and CATI is the safest option to avoid mode effect.

## 9.3.2 HOW TO MIX MODES?

There are no specific rules on how to mix modes. There is no best solution. Experimental research is still ongoing. The aim is to find out evidence for advantages and disadvantages of different approaches. Design decisions should consider the available evidence. Moreover, such decisions will also depend on the specific survey context.

Some practical aspects are discussed here. Different approaches are compared on the basis of costs and response rates. The first aspect is the choice between a sequential mixed-mode survey and a concurrent mixed-mode survey,

A concurrent design has the advantage that it reduces under-coverage errors. It can be implemented in two ways. If the survey researcher decides beforehand which mode is best for every respondent, the survey researcher has control over assigning modes to groups. A proper design makes it possible to distinguish mode effects from selection effects. If respondents decide the mode used for survey participation, no control is in the hands of the researcher. Therefore, mode effects and selection effects may be entangled. This means that under-coverage errors are replaced by selection errors. It is up to the researcher which error to prefer in a practical situation. Concurrent designs ask for a more complex organization, since modes are on the field at the same time and everything must work in parallel and from the beginning of the data collection.

A simple sequential mixed-mode design is one in which recruitment takes place in one mode, and the actual data collection in another mode. For example, sampled individuals are approached by telephone and asked to participate in a web survey. This design has the advantage of the high response rates of the telephone recruitment. Use of a single mode for data collection avoids mode effects.

One step further is a design in which people are approached in a single, interviewer-assisted mode (face-to-face or telephone). At the end of the interview the respondent is invited to participate in a panel. Such a panel may have different modes of data collection. Then there will be not only mode effects within the panel, but also between the recruitment interview and the panel. The situation is complicated by the time lag between recruitment and panel. This makes it impossible to distinguish real changes over time from mode effects.

It is common procedure to follow up nonrespondents in a survey as this helps to increase response rates. The researcher has choice to use the same mode for the follow-up or to use a different mode. For example, nonrespondents can be called by telephone or sent a letter in an attempt to encourage them to complete a web survey questionnaire. If the follow-up is not meant for persuading nonrespondents, but also some additional data are collected, there may be mode effects.

Both in recruitment, follow-up and data collection several modes can be used. Recruitment and follow-up focus on communication and not on real data collection. Therefore, de Leeuw (2005) uses the term mixed-mode system or multimode system to denote either communication with respondents (mixed-mode communication) or data collection (mixed-mode data collection).

Statistics Netherlands carried out some experiments with more complex mixed-mode designs. Some of the findings are described in Example 9.5.

### EXAMPLE 9.5 The Safety Monitor Pilot

Statistics Netherlands has conducted an experiment with its Safety Monitor. It has been studied whether a mixed-mode survey can replace a CAPI or CATI survey without affecting the quality of the results. The Dutch Safety Monitor is an annual survey. It measures actual and perceived safety of the people in the country.

The sample for this survey is selected from the population register. The old Safety Monitor applied to data collection modes. If a telephone number could be found, sampled persons were approached by CATI. If this was not the case, they were approached by CAPI.

The design of this pilot is shown in Figure 9.3. There were four modes of data collection in the pilot for the new Safety Monitor. All sampled individuals received a letter in which they were asked to complete the survey questionnaire on the Internet. The letter also included a postcard that could be used to request a paper questionnaire. Two reminders were sent to those that did not respond by web or mail. If still no response was obtained, nonrespondents were approached by means of CATI, if a listed telephone number was available. If not, these nonrespondents were approached by CAPI.

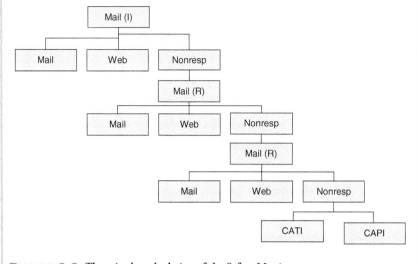

FIGURE 9.3 The mixed-mode design of the Safety Monitor

The data are comparable over the years since 2012.

Large mode effects in the Dutch Safety Monitor have led to a restriction in the modes to web and paper, that is, noninterviewer modes (Schouten et al., 2013).

More detail about this survey experiments is given by Kraan et al. (2010).

Official statistics surveys are on a transition phase due to the need to explore the use of web or mobile web, introducing a mixed-mode design or modifying the one currently used.

The MIMOD survey, run in May 2018, presents the state of the art of the use of mixed-mode strategies by the European NSIs for their social surveys.

Over 31 NSIs, the web mode is used as part of the mix in 23 NSIs, but it is not used at all by 6 NSIs. It is investigated which modes are combined, which sequence of modes are used in sequential designs, whether modes can be chosen by respondents in case of concurrent designs or which auxiliary variables are exploited by the NSIs to assign modes to sample units. The following social surveys are: Labor Force Survey waves 1 and 2 (LFS wave 1 and wave 2), Survey on Income and Living Conditions waves 1 and 2 (EU-SILC wave 1 and wave 2), European Health Interview Survey (EHIS), AES, Survey on Information and Communication Technology (ICT), HBS, and Harmonised European Time Use Survey (HETUS/TUS).

TABLE 9.2    Concurrent versus sequential mixed-mode per survey

| | Percent values | | | |
| | Concurrent mixed-mode | Partly sequential, partly concurrent | Sequential mixed-mode | Total per countries |
| --- | --- | --- | --- | --- |
| LFS wave 1 | 36.4 | 54.5 | 9.1 | 100 |
| LFS wave 2 | 72.2 | 22.2 | 5.6 | 100 |
| EU-SILC wave 1 | 72.7 | 18.2 | 9.1 | 100 |
| EU-SILC wave 2 | 60.0 | 33.3 | 6.7 | 100 |
| EHIS | 41.7 | 33.3 | 25.0 | 100 |
| AES | 23.1 | 53.8 | 23.1 | 100 |
| ICT | 35.3 | 35.3 | 29.4 | 100 |
| HBS | 58.4 | 33.3 | 8.3 | 100 |
| HETUS/TUS | 66.7 | 0.0 | 33.3 | 100 |
| Total surveys | 51.0 | 34.0 | 15.0 | 100 |

Source: Based on Istat, MIMOD project report and deliverables, 2019.

All the EU NSIs, but one, (Romania) use mixed-mode approach. Table 9.2 gives a synthesis of the adopted approaches; the results rely on 112 surveys of the EU countries that are using mixed-mode.

Concurrent mode is the most frequently used. Mixed-mode designs are mainly based on a concurrent approach (51%), meaning that all modes in the mix are in the field at the same time. Modes can be assigned in advance to subgroups of sample units or sample units can choose the mode they prefer. The survey showed that 45.5% of mixed-mode surveys allow respondents' mode choice, while 43.9% assign modes to sample units and 10.5% give the possibility to choose the mode only under certain conditions. Overall, concurrent mode CAWI appears to be less frequently used than interviewer-assisted modes, probably because concurrent designs take advantage from the presence of an interviewer for the management of different modes in the field.

The use of a sequential administration of modes equals 15% of the total examined surveys.

On the contrary, "Partly sequential-partially concurrent" (or hybrid) mixed-mode designs are quite a large number (35% of the considered surveys): This approach represents those strategies where the first data collection mode remains on the field when the second one(s) is (are) made available to respondents.

In mixed-mode surveys, sometimes respondents are given the chance to choose the mode to participate in a survey. The findings of the MIMOD survey show that 45.5% of mixed-mode surveys allows respondents' mode choice, while 40.2% assign modes to sample units. About 14.3% give the possibility to choose mode only under certain conditions (for instance, at the end of LFS wave 1 respondents can choose if they would like a CAPI or CATI interview at wave 2). EHIS presents the highest percentage for allowing mode choice (58.3%); on the contrary, EU-SILC wave 1 has the greatest quote of no choice (54.5%). For HETUS/TUS none of the three countries allows to choose the mode. For more detailed information see MIMOD, 2019, WP1, deliverable1.

### 9.3.3 HOW TO COMPUTE RESPONSE RATES?

A mixed-mode survey is sometimes designed with the objective to reduce survey costs. Another objective can be to increase response rates. Whatever the objective of the design, it is always important to compute the response rate, as it is often seen as a quality indicator: the higher the response rate, the better the quality of the survey. It is however advised that if the search of high response rate is stressed, at the end, quality of the data could decrease due to low quality of the last respondents who were not interested in participating in the survey.

To be able to compare response rates of modes, and also to be able to compare response rates of different surveys, standardized and consistent definitions must be used.

The basic formula for the response rate is

$$(9.1) \qquad RR = \frac{n_R}{n_E} = \frac{n_R}{n_R + n_{NR}},$$

where $n_R$ is the number of (eligible) respondents, $n_E$ is the total number eligible units (individuals, households, businesses, etc.) in the sample and $n_{NR}$ is the number of (eligible) nonrespondents. Eligible units are units that belong to the target population and have been selected in the sample. In practice it may be difficult to compute $n_E$ as it is not always possible to determine whether nonrespondents are eligible or not.

It is important to distinguish different causes of nonresponse. Different causes may have different effects on the outcomes of the survey and may therefore require different treatment. The three basic causes of nonresponse are no contact, refusal and not able. Considering response rate definitions as proposed by AAPOR (2009) and Lynn et al. (2002), expression (9.1) can be rewritten as

$$(9.2) \qquad RR = \frac{n_R}{n_R + n_{IC} + n_{NC} + n_{RF} + n_{NA} + n_{OT}}.$$

Respondents may break off the completion of the questionnaire. This typically happens in web surveys. Respondents get tired or bored of filling in the form and just stop in the middle. The number of such cases is denoted by $n_{IC}$. Partially completed questionnaires are treated as nonresponse in many surveys. That is why this quantity is included in the denominator and not in the numerator of definition 9.2.

The quantities $n_{NC}$, $n_{RF}$, and $n_{NA}$ denote the number of cases of nonresponse caused by noncontact, refusal and not able, respectively. There is a problem with the number of noncontacts $n_{NC}$. This must be the number of eligible noncontacts. However, this number cannot be determined because there was no contact. It is not unlikely that some of the noncontacts may be cases of over-coverage, and those should not be included. In practice, an estimate of $n_{NC}$ will be used.

The quantity $n_{OT}$ denotes the number of other unprocessed cases. This could include units of which the eligibility is unknown. At Statistics Netherlands another cause of nonresponse is distinguished: administrative nonresponse. These are cases that are not processed by interviewers because their workload is too high. This may happen if they have to handle many difficult cases in a limited amount of time. Another reason can be temporary illness of an interviewer.

In case of interviewer-assisted surveys it is usually possible to determine the cause of nonresponse. Interviewers observe this and record the results of their efforts as noncontact, refusal or not-able. This is much more difficult to do in self-administered surveys (web and mail). It is only observed that questionnaire forms are not returned or completed. The reason why is not observed. There can be many reasons. As an example, here are some reasons for nonresponse in web surveys:

- There are various forms of noncontacts. It depends on the way in which sample persons are selected. If the sampling frame is a list with e-mail addresses, noncontact occurs if the e-mail with the invitation to participate in the survey does not reach a selected individual. The e-mail address may be wrong or the e-mail may be blocked by a spam filter. If the sampling frame is a list of postal addresses and letters with an Internet-address sent to selected units, noncontact may be caused by units not receiving the letter. If recruitment for a web survey takes place by means of a face-to-face or telephone survey, noncontact can be due to respondents being not at home or not answering the telephone.

- Nonresponse due to *refusal* can occur after contact has been established with a sampled unit. Refusal to cooperate can have many reasons. In business surveys, a too high perceived or actual response burden, having strategies against surveys, and a lack of interest in providing data can be factors leading to refusal to cooperate. In households or individual surveys, people may not be interested, they may consider it an intrusion of their privacy, they may have no time, etc. Sometimes a refusal can be temporary. In this case it may be attempted to make an appointment for another day and/or time. But often a refusal is permanent. If sample individuals for a web survey are contacted by an e-mail or a letter, they may postpone and forget to complete the questionnaire form. This can be seen as a weak form of refusal. Sending a reminder helps to reduce this form of nonresponse.

- Nonresponse due to *not-able* is a type of nonresponse where respondents may be willing to respond but are not able to do so. Reasons for this type of nonresponse can be, for example, illness, hearing problems or language problems.

If a letter with an Internet address of a web questionnaire is sent to a sampled person, this person receives the letter, and he/she wants to participate in the web survey, but does not have access to the Internet, this can also be seen as a form of nonresponse due to not-able.

If there are self-administered-modes in a mixed-mode survey, and one wants to compare response rates over modes, only simple response rates can be computed (without considering the cause of nonresponse).

Suppose, the mixed-mode survey consists of $M$ concurrent modes, and the survey researcher has preassigned modes to sampled units. Let $n_E^{(h)}$ denote the number of eligible elements in mode $h$, for $h = 1, 2,\ldots, M$. And let $n_R^{(h)}$ be the number of respondents in mode $h$. Then the response rate in mode $h$ is equal to

$$(9.3) \qquad RR^{(h)} = \frac{n_R^{(h)}}{n_E^{(h)}},$$

for $h = 1, 2, \ldots, M$. Consequently, the overall response rate of the survey is equal to

$$(9.4) \qquad RR = \frac{n_R}{n_E} \frac{\sum_{h=1}^{M} n_R^{(h)}}{\sum_{h=1}^{M} n_E^{(h)}} = \sum_{h=1}^{M} \frac{n_E^{(h)}}{n_E} RR^{(h)}.$$

So, the overall response rate is the weighted mean of the mode response rates.

In case of a concurrent mixed-mode survey where people choose their own mode of data collection, it is not possible to compute mode response rates, as the number $n_E^{(h)}$ of eligible units for each mode is unknown. Of course, the overall response rate can be computed as $RR = n_R/n_E$. Also, the rate $n_R^{(h)}/n_E$ can be computed but this rate is not comparable to the response rate that would have been obtained if all respondents were approached in this mode.

In case of a sequential mixed-mode design, the mode specific response rates and the overall response rates can be computed. Note that for this design the number of eligible elements in a mode is equal to the number of nonrespondents that remains after the previous mode. So

$$(9.5) \qquad n_E^{(h)} = n_E^{(h-1)} - n_R^{(h-1)} = n_E - \sum_{i=1}^{h-1} n_R^{(i)}.$$

The importance of using a uniform, standardized response rate definition is stressed once more.

Only then are response rates comparable across studies and survey modes used in different analyses. See also Shih and Fan (2007).

A simple example of the response rates computation is given in Example 9.6.

---

📖 **EXAMPLE 9.6 Computing response rates**

The local authorities of the town of Mudwater in the country of Samplonia conduct a survey about living conditions in the town. It is a simple sequential mixed-mode survey. All people of the age 18 and over are sent a letter with the invitation to complete the questionnaire on the Internet. After two weeks, all nonrespondents are called by telephone in an attempt to fill in the questionnaire in a telephone interview.

The sample size is 1600 persons. The number of persons completing the form on the Internet is equal to 672. The number of respondents in the telephone mode is equal to 278.

The overall response rate of the survey is $100 \times (672 + 278)/1600 = 59.4\%$. The response mode in the web mode is $100 \times 672/1600 = 42\%$. After the web mode, there remain $1600 - 672 = 928$ persons. These are the eligible persons for the telephone mode. The response rate of the telephone mode is $100 \times 278/928 = 30.0\%$.

---

Many survey organizations consider conducting mixed-mode surveys in which one of the data collection modes is the web. The idea is that this will substantially reduce the survey costs, as no interviewers are involved. At the same time the overall response rate must remain at an acceptable level. This is reason for concern as some experiments with web surveys show that response rates are not very high.

In the literature, there is no clear evidence of higher response rates in mixed-mode surveys.

The conclusion is that this does not mean that this approach has no advantages, since many survey quality factors should be considered in choosing the survey mode. For example, in considering changing to a mixed-mode survey with the web as one of the modes, the response rate of the web mode is of crucial importance. The higher the response rate that can be obtained for the web mode, the more cost savings can be realized. There is evidence that starting a mixed-mode sequence with a self-administered mode followed by an interviewer-administered mode can yield potential cost savings (Bianchi, Biffignandi, and Lynn et al., 2017). Jäckle, Lynn, and Burton (2015) comment real cost savings can be obtained, however, avoiding damage to long-term participation rates and to item response rates may prove more challenging. Examining the cost-error tradeoffs associated with mixed-mode designs is an important topic for further exploration and it is an aspect to be considered in designing the survey.

Shih and Fan (2007) carried out a meta-analysis of a large number of mixed-mode surveys. They observed a preference of the mail survey mode over the web survey mode. The response rate for mail survey modes was on average 14% higher than for web survey modes. However, if respondents were offered both the mail and the web option, there were no systematic differences in response rates. They suggest offering the other mode in a follow-up of nonrespondents of the one mode.

How respondents choose the response mode in a mixed-mode survey remains unclear. Factors affecting the choice seem to be:

• Technological background. However, Zhang (2000) suggests the choice not merely depend on respondents' technological backgrounds or on their access to the web. It turns out that experienced and frequent Internet users often choose to reply by mail or fax;
• The modes offered;
• Delivery format (i.e. in which format is the questionnaire initially offered);
• Mode delivery order (i.e. simultaneous mixed-mode or sequential mixed-mode);
• Type of target population;
• Use of incentives;
• Deployment of follow-up reminders.

However, there are several reasons others than costs to apply mixed-mode approach.

There are other aspects also playing a role in the choice of modes to mix. These aspects are:

• Budget restrictions;
• Timeliness;
• Available infrastructure for implementing mixed-mode designs;
• Data quality;
• Specific survey content (types of questions, length of survey, complexity of the questions, and need for visual aids).

Roberts (2007) and Biemer and Lyberg (2003) discuss factors influencing the decision process in choosing modes. They acknowledged that choosing an optimal design is especially difficult in situations where there are a lot of options.

The fact that modes vary with respect to factors such as the cost and speed of fieldwork, their suitability for administering different types of questionnaires and their impact on data quality means that, in principle, mixing modes allows the researcher to minimize both the costs and errors associated with any given single-mode approach. However, while the mixed-mode designs may help reducing survey costs, under-coverage, nonresponse and specific forms of measurement error, there remains a risk of mode effects.

### 9.3.4 AVOIDING AND ADJUSTING MODE EFFECTS FOR INFERENCE

Generally speaking, there are two approaches to inference in survey methodology. They are called the design-based approach and the model-based approach. Both approaches are discussed in the context of web surveys by Couper and Miller (2008).

The *design-based approach* is the classical approach to survey sampling as described in Chapter 4. If samples are based on probability sampling, the theory of probability and statistics can be applied. This results in concepts like unbiased estimators and confidence intervals. The design-based approach can also be used for web surveys. For example, if a random sample is selected from a sampling frame and all selected elements are invited to complete the questionnaire on the web, proper inference to the population is possible.

The *model-based approach* assumes the existence of some kind of model for the relationships between the survey variables. Based on the available data, the parameters of the model are estimated. Next, the model can be used to predict the values of population parameters. The reliability and accuracy of these predictions usually depend on the validity of the model. Unfortunately, it is not always possible to select the validity of the model. If the model is correct, predictions are accurate. If the model is not correct, predictions can be seriously biased.

A typical example of the model-based approach in web survey is recruiting respondents by means of self-selection (see also Chapter 11). The sample is not obtained by means of probability sampling. Selection probabilities are unknown and therefore the design-based approach to inference cannot be applied. A way out of this problem is to model participation in the survey by means of response propensities. See Chapter 13 for a detailed description.

In mixed-mode surveys one of the objectives is to produce reliable and accurate estimates of population characteristics. The inference problem can be more complex for a mixed-mode survey. Alternative models have been proposed. The debate is if they can be used in mixed-mode surveys or whether new methods of inference should be developed. Mode effects may cause estimates to be biased; the surveyor should bear in mind this risk.

There are two categories of effects: mode selection effects, i.e. errors due to coverage and nonresponse, and mode measurement effects relying on the difference among modes in the observed data, i.e. the influence of a survey mode on the answers respondents give, so that one person would give different answers in different modes. Errors may originate from differences in, among others, whether items are presented sequentially or simultaneously to the respondent, interviewer effects and social desirability, primacy and recency effects, recall bias, and acquiescence.

Given that different modes are likely to lead to differences in coverage and in response rates for subgroups, it is advised to attempt to correct for these differences. Weighting adjustment can do this. See Chapter 12 for details. Weighting adjustment is useful to correct for the lack of representativeness of the response. If there are only mode effects in a mixed-mode survey, weighting will not help to solve estimation problems. This is illustrated in Example 9.7 where the composition of the response is representative with respect to age, but the mode effect still causes a bias in the estimator. For this reason, recent literature has paid great attention to the inference problem with respect to selection and measurement bias. Several methodological proposal and empirical analyses have been published. Observe that mode effects are not affecting only the bias. Other aspects relying on the precision, like a bad designed questionnaire, respondent behavior, scarce commitment, noisy or disturbed environment, and so on, are producing an impact on the results. However, this kind of factors should be prevented through a careful questionnaire design and survey administration. Thus, in the present chapter these aspects are not measured and not considered here to study of mixed-mode effect in the inference context.

## EXAMPLE 9.7 Estimation effects in mixed-mode surveys

This small artificial example shows what can happen in a mixed-survey if there are selection effects and mode effects.

The target population consists of two age groups: the young and the elderly. The two groups are of equal size. Objective of the survey is supposed to be estimation of the percentage of voters on the New Internet Party (NIP) at the next elections.

Among the young, the population percentage is 70% and for the elderly it is 10%. So, the overall percentage of voter for the NIP is $0.5 \times 70 + 0.5 \times 10 = 40\%$.

The young are more inclined to participate in a web survey. Their response probability is 0.8 whereas for the elderly it is only 0.2. It is the opposite for a face-to-face survey: the response probability for the young is 0.2 and for the elderly it is 0.8.

Suppose a web survey is conducted. Then the expected value of the percentage of voters for the NIP in the sample would be

$$100 \times \frac{400 \times 70 + 100 \times 10}{500} = 58\%.$$

This percentage is much higher than the population percentage of 40%. This is not surprising as this survey design leads to an overrepresentation of young people and they typically vote for the NIP.

Suppose a face-to-face survey is conducted. Then the expected value of the percentage of voters in the sample would be

$$100 \times \frac{100 \times 70 + 400 \times 10}{500} = 22\%.$$

This percentage is much lower than the population percentage of 40%. Again this is not surprising as this survey design leads to an under-representation of young people and they typically vote for the NIP.

Now suppose a sequential mixed-mode survey is conducted. First the web is offered. Nonrespondents are visited for a face-to-face interview. It is assumed that probability to participate in the face-to-face mode after nonresponse in the web mode remains 0.2 for the young and 0.8 for the elderly. The expected value of the percentage of voters in the sample would be

$$100 \times \frac{400 \times 70 + 20 \times 70 + 100 \times 10 + 320 \times 10}{840} = 40\%.$$

The response rate goes up from 50% to 84%. Moreover, the estimator is now unbiased.

Up until now there were only selection effects and no mode effects. Now a mode effect is introduced. It is assumed that voting for the NIP is a sensitive topic. Therefore, 20% of the NIP-voters will select a different party in the face-to-face survey. They give a truthful answer in the web survey.

The outcome for the web survey will not change. It remains 58%. The expected value of the percentage of NIP-voters in the sample is now equal to

$$100 \times \frac{100 \times 56 + 400 \times 8}{500} = 17.6\%.$$

So the percentage drops from 22% to 17.6% in the face-to-face survey. In case of a mixed-mode design the expected value of the percentage of NIP-voters becomes

$$100 \times \frac{400 \times 70 + 20 \times 56 + 100 \times 10 + 320 \times 8}{840} = 38.9\%.$$

The mixed-mode survey estimator has a bias now. Still the estimates are much better than would have been the case for a single-mode web survey or face-to-face survey.

In a mixed-mode survey, the analysis of the effects to inference can be focused either to the total mode effect, or to measure selection effects, or to disentangle selection and measurement effects.

There are various ways to separate the two sources of mode differences. For example, Tourangeau (2017) lists as approaches to the measurement error: direct assessment of the measurement errors (e.g., by comparing survey reports to a gold standard), different mode groups comparability by statistically weighting or applying regression methods, estimation of the errors using modeling techniques (such as confirmatory factor analysis or latent class modeling; and comparisons of the answers obtained in each mode with records data). See for an extended overview Hox, de Leeuw, and Klausch (2017).

Inference in mixed-mode surveys goes through the assessment of the mode effect, in some case the assessment of the total mode effect is the ultimate task; response rate, analyses of the distribution of the socio-economic characteristics of the participants to the survey and the representativeness of the survey collected data help in the understanding how the mode selectivity affects the results. Traditional adjustments in this context are the same as the ones adopted in single survey mode. Methods commonly used for this purpose include weighting, calibration and regression methods (Bethlehem, Cobben, and Schouten, 2011).

More complete approaches consider measurement of the selection effect or adjustment of estimates trying to disentangle selection and measurement effects.

To estimate mode impact on measurement, control for the selection process can be done computing the difference in likelihood to participate in specific modes. After controlling for the selection process in the modes, any remaining mode differences can be attributed to measurement. This approach hinges on two important assumptions: all aspects of the selection process are controlled, and the variables used to control this process are measured in the same way and are not themselves influenced by the mode.

In assessment studies to deal with the inference problem the survey design has to be carefully planned. Different types of design can be adopted.

Experimental designs allow for the control of the selection effects. The experiments are planned exactly for the mode effect analysis. Therefore, they are more precise, but more costs have to be afforded. There are alternative experimental design approaches. For example, embedded (like the reinterview approach), split sample design and repeated measurement designs are the most used designs. Measurement differences between modes are computed. Usually covariates are considered to separate selection and measurement effects. Methods to adjust for selection effects are: weighting, calibration, and regression methods.

Nonexperimental designs rely on observational studies. They are based on surveys whose primary task is not mode assessment. Therefore, adjustment techniques should be applied to correct survey estimates for bias introduced from different modes. These designs require covariate auxiliary variables to explain the selection mechanism. If covariates are identified, differences between mode groups conditional to the covariates are an estimate of the measurement errors. If the covariates are not adequate to fully explain the selection mechanisms, measurement effects could be incorrect since the total error decomposition is not correct.

In some case, quasi-experimental designs are adopted, too. They consist on basic designs and several variations of them. They lack the element of random assignment to treatment or control. Instead, quasi-experimental designs typically allow the researcher to control the assignment to the treatment condition, but using some criterion other than random assignment.

Auxiliary variables are used as covariates to explain either the selection or the measurement effect. Therefore, appropriated variables for each task should be identified. For example, administrative data, linked frame data and paradata recording the contact data and details about the participation process are useful to get information on selection effects. These types of data are also called in the literature backdoor variables. On the contrary, variables describing control checks and the answering process are informative about measurement effects. This type of data are also called frontdoor variables.

Total bias is given from selection bias and measurement bias. Selection effects and measurement effects are mostly confounded and difficult to separate. Reducing selection bias may increase measurement bias of the mixed-mode estimate. To prevent this problem, questionnaires of the mode determining higher measurement bias can be redesigned when the bias size is known.

Early measurement error studies have been based on imprecise design decisions. For example, different incentives were confounded in the sequence comparison or reference to the total survey error framework was done, thus confounding effects (de Leeuw, 2005).

Recent literature to measurement bias estimation has proposed using statistical adjustment of bias to the level of a measurement benchmark mode in the context of inference from mixed-mode data.

Researchers can use record data as a "gold standard" in order to estimate mode effects (Voogt and Saris, 2005). In some cases the true scores are available from an external source and in other cases a measurement benchmark approach is applied.

A quite recent and rarely used approach (Schouten et al., 2013; Klausch et al., 2017) is the combined use of administrative data or baseline data.

A possible approach is adjusting measurement bias using a benchmark. Estimates (point and variance estimates) are compared to estimates of some benchmark; thus, bias is measured. Comparison between results are usually referred to mean square error (MSE). Cost constrains can be an additional concept to be considered.

To disentangle both mode effects and selection effect it is not adequate to focus only on mixed-mode survey data. Thus, some methodological proposals are focused on comparing mixed-mode survey data with datasets used as a benchmark.

There are different possible ways to identify benchmark data.

- A comparable single-mode survey allows disentangling mode effects to a certain extent. comparing a mixed-mode dataset with a comparable single-mode dataset. It is the so called single mode benchmark (SMB). For instance, Schouten et al. (2013) and Klausch, Schouten, and Hox (2015b) defined differences in bias between a SMB and other modes as "single-mode effects." In this case, the effects are defined between single-mode surveys and a reference survey (the SMB), and not between mode-specific response groups in a mixed-mode design.

- In the relative mode effect approach the effects are between mode-specific response groups in a mixed-mode design. An example is a face-to-face survey as SMB against telephone, mail, and web modes.

- A combined benchmark for selection bias face to face and measurement of web is another proposed approach. It is called hybrid mode benchmark (HMB). This approach is more precise when sensitive questions are involved in the survey and/or when comparing results over time is not a central task. This approach has been experimented by Klausch, Schouten, and Hox (2015b).

- Another approach is based on a reinterview design. In this interview some variables like the survey target variables and some auxiliary variables are collected. Thus, repeated measurements are available in a different mode. A reinterview could be administered to a subset of respondents to collect answers of the main questions included in the main survey questionnaire. Reinterview approach has recently shown that the performance of the estimators strongly depends on the true measurement error model. However, one estimator, called inverse regression estimator, performs particularly well under all considered scenarios. The results suggest that the reinterview method is a useful approach to adjust measurement effects in the presence of non-ignorable selectivity between modes in mixed-mode data (Klausch et al., 2017).

Recent literature has proposed new approaches and comparative studies of adjustment methods of inference in the measurement setting. Most used methods are: calibration, measurement error correction, regression. However, various alternative of these methods and other methods have been applied, too.

Here is a short overview of the abovementioned approaches.

The *calibration method* adjusts the survey weights to balance the response with respect to the survey modes, while the correction approach adjusts measurements using predicted counterfactuals. Calibration methods do not require additional data collection beyond the regular sequential mixed-mode survey; this make them easier for application. They apply to sequential mixed-mode surveys. Mixed-mode calibration uses data only from the mixed-mode group. The standard error of the mode calibration estimator increases rapidly with increasing discrepancies between the distribution in the sample and in the population.

In survey statistics where change over time is strongly confounded with changes in survey mode composition, the calibration and correction methods have a stabilizing effect.

An example is a study by Vannieuwenhuyze and Loosveldt (2013). They apply three calibration methods: mixed-mode calibration, extended mixed-mode comparison, and extended mixed-mode calibration. Comparison is made between simple survey and mixed-mode survey, both samples drawn from the same target population.

Some caution in considering the results of this study is about the use of demographic variables from the survey itself, thus the hypothesis that covariates are not affected by mode is hardly satisfied. This represents a limitation in generalizing conclusions.

One possible way to compute calibration correction is reweighting the survey response to fixed mode distributions. Thus, the survey response is balanced to a prespecified distribution of the respondents over the modes. The calibration of mode proportions to fixed proportions in the case of repeatedly conducted, sequential mixed-mode, cross-sectional surveys has been used by Buelens and Van den Brakel (2015). An experiment linked to the Dutch Crime Victimization Survey is analyzed.

More recently Buelens and Van den Brakel (2017) compared two methods: (a) a calibration adjustment to the survey weights so as to balance the survey response to a prespecified distribution of the respondents over the modes, (b) a prediction method that seeks to correct measurements toward a benchmark mode. The two methods applied to the Labour Force Survey in the Netherlands (a sequential mixed-mode survey combining web, telephone, face-to-face, or other modes) provided similar estimates of the number of unemployed, even if each method has some specific valuable features. They explain this conclusion stating that if the underlying assumptions are met the models are identical for some parametrisation.

*Multiple imputation* is another method to disentangle mode measurement and mode selection effects. Multiple imputation to disentangle mode measurement and mode selection effects has been applied in several papers (see Klausch, Schouten, and Hox, 2017; Kolenikov and Kennedy, 2014).

Different mix of counterfactuals are identified. Prediction of the so-called counterfactuals potential estimates are computed through regression models. The one which minimizes the MSE is chosen. There are examples of imputation of non-observed answers or imputation using logistic regression models or fractional imputation. When the interviewed is choosing the mode, also a method based on fractional imputation can be applied to create imputed values of the unobserved counterfactual outcome variables in the mixed-mode surveys. For example, this approach has been proposed by Park, Kim, and Park (2016). In this study, however, minor measurement differences with respect to a single mode using mixed-mode was found. The application relies on a small-scale simulation study therefore the potential interest of this method should be further investigated.

Kolenikov and Kennedy (2014) investigated different methods for detecting and compensating for measurement differences by mode. They applied a regression adjustment and two methods involving multiple imputation. The regression adjustment incorporated covariates as well as the mode of data collection and adjusted the aggregate value for one mode by subtracting the estimated mode effect. To compare the methods a mixed-mode survey (web followed by telephone) and a single-mode comparison sample (telephone-only) was analysed by Kolenikov and Kennedy (2014). None of the three adjustment methods had much impact on the estimates. In such designs, the populations represented by the two mode groups were presumably similar.

*Measurement error correction approach* uses a prediction method that seeks to correct measurements toward a benchmark mode. The measurement error correction estimator uses additional information by explicitly relying on a specified model to correct the actual observations for a measurement error component. Under the measurement error correction approach, the standard errors increase only slightly, even when the outcomes are corrected to a single mode.

Unlike the calibration method, the measurement error correction method does not have a built-in protection against strong deviations of the sample and population distributions, unless the mixing coefficients are chosen by minimizing the MSE as proposed by Suzer-Gurtekin, or by choosing them close to the observed mode distribution.

Various sophisticated modeling procedures to isolate the measurement effects of different methods of data collection can be applied (Klausch, Schouten, and Hox, 2015a, 2015b, Klausch et al., 2017). Interesting is that it is demonstrated how feasible it is to estimate the measurement effects in presence of missing not at random (NMAR) effects. More precisely, when selection effects cannot be explained by auxiliary information and depend on the target variable is considered.

An early study by Biemer applied an interview reinterview approach and analyzed data with a latent class model to disentangle selection and measurement bias in face-to-face and telephone data collection modes. The reinterview approach is used to obtain the measurement benchmark.

The reinterview approach for statistical adjustment of measurement bias to the level of a measurement benchmark mode in the context of inference from mixed-mode data with web component is used to collect auxiliary information both for the application of a Monte Carlo simulation to compare the performance of six different estimators and to use ordinal multiple-group confirmatory factor analysis to isolate error components and compare modes (web, paper questionnaire, telephone, and face-to face). The reinterview approach is found to be useful to adjust measurement effects in the presence of non-ignorable selectivity between modes in mixed-mode data, as regards measurement error differences, consistent with previous studies, between the two self-administered modes and the two interviewer-administered modes are found.

Other mode effects studied using regression approaches are found for example in Jäckle, Roberts, and Lynn (2010), or latent class analysis Biemer (2010).

A review of the literature is in Hox, de Leeuw, and Klausch (2017) and in Tourangeau (2017).

A new model (MTME) multitrait–multierror (Cernat and Oberski, 2018, 2020) is applied by Cernat and Sakshaug (2020) to estimate multiple types of measurement error simultaneously; social desirability, acquiescence, and method effects are considered. The results show no differences in measurement error between single modes and mode designs with respect to acquiescence and method effect but some difference is found in social desiderability. Thus, scarce relevance of acquiescence effect and importance of mode on social desiderability are consistent with several other studies and can be considered as a typical characteristic in mode effects. A quasi experimental design randomly allocating respondents to either a unimode face-to-face interview or a sequential mixed-mode (web and face-to-face) design has been used to study the abovementioned effects.

A critical issue in estimating bias in a mixed survey is the temporal perspective. That means, if a sequential mixed-mode survey is repeated over time (longitudinal surveys) some different methodological approaches should be considered. Longitudinal mixed-mode surveys are subject to changes in mode composition and estimates can be unstable. The problem of measurement errors is more complex and the question is how the estimation method can be adapted according to the changes over time.

For instance, confounding of true change over time of a survey statistic with change in mode composition limits the usefulness of mixed-mode surveys (Buelens and Van den Brakel, 2015; Cernat, 2015). The overall measurement bias of estimated means and totals of survey variables may change due the mode composition of the mix variations.

A method stemming from the general regression estimator can be applied in sequential mixed-mode design aiming at stabilizing total measurement error in repeated surveys, as proposed in Buelens and Van den Brakel (2015). This method does not require the collection of additional data.

Another method may be derived from the general regression estimation, but without change in the survey weights.

Adjustments to the observed values in order to remove measurement error can be applied.

The abovementioned methods are intended to stabilize the mode distribution in repeated surveys to avoid fluctuations in mode-dependent measurement bias not considering measurements of change over time. Mixed-mode surveys may measure a level different from the true level in the population. As long as the level difference remains constant through time, change over time can be estimated without bias, both in single mode and mixed-mode surveys. It is recommended to choose the distribution for the mode calibration or the mixing proportions for the correction approach close to the observed distribution of the respondents over the modes in the samples. This avoids unnecessary increase of fluctuations in the weights and in the standard errors. The techniques are practically useful as they do not require additional questions, questionnaires, or repeated interviewing. An example about using longitudinal information in the weighting scheme is given in Example 9.8

🔳 EXAMPLE 9.8 **A longitudinal study**

In a mode comparison study in the Netherlands, Schouten et al. (2013) considered four modes of data collection in the first wave of a survey (face-to-face, telephone, paper, or web) and either face-to-face or telephone in the second. The units were assigned randomly to the mode, nonrespondents in the initial wave were retained in the sample for the second wave. Variables to be compared were collected over two survey waves, the researchers combined second wave data for telephone and face-to-face respondents. Incorporating Wave 2 variables into the weighting scheme did not appear to affect the estimated mode effects; weights based on standard population registry variables seemed to allow adequate separation of coverage, nonresponse, and measurement effects.

The weighting model removes mode-dependent selectivity with respect to the survey variables. Survey variables that suffer from large mode-dependent measurement effects, such as attitudes or answers to questions subject to social desirability bias are examples of variables producing this effect. Mode differences could represent greater social desirability bias in the interviewer-administered modes (Tourangeau and Yan, 2007) or more reluctance to give negative answers to an interviewer (Ye, Fulton, and Tourangeau, 2011). In addition, mail and web respondents show less random measurement error than respondents who completed telephone or face-to-face interviews.

Another interesting issue is about how to study longitudinal individual change and the impact of mode effect on random and systematic error and on estimates of change. Cernat (2015) studies mode effects in a longitudinal study, using a quasi-experimental design consisting of random allocation to mixed-mode or single mode, within an existing panel study. The results are analyzed using models which are a type of structural equation models where survey responses to earlier waves are included in the models. In a further study Cernat, Couper, and Ofstedal (2016) use latent class models to study mode effects and find effects predominantly between interviewer administered modes (face-to-face and telephone) and the non-interviewer mode (web).

An approach based on psychometric models can be applied. Overestimation of individual change compared to a single mode design is a risk. An analysis of the Society Innovation Panel (Cernat, 2015) considers a 12 items health scale (SF12 scale). Only one variable out of 12 has systematic differences due to the mixed-mode design. Variance of change in time is overestimated four of the 12 items in the mixed-mode design interviews.

Over the last years a great deal of papers has been devoted to inference and estimation in mixed-mode and literature is still steadily growing. Intensive experimental and methodological effort has been made from the researchers, especially using NSIs survey data. The MIMOD project has been focusing on mixed-mode in official statistics surveys. Since research is continuously going on and new designs are going to be implemented into the future new results are expected to come out. The MIMOD project has deeply analyzed some specific surveys and countries situation with respect to the estimation of selection and measurement error. An example is Istat "Multipurpose Survey on Households—Aspects of daily life—2017" (ADL survey). The objective of the analyses in the MIMOD project is to evaluate first the impact on the survey estimates of the introduction of mixed-mode design with respect to the previous single mode design. The reasons that determine significant differences in the estimates obtained with the two designs are also investigated. The analyses are organized over three steps: (1) (first level) based on the comparison between the two samples single mode and mixed-mode; (2) (second level) addresses the evaluation of the mode effect (selection and measurement) in the samples of respondent web and PAPI in the mixed-mode design; (3) (third level) carried out some experiments to adjust for mode effect. In the following scheme (Table 9.3) the phases and the methods considered in the study are listed.

TABLE 9.3 **Multipurpose survey on households: methods to compare alterntive approaches (Aspects of Daily Life survey-2017)**

| | | Method | Objective | Assumptions/ Conditions |
|---|---|---|---|---|
| First phase | 1 | Tests on the differences in the estimates calculated on the two sample for a set of relevant survey variables | Highlighting the variables for which a suspect of mode effect was significant | Independence between the two samples |
| | 1 | Tests on the response rates in the SM and MM sample | Analysis of the response processes and evaluation of the bias caused by the total nonresponse | Independence between the two samples; |
| | 2 | Indicators of representativeness | | MAR assumption for the response models |
| | 3 | Tests on the differences on estimates of benchmark variables known for selected sample units | | |
| | 1 | Instrumental variable approach | Disentangling measurement and selection effects | Representativity assumption |
| Second phase | 1 | Propensity score | Disentangling measurement and selection effects | MAR assumption for the response models; Balancing assumption |
| | 1 | Multigroup confirmatory factor analysis | Analysis of the equivalence of the measurements in surveys | Identification of the latent structure of the phenomenon |
| Third phase | 1 | Weighting methods as propensity score, calibration | To adjust selection effect | Ignorability of selection mechanism; Measurement error negligible |
| | 1 | Mode calibration | To stabilize the total measurement error | Invariance over time of measurement error |
| | 1 | Multiple imputation (standard) | To adjust measurement effect | MAR assumption |

Aspects of daily life—2017: method to compare alternative mode approaches.

This is an interesting example about the possible use of various methods to achieve different objectives.

## 9.3.5 MIXED-MODE BY BUSINESSES AND HOUSEHOLDS

### 9.3.5.1 Mixed-mode for Business Surveys.

Mixed-mode designs can be applied both in business surveys and in household/individual surveys. Self-administered modes of data collection were already used in businesses early in development of information technology. This is not surprising as businesses were the first to use computers at a wide scale. Therefore, coverage was less of a problem.

The rapid development of the Internet led to new modes of data collection. Already in the 1980s, prior to the widespread introduction of the World Wide Web, e-mail was explored as a new mode of survey data collection. Kiesler and Sproull (1986) describe an early experiment conducted in 1983.

In the first years of the World Wide Web, use of web surveys was limited by the low coverage of the Internet. Clayton and Werking (1998) describe a pilot carried out in 1996 for Current Employment Statistics (CES) program of the US Bureau of Labor Statistics. They expected several advantages like: the lower costs of a web survey, the quick (almost immediate) response to the questions, and the greater flexibility of web survey questionnaires (they could be offered in a form layout or in a question-by-question approach). The drawback was the limited number of respondents having access to the Internet. Only 11% of CES respondents had access to Internet and a compatible browser. This case supports the idea that a mixed-mode approach is needed when using web as a survey tool. In 2004, the situation was much more favorable as shown in Example 9.9.

---

🔲  EXAMPLE 9.9 **Some experiences in the United States and in Europe**

Rosen and Gomes (2004) report on a test conducted in April 2004 by the US Bureau of Labor Statistics. The CES program involves monthly surveys among business establishments. Data are collected on employment, payroll, and working hours. The common way to collect these data is Touchtone Data Entry (TDE). The test investigated whether it was possible to convert TDE respondents to the web respondents. A sample of 3,000 TDE respondents was contacted by telephone, fax and mail (1,000 for each contact method). They were all invited to change to web reporting. The response rate was 74%. All those who agreed to report by web received their initial web account information by mail.

It is worth noting that at the time of this study 71% of the TDE respondents met the criteria imposed for reporting via the web (having access to the Internet, having e-mail at their desk, and using Internet Explorer 6.0 or higher). Of those meeting the eligibility criteria for the web 89% reported that they wanted to switch to web reporting.

An important finding was that offering the web mode had a negative effect on response rates. Initially, the response rate dropped by 8% extensive follow-up procedures were needed to ensure respondents activated their web accounts. As it turned out, fax was the most cost-effective contact method when converting respondents from TDE to web reporting.

As regards Europe, Roos and Wings (2000) conducted a test with Internet data collection at Statistics Netherlands for the Construction Industry. In fact, this was a kind of mixed-mode survey, because respondents could choose between three modes of completing the form:

• Off-line form. The form was sent as an HTML-file that was attached to an e-mail. The form was downloaded, completed off-line, and returned by e-mail;

• Online form. The Internet address of an online web form was sent by e-mail. The form was completed online;

• E-mail form. An e-mail was sent containing the questionnaire in plain text. Respondents clicked the reply button, answered the questions, and sent the e-mail back.

A sample of 1,500 companies was invited to participate in the experiment. About 188 companies were willing and able to participate. Of those, 149 could surf the Internet and 39 only had e-mail. Questionnaire completion times of all three modes were similar to that of a paper form. Respondents preferred the form-based layout over the question-by-question layout. The conclusion of the experiment was that web surveys worked well.

Web surveys began to be run from private firms and marketing research societies to explore various topics of the business behavior. As well, researchers started to administer web surveys to firms in order to study the manufacturing sector. Example 9.10 describes a mixed-mode survey on a sample of businesses.

▣ EXAMPLE 9.10 **A mixed-mode survey of manufacturing firms**

Biffignandi and Fabrizi (2006) conducted a sample survey on manufacturing firms. Data collection was based on a multistage and multimode strategy. Respondents were first contacted by telephone. If contact was established, they were offered the choice to complete the questionnaire on the web or to fill in a paper form that was sent by mail or fax.

Reminders strategies were optimized for each data collection mode separately. There were only two reminders. This decision was based on the analysis of experience in previous surveys: Biffignandi et al. (2004) showed that the effect of successive reminders decreased rapidly.

The decision was also a compromise between costs and response rates. Web respondents received two remainders (after 10 and after 20 days) by e-mail. Fax and mail respondents were reminded by telephone (after 14 and 28 days). For simplicity, reminder periods were taken the same for fax and mail respondents.

The data collection procedure started with telephone recruitment. For the respondents there was a choice of web, mail, or fax. For reminders, e-mail or telephone was used. Figure 9.4 summarized the mixed-mode design of this study.

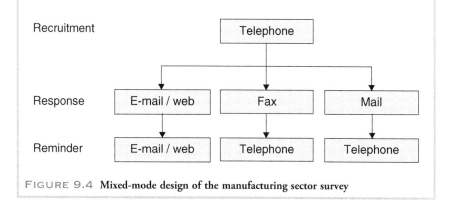

FIGURE 9.4  **Mixed-mode design of the manufacturing sector survey**

In 2005, Meckel, Walters, and Baugh (2005) gave an example of mixed-mode survey with a concurrent design. The target population consisted of small and medium sized companies. Sample companies were sent both a paper questionnaire form and a link to a website. So, they could choose between web and paper. Since then, there has been a constant growing trend in the direction of web surveys and panels in the United States and all around the Europe. This trend has recently included plans to web data collection for the Census. Countries like United Kingdom and Italy, just to mention a few countries, have moved web data collection, mostly in the context of a mixed-mode approach.

The success of the web surveys by businesses relies on the characteristics of the data to be collected (mainly quantitative data already available on a digital support) and to the high penetration of Internet in the businesses. Using the web for self-administered ways of data collected seems an obvious way to go for business survey. Official statistics as well as market research and other research studies are going toward web surveys data collection. Thus, also private research societies need to know and understand the methodological critical aspects  and the trend of this data collection tool.

In 2018, the vast majority (91%) of EU enterprises with at least 10 persons employed used a fixed broadband connection to access the Internet (*Source:* Eurostat (isoc_ciweb)). Breaking down by size, large enterprises present higher penetration than small enterprise. However, the difference is not extremely relevant confirming high diffusion on Internet in every business environment. This share has remained 91–92% since 2014, suggesting that at EU level the uptake of this technology has reached saturation. With almost all enterprises connected to the Internet via broadband, the trend is now businesses have switched to the speed of fixed broadband connections. The share of enterprises using the fastest Internet connections tripled between 2011 and 2018. The share of enterprises using the slowest connection speeds (≥2 Mb/s but <10 Mb/s) fell during successive years between 2011 and 2018 while the share of enterprises using the fastest connections (≥30 Mb/s) constantly increased.

Also the share of employees equipped with company portable devices that allow mobile Internet connection is increasing in the European Union (EU). In 2019, 28% of all persons employed in enterprises in the EU were provided with a company portable computer or other portable device such as a smartphone, which allows Internet connection via mobile telephone networks for business purposes. Among the EU Member States, the Nordic countries and Ireland stood out; more than half of the persons employed in these countries were equipped with company portable devices allowing mobile Internet connection in 2019. This phenomenon is most common in Sweden, with 57% of staff, in Ireland (55%), Finland (54%), and Denmark (52%). In contrast, only 11% of persons employed in enterprises in Bulgaria were provided with such company mobile devices. The shares were also quite low in Cyprus (16%), Greece (17%), Slovakia, Romania, and Portugal (all 18%). Data source is Community survey on ICT usage in enterprise.

The high diffusion of portable devices by the staff of the enterprices is a signal of the importance of staff being able to keep contact with the company as well as with suppliers and customers while on the go. A mobile connection to the Internet also enables staff to check e-mails, access and modify documents or use company software applications when not in office. This allows for a quick communication inside the business to collect by different firm branches the information to be included in the centralized questionnaire to be delivered to the National Statistical Institute. The high diffusion of the Internet by the employeees is also particularly important during the current coronavirus pandemic, with many employees in home office.

All in all, the situation is such that collecting data by enterprise is an already undergoing and positive experience. Moreover, they represent a promising field to experiment new methodological approaches such that advanced measurement error methods and surveys integration with passive data collection tools (sensors, GPS, and so on). For many surveys, all companies have experience with self-completion of forms. Moreover, companies in specific target populations have access to the Internet. So, under-coverage is not a problem. There are additional advantages. One is that companies are asked in many government surveys to provide administrative and financial information. Copying and pasting this information from their systems to a web questionnaire can be a lot easier than having to write it down on a paper form. Another is that checks can be built into the web questionnaire forms. This makes it possible to detect and correct errors while filling the form. Thus, data quality is improved. It should also be noted that many of the questions asked in business surveys are factual questions. This reduces the risk of mode effects in mixed-mode surveys.

It should be noted that the response to a survey may depend on the organization conducting it (a national statistical institute, a private market research company, or an academic researcher). See Snijkers et al. (2013) for an extended discussion of the business surveys issues. Statistical organizations of the government usually have a sampling frame: a business register is available, Moreover, their surveys are often compulsory. So, it is easy to select a sample and contact the businesses in the sample. This simplifies the decision to introduce the web as a data collection mode, even as primary mode. Moreover, businesses are used to completing web forms (for government regulations). As a consequence, they ask for web questionnaires, as they have the perception that this will reduce the burden of survey compliance. Nevertheless, survey design is a crucial aspect to maintain and increase response rates: the communication strategy, the questionnaire, and a web portal are to be carefully planned, whereas a sequential approach focused on web as primary mode is the strategy allowing for cost efficiency as well as satisfactory participation to the survey. With respect to data quality web data collection in business surveys seem to provide improvements, such as consistency among items and reduction in missing values.

Literature shows that mixing mail and web data collection in some case increased response rates. However, if it does not increase response rates, it causes a shift from using the paper version of the questionnaire to the web version of the questionnaire. This is the case also of Example 9.11. To assess quality differences across modes, Thompson, Oliver, and Beck (2015) analyze three quality measures: Unit Response Rate (i.e. unweighted proportion of responding reporting units), Quantity Response Rate (weighted proportion of an estimated total obtained from reported data) and Source Data Item. (i.e. the proportion of responding units that retain their reported data or reported value equals edited value). These indicators help in measuring completeness (URR) and the accuracy with respect to valid reported data (QRR and SDI). The empirical analysis considers two surveys. One is the Annual Capital Expenditure Survey (ACES). This is a mandatory survey collecting data on capital spending for new and used structures and equipment by US businesses. Only larger businesses were considered in the computation of the indicators; quarterly data for the period 2009–2011 have been processed. The second survey is the Quality Services Survey, which estimates the total operating revenue for businesses and the percentage of revenues by class of customers (government, business, consumers and individuals) and other indicators. This is a voluntary survey, based on a mixed mode approach. Different aspects of quality are identified by the indicators and considering the two different surveys; all in all, good quality signals appear.

### EXAMPLE 9.11 Mixed-mode in the Italian SCI survey

Biffignandi and Zeli (2008) investigated response rates in an Italian business survey. This was the SCI survey. SCI stands for "Sistema dei Conti delle

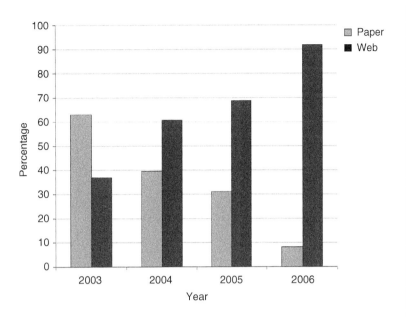

FIGURE 9.5  Response by mode in the SCI survey

Imprese." It is a annual compulsory survey among large enterprises. The survey is conducted by the Italian National Statistical Institute (ISTAT) and collects information about the economic and financial status of companies.

Companies have a choice to respond by paper or via Internet. Starting from 2003 an electronic questionnaire was delivered to businesses and they had the choice to send their answer on a paper questionnaire or via Internet. This study has been a preliminary experiment before moving completely toward a web questionnaire.

An analysis of the response rate by mode, in the first period of introduction of Internet data collection (2003–2006) is interesting. The trend of the response is shown in Figure 9.5. The overall response rate did not change much over the years, but there are considerable changes in the composition of the response by mode.

The share of web mode responses has been constantly growing since 2003. This share increased from a share of 35.7% in 2003 to 92.0% in 2006. This result confirms that companies appreciate this response mode.

The electronic questionnaire form contained consistency checks. There were also facilities to compute totals of amounts. It was expected this would help to improve the quality of the data. Table 9.4 shows the average number of various types of problems by mode. Outliers denote values that differ substantially from the anticipated values. These values are not by definition incorrect. A careful evaluation show whether they are correct and can be left as is, or whether they are wrong and must be corrected.

TABLE 9.4    Detected errors in the SCI survey

| Year | Response mode | Outliers | Substitutions | Corrected errors |
|------|---------------|----------|---------------|------------------|
|      | Paper         | 4.40     | 2.71          | 4.42             |
| 2003 | Web           | 4.03     | 2.40          | 3.68             |
|      | Both modes    | 4.26     | 2.60          | 4.15             |
|      | Paper         | 0.79     | 4.48          | 4.13             |
| 2004 | Web           | 1.03     | 3.30          | 3.15             |
|      | Both modes    | 0.93     | 4.05          | 3.54             |
|      | Paper         | 0.89     | 4.90          | 4.74             |
| 2005 | Web           | 0.85     | 3.39          | 2.88             |
|      | Both modes    | 0.86     | 4.48          | 3.46             |
|      | Paper         | 0.84     | 4.30          | 3.56             |
| 2006 | Web           | 0.85     | 3.31          | 3.05             |
|      | Both modes    | 0.85     | 4.01          | 3.09             |

Substitutions denote incorrect values. Since they are within a predefined threshold from the anticipated value, these values are automatically replaced by the anticipated value. It can be observed in the table that changing to the electronic questionnaire as mode of data collection leads to a reduction of the number of problems.

Private market research companies and academic researchers may have different objectives for their surveys, like market and product analysis, consumer satisfaction, etc. A sampling frame is hardly ever available. And even if a proper sample can be selected, then still the information may be lacking to contact the selected businesses. This is the reason why often the studies are conducted of small target population (where the necessary information can be managed) and why often private companies relate to panel construction for web data collection, instead of sample surveys using mixed-mode. The use of web panels is described in Chapter 14. Because of these reasons, the discussion and examples in this chapter refer to government organization surveys.

The increasing demand for information and the continuing pressure to reduce the administrative burden of companies have encouraged survey organizations all over the world to implement mixed-mode surveys more and more. In the United States, the Bureau of Labor Statistics as well as many federal statistical agencies are conducting mixed-mode surveys where the web is one of the data collection modes. In Europe, several national statistical institutes have also introduced web data collection as a mode. In general, in each country there is an ongoing process of moving respondents from paper to web or mobile surveys. Even for the 2011 census there are examples of mixed-mode data collection including the web.

New ways of data collection are usually welcomed by survey managers, especially when they offer the opportunity to reduce costs or to improve the timeliness of data. The Internet is such an opportunity. Besides the advantages already mentioned, there are other interesting aspects for government survey organizations that conduct large-scale data collection operations, particularly if companies have to participate regularly in several surveys. For example, the US Bureau of Labor Statistics offers companies a common portal or gateway for its web surveys. It is called the "Internet Data Collection Facility" or IDCF. In addition to providing a secure common gateway, the IDCF requires that all survey applications meet internal standards for graphical user interfaces so that on-line questionnaires have the same look and feel. Analogously Italy and many European countries have constructed or have under construction such portals for business web survey management. New approaches in the mixed-mode surveys are expected in the direction of integrating active survey data collection with passive data collection, linking the survey tool to special devices, like sensors, geo-reference tool and other digital forms of measurement relying on passive data collection.

### 9.3.5.2 Mixed-mode for Surveys Among Households and Individuals.

One of the factors standing in the way of large-scale application of the Internet for conducting surveys among households and individuals is the penetration of Internet in the population. Even though the number of persons with Internet access is rapidly growing, population coverage is far from 100%. Moreover, people with Internet access may only have it at their workplace and not at home. This may also prevent participation in web surveys.

Table 9.5 presents an overview of Internet penetration in the world. There is a large variation across regions. In the 2019, penetration is highest in

**TABLE 9.5**   **Internet access in the world**

| World region | Population in 2019 (millions) | Internet users in 2019 (millions) | Internet penetration in 2019 (%) | Internet penetration in 2017 (%) | Share of users in 2019 (%) |
|---|---|---|---|---|---|
| Africa | 132 | 521 | 40 | 27.7 | 12 |
| Asia | 4241 | 2200 | 54 | 45.2 | 51 |
| Europe | 813 | 719 | 88 | 77.4 | 16 |
| Middle East | 212 | 63 | 68 | 59.6 | 4 |
| North America | 364 | 346 | 95 | 56.7 | 7 |
| Latin America and Caribbean | 593 | 205 | 68 | 88.1 | 10 |
| Oceania and Australia | 35 | 21 | 68 | 68.1 | 1 |
| World | 7716 | 4,51 | 58 | 49.6 | 100 |

Source: Based on Internet world stats: usage and Population Statistics, 2017 and 2019.

TABLE 9.6 **Households level Internet access in European countries**

| Country | Internet penetration 2019 |
|---|---|
| Bosnia, Bulgaria, Montenegro, Greece, Serbia | 71–80% |
| Croatia, Portugal, Lithuania, Slovakia, North Macedoia, Romania, Italy, Latvia, Hungary, Malta, Czechia, Poland, Slovenia, Belgium, Estonia, France, Cyprus, Austria | 81–90% |
| United Kingdom, Luxembourg, Finland, Denmark, Netherlands, Ireland, Spain, Kosovo, Germany, Sweden, Iceland | More than 90% |

Source: Based on Eurostat database, Archive:Internet access and use statistics – households and individuals, 2018.

North America, where 95% of the people have access to the Internet (note that in the 2017 the penetration was 56.7%). Penetration is also high in Europe (88% against the 77.4% in 2017). Rather high in Oceania/Australia (68%), in Latin America (68%), and in Middle Est (68%). Penetration is lowest in Asia (54%) and Africa (40%). Comparison with 2017 data shows that a remarkable increase in the penetration has been achieved in the recent years. About 58% of the world population has Internet access in 2019.

Internet penetration in households has a large variation within Europe. See Table 9.6 for details. In several countries penetration is higher than 90%,; in some cases it is very high (98%) like Denmark, The Netherlands, Norway, and Iceland. In most countries, penetration is over 80%, but not more the 90%. On the contrary, in Bosnia, Bulgaria, Montenegro, Greece, Serbia Internet access is still under 81% (anyway higher than 71%, meaning that the diffusion of the Internet is in a growing phase).

Individual usage of the Internet in the European countries has also variability and in general shows level of usage (measured as *Last Internet use: in last 3 months*) very high in a few countries (Sweden, Norway, and Iceland: 98–99%), whereas there are a few countries with the lowest individual Internet usage: Bulgaria (68%) and Bosnia-Herzegovina (70%).

An interesting indicator of the coverage problem related to web or mobile web surveys is the percentage of individuals who have never used Internet. There are some countries (Bulgaria, Greece, Portugal, Montenegro, and Turkey) where in 2019 this percentage is more than 20%. Furthermore, in eight countries this percentage is over 10%.

Summing up, Internet diffusion and usage is growing and quite high, but full coverage is still far at least for most countries. Therefore, web or mobile web surveys are better administered in a mixed-mode to obtain general population estimates. This recommendation becomes especially important in case of international comparisons.

It should be noted that even with (almost) complete Internet coverage there are still sampling problems. Since there is no sampling frame of e-mail address, other modes have to be used to select a sample. Indeed, Couper and Miller (2008) state that "frames of Internet users in a form suitable for sampling do not – and likely will not – exist." Couper (2007) adds that only for specialized populations list may exist, and that there is no analogue for Random Digit Dialing in telephone surveys. There is no algorithm to randomly generate existing e-mail addresses. Further comments on the availability of frames of Internet user are found in Dillman (2017). Currently, even if probability sampling of Internet users would be technically possible, sample will not be representative due to coverage problems (Couper, 2007; Bethlehem, 2009). However, within the framework of probability sampling there are possibilities to conduct mixed-mode surveys. For example, a letter could be sent with a request to go to a survey website, or contact could be made by telephone and an e-mail address asked for in order to send a link to the survey questionnaire.

Keep in mind that there is not much evidence that mixed-mode surveys will increase overall response rates. However, there is clear evidence that including self-administered modes (web, mail) will decrease the costs of the survey (Couper and Miller, 2008). Furthermore, segments of population may be reached only by including mobile web in mixed-mode.

Example 9.12 compares telephone mode with a mixed-mode (web and mail) approach for the ICT survey at Statistics Netherlands.

---

**EXAMPLE 9.12 Response rates in mixed-mode surveys**

Statistics Netherlands carried out some experiments with mixed-mode surveys. One experiment was conducted for the ICT Survey. This survey collects information on the use of computers and Internet in households and by individuals. The regular ICT survey was a CATI survey. The survey was fairly expensive. It also suffered from under-coverage because the sample was selected from the telephone directory. Households with unlisted numbers and mobile-only households could not be selected.

Main objectives of this pilot were:

- to find out what level of response could be obtained;
- to establish whether people without Internet would be properly represented in a mixed-mode survey with the web as the most important mode.

Respondents had the possibility to complete the questionnaire on paper. To prevent those with Internet from responding by paper, the paper questionnaire was not included in the invitation letter. People had to apply for the paper form by returning a stamped return postcard.

The sample was selected from the population register. So, there was no under-coverage. All persons in the sample received an invitation letter by mail. The letter contained the Internet address of the survey and a unique login code.

After one week, a postcard was sent to all nonrespondents with a reminder to complete the survey questionnaire, either by web or mail.

Two weeks after receipt of the invitation letter, the nonrespondents were approached again. They were randomly split into two sub-groups. All people in the first subgroup with a known telephone number were approached by telephone (CATI). The remaining people in the first sub-group received a reminder letter by mail. All people in the other sub-group were reminded by mail. Note that they CATI approach was just to remind people to complete the questionnaire on the Internet, and not to conduct a telephone interview.

Table 9.7 contains the response rates by age, class, and mode. Response rates were substantially lower in the mixed-mode survey than in the single-mode survey. The differences were more extreme for the elderly.

It is also clear from the table that the web response rate decreases with increasing age. The opposite effect can be observed for the mail response rate. These trends can be explained by the lower Internet penetration among the elderly. And even if they have Internet, they may find it easier to complete the paper questionnaire.

The analysis of the results also showed low response rates for single persons, single parents, divorced, widowed people, and ethnic minorities from non-Western countries and people with a low income.

Another conclusion of this experiment was that reminding people by telephone works better than reminding them by mail. The answers to the questions of those reminded by telephone were more accurate than the answer of those reminded by mail. This is probably due to the fact that the personal attention of an interviewer motivates the respondent more. More details can be found in Janssen (2006).

**TABLE 9.7**   **Response rates (%) by age and mode in the ICT-experiment**

|  | Single-mode | | Mixed-mode | | |
| --- | --- | --- | --- | --- | --- |
| Age class | CATI | Web | Mail | Total | Difference |
| 0–25 | 54 | 36 | 6 | 42 | –12 |
| 26–35 | 42 | 26 | 7 | 33 | –9 |
| 36–45 | 52 | 29 | 7 | 36 | –16 |
| 46–55 | 53 | 29 | 9 | 38 | –15 |
| 56–65 | 52 | 23 | 16 | 39 | –13 |
| 66 and older | 59 | 13 | 24 | 37 | –22 |

Problems quoted in Example 9.12 and the rapidly increasing costs of interviewer-assisted surveys caused Statistics Netherlands to look for new ways of collecting data. Mixed-mode surveys seem to be a promising alternative as they may keep response rates at the same level, they can reduce coverage problem, and are cheaper than single-mode CAPI or CATI surveys. Example 9.13 describes the situation of surveys in the Netherlands.

**EXAMPLE 9.13 Surveys in the Netherlands**

Until a few years ago general population surveys were all single-mode surveys. Some like (the first round of) the Labor Force Survey were CAPI surveys, others were CATI surveys. The choice for interviewer-assisted surveys was made, because high data quality was required. Most of the questions in these types of surveys are factual questions, so that there are no problems with socially desirable answers. The sampling frame for both CAPI and CATI surveys is the population register of The Netherlands. Both CAPI and CATI surveys suffer from nonresponse. Moreover, CATI surveys suffer from increasing under-coverage problems, because for increasingly less number of people it is possible to link a telephone number to their addresses. At least 30% of the telephone numbers are unlisted.

Beukenhorst reports on an ongoing program of Statistics Netherlands to move all single-mode surveys to mixed-mode. He states that this did not lead to higher response rates. Respondents who would have responded anyway, have done so again but in a different mode. Face-to-face survey nonrespondents did not change their minds when approached in a follow-up in a different mode. There has also been little success in contacting face-to-face noncontacts by telephone. From field experiments it was concluded that it is possible to use mixed-mode including Internet where the sample consists of individuals, but it is not yet evident if the same applies for household surveys in which all household members have to complete a questionnaire form (for example, the Labor Force Survey).

The redesign program has ended in 2013 and almost all surveys are mixed-mode and have web as the first mode. Currently, there is a lot of interest in adaptive mixed-mode survey design and in multimode case management. The step toward adaptive survey design is logical; modes differ greatly in costs, response rates, and measurement properties. Considering the travel survey. This is now a web only survey that is very likely going to use mobile device applications in the near future.

The ESSnet Project "(DCSS) Data Collection for Social Surveys using Multiple Modes" (see Summary report, 2015) shows the situation of the social survey mixed-mode methodology in several European countries at that time. Some recommendations have been proposed. Anyway, the context, as already explained, is changing fast. MIMOD Project provides a more updated overview of the state of art and trends in the official statistic social surveys. However, discussion and experiments for changes are a currently investigated topic to actualize official statistics surveys by ensuring data quality as well.

An example of a recent study is the Norwegian SILC (Statistics on Income and Living Conditions survey) web pilot held in June 2019. The regular data collection for SILC was currently conducted by telephone only (CATI). The experiment aimed at exploring how using web (CAWI) mode may contribute to cutting costs and increasing representativity of the data. The pilot consisted of three subsamples with differing characteristics in terms of knowledge of the survey and contact mode. For two subsamples the data collection was done by web only, but for one sample we did a short CATI interview and then switched to CAWI (mixedmode). Pilot results were compared to results from the regular CATI SILC from 2018 to 2019. Several aspects are considered in the comparison (see Berg et al., 2019) and problems are highlighted. General conclusion is that the results suggest that Statistics Norway should consider changing to mixed-mode (CATI + CAWI) for SILC. Transition needs several issues be resolved. For example, an important issue is that a case management system and programming software that can handle mixed-mode surveys has to implemented. This example in only one of the numerous experiments in progress by the NSIs.

Summing up decision of moving from one single mode, for example, to mixed-mode survey is a critical one and needs a preliminary study of the impact of the change on quality (response rates, errors, and other characteristics) and costs. The study of the problems and feasibility of the transition to mixed-mode is multifaceted and involves not only official statistics, but also private companies and researchers managing survey data collection.

Another interesting example shows how a cross-national survey have evaluated the possible advantages or disadvantages of moving online or to a mixed-mode approach (Example 9.14).

**EXAMPLE 9.14** Measurement errors in the ESS (European Social Survey) mixed-mode experiment

The ESS has two main objectives. The first one is to explore and explain the interaction between Europe's changing institutions, its political and economic structures, and the attitudes, beliefs and behavior of its people. The second objective is to show that it is possible to conduct a high quality cross-national social survey.

The ESS is funded jointly by the European Commission, the European Science Foundation and Scientific funding bodies in 30 European countries. The survey is conducted bi-annually. In each participating country approximately 2000 respondents have to answer a one-hour questionnaire in a face-to-face interview and to complete a short supplementary questionnaire either face-to-face or on paper.

A major challenge of the European Social Survey is collecting data that allow for comparing countries. This can only be accomplished by applying fundamental methodological principles of survey sampling. The European Social Survey attempts to achieve this by enforcing consistent, standardized procedures and protocols in all participating countries. These include sampling design, questionnaire definition and fieldwork procedures.

Six studies have been carried out recently; three about the effect of mode on measurement and three studies focusing on the challenges of using a different mode and the implications of the errors due the mode design.

As regards measurement error studies, two experiments were focused on measurement equivalence; the first one took place in Hungary in 2003 and considered as modes of data collection (face-to-face, telephone, Internet, and paper self-completion). The second study, conducted in Budapest and Lisbon in 2005, was an experiment designed to investigate the causes of measurement differences between face-to-face and telephone interviewing.

Study 3 was conducted in the UK in 2010, by asking ESS respondents to participate in a short follow-up web survey. The goal was to evaluate measurement differences between face-to-face and web data collection, using a within respondent reinterview design.

As regards the challenges and the errors, the first study assessed the feasibility of using telephone interviews in the European Social Survey, focusing on the effect of varying interview length on respondents' willingness to participate in the survey. The study was carried out in 2006 in five countries: Cyprus, Germany, Hungary, Poland, and Switzerland. The second and third studies tried to evaluate the effects of mixed-mode designs on response rates, representativeness of samples, survey costs, and data quality. The studies were conducted one in the Netherlands in 2008 (parallel to Round 4 of the European Social Survey) and the other one in Estonia, Sweden and the UK in 2012. The Dutch study tested two different mixed-mode (web, telephone, and face-to-face) data collection designs: a "concurrent mode choice design," and a sequential mode choice design. In the other study, countries were allowed to select the ideal mixed-mode design given the country's survey environment. Estonia and the UK focused on a web and face-to-face mix, whereas Sweden tested a telephone and face-to-face design.

The European Social Survey investigates now opportunities and challenges of high-quality data collection by developing a Cross-National Online Survey (CRONOS) probability-based web panel. A harmonized cross-country approach has been used from the recruitment stage to data processing. The effectiveness of the panel recruitment on the back of an existing cross-national survey has been evaluated in terms of costs, sample representativeness, participation (and attrition rates) and data quality.

At present, mixed-mode consists a follow-up online panel based on the same respondents over time. At the moment, there is no intention to replace face-to-face data collection with an online survey.

CRONOS data collection was run alongside ESS Round 8 in 2016. After completing the ESS face-to-face interview, respondents in Estonia, Great Britain and Slovenia, 18 or older, were invited to participate in six 20-minute online surveys over a time period of 12 months. Respondents who did not have Internet access for private use were offered a tablet and an Internet connection. Field activities took place between December 2016 and February 2018. Waves 1 to 6 included around 100 questions on diverse topics, often borrowed from high-standard cross-national surveys (e.g., European Values Study, Generations and Gender Programme, International Social Survey Programme, and European Quality of Life Survey).

For details see Villar and Fitzgerald (2017), Cernat and Revilla (2020).

# 9.4 Application

The ideal situation for a web survey is to have a sampling frame containing e-mail addresses of all members of the target population. This makes it easy to inform sample persons they have been selected in the survey. Moreover, it is also very easy for the sample persons to access the survey questionnaire. It is just a matter of clicking on the link included in the e-mail. Often such sampling frame of e-mail addresses does not exist. Then some other means have to be deployed to get into contact with those selected for the survey. The least expensive way to do this is by mail.

Schonlau, Asch, and Du (2003) conducted a mixed-mode experiment in which they tested this. A sample of high school students was approached by mail. They were invited to complete a survey questionnaire on the web. Objective of this experiment was to obtain answers to three questions:

- Can the majority of respondents be convinced by mail to participate in a web survey?
- Are reminders by telephone effective for increasing response rates?
- Are reminders by mail effective for increasing response rates?

The target population consisted of high school students graduating in 2001. There was a sampling frame with addresses. A sample of 1,750 students was selected.

The fieldwork of the survey consisted of two stages. In the first stage, potential respondents were asked by mail to complete the questionnaire on the Internet. There was a possibility to fill in a paper questionnaire. It was not included in the invitation letter, but it could be requested. Objective of this design was to stimulate the students to respond by web as much as possible, because conducting a web survey is cheaper than conducting a mail survey.

After some time, reminders (without a paper copy of the questionnaire) were sent to all students in the sample. Moreover, a random sample of nonrespondents were contacted and reminded by telephone.

In the second stage of the experiment, all remaining nonrespondents were sent a paper questionnaire form. So, they had a choice to complete either the mail or the web questionnaire. In addition, an incentive was sent to a random sample of nonrespondents. They incentive was a McDonald gift certificate of three dollars.

Two reminders were sent. The first one was just a postcard, and the second one was letter that included a paper copy of the questionnaire. The experimental design is summarized in Table 9.8.

It turned out that of the respondents only 35% of those did so by the web. Although they were not encouraged to do so, 65% responded by mail. Quigley et al. (2000) report on a similar experiment where 73% of the respondents preferred the web. However, there were differences between the experiments. One is that different target populations were approached (students versus the military), and the second is that Quigley et al. offered the paper questionnaires only very late in the fieldwork period.

The fact that only about one-third of the students preferred the web was seen as disappointing. It was expected the students were intensive computer and Internet users. Therefore, it could have been attractive for them to do the survey on the Internet.

The use of incentives increased the response rate. However, this increase only occurred for the mail mode. There was no effect for the web mode. Two factors may explain this. One, by the time the mail response and the incentives were introduced, the survey had been in the field already for a considerable period of time.

**TABLE 9.8** Modes, reminders and incentives in different stages of the survey

| Stage | Mode | Reminder/incentives |
|---|---|---|
| 1 | Invitation letter sent by mail. No paper questionnaire included. | All nonrespondents were reminded by mail (no paper questionnaire included.) |
| | Request to complete the web questionnaire. A paper questionnaire could be requested. | A random subsample of nonrespondents were reminded by telephone. |
| 2 | Letter to stage 1 nonrespondents with a paper copy of the questionnaire. Choice of web or mail response. | First reminder: postcard to all remaining nonrespondents. |
| | | Second reminder: letter with paper questionnaire to all remaining nonrespondents. |
| | | A random subsample of stage 1 nonrespondents received a $3 incentive by mail. The paper questionnaire was included in the letter. |

This may have made the students less aware of the fact that they were expected to complete a web survey questionnaire. Second, the mode used for sending the incentives (mail) may have affected the respondents' choice of the response mode (mail).

Reminding the students by telephone was very effective. There was a substantial increase of the response rate in the web mode. For those receiving the telephone call, the subsequent response rate was 30%. The subsequent response rate of the other students was only 18%.

## 9.5 Summary

A growing interest in conducting mixed-mode surveys where the web is one of the data collection modes is now recognized both in official statistics and in private business and research institutions. Furthermore, referring to mobile web, i.e. the possibility of using mobile devices to complete the questionnaire is becoming a practice the surveyor needs to face and adopt. This opens new possibilities, such as survey integration with alternative tools (GPS for example) or simply to be able to let people participate in any time and in any location to the survey. In this context new challenges have to be considered; however, the clear direction is that the surveyor has to go toward the use of a mixed-mode approach, and a mixed-device approach, and also multisource integration approach. Using mixed-mode with web component has been seen as a potential means for increasing response rates, reducing under-coverage problems, and decreasing survey costs. Literature has shown that mixed-mode surveys might not be very successful in increasing response rates. Particularly, response rates for the web mode are in some cases disappointingly low. Also providing sampled units with a choice of mode does not appear to increase overall response rates. Therefore, mixed-mode surveys are not a solution for declining response rates. There are however other advantages making mixed-mode surveys appealing. For example, they are very successful for reducing survey costs. This is a consequence of replacing (part of) interviewer-assisted interviewing by self-administered completion of questionnaires.

There are two approaches for implementing mixed-mode designs: a sequential design and a concurrent design. In a sequential mixed-mode survey, the sample persons are approached in one mode. Nonrespondents are followed up in a different mode. This process can be repeated for a series of modes. For the concurrent (parallel) approach, the sample is divided into groups and each group is approached with a different mode. Many different modes can be combined in a mixed-mode survey: face-to-face, telephone, mail, fax, IVR, TDE, etc. Therefore, many different designs are possible. Sometimes also sequential modes and concurrent modes are combined.

No best mixed-mode design exists. It depends on the specific research context, the topics of the surveys and the type of questions asked (factual or attitudinal).

Different phenomena may have different effects in different modes. For example, interviewer-assisted surveys may suffer from socially desirable answers.

Mixed-mode surveys may suffer from selection effect, due to the attitude to participate using different modes; considering this possible effect, it is possible to try and obtain the most efficient effect for the survey participation. There is another effect, the measurement effect, i.e. the same question is answered differently in a different mode This produces the mode effect. A good questionnaire design could help in preventing this effect, but it is not possible to remove it. It is not easy to correct for mode effects because they are confounded with other effects such as selection effects.

Recent literature has been focused on preventing and measuring mode effects, Also, attention on the use of mobile web has been growing, and some studies on different mobile devices have also been published. All this aspects have been discussed in this chapter.

The mixed-mode approach is promising both for household/individual surveys and business surveys. Although it may be hard to disentangle some of the methodological problems, it is important to continue research in both application areas. Cross-fertilization of the results of experiments in business and household/individual surveys will help to improve the quality of future mixed-surveys.

In summary, there is an increasing interest in use of mixed-mode surveys for data collection. Although these surveys may help to reduce survey costs, there are still some estimation problems that have not yet been solved. Further research is required in this area. A general recommendation is important: to undertake a survey the critical issues and approaches discussed in this chapter has to be borne in mind, and, eventually, a deeper analysis of the existing experiment should be considered. As well, researchers should have in mind the complex and numerous issues discussed in this chapter which affect the design of a mixed-mode survey and consequently the final results.

Some rules of thumb are useful when designing a mixed-mode survey. The specific research context should drive the final choice of the survey design. Thus, carefully reading the recommendations and experiments presented in this chapter one should define its own rule of type for the specific survey.

The variety of modes and mode combinations will continue to expand as survey researchers adapt to societal and technological changes. There is a constant tension between the need to innovate or invent new methods, whether for market share or peer recognition, and the importance of maintaining comparability across time for key longitudinal estimates and analyses. Some might rush out to try a new method of data collection, while others might wait to assemble the evidence before carefully transitioning from one approach to another. This tension is a healthy one. Both approaches are valuable for the profession. The one constant in survey research seems to be change, and we need to find ways to adapt existing methods and develop new methods, in response to both external changes and methodological research on ways to improve surveys.

## KEY TERMS

**Concurrent mixed-mode**: A mixed-mode design in which respondents can choose among alternative modes of data collection, or in which different groups are approached by different modes.

**Mixed-mode survey**: A survey in which various modes of data collection are combined. Modes can be used concurrently (different groups are approached by different modes) or sequentially (nonrespondents of a mode are reapproached by a different mode).

**Mixed-device survey**: A special form of mixed-mode design is a mixed-device survey. Mobile phones, tablets, and other mobile devices can be used in addition to regular desktop PCs to complete the questionnaire.

**Mixed-mode systems**: Mixed-mode system or multimode system denote either communication with respondents (mixed-mode communication) or data collection (mixed-mode data collection).

**Mode effect**: The phenomenon that a question is answered differently when asked in a different mode. Sometimes this term is used in a wider context, in which it denotes the combined differences of the modes. This includes, for example, differences in coverage and difference in response rates.

**Sequential mixed-mode**: A mixed-mode design in which nonrespondents from one mode are reapproached in another mode.

**Socially desirable answer**: The tendency that respondents give answers that will be viewed as more favorable by others. This particularly happens for sensitive questions in interviewer-assisted surveys. It is a typical cause of mode effects.

**Unimode design:** A mixed-mode survey design in which the same questionnaire is used in each mode. Following the guidelines for unimode questionnaires, the questions are defined in such a way that mode effects are minimized.

## EXERCISES

**Exercise 9.1**   What is a unimode survey?

a. A unimode survey is a special type of mixed-mode survey. Questions are defined in exactly the same way in each mode;

b. A unimode survey is a special type of mixed-mode survey. Questions may be defined differently in each mode, but they measure the same concept;

c. It is a sequential mixed-mode survey in which the first mode produces the highest quality data;

d. It is a different term for a single mode survey, i.e. a survey with only one mode of data collection.

**Exercise 9.2**   Can the bias of estimators due to a mixed-mode approach be corrected by means of a weighting adjustment?

**a.** Yes, weighting adjustment can be successful if the bias is caused by under-coverage and/or nonresponse;

**b.** Yes, weighting adjustment can be successful if the bias is caused by mode effects;

**c.** Yes, weighting adjustment will always be able to reduce a mixed-mode bias;

**d.** No, weighting adjustment will never be able to reduce a mixed-mode bias.

**Exercise 9.3**   A survey was conducted using web and mail for data collection concurrently. An invitation letter was sent by ordinary mail to a sample of 2,000 potential respondents. They were offered a choice of completing a paper questionnaire form or a web form. About 1200 completed questionnaires were returned by mail, and 800 forms were completed on the web. Calculate the overall response rate and the response rate for each mode.

**Exercise 9.4**   A survey was conducted using web and CATI for data collection sequentially. An invitation letter was sent by ordinary mail to a sample of 2,000 potential respondents. They were asked to complete the questionnaire on the web. About 420 persons did so. After two weeks all remaining respondents were called by telephone. About 640 persons completed the questionnaire by telephone. Calculate the overall response rate and response rate by mode.

**Exercise 9.5**   According to the literature, what can be said if the response rates of CATI, web, and IVR are compared?

**a.** In CATI lower than in web;

**b.** In IVR higher than in CATI;

**c.** In CATI higher than in web;

**d.** In IVR lower than in web.

**Exercise 9.6**   According to the literature, what can be said if the item nonresponse rates of CATI, web, and IVR are compared?

**a.** In CATI higher than in web;

**b.** In IVR higher than in CATI;

**c.** In CATI lower than in web;

**d.** In IVR lower than in web.

**Exercise 9.7**   According to the literature, what can be said if the accuracy of the answers of CATI, web, and IVR are compared?

**a.** In CATI higher than in web;
**b.** In IVR higher than in CATI;
**c.** In CATI lower than in web;
**d.** In IVR lower than in web.

**Exercise 9.8**   If a CAPI survey is replaced by a sequential mixed-mode survey, with web as the first mode and CAPI as the second mode. What can be said about the costs and the response rate of the new survey?

**a.** Both response rate and the costs will go up;
**b.** Both response rate and the costs will go down;
**c.** The response rate will go up and the costs will go down;
**d.** The response rate will go down and the costs will go up.

**Exercise 9.9**   Which of the following phenomena cannot cause a mode effect in a mixed-mode survey?

**a.** Straight-lining in matrix questions;
**b.** Memory effects in recall questions;
**c.** Response order effects;
**d.** Socially desirable answers to sensitive questions.

**Exercise 9.10**   Which of the following statements describe advantages of web surveys over mail surveys?

**a.** Checks can be included in the questionnaire;
**b.** The questionnaire can be completed quickly;
**c.** Dynamic routing can be implemented in the questionnaire;
**d.** There are no under-coverage problems.

## REFERENCES

AAPOR (2009), *Standard Definitions: Final Dispositions of Case Codes and Outcome Rates for Surveys*, 6th ed. The American Association for Public Opinion Research, Deerfield, IL.

AAPOR (2019), *Report of the AAPOR Task Force on Transitions from Telephone Surveys to Self-Administered and Mixed-Mode Surveys*. The American Association for Public Opinion Research, Deerfield, IL.

American Community Survey (2014), Design and methodology, chapter 7: data collection and capture for housing units. Retrieved from https://www2.census.gov/programs-surveys/acs/methodology/design_and_methodology/acs_design_methodology_ch07_2014.Pdf.

Antoun, C. & Cernat, A. (2020), Factors Affecting Completion Times: A Comparative Analysis of Smartphone and PC Web Survey. *Social Science Computer Review*, 38, 4, pp. 477–489.

Antoun, C., Couper, M., & Conrad, F. (2017), Effects of Mobile Versus Pc Web on Survey Response Quality. *Public Opinion Quarterly, Special Issue: Survey Research Today and Tomorrow*, 81, pp. 280–306.

Ariel, A., Giesen, D., Kerssemakers, F., & Vis-Visschers, R. (2008), *Literature Review on Mixed-mode Studies. Internal Report DMH-2008-04-16-RVCS*. Statistics Netherlands, Heerlen.

Berg, N., Snellingen, B. L., Rossbach, K., & With, L. (2019), *Report from the Norwegian SILC Web Pilot 2019*. Statistics Norway.

Bethlehem, J. (2009), *Applied Survey Methods – A Statistical Perspective*. Wiley, Hoboken, NJ.

Bethlehem, J., Cobben, F., & Schouten, B. (2011), *Handbook on Nonresponse in Household Surveys*. Wiley, Hoboken, NJ.

Bianchi, A. & Biffignandi, S. (2013), Web Panel Representativeness. In: Giudici, P., Ingrassia, S., & Vichi, M. (eds.), *Statistical Models for Data Analysis*. Springer, Berlin.

Bianchi, A. & Biffignandi, S. (2017), Effects of Targeted Designs in Surveys and Longitudinal Studies,. *Statistical Journal of the International Association for Official Statistics*, 33, pp. 459–467.

Bianchi, A. & Biffignandi, S. (2020), Survey Experiments on Interactions: A Case Study of Incentives and Modes. In: Lavrakas, P. J., de Leeuw, E., Holbrook, A., Kennedy, C., Traugott, M. W., & West, B. T. (eds.), *Experimental Methods in Survey Research: Techniques that Combine Random Sampling with Random Assignment*. Wiley, Hoboken, NJ.

Bianchi, A., Biffignandi, S., & Lynn, P. (2017), Web-CAPI Sequential Mixed-mode Design in a Longitudinal Survey: Effects on Participation Rates, Sample Composition and Costs. *Journal of Official Statistics*, 33, 2, pp. 385–408.

Biemer, P. (2010), *Latent Class Analysis of Survey Error*. Wiley, Hoboken, NJ.

Biemer, P. & Lyberg, L. (2003), *Introduction to Survey Quality*. Wiley, Hoboken, NJ.

Biemer, P., Murphy, J., Zimmer S., Berry, C., Deng, G., & Lewis, K. (2018), Using Bonus Monetary Incentives to Encourage Web Response in Mixed-mode Household Surveys. *Journal of Survey Statistics and Methodology*, 6, 2, pp. 240–261.

Biffignandi, S. & Fabrizi E. (2006), Mixing Web and Paper Based Data Collection Modes in a Survey on Manufacturing Firms. *Proceedings of the European Conference on Quality Survey Statistic – Q2006*, Cardiff, UK.

Biffignandi, S., Pratesi, M., Lozar Manfreda, K., & Vehovar, V. (2004), List Assisted Web Surveys: Quality, Timeliness and Nonresponse in the Steps of the Participation Flow. *Journal of Official Statistics*, 20, pp. 451–465.

Biffignandi, S. & Zeli, A. (2008), Statistical Quality Analysis in a Mixed-mode Survey. *Paper Presented at the Business Data Collection Workshop*, Ottawa, Canada.

Blom, A. G., Bosnjak, M., Cornilleau, A., Cousteaux, A. S., Das, M., Douhou, S., & Krieger, U. (2015), A Comparison of Four Probability Based Online and Mixed-mode Panels in Europe. *Social Science Computer Review*, 34, 1, 8–25.

Bosnjak, M., Dannwolf, T., Enderle, T., Schaurer, I., Struminskaya, B., Tanner, A., & Weyandt, K. (2018), Establishing an Open Probability-Based Mixed-mode Panel of the General Population in Germany: The GESIS Panel. *Social Science Computer Review*, 36, 1, pp. 103–115.

Bowling, A. (2005), Mode of Questionnaire Administration Can Have Serious Effects on Data Quality. *Journal of Public Health*, 27, pp. 281–291.

Buelens, B. & Van den Brakel, J. (2015), Measurement Error Calibration in Mixed-mode Sample Surveys. *Sociological Methods & Research*, 44, 3, pp. 389–426.

Buelens, B. & Van den Brakel, J. (2017), Comparing Two Inferential Approaches to Handling Measurement Error in Mixed-Mode Surveys. *Journal of Official Statistics*, 33, 2, pp. 513–531.

Calinescu, M. & Schouten, B. (2016), Adaptive Survey Designs for Non-response and Measurement Error in Multi-purpose Surveys. *Survey Research Methods*, 10, 1, pp. 35–47.

Callegaro, M. (2013) From Mixed-mode to Multiple Devices. Web Surveys, Smartphone Surveys and Apps: Has the Respondent Gone Ahead of Us in Answering Surveys?. *International Journal of Market Research*, 55, 2, pp. 317–320.

Cernat, A. (2015), Impact of Mixed-modes on Measurement Errors and Estimates of Change in Panel Data. *Survey Research Methods*, 9, 2, pp. 83–99.

Cernat, A., Couper, M., & Ofstedal, M. (2016), Estimation of Mode Effects in the Health and Retirement Study Using Measurement Models. *Journal of Survey Statistics and Methodology*, 4, 4, pp. 501–524.

Cernat, A. & Oberski, D. (2018), Estimating stochastic survey response errors using the multitrait-multierror model. National Centre for Research Methods, NCRM, Working Paper. Retrieved from http://eprints.ncrm.ac.uk/4156/.

Cernat, A. & Oberski, D. (2020), *Extending the Within Persons Experimental Design: The Multitrait-Multierror (MTME) Approach. Experimental Methods in Survey Research*, New York, NY: John Wiley & Sons.

Cernat, A. & Revilla, M. (2020), Moving From Face-To-Face To A Web Panel: Impacts On Measurement Quality. *Journal of Survey Statistics and Methodology*, 0, pp. 1–19.

Cernat, A. & Sakshaug, J. (2020), The Impact of Mixed Modes on Multiple Types of Measurement Error. *Survey Research Methods*, 14, 1, pp. 79–91.

Christian, L., Dillman, D., & Smyth, J. (2005), Instructing Web and Telephone Respondents to Report Date Answers in Format Desired by the Surveyor. *Social and Economic Sciences Research Center Technical Report 05-067*, Washington State University, Pullman, WS.

Clayton, R. & Werking, G. (1998), Business Surveys of the Future: The World Wide Web as a Data Collection Methodology. In: Couper, M., Baker, R., Bethlehem, J., Clark, C., Martin, J., Nicholls II, W., & O'Reilly, J. (eds.), *Computer Assisted Survey Information Collection*. Wiley, New York, NY, pp. 543–562.

Conrad, F., Schober, M., Antoun, C., Hupp, A. L., & Yan, H. (2017a), Text Interviews on Mobile Devices. In Biemer, P., de Leeuw, E., Eckman, S., Edwards, B., Krauter, F., Lyberg, L., Tucker, N. & West, B. (Eds.), *Total Survey Error in Practice* (pp. 299–318). Hoboken, NJ: John Wiley & Sons.

Conrad, F., Schober, M., Antoun, C., Yan, H., Hupp, A., Johnston, M., Ehlen, P., Vickers, L., & Zhang, C. (2017b), Respondent Mode Choice in a Smartphone Survey, *Public Opinion Quarterly* 81, Special Issue, pp. 307–337.

Couper, M. (2007), Web Surveys in a Mixed-mode World. *Paper presented at the ESRC Conference 'Survey Research in the 21st Century: Challenges and Opportunities'*, London, UK.

Couper, M. (2011), The Future of Modes of Data Collection, *Public Opinion Quarterly*, 75, pp. 889–908.

Couper, M., Antoun, C., & Mavletova, A. (2017), Mobile Web Surveys. In Biemer, P., de Leeuw, E., Eckman, S., Edwards, B., Krauter, F., Lyberg, L., Tucker, N., & West, B. (eds.), *Total Survey Error in Practice*. Wiley, Hoboken, NJ, pp. 133–154.

Couper, M. & Miller, P. (2008), Web Survey Methods Introduction. *Public Opinion Quarterly*, 72, pp. 831–835.

Daikeler, J., Bošnjak, M., & Lozar Manfreda, K., (2020), Web Versus Other Survey Modes: An Updated and Extended Meta-Analysis Comparing Response Rates. *Journal of Survey Statistics and Methodology*, 8, 3, pp. 513–539.

de Bruijne, M. & Wijnant, A. (2013), Comparing Survey Results Obtained via Mobile Devices and Computer: An Experiment with a Mobile Web Survey on a Heterogeneous Group of Mobile Devices Versus A Computer-Assisted Web Survey. *Social Science Computer Review*, 31, 4, pp. 482–504.

de Leeuw, E. D. (1992), *Data Quality in Mail, Telephone, and Face-to-Face Surveys*. TT-Publications, Amsterdam.

de Leeuw, E. D. (2005), To Mix or Not to Mix Data Collection Modes in Surveys. *Journal of Official Statistics*, 21, pp. 233–255.

de Leeuw, E. D. (2008), Choosing the Method of Data Collection. In: de Leeuw, E., Hox, J. & Dillman, D. (eds.), *International Handbook of Survey Methodology*. Lawrence Erlbaum Associates, New York, NY, pp. 113–135.

de Leeuw, E. D. (2018), Mixed-Mode: Past, Present, Future. *Survey Research Methods*, 12, 2, pp. 9999–10013.

de Leeuw, E. D. & De Heer W. (2002), Trends in Household Survey Nonresponse: A Longitudinal and International Comparison. In: Groves, R. M., Dillman, D. A., Eltinghe, J. I., & Little, R. J. A. (eds.), *Survey Nonresponse*. Wiley, New York, NY.

de Leeuw, E. D., Hox, J., and Luiten, A. (2018), International Nonresponse Trends across Countries and Years: An analysis of 36 years of Labour Force Survey data. Survey Insights: Methods from the Field. https://dspace.library.uu.nl/handle/1874/375517.

de Leeuw, E. D. & Toepoel, V. (2018), Mixed-Mode and Mixed-Device Surveys. In: Vannette, D., Krosnick, J. (eds.), *The Palgrave Handbook of Survey Research*. Palgrave Macmillan, Cham.

Dillman, D. (2007), *Mail and Internet Surveys: The Tailored Design Method*, 2nd edition. Wiley, New York, NY.

Dillman, D. (2017), The Promise and Challenges of Pushing Respondents to the Web in Mixed-mode Surveys. *Survey Methodology*, 43, 1, pp. 3–30.

Dillman D., Clark, J., & West, K. (1995), Influence of an Invitation to Answer by Telephone on Response to Census Questionnaire. *Public Opinion Quarterly*, 58, pp. 557–568.

Dillman D. & Edwards, M. (2016), Designing a Mixed-Mode Survey. In: Wolf, C., Joye, D., Smith, T., & Fu Y. (eds.), *The SAGE Handbook of Survey Methodology*. Sage Publications, London.

Dillman, D., Smyth, J., & Christian, L. (2014), *Internet, Mail and Mixed-mode Surveys: The Tailored Design Method*. Wiley, Hoboken, NJ.

Griffin, D., Fischer, D., & Morgan, M. (2001), Testing an Internet Response Option for the American Community Survey. *Paper presented at the Annual Conference of the American Association for Public Opinion Research*, Montreal, Canada.

Groves, R., Fowler, F. J., Couper, M. D., Lepkowski, P., Singer, J. M., & Tourangeau, R. (2004), *Survey Methodology*. Wiley, New York, NY.

Groves, R. & Heringa, S, (2006), Responsive Design for Household Surveys: Tools for Actively Controlling Errors and Costs. *Journal of Royal Statistical Society, Series A Statistics in Society*, 169, 3, pp. 439–457.

Groves, R. & Lyberg, L. (2010), Total Survey Error Past, Present and Future, *Public Opinion Quarterly*, 74, 5, pp. 849–879.

Haan, M., Lugtig, P., & Toepoel, V., (2019), Can We Predict Device Use? An Investigation into Mobile Device Use in Surveys *International Journal of Social Research Methodology* 22, 5, 517–531.

Hochstim, J. (1967), A Critical Comparison of Three Strategies of Collecting Data From Households. *Journal of American Statistical Association*, 62, pp. 976–989.

Hox, J., de Leeuw, E., & Klausch, T. (2017), Mixed-mode Research. Biemer, P., de Leeuw, E., Eckman, S., Edwards, B., Krauter, F., Lyberg, L., Tucker, N., & West, B., *Total Survey Error in Practice*. Wiley, New Jersey, NJ.

Jäckle, A. Lynn, P., & Burton, J. (2015), Going Online with a Face-to-Face Household Panel: Effects of a Mixed-mode Design on Item and Unit Non-Response. *Survey Research Methods*, Vol. 9, No. 1, pp. 57–70.

Jäckle, A, Roberts, C., & Lynn, P. (2010), Assessing the Effect of Data Collection Mode on Measurement. *International Statistical Review*, 8, 1, pp. 3–20.

Janssen, B. (2006), Web Data Collection in a Mixed-mode Approach: An Experiment. *Paper presented at the European Conference on Quality in Official Statistics (Q2006)*, Cardiff, UK.

Kappelhof, J. (2017), Survey Research and the Quality of Survey Data Among Ethnic Minorities. In: Biemer, P., de Leeuw, E., Eckman, S., Edwards, B., Krauter, F., Lyberg, L., Tucker, N., & West, B. (eds.), *Total Survey Error in Practice*. Wiley Series in Survey Methodology. Wiley, Hoboken, NJ, pp. 235–252.

Kiesler, S. & Sproull, L. (1986), Response Effects in the Electronic Survey. *Public Opinion Quarterly*, 50, pp. 402–413.

Klausch, T., Hox, J. J., & Schouten, B. (2013), Measurement Effects of Survey Mode on the Equivalence of Attitudinal Rating Scale Questions. *Sociological Methods & Research*, 42, pp. 227–263.

Klausch, T., Hox, J. J., & Schouten, B. (2015a), Selection Error in Single and Mixed-mode Surveys of the Dutch General Population. *Journal of the Royal Statistical Society: Series A*, 178, 4, pp. 945–961.

Klausch T., Schouten, B., Buelens B., & Van Den Brakel, J. (2017), Adjusting Measurement Bias in Sequential Mixed-Mode Surveys Using Re-Interview Data. *Journal of Survey Statistics and Methodology*, 5, pp. 409–432.

Klausch, T., Schouten, B., & Hox, J. (2015b), Evaluating Bias of Sequential Mixed-mode Designs Against Benchmark. *Surveys Sociological Methods & Research*, 46, 3, pp. 456–489.

Knapp, H. & Kirk, S. (2003), Using Pencil and Paper, Internet and Touch-Tone Phones for Self-Administered Surveys: Does Methodology Matter?. *Computers in Human Behavior*, 19, pp. 117–134.

Kolenikov, S. & Kennedy, C. (2014), Evaluating three Approaches to Statistically Adjust for Mode Effects. *Journal of Survey Statistics and Methodology*, 2, 2, pp. 126–158.

Kraan, T., Van den Brakel, J., Buelens, B., & Huys, H. (2010), Social Desirability Bias, Response Order Effect and Selection Effects in the New Dutch Safety Monitor. Discussion Paper 10004, Statistics Netherlands, The Hague/Heerlen, The Netherlands.

Kreuter, F., Presser, S., & Tourangeau, R. (2008), Social Desirability Bias in CATI, IVR, and Web Surveys, The Effects Of Mode And Question Sensitivity. *Public Opinion Quarterly*, 72, pp. 847–865.

Lugtig, P. & Toepoel, V. (2016), The Use of PCs, Smartphones, and Tablets in a Probability-Based Panel Survey: Effects on Survey Measurement Error. *Social Science Computer Review*, 34, pp. 78–94.

Luiten, A., Hox, J. & de Leeuw, E. (2020). *Journal of Official Statistics*, 6, 3, 2020, pp. 469–487.

Luiten, A. & Schouten, B. (2013), Tailored Fieldwork Design to Increase Representative Household Survey Response: An Experiment in the Survey of Consumer Satisfaction, *Journal of the Royal Statistical Society, Series A*, 176, 1, pp. 169–189.

Lynn, P. (2017), From Standardized to Targeted Survey Procedure for Tackling Nonresponse and Attrition. *Survey Research Methods*, 11, 1, pp. 93–103.

Lynn, P., Beerten, R., Laiho, J., & Martin, J. (2002), Towards Standardisation of Survey Outcome Categories and Response Rate Calculations. *Research in Official Statistics* 1, pp. 63–86.

Mariano, L. & Elliot, M. (2017), An Item Response Theory Approach to Estimating Survey Mode Effects: Analysis of Data from a Randomized Mode. *Experimental Journal of Survey Statistics and Methodology* 5, 2, pp. 233–253.

Maslovskaya, O., Durrant, G., Smith, P., Hanson, T., & Villar, A. (2019), What Are the Characteristics of Respondents Using Different Devices in Mixed-device Online Surveys? Evidence from Six UK Surveys. *International Statistical Review*, 87, 2, pp. 326–346.

Meckel, M., Walters, D., & Baugh, P. (2005), Mixed-Mode Surveys Using Mail and Web Questionnaires. *The Electronic Journal of Business Research Methodology*, 3, pp. 69–80.

Medway, R. & Fulton, J., (2012), When More Gets You Less. A Meta-analysis of the Effect of Concurrent Web Options on Mail survey Response Rates. *Public Opinion Quarterly*, 76, 4, pp. 733–746.

MIMOD (Mixed-mode Design on Social Surveys) ESSnet Project (2017–2019), Final Report and Other Deliverables. Retrieved from https://www.istat.it/it/ricerca-in-istat/ricerca-internazionale/essnet-e-grants.

Park, S., Kim, J., & Park, S. (2016), An Imputation Approach for Handling Mixed-Mode Surveys. *The Annals of Applied Statistics*, 10, 2, pp. 1065–1085.

Pierzchala, M. (2006), Disparate Modes and Their Effect on Instrument Design. *Proceedings of the 10th International Blaise Users Conference*, Arnhem, The Netherlands. http://www.blaiseusers.org/2006/Papers/209.pdf.

Quigley, B., Riemer, R., Cruzen, D., & Rosen, S. (2000), Internet Versus Paper Survey Administration: Preliminary Finding on Response Rates. *Paper presented at the 42nd Annual Conference of the International Military Testing Association*, Edinburgh, Scotland.

Read, B. (2019), The Influence of Device Characteristics on Data Collection Using a Mobile App. *Understanding Society Working Paper Series, n. 2019-01.*

Revilla, M., Couper, M., & Ochoa, C. (2018), Giving Respondent Voice? The Feasibility of Voice Input for Mobile Web Surveys. *Survey Practice*, 11, 2, pp. 1–11.

Roberts, C. (2007), *Mixing Modes of Data Collection in Surveys: A Methodological Review. Report, Economic and Social Research Council, National Centre for Research Methods*, City University, London, UK.

Roos, M. & Wings, H. (2000), Blaise Internet Services Put to the Test: Web-surveying the Construction Industry. *Proceedings of the 6th International Blaise Users Conference*, Kinsale, Ireland.

Rosen, R. & Gomes, T. (2004), Converting CES Reporters from TDE to Web Data Collection. *Paper presented at the Joint Statistical Meeting*, Toronto, Canada.

Scherpenzeel, A. & Das, M. (2011), True Longitudinal Based Probability Based. In Das, M., Ester, P. & Kaczmirek, L., *Social and Behavioral Research and the Internet: Advances in Applied Methods and Research Strategies*. Routledge, Boca Raton, FL, pp. 74–104.

Schonlau, M., Asch, B.J. & Du, C. (2003), Web Surveys as Part of a Mixed-mode Strategy for Populations than Cannot Be Contacted by E-mail. *Social Science Computer Review* 21, pp. 218–222.

Schouten, B., van den Brakel, J., Buelens, B., van der Laan, J., & Klausch, T. (2013), Disentangling Mode-specific Selection and Measurement Bias in Social Surveys. *Social Science Research*, 42, 6, pp. 1555–1570.

Shih, T. & Fan, X. (2007), Response Rates and Mode Preferences in Web-Mail Mixed-Mode Surveys: A Meta-Analysis. *International Journal of Internet Science*, 2, pp. 59–82.

Snijkers, G., Haraldsen, G., Jones, J., & Willimack D. (2013), *Designing and Conducting Business Surveys*. Wiley, New Jersey, NJ.

Struminskaya, B., Weyandt, K., & Bosnjak, M. (2015), The Effects of Questionnaire Completion Using Mobile Devices on Data Quality. *Methods, Data, Analysis* 9, 261–292.

Thompson, K., Oliver, B., & Beck, J. (2015), An Analysis of the Mixed Collection Modes for Two Business Surveys Conducted by the US Census Bureau. *Public Opinion Quarterly*, 79, 3, pp. 769–789.

Toepoel, V. & Lugtig, P. (2015), Online Surveys are Mixed-device Surveys. *Methods, Data, Analysis*, 9, 2, pp. 155–162.

Toepoel, V. & Lugtig, P. (2018), Modularization in an Era of Mobile Web: Investigating the Effects of Cutting a Survey into Smaller Pieces on Data Quality. *Social Science Computer Review*, 1, pp. 1–15.

Tourangeau, R. (1984), Cognitive Sciences and Survey Methods. In: Janine, T. T., Lofts, G., Strafe, M., Tanner, J., & Tourangeau, R. (eds.), *Cognitive Aspects of Survey Methodology: Building a Bridge Between Disciplines*. National Academy of Science, Washington, DC, pp. 73–100.

Tourangeau, R. (2017), Mixing Modes. Tradeoffs Among Coverage, Nonresponse, and Measurement Error. In: Biemer, P., de Leeuw, E., Eckman, S., Edwards, B., Krauter, F., Lyberg, L., Tucker, N., & West, B. (eds.), *Total Survey Error in Practice*. Wiley, New Jersey, NJ.

Tourangeau, R., Maitland, A., Rivero, G., Sun, H., Williams, D., & Yan, T. (2017), Web Surveys by Smartphone and Tablets: Effects on Survey Responses. *Public Opinion Quarterly*, 81, 896–929.

Tourangeau, R., Sun, H., Yan, T., Maitland, A., Rivero, A., & Williams, D. (2018), Web Surveys by Smartphones and Tablets: Effects on Data Quality. *Social Science Computer Review* 36, 5, pp. 542–556.

Tourangeau, R. & Yan, T. (2007), Sensitive Questions in Surveys. *Psychological Bulletin*, 133, pp. 859–883.

Vannieuwenhuyze, J. & Loosveldt, G. (2013), Evaluating Relative Mode Effect in Mixed-Mode Surveys: Three Methods to Disentangle Selection and Measurement Effects *Sociological Method Research*, 42, pp. 82–104.

Villar, A. & Fitzgerald, R. (2017), Using Mixed-modes in Survey Data Research: Results from Six Experiments. In: Breen, M. (ed.), *Values and Identities in Europe: Evidence from the European Social Survey*. Routledge, New York, NY, pp. 273–310.

Voogt, R. J. & Saris, W. (2005), Mixed-mode Designs: Finding the Balance Between Nonresponse Bias and Mode Effects. *Journal of Official Statistics*, 21, pp.367–387.

Wells, T. (2015), What Market Researchers Should Know about Mobile Surveys, *International Journal of Market Research*, 57, 4, pp. 521–532.

Wells, T., Bailey, J., & Link, M. (2014), Comparison of Smartphones and Online Computer Survey Administration. *Social Science Computer Review*, 32, pp. 238–255.

Ye, C., Fulton, J., & Tourangeau, R. (2011), More Positive or More Extreme? A Meta-analysis of Mode Differences in Response Choice. *Public Opinion Quarterly*, 75, 2, pp. 349–365.

Zhang, Y. (2000), Using the Internet for Survey Research: A Case Study. *Journal of the American Society for Information Science*, 51, pp. 57–68.

# CHAPTER TEN

# The Problem of Under-coverage

## 10.1 Introduction

Collecting data with a survey is often a complex, costly, and time-consuming process. Not surprisingly, continuous attempts have been made all through the history of survey research to improve timeliness and reducing costs while at the same time maintaining a high level of data quality.

Developments in information technology in the last decades of the previous century made it possible to use microcomputers for data collecting. This led to the introduction of computer-assisted interviewing (CAI). Replacing the paper questionnaire by an electronic one turned out to have many advantages, among which were considerably shorter survey processing times and higher data quality. More on the benefits of CAI can be found in Couper et al. (1998). The next important development in the area of survey research was the fast rise of the Internet in the 1990s. This made it possible to conduct surveys online (Bethlehem and Hofman, 2006; File and Ryan, 2014). Web surveys seem to have some attractive advantages in terms of costs and timeliness:

- Nowadays many people are connected to the Internet. Therefore, a web survey is a simple means to get access to a large group of potential respondents.

*Handbook of Web Surveys*, Second Edition. Silvia Biffignandi and Jelke Bethlehem.
© 2021 John Wiley & Sons, Inc. Published 2021 by John Wiley & Sons, Inc.

- Web survey questionnaires can be distributed at very low costs. No interviewers are needed. This makes it cheaper than face-to-face or telephone surveys. A web survey is also cheaper in terms of mailing and printing costs.
- Web surveys can be set up very quickly. Fieldwork can start immediately after the questionnaire has been installed on the Internet.

So, a web survey is a fast and cheap means of collecting large amounts of data. Not surprisingly, many survey organizations (and other organizations, too) started conducting such surveys. However, costs and timeliness are not the only aspect. More important is the question whether web surveys can produce reliable and precise estimates of population characteristics.

When conducting a survey, a researcher is confronted with all kinds of phenomena that may have a negative impact on the quality, and therefore the validity, of the outcomes (Bethlehem, 2009). Some of these disturbances are almost impossible to prevent. So, efforts will have to be aimed at reducing their impact as much as possible. Nevertheless, notwithstanding all these efforts, final estimates of population characteristics may be affected. One of these phenomena is under-coverage. This is the topic of this chapter.

To be able to select a sample from a target population, a sampling frame is required. A *sampling frame* is a list of all elements in the target population. For every element in the list, there must be information on how to contact that element. Such contact information can consist of, for example, name and address, telephone number, or e-mail address. Such lists can exist on paper (a card-index box for the members of a club, a telephone directory) or in a computer (a database containing a register of all companies). If such lists are not available, detailed geographical maps are sometimes used.

For a face-to-face survey among persons, the sampling frame could consist of names and addresses. Some countries (for example, the Netherlands and the Scandinavian countries) have population registers. These registers contain names and addresses of all permanent residents in the country. In Italy, the ANPR (Anagrafe nazionale della popolazione residente) starting from 2020 is a centralized database which includes and harmonizes the LAC (Liste Anagrafiche Comunali) data collected from the municipaities. This centralized harmonization project is a target frame for the general population sampling surveys and for the sample-based Permanent Census.

If a population register is not available, an address list could be an alternative. An example of such a list is a Postal Address File (PAF). Postal service organizations in several countries maintain database of all postal delivery points in the country. Examples of such countries are the United Kingdom, Australia, New Zealand, and the Netherlands. Such databases contain addresses of both private houses and companies. Typically, a PAF can be used to select a sample of addresses and therefore of households. It is sometimes not clear whether an address in a PAF belongs to a private address or a company. So, if the aim is to select a sample of households, there can be over-coverage caused by companies in the file.

For example, PostNL, the postal service company in the Netherlands, has a PAF. This is a computer file containing all addresses where to deliver post. There are separate files for households and companies. Typically, the use of this file is to draw a sample of households. If required, a person can be randomly drawn from each selected address.

For a telephone survey, the sampling frame could be telephone directory. A disadvantage of this sampling frame is that many people have unlisted telephone numbers. There are many people with unlisted landline telephones. Moreover, almost all mobile telephone numbers are missing. An alternative could be to apply random digit dialing (RDD), where valid telephone numbers generation is by some computer algorithm.

The obvious sampling frame for a web survey would be a list of e-mail addresses. Sometimes such a sampling frame exists. For example, all employees of a large company may have a company e-mail address. Similarly, all students of university usually have an e-mail address. The situation is more complicated for a general population survey. Unfortunately, there does not exist (yet) a list of e-mail addresses of everybody in the country. It is also not possible to generate random e-mail address in a fashion similar to RDD.

The sampling frame should be an accurate representation of the population. There is a risk of drawing wrong conclusion from the survey if the sample selection was from a sampling frame that differs from the population. Figure 10.1 shows what can go wrong.

The first problem is *under-coverage*. This occurs if the target population contains elements that do not have a counterpart in the sampling frame. Such elements can never be selected in the sample. An example of under-coverage is survey where the sample selection is from a population register. Illegal immigrants are to consider part of the population, but they are not in the sampling frame. Another example is a general population survey, where respondent selection is via the Internet. Then there will be under-coverage due to people without Internet access. Under-coverage can have serious consequences. If the elements outside the sampling frame systematically differ from the elements in the sampling frame, estimates of population parameters may be seriously biased. A complicating factor is that it is often not very easy to detect the existence of under-coverage.

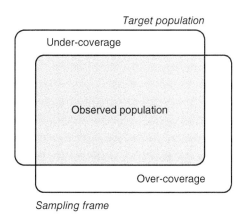

FIGURE 10.1 **The target population and the sampling frame**

The second sampling frame problem is *over-coverage*. This refers to the situation where the sampling frame contains elements that do not belong to the target population. If such elements end up in the sample and their data used in the analysis, estimates of population parameters may be affected. It should be rather simple to detect over-coverage in the field. This should become clear from the answers to the questions. Example 10.1 is about the under-coverage from a telephone directory.

EXAMPLE 10.1 **A web survey with telephone recruitment**

The local authorities of a town intend to set up a web panel of citizens. To recruit people for this panel, a simple random sample selection is drawn from the telephone directory of the town. People are called, invited to join the panel. If they agree, they are asked for their e-mail address. A link to the questionnaire on the Internet and a unique access code is sent to each respondent by e-mail.

At first sight, it might be a good idea to use the telephone directory of the town as a sampling frame. However, it suffers from serious coverage problems. Under-coverage occurs because many people have unlisted numbers and some will have no telephone at all. Moreover, there are many people with only a mobile telephone. In many countries, mobile phone numbers are not listed in directories.

The telephone directory also suffers from over-coverage, because it contains the telephone numbers of shops, companies, etc. Hence, it may happen that people are contacted that do not belong to the target population. Moreover, some people may have a higher than assumed contact probability, because they can be contacted both at home and in the office.

Even if people agree to participate, there may still be problems, because these people do not necessarily have Internet at home. This would be a case of nonresponse. This problem could be solved by advising them to go to a public library with Internet access, to complete the questionnaire at work, or by simply giving people Internet access. Another solution could be to use a different mode of data collection, for example, a paper questionnaire that is sent by ordinary mail. This would be an example of a mixed-mode survey. See also Chapter 9.

The set of all elements that can be contacted through the sampling frame is called the *frame population*. The sample is always selected from the sampling frame. Consequently, the conclusions drawn from the survey will apply to the frame population. Only if the frame population coincides with the target population will the results also apply to the target population.

The frame population for a web survey is by definition restricted to those having Internet. This can be a problem if the target population is the general population. Not everyone has access to the Internet.

By 2018, the share of EU-28 households with Internet access had risen to 89%, some 29 percentage points higher than in 2008. The penetration differs by country. It is however generally rather high since even countries where Internet access by households was in 2009 low registered an important increase. It should be observed that the Netherlands has the highest penetration rate (98%). Figure 10.2 shows the distribution of household Internet penetration in the European Union (EU) countries by household.

At the beginning of 2018, 85% of all individuals in the EU-28 used the Internet (at least once within the three months prior to the survey date). The proportion of the EU-28's population that had never used the Internet is 11% in 2018. This share was 33% in 2008; the decrease is high, and the recent trend patterns have shown a constant falling.

Figure 10.3 gives an overview of Internet access by individuals in the EU in 2018. It is clear there are differences across countries; however a rather high level of Internet diffusion is reached in several countries. Internet access is over 90% in seven countries: Denmark, Luxembourg, the Netherlands, the United Kingdom, Finland, Germany, and Sweden. In four of them, the Internet access is extremely high (95% or more). These are mainly North European countries. In other 13 countries the percentage is between 89 and 80. A few countries of the EU (Poland, Hungary, Croatia, Portugal, Italy, Greece, Romania, Bulgaria) are still between 78 and 70, except Bulgaria, which has 65% of individuals with Internet access. On the other side, almost full penetration in individual access is registered in Europe in countries like Iceland and Norway.

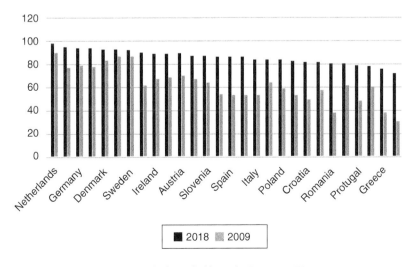

FIGURE 10.2 Internet access by households in the European Union in 2018

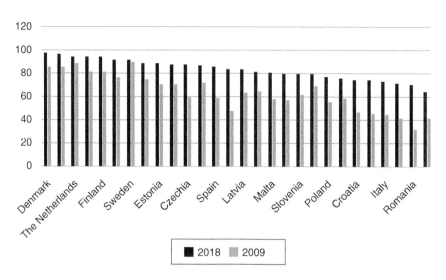

FIGURE 10.3 **Percentage of individuals Internet access (last three months) by country: 2009 and 2018**

Under-coverage in web surveys would not be a problem if those with Internet did not differ systematically from those without it. If people with Internet could be seen as a random sample from the population, still valid conclusions could be drawn from web surveys. Unfortunately, this is not the case. There are substantial differences between those with and without Internet.

Analysis of data (year 2018) about Internet access (Eurostat, 2020) registers some patterns common to most countries. Clearly, Internet access decreases with age. The distribution of Internet access by gender shows that more males than females have access to the Internet and by age young people is in most countries almost totally on Internet (percentage of individuals accessing in the last three months over 90%). Older people (aged 55 or more) are in general less on the Internet, especially women. However, in countries where the diffusion of Internet access is especially high like the northern countries, male aged people show high percentages, over 80%.

The use of the Internet while on the move, in other words when away from home or work, for example, using the Internet on a portable computer or handheld device via a mobile or wireless connection, has been highly increased during the last five years. Figure 10.4 compares 2013 data. In 2013, the percentage of individuals aged 16–74 within the EU-28 using a mobile device to connect to the Internet was 43%, whereas the similar share in 2018 had risen to 69%. The most common mobile devices for Internet connections were mobile or smartphones, laptops, and tablet computers.

Denmark, Sweden, the Netherlands, Luxembourg, and the United Kingdom recorded the highest proportions of mobile Internet use in 2018, with more than four-fifths of individuals aged 16–74 using the Internet while on the move,

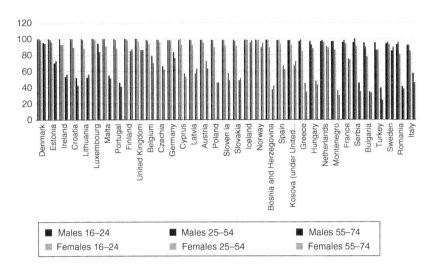

FIGURE 10.4 **Percentage of individuals Internet access (daily access) by age in different countries: 2018**

peaking at 88% in Denmark and Sweden; with one percentage point more, the maximum share was recorded in Norway (89%). By comparison, between 50% and 60% of individuals aged 16–74 in Greece, Bulgaria, Romania, Latvia, and Portugal used the Internet away from home or work, with this proportion as low as 49% in Poland and 40% in Italy.

It is clear that use of the Internet as a sampling frame can cause problems, because certain specific groups are substantially under-represented. Specific groups in the target population will not be able to fill in the (electronic) questionnaire form. It is however true that the diffusion of Internet both at household and individual level is a constantly increasing phenomenon and surveys have to handle with the new characteristics of the sampling frame.

## 10.2 Theory

### 10.2.1 THE INTERNET POPULATION

Let the target population $U$ of the survey consist of $N$ identifiable elements, which are labeled 1, 2, ..., $N$. Therefore, the target population can be denoted by

$$(10.1) \qquad U = \{1, 2, ..., N\}.$$

Associated with each element $k$ is a value $Y_k$ of the target variable $Y$. The aim of the web survey is assumed to be estimation of the population mean:

(10.2)
$$\overline{Y} = \frac{1}{N} \sum_{k=1}^{N} Y_k$$

of the target variable $Y$.

The population $U$ is divided into two subpopulations. There is a subpopulation $U_I$ of elements having access to the Internet. There is also a subpopulation $U_{NI}$ of elements not having access to the Internet. Associated with each element $k$ is an indicator $I_k$, where $I_k = 1$ if element $k$ has access to the Internet (and thus is an element of subpopulation $U_I$) and $I_k = 0$ otherwise. The subpopulation $U_I$ will be called the *Internet population* and $U_{NI}$ is the *non-Internet population*. The number of elements in the Internet population $U_I$ is equal to

(10.3)
$$N_I = \sum_{k=1}^{N} I_k.$$

Likewise,

(10.4)
$$N_{NI} = \sum_{k=1}^{N} (1 - I_k)$$

denotes the size of the non-Internet population $U_{NI}$, where $N_I + N_{NI} = N$.

The mean of the target variable for the elements in the Internet population is equal to

(10.5)
$$\overline{Y}_I = \frac{1}{N_I} \sum_{k=1}^{N} I_k Y_k.$$

Likewise, the mean of the target variable for the non-Internet population is denoted by

(10.6)
$$\overline{Y}_{NI} = \frac{1}{N_{NI}} \sum_{k=1}^{N} (1 - I_k) Y_k.$$

## 10.2.2 A RANDOM SAMPLE FROM THE INTERNET POPULATION

The more or less ideal case is considered now in which it is possible to select a random sample without replacement from the Internet population. This would require a sampling frame listing all elements having access to the Internet. Often there is no such list. A solution could be to select a random sample from a larger sampling frame. Examples are a population register and an address list. People selected from such a list are recruited for the web survey by sending them a letter, by calling them on the telephone, or by visiting them at home. Only those with access to Internet are selected for the web surveys. These persons are provided with a link to the electronic questionnaire and possibly a unique entry code.

A random sample selected without replacement from the Internet population is denoted by a series

$$(10.7) \qquad\qquad a_1, a_2, \ldots, a_N$$

of $N$ indicators. The $k$th indicator $a_k$ assumes the value 1 if element $k$ is selected, and otherwise it assumes the value 0, for $k = 1, 2, \ldots, N$. Note that always $a_k = 0$ for elements $k$ in the non-Internet population. The sample size is denoted by

$$(10.8) \qquad\qquad n_I = a_1 + a_2 + \cdots + a_N.$$

Horvitz and Thompson (1952) have shown that always an unbiased estimator of a population mean can be defined if all elements in the population have known positive probability of being selected. The Horvitz–Thompson estimator for the mean of the Internet population is defined by

$$(10.9) \qquad\qquad \bar{y}_{HT} = \frac{1}{N_I} \sum_{k=1}^{N} a_k I_k \frac{Y_k}{\pi_k}.$$

The quantity $\pi_k$ is called the *first-order inclusion probability* of element $k$. It is defined as the expected value

$$(10.10) \qquad\qquad \pi_k = E(a_k)$$

if the indicator $a_k$. Note that by definition $Y_k/\pi_k = 0$ for all elements outside the Internet population. The values of the $\pi_k$ are determined by the sampling design. For a simple random sample of size $n$ (with equal probabilities and without replacement) from a population of size $N$, all $\pi_k$ are equal to $n/N$.

In the case of a simple random sample from the Internet population, all first-order inclusion probabilities are equal to $n/N_I$. Therefore expression (10.9) reduces to

$$(10.11) \qquad\qquad \bar{y}_I = \frac{1}{n} \sum_{k=1}^{N} a_k I_k Y_k.$$

This estimator is an unbiased estimator of the mean $\overline{Y}_I$ of the Internet population, but not necessarily of the mean $\overline{Y}$ of the target population. The bias is written as

(10.12)          $B(\bar{y}_{HT}) = E(\bar{y}_{HT}) - \overline{Y} = \overline{Y}_I - \overline{Y} = \dfrac{N_{NI}}{N}(\overline{Y}_I - \overline{Y}_{NI}).$

The magnitude of this bias is due to two factors. The first factor is the relative size $N_{NI}/N$ of the non-Internet population. The bias is increasing if a larger proportion of the population does not have access to the Internet. The second factor is the *contrast* $\overline{Y}_I - \overline{Y}_{NI}$ between the Internet population and the non-Internet population. It is the difference between the population means of the two subpopulations. The more the mean of the target variable differs for the two subpopulations, the larger the bias will be.

The relative size of the non-Internet population cannot be neglected in many countries (see Figure 10.2). Furthermore, there are substantial differences between those with and without Internet. Specific groups are underrepresented in the Internet population, for example, the elderly, those with a low level of education, and ethnic minority groups. So, the conclusion is that generally a random sample from an Internet population will lead to biased estimates for the parameters of the target population.

It is to be expected that Internet coverage will increase over time. The factor $N_{NI}/N$ will become smaller, and this will reduce the bias. It is unclear, however, whether the contrast will also become smaller over time. It is even possible that it increases, as the remaining group of people without Internet access may differ more and more from the Internet users. So, the combined effect of a smaller non-Internet population and a larger contrast need not necessarily lead to a smaller bias.

It is important to note that the value of expression (10.12) does not depend on the sample size. Increasing the sample size will not reduce the bias. So, the problem of under-coverage in web surveys does not diminish by collecting a larger number of observations.

The precision of an estimator is often set at a 95% confidence interval. Suppose, a simple random sample is selected from the target population. Then the sample mean $\bar{y}$ can be computed. This is an unbiased estimator for the population mean $\overline{Y}$. Since the sample mean is (approximately) normally distributed, the 95% confidence interval for the population mean is equal to

(10.13)          $I = (\bar{y} - 1.96 \times S(\bar{y}); \bar{y} + 1.96 \times S(\bar{y})),$

where $S(\bar{y})$ is the standard error of the sample mean. The probability that this interval contains the true value is by definition (approximately) equal to

(10.14)          $P(\overline{Y} \in I) = 0.95.$

The standard error will decrease if the sample size increases. This will lead to a smaller confidence interval. The estimator is more precise. If a simple random sample is selected from the Internet population, the sample mean $\bar{y}_I$ is used to estimate the population mean $\overline{Y}$. Analogous to expression (10.13), the confidence will be computed as

$$(10.15) \qquad I_I = \left(\bar{y}_I - 1.96 \times S\left(\bar{y}_I\right); \bar{y}_I + 1.96 \times S\left(\bar{y}_I\right)\right),$$

The confidence level of this interval is not by definition equal to 0.95. It can be shown that

$$(10.16) \qquad P\left(\overline{Y} \in I_I\right) = \Phi\left(1.96 - \frac{B\left(\bar{y}_I\right)}{S\left(\bar{y}_I\right)}\right) - \Phi\left(-1.96 - \frac{B\left(\bar{y}_I\right)}{S\left(\bar{y}_I\right)}\right),$$

in which $\Phi$ is the standard normal distribution function. The quantity $B\left(\bar{y}_I\right)/S\left(\bar{y}_I\right)$ is called the *relative bias*. Apparently, the confidence level depends on the value of this relative bias. Figure 10.5 contains a plot of the confidence level as a function of the relative bias.

It is clear that the confidence level can be much lower than expected. A larger sample size will lead to a smaller standard error, but the bias will remain the same. So, the relative bias increases. If the bias is equal to the standard error, i.e., the relative bias is 1, the confidence level is only 0.83. As the relative bias increases, the situation becomes worse. The confidence level is even less than 0.5 for a

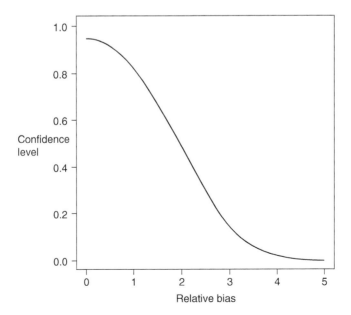

FIGURE 10.5 **The confidence level of the 95% confidence interval as a function of the relative bias**

relative bias of 2. The conclusion is that under-coverage may lead to incorrect confidence interval.

## 10.2.3 REDUCING THE NON-COVERAGE BIAS

There are several ways in which the negative effects of under-coverage can be reduced. Three approaches are discussed here.

The first approach is to give Internet access to persons in the sample without it. The Dutch LISS panel is the result of such an attempt. See Scherpenzeel (2008) for more information. This web panel has been constructed by selecting a random sample of households from the population register of the Netherlands. Selected households were recruited for this panel by means of a face-to-face interview (CAPI) or a telephone interview (CATI). Cooperative households without Internet access were provided with equipment giving them access to Internet. It should be noted, however, that there is always a (small) group that refuses to use the Internet. These are particularly the elderly. Analysis by Scherpenzeel and Bethlehem (2011) showed nevertheless that this approach reduces the under-coverage bias.

Another example is the ELIPSS Panel. This web panel is representative of the population living in metropolitan France (France without overseas regions). The panel has approximately 3,500 members. They were randomly selected by The French National Institute of Statistics and Economic Studies (INSEE). All members of the panel (whether they already had Internet) were provided with the same touchscreen tablet and a 3G Internet access subscription. This guarantees that the respondents all used the same device. See Chevallier and Olivier (2016).

A second approach is to conduct a mixed-mode survey. This is a survey in which different modes of data collection are used. Possible modes are face-to-face interviewing (CAPI), telephone interviewing (CATI), and self-administered modes like mail and web. de Leeuw (2005) describes two-mixed mode designs. The first design is the use of different modes *concurrently*. The sample is divided into groups, and each group is approached by a different mode. The other approach is the use of different modes *sequentially*. All sample persons are approached by one mode. The nonrespondents are then followed up by a different mode than the one used in the first approach. This process can be repeated for a number of modes.

The under-coverage problem of web surveys could be addressed by a sequential mixed-mode design. First, a sample is selected. If it turns out that selected persons are willing to participate in a survey but do not have access to the Internet, they are, for the time being, considered to be nonrespondents. In the second phase of the fieldwork, these nonrespondents are approached with a different node of data collection. The cheapest one would be a mail survey, but if quality is vital, CATI or CAPI should be preferred. The collected data of the two phases of the fieldwork are combined into an estimate for the target population.

A third approach is to apply adjustment weighting. This is a family of techniques that attempt to reduce the bias of survey estimates by assigning weights to responding elements. These weights correct for the over- or underrepresentation of specific groups in the response. Adjustment weighting is treated in detail in Chapter 12. Here one such technique is summarized. It is called post-stratification.

*Post-stratification* requires one or more auxiliary variables. An auxiliary variable is a variable that has been measured in the survey and for which the distribution in the target population is available. Typical target variables are gender, age, marital status, and region. By comparing the response distribution of an auxiliary variable with its population distribution, it can be assessed whether or not the survey response is representative for the population (with respect to this variable). If these distributions differ considerably, one must conclude that the response is selective. To correct this, adjustment weights are computed. Weights are assigned to all records of observed elements. Estimates of population characteristics can now be obtained by using the weighted values instead of the unweighted values.

To carry out post-stratification, one or more categorical auxiliary variables are needed. Here, only one such variable is considered. The situation for more variables is not essentially different. Suppose there is an auxiliary variable $X$ having $L$ categories. So, it divides the target population into $L$ subsets called strata. The strata are denoted by $U_1, U_2, \ldots, U_L$. The number of target population elements in stratum $U_h$ is denoted by $N_h$, for $h = 1, 2, \ldots, L$. The population size $N$ is equal to $N = N_1 + N_2 + \cdots + N_L$. This is the population information assumed to be available.

Suppose a simple random sample of size $n$ is selected from the Internet population. If $n_h$ denotes the number of sample elements in stratum $h$, then $n = n_1 + n_2 + \cdots + n_L$. The values of the $n_h$ are the result of a random selection process, so they are random variables. Note that since the sample is selected from the Internet population, only elements in the substrata $U_I \cap U_h$ are observed (for $h = 1, 2, \ldots, L$).

Post-stratification assigns identical adjustment weights to all elements in the same stratum. The weight $w_k$ for an element $k$ in stratum $h$ is equal to

$$(10.17) \qquad w_k = \frac{N_h/N}{n_h/n}.$$

Post-stratification comes down to replacing the simple sample mean

$$(10.18) \qquad \bar{y}_I = \frac{1}{n}\sum_{k=1}^{N} a_k I_k Y_k$$

by the weighted sample mean

$$(10.19) \qquad \bar{y}_{I,PS} = \frac{1}{n}\sum_{k=1}^{N} a_k w_k I_k Y_k.$$

Substituting the weights and working out this expression leads to the *post-stratification estimator*

$$(10.20) \qquad \bar{y}_{I,PS} = \frac{1}{N} \sum_{h=1}^{L} N_h \bar{y}_I^{(h)} = \sum_{h=1}^{L} W_h \bar{y}_I^{(h)},$$

where $\bar{y}_I^{(h)}$ is the sample mean in stratum $h$ and $W_h = N_h/N$ is the relative size of stratum $h$. The expected value of this post-stratification estimator is equal to

$$(10.21) \qquad E\left(\bar{y}_{I,PS}\right) = \frac{1}{N} \sum_{h=1}^{L} N_h E\left(\bar{y}_I^{(h)}\right) = \sum_{h=1}^{L} W_h \overline{Y}_I^{(h)} = \widetilde{Y}_I,$$

where $\overline{Y}_I^{(h)}$ is the mean of the target variable in stratum $h$ of the Internet population. Generally, this mean will not be equal to the mean $\overline{Y}^{(h)}$ of the target variable in stratum $h$ of the target population. The bias of this estimator is equal to

$$(10.22) \qquad \begin{aligned} B\left(\bar{y}_{I,PS}\right) &= E\left(\bar{y}_{I,PS}\right) - \overline{Y} = \widetilde{Y}_I - \overline{Y} = \sum_{h=1}^{L} W_h \left(\overline{Y}_I^{(h)} - \overline{Y}^{(h)}\right) \\ &= \sum_{h=1}^{L} W_h \frac{N_{NI,h}}{N_h} \left(\overline{Y}_I^{(h)} - \overline{Y}_{NI}^{(h)}\right), \end{aligned}$$

where $N_{NI,h}$ is the number of elements in stratum $h$ of the non-Internet population.

The bias will be small if there is (on average) no difference between elements with and without Internet within the strata. This is the case if there is a strong relationship between the target variable $Y$ and the stratification variable $X$. The variation in the values of $Y$ will manifest itself in this case between strata but not within strata. In other words, the strata are homogeneous with respect to the target variable. In nonresponse correction terminology, this situation comes down to Missing at Random (MAR).

Application of post-stratification will successfully reduce the bias of the estimator if proper auxiliary variables can be found. Such variables should satisfy three conditions:

- They must have been measured in the survey (or complete sample).
- Their population distribution $(N_1, N_2, ..., N_L)$ must be known.
- They must be strongly correlated with all target variables.

Unfortunately, such variables are not very often available, or there is only a weak correlation.

## 10.2.4 MIXED-MODE DATA COLLECTION

The fundamental problem of a web survey is that persons without Internet are excluded from the survey. This problem could be solved by selecting a stratified sample. The target population is assumed to consist of two strata: Internet population $U_I$ of size $N_I$ and the non-Internet population $U_{NI}$ of size $N_{NI}$.

To be able to compute an unbiased estimate, a simple random sample selection from both strata is needed. The survey provides the data about the Internet stratum. If this is a random sample with equal probabilities, the sample mean

$$(10.23) \qquad \bar{y}_I = \frac{1}{n} \sum_{k=1}^{N} a_k I_k Y_k$$

is an unbiased estimator of the mean of the Internet population.

Now suppose a random sample (with equal probabilities) of size $m$ is selected from the non-Internet stratum. Of course, there is no sampling frame for this population. This problem could be avoided by selecting a sample from the complete target population (a reference survey) and only using people without Internet access. Selected people with Internet access can be added to the large online sample, but this will have no substantial effect on estimators. The sample mean of the non-Internet sample is denoted by

$$(10.24) \qquad \bar{y}_{NI} = \frac{1}{m} \sum_{k=1}^{N} b_k (1 - I_k) Y_k,$$

where the indicator $b_k$ denotes whether or not element $k$ is selected in the non-Internet survey, and

$$(10.25) \qquad m = \sum_{k=1}^{N} b_k (1 - I_k).$$

The stratification estimator is now defined by

$$(10.26) \qquad \bar{y}_{ST} = \frac{N_I}{N} \bar{y}_I + \frac{N_{NI}}{N} \bar{y}_{NI}.$$

This is an unbiased estimator for the mean of the target population. Application of this estimator assumes the size $N_I$ of the Internet population and the size $N_{NI}$ of the non-Internet population to be known. The variance of the estimator is equal to

$$(10.27) \qquad V(\bar{y}_{ST}) = \left(\frac{N_I}{N}\right)^2 V(\bar{y}_I) + \left(\frac{N_{NI}}{N}\right)^2 V(\bar{y}_{NI}).$$

The variance of the sample mean in the Internet stratum is of order $1/n$, and the variance in the non-Internet stratum is of order $1/m$. Since $m$ will be much smaller than $n$ in practical situation and the relative sizes of the Internet population and the non-Internet population do not differ that much, the second term will determine the magnitude of the variance. So, the advantages of the large sample size of the web survey are for a great part lost by the bias correction.

Note that the sizes of the Internet and non-Internet population are usually unknown. In this case they have to be estimated. This can, for example, be done using data from the non-Internet survey.

# 10.3 Application

The possible consequences of under-coverage and the effectiveness of correction techniques are now illustrated using a simulation experiment. A fictitious population was constructed. For this population, reported voting behavior in an election survey was simulated and analyzed. The relationship between variables involved was such that it could resemble more or less a real-life situation. This relationship is shown graphically in Figure 10.6.

With respect to the Internet population, both *Missing At Random* (MAR) and *Not Missing at Random* (NMAR) were introduced. The characteristics of estimators (before and after correction) were computed based on a large number of simulations.

First, the distribution of the estimator was determined in the ideal situation of a simple random sample from the target population. Then, it was explored how the characteristics of the estimator change if a simple random sample is selected just from the Internet population. Finally, the effects of weighting (post-stratification and reference survey) were analyzed.

A fictitious population of 30,000 individuals was constructed. There were five variables:

- Age in three categories: Young (with probability 0.40), Middle-aged (with probability 0.35), and Old (with probability 0.25).

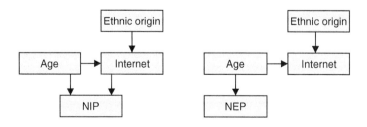

FIGURE 10.6 **Relationships between variables**

- Ethnic origin in two categories: Native (with probability 0.85) and Nonnative (with probability 0.15).

- Having access to Internet with two categories: Yes and No. The probability of having access to Internet depended on the two variables: Age and Ethnic origin. For natives, the probabilities were 0.90 (for Young), 0.70 (for Middle-aged), and 0.50 (for Old). So, Internet access decreased with age. For nonnatives, these probabilities were 0.20 (for Young), 0.10 (for Middle-aged), and 0.00 (for Old). These probabilities reflected the much lower Internet access among nonnatives.

- Voted for the National Elderly Party (NEP). The probability to vote for this party depended on age only. Probabilities were 0.00 (for Young), 0.40 (for Middle-aged), and 0.60 (for Old).

- Voted for the New Internet Party (NIP). The probability to vote for this party depended on both age and having Internet. For people with Internet access, the probabilities were 0.80 (for Young), 0.40 (for Middle-aged), and 0.20 (for Old). For people without Internet access, all probabilities were equal to 0.10. So, for people with Internet voting decreased with age. The voting probability was low for people without Internet.

In the experiment the variable NEP suffered from missingness due to MAR. There was direct relationship between voting for this party and age, and also there was a direct relationship between age and having Internet. This will cause estimates to be biased. It should be possible to correct for this bias by weighting using the variable age.

The variable NIP suffered from NMAR. There existed (among other relationships) a direct relationship between voting for this party and having Internet. As a result, estimates will be biased, and there is no correction possible.

The distribution of estimators for the percentage of voters for both parties was determined in various situations by repeating the selection of the sample 1,000 times. In all cases the sample size was $n = 2,000$.

Figure 10.7 contains the results for the variable NEP (voted for the NEP). The distributions of the estimator are displayed by means of box plots. The upper box plot shows the distribution of the estimator for simple random sample from the complete target population. The vertical line denotes the population value to be estimated (25.4%). The estimator has a symmetric distribution around this value. The estimator is clearly unbiased.

The middle box plot shows the distribution of the estimator if samples are not selected from the complete target population, but just from the Internet population. The shape of the distribution remains the same, but the distribution as a whole has shifted to the left. All values of the estimator are systematically lower. The expected value of the estimator is only 20.3%. The estimator is biased. The explanation of this bias is simple: relative few elderly have Internet access. Therefore, they are underrepresented in samples selected from the Internet. These persons typically vote for the NEP.

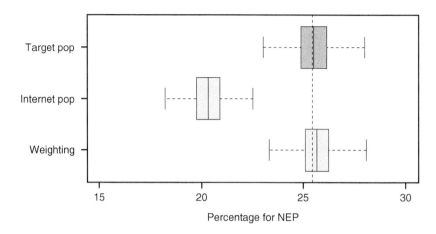

FIGURE 10.7   **Results of the simulations for variable NEP (National Elderly Party)**

The lower box plot shows the distribution of the estimator in case of post-stratification by age. The bias disappears. This was possible because this is a case of MAR.

The simulation results are summarized in Table 10.1. Sampling from the Internet results in a large relative bias of −6.4. Weighting eliminates the bias, and therefore the relative bias is 0. In case of sampling from the Internet, the confidence level of the 95% confidence interval is close to 0. This implies that almost certainly wrong conclusion will be drawn from this web survey.

Figure 10.8 contains the results for the variable NIP (voted for the NIP). The upper box plot shows the distribution of the estimator for simple random samples from the complete target population. The vertical line denotes the population value to be estimated (39.5%). Since the estimator has a symmetric distribution around this value, it is clear that the estimator is unbiased.

The middle box plot shows what happens if samples are not selected from the complete target population, but just from the Internet population. The distribution has shifted to the right considerably. All values of the estimator are systematically too high. The expected value of the estimator is now 56.5%. The estimator

TABLE 10.1   **Summary of simulation results for the variable NEP**

| Simulation | Mean | Standard error | Bias | Relative bias |
|---|---|---|---|---|
| Samples from the target population | 25.4 | 0.9 | 0.0 | 0.0 |
| Samples from the Internet population | 20.3 | 0.8 | −5.1 | −6.4 |
| Samples from the Internet population, with weighting adjustment | 25.4 | 0.8 | 0.0 | 0.0 |

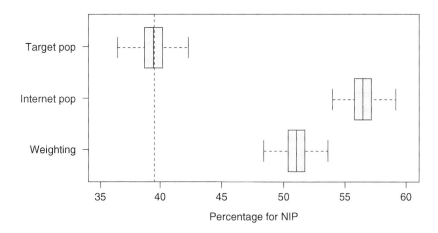

FIGURE 10.8 **Results of the simulations for variable NIP (New Internet Party)**

TABLE 10.2 **Summary of simulation results for the variable NIP**

| Simulation | Mean | Standard error | Bias | Relative bias |
|---|---|---|---|---|
| Samples from the target population | 39.5 | 1.1 | 0.0 | 0.0 |
| Samples from the Internet population | 56.5 | 1.1 | 17.0 | 15.5 |
| Samples from the Internet population, with weighting adjustment | 51.1 | 1.0 | 11.6 | 11.6 |

is severally biased. The explanation of this bias is straightforward: voters for the NIP are overrepresented in Internet samples.

The lower box plot in Figure 10.8 shows the effect of post-stratification by age. Only a small part of the bias is removed. This is not surprising as there is a direct relationship between voting for the NIP and having access to the Internet. This is a case of NMAR.

The simulation results are summarized in Table 10.2. Sampling from the Internet results in a very large relative bias of 15.5. Weighting can reduce this relative bias only to 11.6. In both cases, the confidence level of the 95% confidence interval is close to 0. This implies that almost certainly wrong conclusions will be drawn from this web survey.

# 10.4 Summary

A web survey may suffer from under-coverage. If not all elements in the population have access to the Internet, this happens. Elements without Internet access will never be selected in the sample.

Under-coverage may cause estimates of population characteristics to be biased. The magnitude of the bias is determined by two factors:

1. The relative size of the group without Internet. The larger the group, the larger the bias.
2. The difference between those with and without Internet. The larger the difference (on average). The larger the difference, the larger the bias.

There are several approaches that may help to reduce a bias due to under-coverage. A first approach is to provide those in the sample without Internet with Internet access. This may not completely solve the problem as there may still be persons refusing to work with Internet.

A second approach is to conduct a mixed-mode survey. A different mode of data collection (face-to-face, telephone, mail) can be used for those without Internet.

A third approach is to carry out some kind of adjustment weighting. By assigning weights the response is corrected for under- or overrepresented groups. There is no guarantee that weighting will completely remove the bias.

Even if Internet diffusion is highly increasing, it has to be advised that higher coverage in the individuals or household doesn't mean that an adequate sampling frame to draw a probability-based sample will be available and bias avoided.

## KEY TERMS

**Contrast**: The difference between the average of the target variable in the Internet population and the average of this variable in the non-Internet population.

**First-order inclusion probability**: The probability that a population element is selected in the sample. The first-order inclusion probability is determined by the sampling design.

**Frame population**: All elements that are represented in the sampling frame.

**Internet population**: The subpopulation of the target population consisting of elements that have access to Internet.

**Missing at Random (MAR)**: Nonresponse depends on auxiliary variables only. Estimators will be biased, but a correction is possible if some technique is used that takes advantage of this auxiliary information.

**Mixed-mode survey**: A survey in which various modes of data collection (for example, face to face, telephone, mail, web) are combined.

**Non-Internet population**: The subpopulation of the target population consisting of elements that do not have access to the Internet.

**Not Missing at Random (NMAR)**: Nonresponse depends directly on the target variables of the survey. Estimators will be biased, and correction techniques will not be successful.

**Over-coverage**: The phenomenon that the sampling frame contains elements that do not belong to the target population.

**Post-stratification**: A weighting adjustment technique that divides the population in strata and subsequently assigns the same weight to all observed elements within a stratum.

**Sampling frame**: A list (electronic or on paper) of all elements in the target population. Using the information in the sampling frame, it must be possible to actually contact each element.

**Under-coverage**: The phenomenon that not all elements in the population are represented in the sampling frame.

# EXERCISES

**Exercise 10.1** What is the difference between under-coverage and nonresponse?

**a.** In case of nonresponse, persons are selected in the sample, and in case of under-coverage, they are never selected.

**b.** In case of under-coverage, persons are selected in the sample, and in case of nonresponse they are never selected.

**c.** Nonresponse is caused by the persons selected in the sample, and under-coverage by the researcher.

**d.** There is no difference between under-coverage and nonresponse.

**Exercise 10.2** Can the bias due to under-coverage always be corrected by means of a weighting adjustment?

**a.** Yes, weighting adjustment will always remove the bias.

**b.** No, weighting adjustment will only work if persons are missing due to Missing at Random.

**c.** No, weighting adjustment will only work if persons are missing due to Missing at Random and the proper auxiliary variables are included in the weighting model.

**d.** No, weighting adjustment will never remove or reduce such a bias.

**Exercise 10.3**   What happens to the bias due to under-coverage in web surveys if Internet access increases in the target population?

**a.** The bias will increase.

**b.** The bias will decrease.

**c.** The bias will not change.

**d.** It depends on the average difference between persons with and without Internet access.

**Exercise 10.4**   What happens to the bias due to under-coverage in web surveys if the sample size is increased?

**a.** The bias will increase.

**b.** The bias will decrease.

**c.** The bias will not change.

**d.** It depends on the average difference between persons with and without Internet access.

**Exercise 10.5**   A researcher wants to estimate the average number of hours per week the adult inhabitants of Samplonia spend on the Internet? He draws a simple random sample of Internet users. There is no nonresponse. The sample mean turns out to be five hours.

**a.** Given that only three out of five inhabitants have access to the Internet, compute an estimate of the bias of the sample mean.

**b.** Compute a better estimate for the average number of hours an inhabitant spends on the Internet.

**Exercise 10.6**   A town council wants to know what percentage of the population is engaged in some form of voluntary work. Since there is only a limited budget available, it is decided to conduct a web survey. The target population consists of 1,000,000 persons. Only 70% of these persons have access to the Internet. It turns out that 10,000 persons participate in the survey. Of these respondents, 70% does some voluntary work.

**a.** Assuming that the 10,000 respondents are a simple random sample without replacement from the target population, compute the 95% confidence interval of the percentage of persons in the population doing voluntary work.      There is a strong suspicion that the survey estimates may be biased because only people with Internet access can participate. Therefore, a follow-up survey is conducted among people without Internet access. It turns out to be possible to draw a simple random sample of size 100 from this non-Internet population. The result is that 40% of the respondents in the follow-up survey do voluntary work.

**b.** Compute an improved estimate for the population percentage of people involved in voluntary work.

**c.** Compute a new 95% confidence interval of the percentage of persons in the population doing voluntary work.

**d.** Compare both confidence intervals and explain any differences.

## REFERENCES

Bethlehem, J. G. (2009), *Applied Survey Methods, A Statistical Perspective.* Wiley, Hoboken, NJ, USA.

Bethlehem, J. G. & Hofman, L. P. M. B. (2006), Blaise—Alive and Kicking for 20 Years. *Proceedings of the 10-th Blaise, Users Meeting, Statistics Netherlands*, Voorburg/Heerlen, the Netherlands, pp. 61–88.

Chevallier, A. & Olivier, M. (2016), Blaise on Touch-Screen Tablets: The ELIPSS Example. *Paper Presented at the 17th International Blaise Users Conference*, The Hague, the Netherlands.

Couper, M. P., Baker, R. P., Bethlehem, J. G., Clark, C. Z. F., Martin, J., Nicholls II, W. L., & O'Reilly, J. M. (eds.) (1998), *Computer Assisted Survey Information Collection.* Wiley, New York, USA.

de Leeuw, E. D. (2005), To Mix or Not To Mix Data Collection Modes in Surveys. *Journal of Official Statistics*, 21, pp. 233–255.

Eurostat (2020), *Level of Internet Access—Households.* Eurostat, Luxembourg. Retrieved from: http://ec.europa.eu/eurostat/tgm/table.do?tab=table&init=1&language=en&pcode=tin00134&plugin=1.

File, T. & Ryan, C. (2014), *Computer and Internet Use in the United States: 2013. American Community Survey Reports.* U.S. Census Bureau, Washington, DC.

Horvitz, D. G. & Thompson, D. J. (1952), A Generalization of Sampling Without Replacement from a Finite Universe. *Journal of the American Statistical Association*, 47, pp. 663–685.

Scherpenzeel, A. (2008), An Online Panel as a Platform for Multi-disciplinary Research. In: Stoop, I. & Wittenberg, M. (eds.), *Access Panels and Online Research, Panacea or Pitfall?* Aksant, Amsterdam, pp. 101–106.

Scherpenzeel, A. & Bethlehem, J. (2011), How Representative Are Online-Panels? Problems of Coverage and Selection and Possible Solutions. In: Das, M., Ester, P., Kaczmirek, L., & Mohler, P. (eds.), *Social Research and the Internet: Advances in Applied Methods and New Research Strategies.* Routledge Academic, New York, USA, pp. 105–132.

# CHAPTER ELEVEN

# The Problem of Self-Selection

## 11.1 Introduction

Web surveys are a fast, cheap, and attractive means of collecting large amounts of data. Not surprisingly, many survey organizations are conducting such surveys. The question is, however, whether a web survey is also attractive from a quality point of view, because there are methodological problems. One of these problems is *self-selection*. This is the phenomenon that the sample is not a probability sample. Instead, it is left to Internet users themselves to participate in a mobile web survey. Estimation problems caused by self-selection are the topic of this chapter. After an introduction, some theory is described. It is also explored whether weighting adjustment techniques can help to solve the problem. Practical implications are shown using simulated samples from a fictitious population.

Objective of a survey is to collect information about a well-defined target population. To this end a sample is selected from this population. The methodology of survey sampling has been developed over a period of more than 100 years. It is based on the fundamental principle of probability sampling. Selecting random samples makes it possible to apply probability theory. Unbiased estimators can be defined, and the accuracy of these estimators can be quantified and controlled. The probability sampling principle has been successfully applied in official and academic statistics since the 1940s and to a lesser extent also in more commercial market research. See Chapter 1 for a historical overview of the development of survey sampling.

*Handbook of Web Surveys*, Second Edition. Silvia Biffignandi and Jelke Bethlehem.
© 2021 John Wiley & Sons, Inc. Published 2021 by John Wiley & Sons, Inc.

Horvitz and Thompson (1952) showed in their seminal paper that unbiased estimates of population characteristics can be computed only if a real probability sample has been selected, every element in the population has a nonzero probability of selection, and all these probabilities are known to the researcher. Furthermore, the accuracy of estimates can be computed only under these conditions.

At first sight, web surveys seem to have much in common with other types of surveys. It is just another mode of data collection. Questions are not asked face-to-face, by telephone, or on paper, but over the Internet. What is different, however, is that many web surveys are self-selection surveys. The principles of probability sampling have not been applied. Samples are not constructed by means of probability sampling but instead rely on self-selection of respondents. This can have a major impact on survey results.

Web surveys appear in many different forms, from simple e-mail surveys to professionally designed interactive forms. Of course, web surveys can be based on probability sampling. An example is a survey among students of a university, where every student has an e-mail address. So a random sample can be selected from the list of all e-mail addresses. Unfortunately, many web surveys, particularly those conducted by market research organizations, are not based on probability sampling. The survey questionnaire is simply put on the web. Respondents are those people who happen to have Internet, visit the website, and spontaneously decide to participate in the survey. The survey researcher is not in control over the selection process. Therefore, the selection probabilities are unknown, which implies that no unbiased estimates can be computed nor can the accuracy of estimates be determined. Some cases of self-selection are described in Example 11.1.

### ▣ EXAMPLE 11.1  Opinion polls in the Netherlands

All major opinion polls in the Netherlands use web panels that have been set up by means of self-selection. Examples are the *Politieke Barometer* and *Peil.nl*. The values of some demographic variables are recorded during the recruitment phase. Therefore the distribution of these variables in a poll can be compared with their distribution in the population. Weighting adjustment techniques can be applied in an attempt to correct for over- or underrepresentation of specific groups.

Another example of large self-selection web survey in the Netherlands was *21minuten.nl*, a survey supposed to supply answers to questions about important problems in Dutch society. The first edition of this survey was conducted in 2006. Within a period of six weeks, about 170,000 people completed the online questionnaire. A similar survey was conducted in Germany (*Perspektive Deutschland*).

Vonk, Van Ossenbruggen, and Willems (2006) describe a study across 19 online panels of Dutch market research organizations. It shows that most of them use self-selection.

There was an intensive political discussion in the Netherlands in January 2010 about the introduction of a system of road pricing. An important participant in this discussion was the Dutch Automobile Association (ANWB). This organization conducted a poll on its website.

It was a self-selection survey. Everyone could participate. Everyone could participate even more than once. There was no check on this. Within a period of a few weeks, the questionnaire was completed more than 400,000 times. It turned out that 68% was in favor of road pricing. In the same period, the Dutch newspaper *De Telegraaf* conducted a simple self-selection web survey on the same topic on its website. In one weekend the questionnaire was completed about 196,000 times. Since this newspaper is known to support the interests of car owners, it was not surprising that the great majority (89%) turned out to be against road pricing.

So there is a very large difference between the outcome of the ANWB survey and the Telegraaf survey. Brüggen, Van den Brakel, and Krosnick (2016) studied 18 online self-selection panels of Dutch market research organizations. They compared these panels with the LISS panel (based on probability sampling), with two CAPI surveys (also based on probability sampling), and with data from the Dutch population register. They show that the self-selection web panels behave inconsistently. Panels may perform well for some variables, but not for other variables. Some variables are estimated incorrectly by only some panels, whereas other variables are estimated incorrectly by most panels. Some panels simply perform poorly for most variables. Moreover, there are also problems with estimating relationships between variables. Panels differ significantly from each other with respect to the strength and even sign of relationships. Unfortunately, weighting adjustment does not improve the situation.

Self-selection web survey results are sometimes claimed to be "representative" in the media because of the large number of respondents or as a result of advanced adjustment weighting procedures. This claim was typically made for some of the web surveys mentioned in Example 11.1. Unfortunately, such claims are not based on methodological knowledge.

The term representative is rather confusing. Kruskal and Mosteller (1979a, 1979b, 1979c) show that it can have many meanings and it is often used in a very loose sense to convey a vague idea of good quality. It is even sometimes claimed that a large number of respondents ensure validity and reliability. Unfortunately, it is a well-known fact in the survey methodology literature that this is not the case. It is shown again in this chapter.

The essential problem of self-selection is that the selection probabilities are unknown. Some of these probabilities may even be equal to 0. This makes it impossible to construct unbiased estimators using the theory of Horvitz and Thompson (1952).

The problem of self-selection is illustrated using survey results related to the general election in the Netherlands in 2012. Various market research organizations carried out opinion polls in an attempt to predict the outcome of this election. The results of the four major polls are summarized in Table 11.1. The polls were conducted one day before the elections. All polls are based on self-selection web panels. The table contains numbers of seats. There are 150 seats in total in parliament.

The largest difference between a prediction and the true election outcome is found for the SP in the poll of De Stemming. The prediction was 22 seats, whereas the election result was only 15 seats. This comes down to a difference of seven seats. In four cases, there was a difference of six seats, and in two cases there was a five-seat difference.

All differences of five seats or more are printed in boldface in the table. These are significant differences. They are larger than the margin of error. It is not unlikely that significant differences were caused by self-selection sampling. There could be other explanations. One unlikely explanation is that people could have changed their mind in the night after the last poll and before election day.

TABLE 11.1   **Predictions (seats in parliament) for the Dutch parliamentary elections of September 12, 2012**

| Party | Election result | Peil. nl | Politieke Barometer | TNS NIPO | De Stemming |
|---|---|---|---|---|---|
| VVD (Liberals) | 41 | **36** | 37 | **35** | **35** |
| PvdA (Social Democrats) | 38 | 36 | 36 | 34 | 34 |
| PVV (Populists) | 15 | 18 | 17 | 17 | 17 |
| CDA (Christian Democrats) | 13 | 12 | 13 | 12 | 12 |
| SP (Socialists) | 15 | **20** | **21** | **21** | **22** |
| D66 (Liberal Democrats) | 12 | 11 | 10 | 13 | 11 |
| GroenLinks (Green) | 4 | 4 | 4 | 4 | 4 |
| ChristenUnie (Christian) | 5 | 5 | 5 | 6 | 7 |
| SGP (Christian) | 3 | 3 | 2 | 2 | 3 |
| PvdD (Animals) | 2 | 3 | 3 | 2 | 2 |
| 50Plus (Elderly) | 2 | 2 | 2 | 4 | 3 |
| Mean absolute difference | | 1.6 | 1.6 | 2.2 | 2.2 |

It was already explained why probability sampling should be preferred over self-selection sampling. There are more issues with self-selection web surveys. Three issues are discussed here:

1. The target population is unclear. If a questionnaire is put on the Internet, everybody can fill it in, even people outside the intended target population.
2. Sometimes the questionnaire can be completed more than once by the same person. It is possible to check the IP address of the computer of the respondents, but such a check is not always implemented.
3. There is a risk that a group of people attempt to manipulate the outcome of the survey by actively participating in the survey.

These issues are discussed in some more detail. The first issue is the target population. The target population must always be clearly defined by the researcher. It must always be possible to determine whether a person belongs to this target population or not. The sample should only consist of people who belong to the target population. In case of probability sampling, the researcher is in control of the selection process. In this case, he/she can guarantee that selected persons are from the target population. The situation is unclear for a self-selection survey. Everybody can complete the questionnaire. The researcher can include some questions allowing him/her to detect and remove people who are not in the target population. Of course, this only works if people answer correctly and not try to circumvent this filter by faking someone in the target population.

An example was a survey about road pricing that was conducted by the Dutch Automobile Association (ANWB). See also Example 11.1. The survey was announced as a consultation of the members of the association about introducing road pricing in the Netherlands. The survey was implemented as a self-selection web survey. Everybody could go to the website of the ANWB and fill in the questionnaire. Also nonmembers could participate. The questionnaire included a question about membership. In the end, about 400,000 people completed the questionnaire, of which 50,000 said they were not a member. It is not clear to what extent all respondents told the truth. Another unclear aspect was that there was a question asking whether the respondents owned a car. This adds to the confusion about what the target population really is. All Dutch? Or all Dutch who are a member of the ANWB? Or maybe all Dutch car owners who are a member of the ANWB? Note that if it was the objective of the ANWB to consult its members, it would have been easy to draw a random sample from its membership administration, resulting in a representative sample.

The second issue with a self-selection is that sometimes it is possible to fill in the questionnaire more than once. If some respondents complete the questionnaire only once and others multiple times, the representativity of the sample is affected. It is possible to check respondents for multiple completion of questionnaires. One way to do this is to store the unique IP address of the device used by the respondent and to block filling in a questionnaire if the IP address was already used. There are also systems that put cookies on the computers of respondents to record it was already used.

Not every web survey has a check on multiple completion of questionnaires. Sometimes researchers seem to think it is more important to have a large response than a representative response. A web survey can also be used in a discussion about a political issue as an instrument giving everybody the opportunity to express his or her opinion.

Here is an example of problems that may occur if a web survey is not protected against multiple completion. The self-selection online survey was an opinion poll conducted during the campaign for the local elections in the Netherlands in 2012. A public debate was organized between local party leaders in Amsterdam. A local newspaper, *Het Parool*, conducted the poll to find out who won the debate. Campaign teams of two parties (the Socialist Party and the Liberal Democrats) discovered that after disabling cookies on their computer, it was possible to fill in the questionnaire repeatedly. So, the campaign teams stayed up all night and voted as many times as possible. In the morning, the leaders of these two parties had a disproportionally large number of votes. The newspaper realized that something was wrong and cancelled the poll. It accused the two political parties of manipulating the poll. It was the newspaper, however, that had set up a bad poll.

A third issue with self-selection is the risk of manipulation of a survey. Probability sampling has the advantage that it provides protection against certain groups in the population attempting to manipulate the outcomes of the survey. This may typically play a role in opinion polls. Self-selection does not have this safeguard.

There are several examples where organizations attempted to influence the outcomes of surveys by advising their members to participate in it. One example is the election of the 2005 Book of the Year Award (Dutch: NS Publieksprijs), a high-profile literary prize in the Netherlands. The winning book was determined by means of a poll on a website. People could vote for one of the nominated books or mention another book of their choice. More than 90,000 people participated in the survey. The winner turned out to be the new interconfessional Bible translation published by the Netherlands and Flanders Bible Societies. This book was not nominated, but nevertheless an overwhelming majority (72%) voted for it. This was due to a campaign launched by (among others) Bible societies, a Christian broadcaster, and Christian newspaper. Although this was all completely within the rules of the poll, the group of voters was clearly not representative for the Dutch population.

Another example of survey manipulation is a single-question poll carried out during the campaign of the presidential election in the United States in 2016. *Single-question* polls are becoming more and more popular. They can typically be encountered on the front page of websites of news media. They have just one question. It is not clear what the target population of a single-question poll is. Everybody can answer this question. For example, people from Europe could participate in a poll about the presidential elections in the United States.

There is an increasing risk that a single-question poll is manipulated by a votebot. A votebot is a special kind of Internet bot. An *Internet bot* is a software application that carries out automated tasks (scripts) over the Internet. A *votebot* aims to automatically answer the questions in online polls, often in a malicious manner. A votebot attempts to act like a human, but answers the questions in an automated manner in order to manipulate the result of the poll. Votebots are sold on the Internet, but simple votebots are easy to code and deploy. Developers of web survey application can protect their software against attacks of votebots, but this requires extra efforts.

Many single-question polls were conducted during the campaign for the presidential election on November 8, 2016, in the United States. They were also used for measuring the performance of the candidates in the debates. There were three debates between Hillary Clinton and Donald Trump. After the third debate, there was a single-question poll on the website of *Breitbart News Network*. This is a politically conservative American news, opinion, and propaganda website. The poll question was: "Who won the debate?" There were two possible answers: Donald Trump and Hillary Clinton. The poll was activated immediately after the end of the debate. Since Breitbart is a conservative news network, many visitors will be Republicans. Therefore, it was to be expected that the Republican Donald Trump would get more votes than the Democrat Hillary Clinton. After a couple of hours, 150,000 people had answered the poll question. Surprisingly, Hillary Clinton was in the lead. According to over 60% of the respondents, she had won the debate.

Breitbart claimed that the poll was infiltrated by votebots that were operated from countries such as Romania, Germany, and South Korea. Based on these votes, Hillary Clinton was overwhelmingly the winner of the final presidential debate. Later, after 300,000 people had participated in the poll, the situation changed, and Donald Trump took the lead with 54% of the votes. Probably there were also votebots working for Donald Trump. This example clearly shows that representativity of this single-question poll is seriously affected by self-selection, manipulation by votebots. Therefore, the outcomes of this poll should not be trusted. See Example 11.2 for a detailed description.

▓ EXAMPLE 11.2 **The presidential elections in the United States in 2016**

On November 8, 2016, Donald Trump was elected president of the United States. For many, the victory of Donald Trump came as a surprise. They thought Hillary Clinton would be the new president. Apparently many polls were wrong.

Table 11.2 contains an overview of polls that were conducted in the last two days before the election. People could choose between four candidates, of which Hillary Clinton and Donald Trump were by far the most important ones.

TABLE 11.2 **Predictions for the presidential election in the United States on November 8, 2016**

| Organization | Sample | Mode | % Clinton | % Trump |
|---|---|---|---|---|
| YouGov | 3,677 | Online | 45 | 41 |
| Insights West | 940 | Online | 49 | 45 |
| Ipsos | 2,195 | Online | 42 | 39 |
| NBC News | 70,194 | Online | 47 | 41 |
| Bloomberg | 799 | Telephone | 44 | 41 |
| ABC News | 2,220 | Telephone | 47 | 43 |
| Fox News | 1,295 | Telephone | 48 | 44 |
| IBD/TIPP | 1,026 | Telephone | 41 | 43 |
| Monmouth | 802 | Telephone | 50 | 44 |
| CBS News | 1,426 | Telephone | 45 | 41 |
| Gravis | 16,639 | Robopoll | 47 | 43 |
| Rasmussen | 1,500 | Robopoll | 45 | 43 |
| Election | | | 47.8 | 47.3 |

Four of the 12 polls were online polls. Their samples were selected from online panels. The other eight polls were telephone polls, for which samples were selected by means of random digit dialing.

Two telephone polls were robopolls. Robopolls are completely automated telephone polls. A computer system generates random telephone numbers and calls these numbers. Respondents answer the recorded voice using the keypad on their telephone. A federal law in the United States prohibits robopolls for mobile telephones. To get people with mobile telephones in the sample, the robopolls in the table have additional online polls.

Figure 11.1 compares the poll estimates with the election result for Donald Trump. The dots represent the prediction of the percentage of votes for Trump. The vertical line denotes the election result (46.3%). The estimates of all polls are smaller than the election result. The horizontal line segments indicate the margins of error. For nine of the 12 polls, the estimates are outside the margins of error. So it can be concluded that these differences are significant.

Most polls underestimate the percentage of votes for Donald Trump. One of the possible explanations is that the samples are not representative. The lack of representativity of the online polls could be caused by the use of self-selection web panels. For the telephone polls, problems could be caused by the high nonresponse rates.

Another possible cause of the lack of representativity could be a "Shy Trump Factor." People may not want to admit the intent to vote for Donald Trump and therefore do not respond in the poll. Further research should make clear what really went wrong.

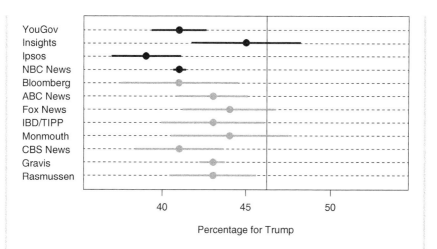

FIGURE 11.1 **The difference between the polls and the election for Donald Trump**

This example also shows that increasing the sample does not help to reduce the bias of polls. The poll of NBC News had a sample size of no less than 70,194 people. Still, this does not help to reduce the bias. So the guideline is: if an estimate is biased, do not increase the sample size, but improve the composition of the sample.

## 11.2 Theory

### 11.2.1 BASIC SAMPLING THEORY

Let the target population $U$ of the survey consist of $N$ identifiable elements, which are labeled 1, 2, ..., $N$. Therefore, the target population can be denoted by

$$(11.1) \qquad U = \{1, 2, ..., N\}.$$

Associated with each element $k$ is a value $Y_k$ of the target variable $Y$. The aim of the web survey is assumed to be estimation of the population mean

$$(11.2) \qquad \overline{Y} = \frac{1}{N} \sum_{k=1}^{N} Y_k$$

of the target variable $Y$.

Suppose a probability sample is selected without replacement. It means that each element can appear at most once in the sample. Therefore the sample can be represented by a set of indicators:

$$(11.3) \qquad a = a_1, a_2, \ldots, a_N.$$

The $k$th indicator $a_k$ assumes the value 1 if element $k$ is selected in the sample, and otherwise it assumes the value 0. The expected value, i.e., the mean value over all possible samples, of $a_k$ is denoted by

$$(11.4) \qquad \pi_k = E(a_k).$$

The quantity $\pi_k$ is called the *first-order inclusion probability* of element $k$ (for $k = 1$, 2, ..., $N$). For deriving variance formulas, also second-order inclusion probabilities are required. The *second-order inclusion probability* of elements $k$ and $l$ (with $k \neq l$) is equal to

$$(11.5) \qquad \pi_{kl} = E(a_k a_l),$$

and by definition $\pi_{kk} = \pi_k$. The *sample size*, i.e., the number of selected elements, is denoted by $n$. Since the indicators $a_k$ have the value 1 for all elements in the sample and the value 0 for all other elements, the sample size can be written as the sum of the values of the indicators:

$$(11.6) \qquad n = \sum_{k=1}^{N} a_k.$$

The Horvitz–Thompson estimator is defined by

$$(11.7) \qquad \bar{y}_{HT} = \frac{1}{N} \sum_{k=1}^{N} a_k \frac{Y_k}{\pi_k}.$$

The indicators $a_k$ filter out the sample values of the target variable. Note that each value $Y_k$ is weighted with its inverse selection probability $\pi_k$. Thus, the estimator is corrected for the fact that elements with a large inclusion probability are overrepresented in the sample.

The Horvitz–Thompson estimator is an unbiased estimator of the population mean. The variance of this estimator is equal to

$$(11.8) \qquad V(\bar{y}_{HT}) = \frac{1}{N^2} \sum_{k=1}^{N} \sum_{l=1}^{N} (\pi_{kl} - \pi_k \pi_l) \frac{Y_k}{\pi_k} \frac{Y_l}{\pi_l}.$$

For without replacement samples of fixed size $n$, the variance can be rewritten in the form

$$(11.9) \qquad V\left(\bar{y}_{HT}\right) = \frac{1}{2N^2} \sum_{k=1}^{N} \sum_{l=1}^{N} (\pi_k \pi_l - \pi_{kl}) \left(\frac{Y_k}{\pi_k} - \frac{Y_l}{\pi_l}\right)^2.$$

This expression shows that the variance can be reduced by taking the first-order inclusion probabilities as much as possible proportional to the values of the target variable.

The *variance of the estimator* is just one way to quantify the precision of an estimator. A small variance means a high precision and a large variance a small precision. Another way to quantify the precision is the *standard error*. It is defined by

$$(11.10) \qquad S\left(\bar{y}_{HT}\right) = \sqrt{V\left(\bar{y}_{HT}\right)}.$$

The standard error is required to compute the confidence interval. The *confidence interval* is a range of possible values of the population mean. If an estimator is unbiased, the interval encompasses the true value of the population mean with a high probability. This probability is called the *confidence level*. It is denoted by $(1 - \alpha)$ where $\alpha$ is a small probability. Often the value $\alpha = 0.05$ is used, corresponding to a confidence level of 95%.

The distribution of many estimators (including the sample mean) can for large probability samples be approximated by a normal distribution. This makes it easier to compute confidence intervals. Only the standard error of the estimator is required. The 95% confidence interval of the Horvitz–Thompson estimator is equal to

$$(11.11) \qquad \left(\bar{y}_{HT} - 1.96 \times S\left(\bar{y}_{HT}\right); \bar{y}_{HT} + 1.96 \times S\left(\bar{y}_{HT}\right)\right).$$

The best known and probably most often used type of probability sample is a *simple random sample without replacement*. First-order inclusion probabilities of all elements are equal for this type of sample. It can be shown that all first-order inclusion probabilities are equal to $n/N$. Furthermore, all second-order inclusion probabilities are equal to $n(n-1)/N(N-1)$. Substitution of these values of the inclusion probabilities in expression in (11.7) results in a simple estimator, the *sample mean*

$$(11.12) \qquad \bar{y} = \frac{1}{n} \sum_{k=1}^{N} a_k Y_k = \frac{1}{n} \sum_{i=1}^{n} y_i,$$

where $y_1, y_2, \ldots, y_n$ denote the $n$ observations that have become available in the sample. This is an unbiased estimator with variance

(11.13)
$$V(\bar{y}) = \frac{1-f}{n} S^2,$$

where $f = n/N$ is the *sampling fraction* and $S^2$ is the population variance, defined by

(11.14)
$$S^2 = \frac{1}{N-1} \sum_{k=1}^{N} \left(Y_k - \overline{Y}\right)^2.$$

From expression (11.13), it is clear that an increased sample size produces more precise estimators.

## 11.2.2 A SELF-SELECTION SAMPLE FROM THE INTERNET POPULATION

The population $U$ is divided into two subpopulations: a subpopulation $U_I$ of elements having access to the Internet and a subpopulation $U_{NI}$ of elements not having access to the Internet. Associated with each element $k$ is an indicator $I_k$, where $I_k = 1$ if element $k$ has access to the Internet ($k \in U_I$) and $I_k = 0$ otherwise ($k \in U_{NI}$). The subpopulation $U_I$ is called the *Internet population*, and the subpopulation $U_{NI}$ is called the *non-Internet population*. Let

(11.15)
$$N_I = \sum_{k=1}^{N} I_k$$

denote the size of the Internet population $U_I$. The mean of the values of the target variable in the Internet population is defined by

(11.16)
$$\overline{Y}_I = \frac{1}{N_I} \sum_{k=1}^{N} I_k Y_k.$$

Likewise, $N_{NI} = N - N_I$ denotes the size of the subpopulation without Internet, where $N_I + N_{NI} = N$. The mean of the values of the target variable in the non-Internet population is defined by

(11.17)
$$\overline{Y}_{NI} = \frac{1}{N_{NI}} \sum_{k=1}^{N} (1 - I_k) Y_k.$$

What happens if a self-selection sample is selected from the Internet population? This section shows that estimators can be substantially biased and also that this bias can be larger than the bias caused by nonresponse in surveys based on probability samples.

Participation in a self-selection survey requires respondents to be aware of the existence of the survey. They have to accidentally visit the website, or they have to follow up a banner or an e-mail message. They also have to decide to fill in the questionnaire on the Internet. This means that each element $k$ in the Internet population has unknown probability $\rho_k$ of participating in the survey, for $k = 1, 2, \ldots, N_I$.

The responding elements are denoted by a set of indicators:

$$(11.18) \qquad R_1, R_2, \ldots, R_N,$$

where the $k$th indicator $R_k$ assumes the value 1 if element $k$ participates and otherwise it assumes the value 0, for $k = 1, 2, \ldots, N$. The expected value $\rho_k = E(R_k)$ is called the *response probability* of element $k$. For sake of convenience, response probabilities are also introduced for elements in the non-Internet population. By definition the values of all these probabilities are 0. The realized sample size is denoted by

$$(11.19) \qquad n_S = \sum_{k=1}^{N} R_k.$$

Lacking any knowledge about the values of the response probabilities, researchers usually implicitly assume all these probabilities to be equal. In other words, simple random sampling is assumed. Consequently, the sample mean

$$(11.20) \qquad \bar{y}_S = \frac{1}{n_S} \sum_{k=1}^{N} R_k Y_k$$

is used as an estimator for the population mean. The expected value of this estimator is approximately equal to

$$(11.21) \qquad E\left(\bar{y}_S\right) \approx \tilde{Y} = \frac{1}{N_I \bar{\rho}} \sum_{k=1}^{N} \rho_k I_k Y_k$$

where $\bar{\rho}$ is the mean of all response propensities in the Internet population. This expression was derived by Bethlehem (1988).

Using an approach similar to Cochran (1977, p. 31), it can be shown that the variance of estimator (11.20) is approximately equal to

$$(11.22) \qquad V(\bar{y}) \approx \frac{1}{(N\bar{\rho})^2} \sum_{k=1}^{N} \rho_k (1 - \rho_k) \left(Y_k - \tilde{Y}\right)^2.$$

Note that this expression for the variance does not contain the sample size $n$ (because no fixed size sample was drawn), but the expected sample size $N\bar{\rho}$. Not surprisingly, the variance decreases as the expected sample size increases.

It is clear from expression (11.21) that, generally, the expected value of the sample mean is not equal to the population mean of the Internet population. One situation in which the bias vanishes is that in which all response probabilities in the Internet population are equal. In terms of nonresponse correction theory, this comes down to Missing Completely Missing at Random (MCAR). This is the situation in which the cause of missing data is completely independent of all variables measured in the survey. For more information on MCAR and other missing data mechanisms, see Little and Rubin (2002). Indeed, in the case of MCAR, self-selection does not lead to an unrepresentative sample because all elements have the same selection probability.

Bethlehem (2002) shows that the bias of the sample mean (11.20) can be written as

$$(11.23) \qquad B\left(\bar{y}_S\right) = E\left(\bar{y}_S\right) - \overline{Y}_I \approx \widetilde{Y} - \overline{Y}_I = \frac{C_{\rho Y}}{\bar{\rho}} = \frac{R_{\rho Y} S_\rho S_Y}{\bar{\rho}},$$

in which

$$(11.24) \qquad C_{\rho Y} = \frac{1}{N_I}\sum_{k=1}^{N} I_k(\rho_k - \bar{\rho})\left(Y_k - \overline{Y}_I\right)$$

is the covariance between the values of target variable and the response probabilities in the Internet population and $\bar{\rho}$ is the average response probability. Furthermore, $R_{\rho Y}$ is the correlation coefficient between the target variable and the response behavior, $S_\rho$ is the standard deviation of the response probabilities, and $S_Y$ is the standard deviation of the target variable. The bias of the sample mean (as an estimator of the mean of the Internet population) is determined by three factors:

- The average response probability. If people are more likely to participate in the survey, the average response probability will be higher, and thus the bias will be smaller.

- The relationship between the target variable and response behavior. A strong correlation between the values of the target variable and the response probabilities will lead to a large bias.

- The variation in the response probabilities. The more these values vary, the larger the bias will be.

There are three situations in which this bias vanishes:

1. All response probabilities are equal. Again, this is the case in which the self-selection process can be compared with a simple random sample.
2. All values of the target variable are equal. This situation is very unlikely to occur in practice. No survey would be necessary in this case. One observation would be sufficient.
3. There is no relationship between the target variable and the response behavior. It means participation does not depend on the value of the target variable.

Expression (11.23) for the bias of the estimator can be used to compute an upper bound for the bias (the worst case). Given the mean response probability $\overline{\rho}$, there is a maximum value the standard deviation $S_\rho$ of the response probabilities cannot exceed:

$$(11.25) \qquad S(\rho) \leq \sqrt{\overline{\rho}(1-\overline{\rho})}.$$

This implies that in the worst case $S_\rho$ assumes its maximum value and the correlation coefficient $R_{\rho Y}$ is equal to either $+1$ or $-1$. Then the absolute value of the bias will be equal to

$$(11.26) \qquad |B_{\max}| = S_Y \sqrt{\frac{1}{\overline{\rho}} - 1}.$$

This worst-case expression of the value of the bias also applies to the situation in which a probability sample has been drawn and subsequently nonresponse occurs in the fieldwork. Therefore, expression (11.26) provides a means to compare potential biases in various surveys. See Example 11.3 for an empirical analysis.

### EXAMPLE 11.3 The bias worst case in Dutch surveys

Around 2006, general population surveys of Statistics Netherlands had response rates of around 70%. This means the absolute maximum bias was equal to

$$0.65 \times S_Y.$$

A large self-selection web survey in the Netherlands was *21minuten.nl*. This survey was supposed to provide answers to questions about important problems in the Dutch society. Within a period of six weeks in 2006, about

> 170,000 people completed the questionnaire (which took about 21 minutes). As everyone could participate in the survey, the target population was not defined properly. If it is assumed the target population consisted of all Dutch citizens from the age of 18, the average response probability was $170,000/12,800,000 = 0.0133$. Hence the absolute maximum bias is equal to
>
> $$8.61 \times S_Y.$$
>
> The conclusion is that the bias of the large web survey could have been a factor 13 larger than the bias of the small probability survey.

It is important to note that the value of expression (11.22) does not depend on the sample size. Consequently, increasing the sample size will not reduce the bias. So the problem of self-selection bias in web surveys does not diminish by having more people completing the survey questionnaire.

The precision of an estimator is often quantified by a 95% confidence interval. Suppose a simple random sample is selected from a target population. Then the sample mean $\bar{y}$ can be computed.

This is an unbiased estimator for the population mean $\bar{Y}$. Since the sample mean is (approximately) normally distributed, the 95% confidence interval for the population mean is equal to

$$(11.27) \qquad I = (\bar{y} - 1.96 \times S(\bar{y}); \bar{y} + 1.96 \times S(\bar{y})),$$

where $S(\bar{y})$ is the standard error of the sample mean. The probability that this interval contains the true value is by definition (approximately) equal to

$$(11.28) \qquad P(\bar{Y} \in I) = 0.95.$$

The standard error decreases with an increasing sample size. Therefore, the width of interval in expression (11.27) is smaller for a larger sample.

If a self-selection sample is selected from the Internet population, the sample mean $\bar{y}_S$ is used to estimate the population mean $\bar{Y}$. Analogous to expression (11.27), the confidence will be computed as

$$(11.29) \qquad I_S = (\bar{y}_S - 1.96 \times S(\bar{y}_S); \bar{y}_S + 1.96 \times S(\bar{y}_S)).$$

The confidence level of this interval is not by definition equal to 0.95. It can be shown that

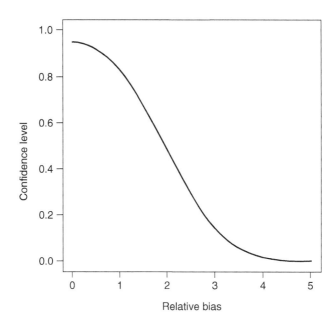

**FIGURE 11.2** The confidence level of the 95% confidence interval as a function of the relative bias

$$(11.30) \qquad P\left(\overline{Y} \in I_S\right) = \Phi\left(1.96 - \frac{B\left(\overline{y}_S\right)}{S\left(\overline{y}_S\right)}\right) - \Phi\left(-1.96 - \frac{B\left(\overline{y}_S\right)}{S\left(\overline{y}_S\right)}\right),$$

in which $\Phi$ is the standard normal distribution function. The quantity $B\left(\overline{y}_S\right)/S\left(\overline{y}_S\right)$ is called the *relative bias*. Apparently, the confidence level depends on the value of this relative bias. Figure 11.2 contains a plot of the confidence level as a function of the relative bias.

An increased sample size will reduce the standard error, but the bias remains the same. Hence, the relative bias will increase. It is clear that the confidence level can be much lower than expected. If the bias is equal to the standard error, i.e., the relative bias is 1, the confidence level is only 0.83. As the relative bias increases, the situation becomes worse. The confidence level is even less than 0.5 for a relative bias of 2. The conclusion is that self-selection may lead to an incorrect confidence interval.

## 11.2.3 REDUCING THE SELF-SELECTION BIAS

There are several ways in which the negative effects of self-selection can be reduced. Three approaches are discussed here.

The first approach is to avoid self-selection by selecting a proper probability sample. It is possible to conduct a web survey that is based on probability sampling. This requires a sampling frame. Sometimes such sampling frames are available. An example is a survey among employees of a company, where every employee has a company-assigned e-mail address. The sampling frame for this example consists of the list of e-mail addresses. The situation is not so straightforward for a general population survey. Unfortunately, there are no population registers containing e-mail addresses. A solution can be to approach sampled persons by some other mode. One option is to send them a letter with the request to go to a specific website, where the online questionnaire can be completed. Such a letter should also contain a unique identification code that has to be entered. Use of such identifying codes guarantees that only sampled persons respond and that they respond only once. Another option is to approach sampled persons face-to-face (CAPI) or by telephone (CATI) and asking them for their e-mail address (if they want to participate). If they cooperate, they are sent a link to the online questionnaire form.

An interesting approach is the one used in the LISS panel, as explained in Example 11.4.

---

■ EXAMPLE 11.4 **The LISS panel**

The Dutch LISS panel is the result of an attempt to set up a web panel where the panel members are recruited by means of probability sampling. See Scherpenzeel (2008) for a detailed description of this panel.

The panel has been constructed by selecting a random sample of households from the population register of the Netherlands. Selected households were recruited for this panel by means of a face-to-face interview (CAPI) or a telephone interview (CATI). Cooperative households without Internet access were provided with equipment giving them access to the Internet. Analysis by Scherpenzeel and Bethlehem (2011) showed that this panel produced better estimates than panels based on self-selection.

---

A second approach to reduce the negative effects of self-selection is applying some form of adjustment weighting. Adjustment weighting is a family of techniques that attempt to reduce the bias of survey estimates by assigning weights to responding elements. These weights correct for the over- or underrepresentation of specific groups in the response. Adjustment weighting is treated in detail in Chapter 12. Here one such technique is summarized. It is called post-stratification.

*Post-stratification* requires one or more auxiliary variables. An auxiliary variable is a variable that has been measured in the survey and for which the distribution in the target population is available. Typical auxiliary variables are gender, age, marital status, and region. By comparing the response distribution of an auxiliary variable with its population distribution, it can be assessed whether or not the survey response is representative for the population (with respect to this variable). If these distributions differ considerably, one must conclude that the response is selective. To correct this, adjustment weights are computed. Weights are assigned to all records of observed elements. Estimates of population characteristics can now be obtained by using the weighted values instead of the unweighted values.

To carry out post-stratification, one or more qualitative auxiliary variables are needed. Here, only one such variable is considered. The situation for more variables is not essentially different. Suppose there is an auxiliary variable $X$ having $L$ categories. So it divides the target population into $L$ strata. The strata are denoted by the subsets $U_1$, $U_2$, ..., $U_L$ of the population $U$. The number of target population elements in stratum $U_h$ is denoted by $N_h$, for $h = 1, 2, ..., L$. Hence, the population size $N$ is equal to $N = N_1 + N_2 + ... + N_L$. This is the population information assumed to be available.

Suppose a sample of size $n$ is selected from the Internet population. If $n_h$ denotes the number of sample elements in stratum $h$, then $n = n_1 + n_2 + \cdots + n_L$. Note that since the sample is selected from the Internet population $U_I$, only elements in the substrata $U_I \cap U_h$ are observed (for $h = 1, 2, ..., L$).

Post-stratification assigns identical adjustment weights to all elements in the same stratum. The weight $w_k$ for an element $k$ in stratum $h$ is equal to

$$(11.31) \qquad w_k = \frac{N_h/N}{n_h/n}.$$

Post-stratification comes down to replacing the simple sample mean

$$(11.32) \qquad \bar{y}_S = \frac{1}{n_S} \sum_{k=1}^{N} R_k Y_k$$

by the weighted sample mean

$$(11.33) \qquad \bar{y}_{S,PS} = \frac{1}{n_S} \sum_{k=1}^{N} w_k R_k Y_k.$$

Substituting the weights and working out this expression leads to the post-stratification estimator

$$(11.34) \qquad \bar{y}_{S,PS} = \frac{1}{N} \sum_{h=1}^{L} N_h \bar{y}_S^{(h)} = \sum_{h=1}^{L} W_h \bar{y}_S^{(h)},$$

where $\bar{y}_S^{(h)}$ is the sample mean in stratum $h$ and $W_h = N_h/N$ is the relative size of stratum $h$. The expected value of this post-stratification estimator is equal to

$$(11.35) \qquad E(\bar{y}_{S,PS}) = \frac{1}{N} \sum_{h=1}^{L} N_h E(\bar{y}_S^{(h)}) = \sum_{h=1}^{L} W_h \tilde{Y}^{(h)} = \tilde{Y}^*,$$

where

(11.36)
$$\tilde{Y}^* = \frac{1}{N_h} \sum_{k=1}^{N_h} \frac{\rho_{k,h}}{\overline{\rho}_h} Y_{k,h}$$

is the weighted mean of the target variable in stratum $h$. The subscript $k$, $h$ denotes the $k$th element in stratum $h$, and $\overline{\rho}_h$ is the average response probability in stratum $h$.

Generally, this mean will not be equal to the mean $\overline{Y}_h$ of the target variable in stratum $h$ of the target population. The bias of this estimator is equal to

(11.37)
$$B(\overline{y}_{S,PS}) = E(\overline{y}_{S,PS}) - \overline{Y} = \tilde{Y}^* - \overline{Y} = \sum_{h=1}^{L} W_h \left( \tilde{Y}^{(h)} - \overline{Y}^{(h)} \right)$$
$$= \sum_{h=1}^{L} W_h \frac{R_{\rho Y}^{(h)} S_{\rho}^{(h)} S_Y^{(h)}}{\overline{\rho}^{(h)}},$$

where the subscript $h$ indicates that the respective quantities are computed just for stratum $h$ and not for the complete population. The bias (11.37) will be small if

- The response propensities are similar within strata;
- The values of the target variable are similar within strata;
- There is no correlation between response behavior and the target variable within strata.

These conditions can be realized if there is a strong relationship between the target variable $Y$ and the stratification variable $X$. Then the variation in the values of $Y$ manifests itself between strata and not within strata. In other words, the strata are homogeneous with respect to the target variable. Also if the strata are homogeneous with respect to the response propensities, the bias will be reduced. In terms of missing data terminology, this situation comes down to Missing at Random (MAR).

It can be shown that, in general, the variance of the post-stratification estimator is approximately equal to

(11.38)
$$V(\overline{y}_{PS}) = \sum_{h=1}^{L} W_h^2 V(\overline{y}_h).$$

In the case of a self-selection web survey, the variance $V(\overline{y}_h)$ of the sample mean in a stratum is the analogue of variance (11.22) but restricted to observations in that stratum. Therefore, the variance of the post-stratification estimator is approximately equal to

$$(11.39) \qquad V\left(\bar{y}_{S,PS}\right) = \sum_{h=1}^{L} W_h^2 \frac{1}{\left(N_h \bar{\rho}_h\right)^2} \sum_{k \in U_h} \rho_k (1 - \rho_k) \left(Y_k - \widetilde{Y}^{(h)}\right)^2.$$

This variance is small if the strata are homogeneous with respect to the target variable. So, a strong correlation between the target variable $Y$ and the stratification variable $X$ will reduce both the bias and the variance of the estimator.

The conclusion can be that application of post-stratification will successfully reduce the bias of the estimator if proper auxiliary variables can be found. Such variables should satisfy three conditions:

- They have to be measured in the survey.
- Their population distribution $(N_1, N_2, ..., N_L)$ must be known.
- They must produce homogeneous strata.

Unfortunately, such variables are rarely available, or there is only a weak correlation. One way to solve this problem is to carry out a *reference survey*. The objective of such a survey is to measure just the auxiliary variables required for weighting purposes. To obtain unbiased estimates of the population distributions of these variables, data should preferably be collected with CAPI or CATI. The reference survey is discussed in more detail in Chapter 12.

A third approach to reduce a self-selection bias is to apply *propensity weighting*. This technique is particularly used by market research organizations. See also Börsch-Supan et al. (2004) and Duffy et al. (2005). The original idea behind propensity weighting goes back to Rosenbaum and Rubin (1983, 1984).

*Propensity scores* are obtained by modeling a variable that indicates whether or not someone participates in the survey. Usually a logistic regression model is used where the indicator variable is the dependent variable and attitudinal variables are the explanatory variables. These attitudinal variables are assumed to explain why someone participates or not. Fitting the logistic regression model comes down to estimating the probability (propensity score) of participating, given the values of the explanatory variables.

Each person $k$ in the population is assumed to have a certain, unknown probability $\rho_k$ of participating in the survey, for $k = 1, 2, ..., N$. Let $R_1, R_2, ..., R_N$ denote indicator variables, where $R_k = 1$ if person $k$ participates in the survey and $R_k = 0$ otherwise. Consequently, $P(R_k = 1) = \rho_k$.

The *propensity score* $\rho(X)$ is the conditional probability that a person with observed characteristics $X$ participates, i.e.,

$$(11.40) \qquad \rho(X) = P(R = 1 \mid X).$$

It is assumed that within the strata defined by the values of the observed characteristics $X$, all persons have the same participation propensity. This is the MAR assumption. The propensity score is often modeled using a logit model:

$$(11.41) \qquad \log \left( \frac{\rho(X_k)}{1 - \rho(X_k)} \right) = \alpha + \beta' X_k.$$

The model is fitted using maximum likelihood estimation. Once propensity scores have been estimated, they are used to stratify the population. Each stratum consists of elements with (approximately) the same propensity scores. If indeed all elements within a stratum have the same response propensity, there will be no bias if just the elements in the Internet population are used for estimation purposes. Cochran (1968) claims that five strata are usually sufficient to remove a large part of the bias.

From a theoretical point of view, propensity weighting should be sufficient to remove the bias. However, in practice the propensity score variable will often be combined with other (demographic) variables in a more extended weighting procedure; see, e.g., Schonlau et al. (2004). The use of propensity scores is described in more detail in Chapter 13.

# 11.3  Applications

Two applications of self-selection are discussed in this section. The first one is an experiment in which self-selection is simulated. Since the target population is completely known, the effects of self-selection can be shown. The second application describes an experiment that was carried out in the town of Alphen a/d Rijn in the Netherlands. The same poll was carried out with the same questionnaire, at the same time, but with three different modes of data collection, one of them being self-selection. So a self-selection poll could be compared with a poll based on probability sampling.

## 11.3.1  APPLICATION 1: SIMULATING SELF-SELECTION POLLS

The possible consequences of self-selection and the effectiveness of correction techniques are illustrated using a simulation experiment. A fictitious population was constructed. For this population, reported voting behavior in a self-selection survey was simulated and analyzed. The relationships between the variables involved were modeled somewhat stronger than they probably would be in a real-life situation. Effects are therefore more pronounced, making it clearer what the pitfalls are.

The characteristics of estimators (before and after correction) were computed based on a large number of simulations. First, the distribution of the estimator was determined in the ideal situation of a simple random sample from the target population. Then, it was explored how the characteristics of the estimator changed if self-selection was applied. Finally, the effect of weighting (post-stratification) was analyzed.

A fictitious population of 100,000 individuals was constructed. There were five variables:

- The variable *Internet* indicated how active a person was on the Internet. There were two categories: very active users and passive users. The population consisted for 1% of very active users and for 99% of passive users. Active users had a response probability of 0.99, and passive users had a response probability of 0.01.

- The variable *Age* in three categories: young, middle-aged, and old. The active Internet users consisted for 60% of young people, for 30% of middle-aged people, and for 10% of old people. The age distribution for passive Internet users was 40% young people, 35% middle-aged people, and 25% old people. Typically younger people were more active Internet users.

- Voted for the National Elderly Party (NEP). The probability to vote for this party only depended on age. Probabilities were 0.00 (for young), 0.30 (for middle-aged), and 0.60 (for old).

- Voted for the New Internet Party (NIP). The probability to vote for this party depended on both age and use of Internet. For active Internet users, the probabilities were 0.80 (for young), 0.40 (for middle-aged), and 0.20 (for old). For passive Internet users, all probabilities were equal to 0.10. So, for active users, voting for the NIP decreased with age. Voting probability was always low for passive users.

Figure 11.3 shows the relationships between the variables in a graphical way. The decision not to participate in a self-selection survey can be seen as a form of nonresponse. Nonresponse theory distinguishes three nonresponse generating mechanisms:

- *Missing Completely at Random* (MCAR). There is no relationship at all between the mechanism causing data to be missing and the target variables of the survey. This situation causes no problems. The mechanism only leads to a reduced number of observations. Estimators will not be biased.

- *Missing at Random* (MAR). There is an indirect relationship between the mechanism causing data to be missing and the target variables of the survey. The relationship runs through a third variable, and this variable is measured in the survey as an auxiliary variable. Estimates are biased in this case, but it is

FIGURE 11.3 **Relationships between variables**

possible to correct for this bias. For example, if the auxiliary variable is used to construct strata, there will be no bias within strata, and post-stratification will remove the bias.

• *Not Missing at Random* (NMAR). There is a direct relationship between the mechanism causing data to be missing and the target variables of the survey. This is the worst case. Estimators will be biased, and it is not possible to remove this bias.

The variable NEP (National Elderly Party) suffers from MAR. There is a direct relationship between voting for this party and age, and also there is a direct relationship between age and the probability to participate in the survey. This will cause estimates to be biased. It should be possible to correct for this bias by weighting using the variable age.

The variable NIP (National Internet Party) suffers from NMAR. There exists a direct relationship between voting for this party and the response probability. Estimates will be biased, and there is no correction possible.

The distribution of estimators for the percentage of votes for both parties was determined in various situations by repeating selection of the sample 1,000 times. The average response probability in the population was 0.01971. Therefore, the expected sample size in a self-selection survey was equal to $100,000 \times 0.01971 = 1,971$.

Figure 11.4 contains the results for the variable NEP (voted for National Elderly Party). The upper box plot shows the distribution of the estimator for simple random samples of size $n = 1,971$ from the target population. The vertical line denotes the population value to be estimated (25.6%). The estimator has a symmetric distribution around this value. This is a clear indication that the estimator is unbiased.

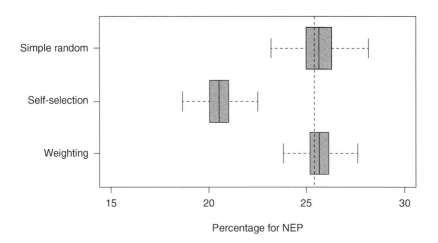

FIGURE 11.4 **Results of the simulations for variable NEP (National Elderly Party)**

The middle box plot shows what happens if samples are selected by means of self-selection. The shape of the distribution remains more or less the same, but the distribution as a whole has shifted to the left. All values of the estimator are systematically too low. The expected value of the estimator is only 20.4%. The estimator is biased. The explanation of this bias is simple: relative few elderly are active Internet users. Therefore, they are underrepresented in the samples. These are typically people who will vote for the NEP.

The lower box plot shows the distribution of the estimator in case of poststratification by age. The bias is removed. This was possible because this is a case of MAR.

The simulation results are summarized in Table 11.3. Self-selection results in a large relative bias of −7.4. Weighting eliminates the bias, and therefore the relative bias is 0. In case of self-selection, the confidence level of the 95% confidence interval is close to 0. This implies that almost certainly wrong conclusion will be drawn from this web survey.

Figure 11.5 contains the results for the variable NIP (voted for the New Internet Party). The upper box plot shows the distribution of the estimator for simple random samples of size 1,971 from the target population. The vertical line denotes the population value to be estimated (10.4%). Since the estimator has a symmetric distribution around this value, it is clear that the estimator is unbiased.

TABLE 11.3    **Summary of simulation results for the variable NEP**

| Simulation | Mean | Standard error | Bias | Relative bias |
|---|---|---|---|---|
| Random | 25.6 | 1.0 | 0.0 | 0.0 |
| Self-selection | 20.4 | 0.7 | −5.2 | −7.4 |
| Self-selection, with weighting adjustment | 25.6 | 0.7 | 0.0 | 0.0 |

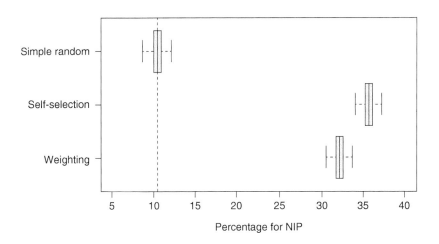

FIGURE 11.5    **Results of the simulations for variable NIP (New Internet Party)**

TABLE 11.4   **Summary of simulation results for the variable NIP**

| Simulation | Mean | Standard error | Bias | Relative bias |
|---|---|---|---|---|
| Random samples | 10.4 | 0.7 | 0.0 | 0.0 |
| Self-selection | 35.6 | 0.6 | 25.2 | 42.0 |
| Self-selection, with weighting adjustment | 32.1 | 0.6 | 21.7 | 36.2 |

The middle box plot shows what happens if samples are obtained by means of self-selection. The distribution has shifted to the right considerably. All values of the estimator are systematically too high. The expected value of the estimator is now 35.6%. The estimator is severally biased. The explanation of this bias is straightforward: voters for the NIP are overrepresented in Internet samples.

The lower box plot in Figure 11.5 shows the effect of post-stratification by age. Only a small part of the bias is removed. This is not surprising as there is a direct relationship between voting for the NIP and having access to the Internet. This is a case of NMAR.

The simulation results are summarized in Table 11.4. Self-selection results in a very large relative bias of 42.0. Weighting can reduce this relative bias only to 36.2. In both cases, the confidence level of the 95% confidence interval is close to 0. This implies that almost certainly wrong conclusions will be drawn from this web survey.

## 11.3.2 APPLICATION 2: SUNDAY SHOPPING IN ALPHEN A/D RIJN

Should shops be open on Sunday, or should they be closed on this day? This was the question that had to be answered by local politicians in the municipality of *Alphen a/d Rijn* in the western part of the Netherlands. They had opposing views. On the one hand, liberals believed that shopkeepers should be able to decide for themselves whether their shop should be open on Sunday or not. On the other hand, Christian parties wanted the shops closed, because Sunday is the day of the Lord. It is a day of rest, of going to church, and it is not a day for economic activities. The local politicians were not able to find a compromise. So in the end they decided to ask the inhabitants of the municipality for their opinion.

Lacking knowledge about the methodological aspects of good polls and surveys, the local politicians decided to do face-to-face interviews in the shopping centers on Saturday afternoon. After it was pointed out to them that this would not result in a representative sample, they decided to use the *Alphenpanel* for another survey. This was a web panel of inhabitants of the town of Alphen a/d Rijn. The approximately 1,600 panel members were recruited for a large part by means of a random sample from the population register of the municipality. Moreover, the politicians decided to also conduct a third survey. The idea was to offer a questionnaire on the Internet. There were no restrictions. Everybody could complete the questionnaire, even more than once. This self-selection survey was mainly offered as a means to give all inhabitants the possibility to express their opinion.

So the interesting situation occurred in which three surveys were carried out at the same time, with the same target population, and with the same questionnaire, but with a different mode of data collection. This made it possible to compare the three modes of data collection.

The results of the three surveys were published on March 3, 2015. It turned out that 754 people had completed the questionnaire of the face-to-face survey in the shopping centers. The self-selection survey produced 1,550 completed forms. A total of 1,600 members of Alphenpanel were invited to complete the questionnaire. In the end, 857 members did so. This comes down to a response rate of 54%. Taken into account the topic of the survey and the fact that all panel members agreed to participate in surveys, the response rate could have been higher.

The municipality of Alphen a/d Rijn had 107,000 inhabitants at the time. Approximately 66% of the people live in the urban town with the same name. The other 34% live in seven small rural towns around the urban area. For the panel survey and the self-selection survey, the town in which the respondents lived was recorded. This made it possible to compare the distribution over the towns in these surveys with the distribution in the population. Table 11.5 contains the data.

The response distribution in the panel resembles the distribution in the population. The largest difference is with respect to the percentage of people in the town of Alphen a/d Rijn. Sixty-six percent of the population lives in this town, whereas 70% of the panel respondents come from this town. This is a difference of 4 percentage points. For all other towns the difference is at most 1 percentage point. It can be concluded that the panel survey is reasonably representative with respect to town of residence.

TABLE 11.5 **Distribution of the respondents over the towns of Alphen a/d Rijn**

| Town | Panel (%) | Self-selection (%) | Population (%) |
|---|---|---|---|
| Aarlanderveen | 2 | 1 | 1 |
| Alphen a/d Rijn | 70 | 55 | 66 |
| Benthuizen | 3 | 13 | 3 |
| Boskoop | 13 | 18 | 14 |
| Hazerswoude-Dorp | 4 | 8 | 5 |
| Hazerswoude-Rijndijk | 4 | 2 | 5 |
| Koudekerk a/d Rijn | 3 | 2 | 4 |
| Zwammerdam | 1 | 1 | 2 |
| Total | 100 | 100 | 100 |

There are problems with the representativity of the self-selection survey. There are substantial differences between the percentages in the population and the percentages in the survey. For example, only 3% of the population lives in the small town of Benthuizen, but no less than 13% of the self-selection respondents are from this town. There is also an overrepresentation of people from the towns of Boskoop (18% instead of 14%) and Hazerswoude-Dorp (8% instead of 5%). A logical consequence of the overrepresentation of these three towns is that one or more other towns are underrepresented. This is indeed the case for the town of Alphen a/d Rijn. Sixty-six percent of the population lives in this town, but in the self-selection survey it is only 55%.

Why are the towns Benthuizen, Hazerswoude-Dorp, and Boskoop overrepresented in the self-selection survey? A plausible explanation is that these three towns are part of or close to the Dutch Bible Belt. This is a strip of land across the country that is inhabited by a high percentage of conservative Protestants. For example, in the town of Benthuizen, almost 50% voted for conservative Protestant parties in the local elections of 2010, while the average in the country was around 6%. There are many conservative Protestants in Benthuizen, Hazerswoude-Dorp, and Boskoop, and they were asked by their churches to participate in the self-selection survey. So one can expect these people to be overrepresented in this survey.

The objective of the surveys was to obtain more insight in the opinion of the inhabitants about shopping Sundays. So one of the questions was whether one favored or opposed shopping on Sunday. Figure 11.6 shows the percentages of opponents of shopping on Sundays. Note that weighting adjustment was applied to the panel survey and the self-selection survey and not to the face-to-face survey in the shopping centers.

The estimates differ substantially. They range from 22% to 43%. Given the ways in which the three surveys were conducted, one can expect the 22% of the panel survey closer to the true value in the population than the other two estimates. Therefore, 22% is the best guess for the percentage of opponents. Of course, it must be taken into account that the estimate is based on a sample from the

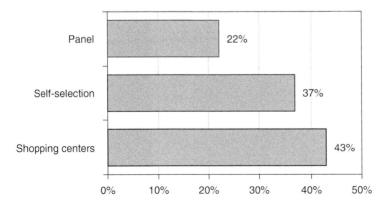

FIGURE 11.6  **Opponents of shopping Sundays by mode of data collection (percentages)**

population. So there are margins of error. For the estimate of 22%, the margin is approximately 3 percentage points. This means that with a high probability the true value will be between 19% and 25%. It must also be taken into account that the panel suffered from nonresponse in the recruitment process. This may cause the estimate to have some bias.

The estimate based on the self-selection survey was much higher: 37% instead of 22%. So there was a difference of 15 percentage points. Note that no statistical test was carried out to determine whether the difference was significant. This is not possible because the underlying distributions of the self-selection poll were unknown. However, the difference was so large that it could not be attributed to sampling error. Even after weighting adjustment, this estimate was much higher. Apparently, the conservative Protestants were still overrepresented. Weighting by region did not sufficiently help to reduce the overrepresentation of towns like Benthuizen and Hazerswoude-Dorp.

The face-to-face survey in the shopping centers produced an even larger estimate: 43%. It was almost double the value in the panel survey (22%). Without more research no clear explanation can be given for this large value. Maybe Saturday shoppers did not have a need to shop on Sundays. Further, people who like to shop on Sundays because it was not possible for them to shop on Saturday will not be included in the survey.

The differences between the three surveys are too large to be able to attribute them to random sample fluctuations. There are significant systematic differences. The only conclusion that can be drawn is that the self-selection survey and the face-to-face survey in the shopping centers are wrong. Their results should not be used.

More about the experiment in Alphen a/d Rijn can be found in Bethlehem (2015).

# 11.4 Summary

A web survey may suffer from self-selection. The survey questionnaire is simply put on the web. Respondents are those people who happen to have Internet access, visit the website, and decide to participate in the survey. The survey researcher is not in control over the selection process.

Since selection probabilities are unknown, it is not possible to compute unbiased estimates. It is also impossible to compute the precision of estimates.

It is usually assumed that a self-selection sample can be treated as a simple random sample. Hence, sample means are used as estimates as population means. Such an estimator can be seriously biased. The magnitude of the bias depends on:

1. The average response probability. If people are more likely to participate in the survey, the average response probability will be higher, causing the bias to be smaller.

2. The relationship between the target variable and response behavior. A strong correlation between the values of the target variable and the response probabilities will lead to a large bias.

3. The variation in the response probabilities. The more these values vary, the larger the bias will be.

The bias of the estimator is independent of the sample size. The bias does not go away for large samples. Particularly for large samples, confidence intervals give a wrong picture. The confidence level of this interval is often closer to 0% than to 95%

There are several approaches that may help to reduce a bias due to self-selection. A first approach is to select a proper probability sample from a sampling frame. A second approach is to carry out some kind of adjustment weighting. By assigning weights the response is corrected for under- or overrepresented groups. There is no guarantee that weighting will completely remove the bias.

## KEY TERMS

**First-order inclusion probability**: The probability that a population element is selected in the sample. The first-order inclusion probability is determined by the sampling design.

**Internet population**: The subpopulation of the target population consisting of elements that have access to the Internet.

**Missing at Random (MAR)**: Nonresponse depends on auxiliary variables only. Estimators will be biased, but correction is possible if some technique is used that takes advantage of this auxiliary information.

**Missing Completely at Random (MCAR)**: Nonresponse happens completely independent of all survey variables. Estimators will not be biased.

**Non-Internet population**: The subpopulation of the target population consisting of elements that do not have access to the Internet.

**Not Missing at Random (NMAR)**: Nonresponse depends directly on the target variables of the survey. Estimators will not be biased and correction techniques will not be successful.

**Post-stratification**: A weighting adjustment technique that divides the population in strata and subsequently assigns the same weight to all observed elements within a stratum.

**Self-selection survey**: A survey for which the sample has been recruited by means of self-selection. It is left to the persons themselves to decide to participate in a survey. The selection probabilities are unknown.

## EXERCISES

**Exercise 11.1**   Which of the following reasons may cause a researcher to use self-selection for a web survey instead of a probability sample?

**a.** A self-selection survey is cheaper.
**b.** The sample will be larger.
**c.** There is no nonresponse.
**d.** No sampling frame is needed.

**Exercise 11.2**   Can the bias due to self-selection always be corrected by means of a weighting adjustment?

**a.** Yes, weighting adjustment will always remove the bias.
**b.** No, weighting adjustment will only work if persons are missing due to Missing at Random.
**c.** No, weighting adjustment will only work if persons are missing due to Missing at Random, and the proper auxiliary variables are included in the weighting model.
**d.** No, weighting adjustment will never remove or reduce such a bias.

**Exercise 11.3**   What happens to the bias due to self-selection in web surveys if Internet access increases in the target population?

**a.** The bias will increase.
**b.** The bias will decrease.
**c.** The bias will not change.
**d.** It depends on the response probabilities of the people getting Internet access.

**Exercise 11.4**   Suppose all response probabilities in a specific population are less than 0.5. What would happen to the bias due to self-selection if all response probabilities were double as large?

**a.** The bias will be twice as large.
**b.** The bias will be halved.
**c.** The bias will not change.
**d.** The bias will vanish.

**Exercise 11.5**   The target variable $Y$ of a survey indicates whether persons have ($Y = 1$) or have not ($Y = 0$) a specific property. There is also an auxiliary variable $X$ with two categories ($X = 1$ and $X = 2$). The population consists of 2,400 people. The distribution of these people in the table obtained by crossing $X$ and $Y$ is given below. Within each cell of the table, the response probabilities are the same. The values of these probabilities are also given in the table below.

|         | $Y = 0$        | $Y = 1$        |
|---------|----------------|----------------|
| $X = 1$ | $N = 1,000$    | $N = 200$      |
|         | $\rho = 0.4$   | $\rho = 0.4$   |
| $X = 2$ | $N = 200$      | $N = 1,000$    |
|         | $\rho = 0.8$   | $\rho = 0.8$   |

**a.** Suppose a simple random sample is selected. Assume that all selected persons will respond. What will be expected value of the percentage of people having the specific property?

**b.** Suppose sampling relies on self-selection and the response probabilities are as indicated in the table. What will be the expected value of the estimated percentage?

**c.** Suppose sampling relies on self-selection and the response probabilities are as indicated in the table. If the variable $X$ is used for adjustment weighting (post-stratification), what will be the expected value of the estimated percentage?

**d.** Explain why or why not the self-selection bias is removed by weighting by $X$.

**Exercise 11.6**   A researcher wants to investigate transport behavior of commuters in a town. He/she intends to conduct a survey in which he/she asks commuters what means of transport they use to go to their work. The researcher thinks it is a good idea to have information about many commuters. He/she does not have the money and time to draw a proper random sample. He/she decides to go to the main railway station in the rush hour. He/she succeeds in interviewing a lot of people in the station. Not surprisingly, it turns out that many people use the train to commute.

**a.** Explain what is wrong with this sampling design?

The researcher observes that he/she has many young people in his survey and only a very few elderly. He/she decides to carry out a weighting adjustment by post-stratifying by the variable age.

**b.** Explain why or why not this will help to improve his estimators.

# REFERENCES

Bethlehem, J. G. (1988), Reduction of the Nonresponse Bias Through Regression Estimation. *Journal of Official Statistics*, 4, pp. 251–260.

Bethlehem, J. G. (2002), Weighting Nonresponse Adjustments Based on Auxiliary Information. In: Groves, R. M., Dillman, D. A., Eltinge, J. L., & Little, R. J. A. (eds.), *Survey Nonresponse*. Wiley, New York, USA.

Bethlehem, J. G. (2015), Essay: Sunday Shopping—The Case of Three Surveys. *Survey Research Methods*, 9, 3, pp. 221–230.

Börsch-Supan, A., Elsner, D., Faßbender, H., Kiefer, R., McFadden, D. & Winter, J. (2004), *Correcting the Participation Bias in an Online Survey*. Report, University of Munich, Germany.

Brüggen, E., Van Den Brakel, J., & Krosnick, J. (2016), *Establishing the Accuracy of Online Panels for Survey Research*. Report 2016|04, Statistics Netherlands, The Hague, the Netherlands.

Cochran, W. G. (1968), The Effectiveness of Adjustment by Subclassification in Removing Bias in Observational Studies. *Biometrics*, 24, pp. 205–213.

Cochran, W. G. (1977), *Sampling Tecniques*, 3rd edition. Wiley.

Duffy, B., Smith, K., Terhanian, G., & Bremer, J. (2005), Comparing Data from Online and Face-to-Face Surveys. *International Journal of Market Research*, 47, pp. 615–639.

Horvitz, D. G. & Thompson, D. J. (1952), A Generalization of Sampling Without Replacement from a Finite Universe. *Journal of the American Statistical Association*, 47, pp. 663–685.

Kruskal, W. & Mosteller, F. (1979a), Representative Sampling, I: Non-scientific Literature. *International Statistical Review*, 47, pp. 13–24.

Kruskal, W. & Mosteller, F. (1979b), Representative Sampling, II: Scientific Literature, Excluding Statistics. *International Statistical Review*, 47, pp. 111–127.

Kruskal, W. & Mosteller, F. (1979c), Representative Sampling, III: The Current Statistical Literature. *International Statistical Review*, 47, pp. 245–265.

Little, R. J. A. & Rubin, D. B. (2002), *Statistical Analysis with Missing Data*, 2nd edition. Wiley, New York, USA.

Rosenbaum, P. R. & Rubin, D. B. (1983), The Central Role of the Propensity Score in Observational Studies for Causal Effects. *Biometrika*, 70, pp. 41–55.

Rosenbaum, P. R. & Rubin, D. B. (1984), Reducing Bias in Observational Studies Using Subclassification on the Propensity Score. *Journal of the American Statistical Association*, 79, pp. 516–524.

Scherpenzeel, A. (2008), An Online Panel as a Platform for Multi-disciplinary Research. In: Stoop, I. & Wittenberg, M. (eds.), *Access Panels and Online Research, Panacea or Pitfall?* Aksant, Amsterdam, pp. 101–106.

Scherpenzeel, A. & Bethlehem, J. (2011), How Representative Are Online-Panels? Problems of Coverage and Selection and Possible Solutions. In: Das, M., Ester, P., Kaczmirek, L., & Mohler, P. (eds.), *Social Research and the Internet: Advances in Applied Methods and New Research Strategies*. Routledge Academic, New York, USA, pp. 105–132.

Schonlau, M., Zapert, K., Payne Simon, L., Haynes Sanstad, K., Marcus, S., Adams, J., Kan, H., Turber, R., & Berry, S. (2004), A Comparison Between Responses from Propensity-Weighted Web Survey and an Identical RDD Survey. *Social Science Computer Review*, 22, pp. 128–138.

Vonk, T., Van Ossenbruggen, R., & Willems, P. (2006), The Effects of Panel Recruitment and Management on Research Results, A Study among 19 Online Panels. *Panel Research 2006, ESOMAR World Research*, ESOMAR Publication Services, Vol. 317, pp. 79–99.

# CHAPTER TWELVE

# Weighting Adjustment Techniques

## 12.1 Introduction

It is the basic idea of survey sampling that observations on only a part of the elements in a population allow for drawing valid and accurate conclusion about the population as a whole. Horvitz and Thompson (1952) have shown in their seminal paper that this is possible provided a probability sample has been selected and each element in the population has a positive probability of selection in the sample. If these conditions are satisfied, unbiased estimates of population characteristics can be computed. Moreover, also the accuracy of these estimates can be computed.

Let $U = \{1, 2, \ldots, N\}$ denote the population to be surveyed. Let $Y$ denote a target variable of the survey. The value of $Y$ for element $k$ is denoted by $Y_k$, for $k = 1, 2, \ldots, N$. Let the aim of the survey be estimation of the population mean

$$(12.1) \qquad \overline{Y} = \frac{1}{N} \sum_{k=1}^{N} Y_k.$$

A sample design must be chosen to select a sample from this population. Only sampling designs are considered here that draw a sample without replacement. This implies a sample can be represented by a series of indicators

$$(12.2) \qquad a_1, a_2, \ldots, a_N,$$

*Handbook of Web Surveys*, Second Edition. Silvia Biffignandi and Jelke Bethlehem.
© 2021 John Wiley & Sons, Inc. Published 2021 by John Wiley & Sons, Inc.

where the indicator assumes the value 1 if element $k$ is selected in the sample and otherwise it assumes the value 0. The expected value of $a_k$ is denoted by

$$(12.3) \qquad \pi_k = E(a_k).$$

This quantity $\pi_k$ is called the *first-order inclusion probability* of element $k$. It is equal to the probability that this element is selected in the sample. The second-order inclusion probability of two elements $k$ and $l$ (with $k \neq l$) is defined as

$$(12.4) \qquad \pi_{kl} = E(a_k a_l).$$

It is the probability that elements $k$ and $l$ are selected together in the sample.

The most common sampling design is a *simple random sample* (without replacement). This sampling design assigns the same probability of selection to each element in the population. This implies that $\pi_k = n/N$ for all $k$. Furthermore, all second-order inclusion probabilities are equal to $\pi_k = n(n-1)/N(N-1)$.

Horvitz and Thompson (1952) show that always an unbiased estimator can be constructed. Their estimator can be written as

$$(12.5) \qquad \bar{y}_{HT} = \frac{1}{N} \sum_{k=1}^{N} a_k \frac{Y_k}{\pi_k}.$$

The Horvitz–Thompson estimator (12.5) is an unbiased estimator of the population mean $\bar{Y}$ provided that $\pi_k > 0$ for all $k$. The variance of this estimator is equal to

$$(12.6) \qquad V(\bar{y}_{HT}) = \frac{1}{2N^2} \sum_{k=1}^{N} \sum_{l=1}^{N} (\pi_k \pi_l - \pi_{kl}) \left( \frac{Y_k}{\pi_k} - \frac{Y_l}{\pi_l} \right)^2.$$

A closer look at estimator (12.5) makes clear that proper estimation requires the sample elements to be weighted. Elements with a larger inclusion probability will be overrepresented in the sample. This is corrected in the estimator by dividing by the inclusion probability.

If the *design weight* $d_k$ for element $k$ is defined as $d_k = 1/\pi_k$ (for $k = 1, 2, \ldots, N$), then the Horvitz–Thompson estimator can be written as

$$(12.7) \qquad \bar{y}_{HT} = \frac{1}{N} \sum_{k=1}^{N} a_k d_k Y_k.$$

If a sample of size $n$ is selected, the values of $Y$ for the selected elements are denoted by $y_1, y_2, \ldots, y_n$, and $d_1, d_2, \ldots, d_n$ are the corresponding design weights, the Horvitz–Thompson estimator can also be written as

$$(12.8) \qquad \bar{y}_{HT} = \frac{1}{N} \sum_{i=1}^{n} d_i y_i.$$

In case of simple random sampling, all design weights are equal to $N/n$. Consequently, the Horvitz–Thompson estimator reduces to the simple sample mean

$$(12.9) \qquad \bar{y} = \frac{1}{n} \sum_{i=1}^{n} y_i.$$

See Example 12.1 for further discussion.

EXAMPLE 12.1  **A web survey based on an address sample**

Suppose a web survey is conducted among all adult inhabitants of a town. To select a random sample of persons, addresses are randomly selected from a list of all addresses in a town. One person is randomly drawn at each selected address by determining the adult who is the first to have his/her birthday. This person is invited to complete the online questionnaire. This is not an equal probability sample, and therefore design weights have to be included in the estimator.

Let the size of the adult population be denoted by $N$. Suppose this population is distributed over $M$ addresses, where the number of adults at address $h$ is denoted by $N_h$, for $h = 1, 2, \ldots, M$. The consequence of this sampling design is that the inclusion probability for an element $k$ at address $h$ is equal to

$$\pi_k = \frac{n}{M} \frac{1}{N_h}.$$

Let the indicators $a_1, a_2, \ldots, a_M$ denote the selected addresses. Furthermore, let the indicators $b_{hk}$ denote which persons are selected at address $h$ (for $h = 1, 2, \ldots, M$ and $k = 1, 2, \ldots, N_h$). By substituting all these quantities in expression (12.5), the Horvitz–Thompson estimator becomes

$$\bar{y}_{HT} = \frac{M}{N} \frac{1}{n} \sum_{h=1}^{M} a_h N_h \sum_{k=1}^{N_h} b_{hk} Y_{hk}.$$

If the measured value of $Y$ for selected address $i$ is denoted by $y_i$, the estimator can be rewritten as

$$\bar{y}_{HT} = \frac{M}{N}\frac{1}{n}\sum_{i=1}^{n} N_i y_i.$$

It is clear that this expression is not equal to the sample mean of the $y_i$.

It is convenient to assume that a web survey sample is a random sample selected with equal probabilities. Under this assumption the sample mean is an unbiased estimator of the population mean, and a population percentage is an unbiased estimator of a population percentage.

However, such an assumption can lead to a serious bias. Suppose the sample is in fact selected with first-order inclusion probabilities $\pi_1$, $\pi_2$, ..., $\pi_N$. Then the expected value of the sample mean is equal to

$$(12.10) \qquad E(\bar{y}) = \frac{1}{n}\sum_{k=1}^{N} E(a_k) Y_k = \frac{1}{n}\sum_{k=1}^{N} \pi_k Y_k.$$

The bias of this estimator turns out to be equal to

$$(12.11) \qquad B(\bar{y}) = E(\bar{y}) - \overline{Y} = \frac{1}{n}\sum_{k=1}^{N} (\pi_k - \bar{\pi})(Y_k - \overline{Y}) = \frac{N}{n} C_{\pi Y}$$

where $C_{\pi Y}$ is the covariance between the inclusion probabilities and the target variable. The stronger the (linear) relationship between inclusion probabilities and the target variable, the larger the bias will be.

The theory above shows that if a sample is not selected with equal probabilities, weighting is always required to obtain unbiased estimates of population characteristics. The design weights have to be computed for this.

A second type of weighting may be applied to improve the precision of estimators. Aim is not reducing or removing a bias, but reducing the variance of the estimator. An additional advantage of such weighting techniques is that the weighted sample becomes representative with respect to some auxiliary variables. The weighting techniques described in Sections 12.2.2 (post-stratification), 12.2.3 (the generalized regression estimator), and 12.2.4 (raking ratio estimation) all can do this if the proper auxiliary information is available.

A third type of weighting is often used to correct for bias caused by nonresponse. According to the *random response model* (see, e.g., Bethlehem, 2009), each element $k$ in the population has an (unknown) response probability $\rho_k$. If element $k$ is selected in the sample, a random mechanism is activated that results with probability $\rho_k$ in response and with probability $1 - \rho_k$ in nonresponse. If nonresponse occurs in a simple random sample, the response mean $\bar{y}_R$ is not unbiased anymore. Bethlehem (2009) shows that the bias is equal to

$$(12.12) \qquad B\left(\bar{y}_R\right) = \frac{R_{\rho Y} S_\rho S_Y}{\bar{\rho}},$$

where $R_{\rho Y}$ is the correlation between the values of the target variable and the response probabilities, $S_\rho$ is the standard deviation of the response probabilities, $S_Y$ is the standard deviation of the variable $Y$, and $\bar{\rho}$ is the population mean of the response probabilities. The bias will be large if:

- The relationship between the target variable and response behavior is strong;
- The variation in the response probabilities is large;
- The average response probability is low.

The weighting techniques described in the following sections can reduce the non-response bias provided proper auxiliary information is available.

The three reasons for weighting described above apply to any survey, whatever the mode of data collection. There are two more reasons for weighting that are particularly important for many web surveys. These reasons are under-coverage and self-selection.

Under-coverage problems were described in detail in Chapter 10. *Under-coverage* occurs if the target population contains elements that do not have a counterpart in the sampling frame. Such elements can never be selected in the sample. Under-coverage occurs in a web survey if the target population is wider than just persons with Internet access and respondents are selected via the Internet. There is under-coverage because it is impossible for people without Internet to participate in the survey. This type of under-coverage can have serious consequences. If people without Internet access systematically differ from persons with access to the Internet, estimates of population parameters may be seriously biased. It is shown in Chapter 10 that the bias of the sample mean $\bar{y}_I$ is equal to

$$(12.13) \qquad B\left(\bar{y}_I\right) = \overline{Y}_I - \overline{Y} = \frac{N_{NI}}{N}\left(\overline{Y}_I - \overline{Y}_{NI}\right).$$

The magnitude of this bias is determined by two factors.

The first factor is the relative size $N_{NI}/N$ of the non-Internet population. The bias will increase as a larger proportion of the population does not have access to Internet. The second factor is the *contrast* $\overline{Y}_I - \overline{Y}_{NI}$ between the Internet population and the non-Internet population. It is the difference between the population means of the two subpopulations. The more the mean of the target variable differs for these two subpopulations, the larger the bias will be.

The weighting techniques in the following sections can be attempted to reduce the bias due to under-coverage. There is no guarantee that this will be successful, as will be shown in the subsequent section.

Self-selection problems are described in detail in Chapter 11. *Self-selection* is the phenomenon that the sample is not selected by means of a probability sample. Instead, it is left to the Internet users themselves to participate in a web survey. The survey questionnaire is simply put on the web. Respondents are those people who happen to have Internet, visit the website, and decide to participate in the survey. The survey researcher is not in control of the selection process.

Participation in a self-selection web survey requires respondents to be aware of the existence of the survey. They have to accidentally visit the website, or they have to follow up a banner or an e-mail message. They also have to decide to fill in the questionnaire on the Internet. This means that each element $k$ in the Internet population has unknown probability $\rho_k$ of participating in the survey, for $k = 1, 2, \ldots, N_I$, where $N_I$ is the size of the population of persons having access to the Internet.

Assuming it is the objective of the web survey to estimate the mean $\overline{Y}_I$ of the Internet population, Bethlehem (2002) shows that the bias of the sample mean $\overline{y}_S$ can be written as

$$(12.14) \qquad B(\overline{y}_S) = E(\overline{y}_S) - \overline{Y}_I = \frac{R_{\rho Y} S_\rho S_Y}{\overline{\rho}},$$

where $R_{\rho Y}$ is the correlation coefficient between target variable and the response behavior, $S_\rho$ is the standard deviation of the participation probabilities, $S_Y$, is the standard deviation of the target variable, and $\overline{\rho}$ is the population mean of the participation probabilities. This bias is large if:

- The relationship between the target variable and the participation probabilities is strong;
- The variation in the participation probabilities is large;
- The average participation probability is low.

The weighting techniques in the following sections can also attempt to reduce the bias due to self-selection. Again, there is no guarantee that this will be successful, as will be shown in this chapter.

Three types of weighting adjustment techniques will be described: post-stratification, generalized regression estimation, and raking ratio estimation. Note that Chapter 13 is devoted to the use of so-called propensity scores. These propensity scores can also be used for weighting.

It will be made clear in this chapter that an effective weighting adjustment procedure requires proper auxiliary information.

## 12.2 Theory

### 12.2.1 THE CONCEPT OF REPRESENTATIVITY

The principles of weighting adjustment are closely related to the concept of *representativity*. This concept is often used in survey research, but usually it is not clear what it means. Kruskal and Mosteller (1979a, 1979b, 1979c) present an extensive overview of what *representative* is supposed to mean in nonscientific literature, in scientific literature excluding statistics, and in the statistical literature. They found the following meanings for "representative sampling":

- General acclaim for data;
- Absence of selective forces;
- Miniature of the population;
- Typical or ideal case(s);
- Coverage of the population;
- A vague term, to be made precise;
- Representative sampling as a specific sampling method;
- As permitting good estimation;
- Good enough for a particular purpose.

To avoid confusion, Kruskal and Mosteller recommended not using the word *representative*, but instead to specify what one means. In this chapter, the concept of representativity with respect to a variable is used. A survey data set is defined to be *representative with respect to a variable $X$* if the distribution of $X$ in the data set is equal to the distribution of this variable in the population.

Weighting adjustment is based on the use of *auxiliary information*. Auxiliary information is defined here as a set of variables that have been measured in the survey and for which the distribution in the population is available. By comparing the population distribution of an auxiliary variable with its response distribution, it can be assessed whether or not the response is representative for the population (with respect to this variable). If these distributions differ considerably, one must conclude that the survey response is not representative.

The next step is to use the auxiliary information to compute *adjustment weights*. Weights are assigned to all observed elements. Estimates of population characteristics can now be obtained by using weighted values instead of the unweighted values. The weights are defined in such a way that population characteristics for the auxiliary variables can be computed without error. So, the weighted sample is forced to be *representative with respect to the auxiliary variables* used.

Recall that, whatever sampling design is used, always an unbiased estimator can be constructed. This is the Horvitz–Thompson estimator. It can be written as

(12.15)
$$\bar{y}_{HT} = \frac{1}{N}\sum_{i=1}^{n} d_i y_i,$$

where $d_i = 1/\pi_i$ is the *design weight* and $y_i$ is observed value of sample element $i$, for $i = 1, 2, ..., n$.
Adjustment weighting replaces this estimator by a new estimator:

(12.16)
$$\bar{y}_W = \frac{1}{N}\sum_{i=1}^{n} w_i y_i.$$

where the weight $w_i$ is equal to

(12.17)
$$w_i = c_i \times d_i$$

and $c_i$ is a *correction weight* produced by a weighting adjustment technique.

Weighting adjustment techniques impose the condition of representativity with respect to one or more selected auxiliary variables. Suppose $X$ is such an auxiliary variable. Representativity with respect to $X$ implies that the weights $w_i$ have to be such that

(12.18)
$$\frac{1}{N}\sum_{i=1}^{n} w_i x_i = \overline{X}.$$

This means that if the weights are used to estimate the population mean of the auxiliary variable, the estimate is exactly equal to the population mean.

If the response can be made representative with respect to several auxiliary variables, and if all these variables have a strong relationship with the phenomena to be investigated, then the weighted sample will also be (approximately) representative with respect to these phenomena, and hence estimates of population characteristics will be more accurate.

Several weighting techniques will be described in this section. It starts with the simplest and most commonly used one: *post-stratification*. Next, *generalized regression estimation* is described, which is more general than post-stratification. This technique can be applied in situations where the auxiliary information is inadequate for post-stratification. Furthermore, *raking ratio estimation* is discussed as an alternative for generalized regression estimation. Also, an introduction into *calibration* is given. This can be seen as an even more general theoretical framework for adjustment weighting that includes generalized regression estimation and raking ratio estimation as special cases.

## 12.2.2 POST-STRATIFICATION

The concept of *stratification* has a long history in survey methodology. Stratification means that the target population of the survey is divided into a number of groups. These groups are called *strata*. A sample is selected from each group so that estimates can be computed for each group separately. The next step is to combine the group estimates into an estimate for the whole population.

Stratification played already a role in the first ideas that emerged about sampling. It was Anders Kiaer, director of the Norwegian statistical institute, who proposed at a meeting of the ISI (International Statistical Institute) in Bern in 1895 to use sampling instead of complete enumeration. He argued that good results could be obtained with his *representative method*. His idea was to select a sample that should reflect all aspects of the population as much as possible. One way to realize such a sample was the "balanced sample." He divided the population into groups using variables like gender, age, and region. The sizes of the groups were supposed to be known. The same percentage of persons was taken in each group. Selection of samples took place in some haphazard way (probability sampling had not yet been invented). As a result, the sample distribution of variables like gender, age, and region was similar to the distribution in the population. Hence, the sample was representative with respect to these variables.

Holt and Smith (1979) noted the wide use of stratification in the 1970s as it has two attractive properties: (1) it leads to representative samples, and (2) it improves the precision of estimators. There is, however, also a drawback. To be able to draw a stratified sample, a sampling frame is required for each group separately. This is not always the case. There are many situations in which membership of a group can be established only after inspection of the sampled data. For example, to obtain a sample that is representative with respect to age groups, a sample would have to be drawn from each age group separately. Usually, there is no sampling frame for each age group.

*Post-stratification* is an estimation technique that attempts to make the sample representative after the data has been collected. It comes down to assigning stratum weights. Respondents in underrepresented groups get a weight larger than 1, and respondents in overrepresented groups get a weight smaller than 1. By using weighted values as in expression (12.16), properties of estimators will be improved.

Using a single auxiliary variable or by crossing several auxiliary variables, strata are obtained. Post-stratification is particularly effective if the strata are *homogeneous*. This means the people within strata resemble each other. If this is the case, post-stratification will not only improve the precision (as measured by the variance or the standard error of estimators) but also reduce a possible bias.

First, in the ideal case of a simple random sample without nonresponse, under-coverage, and self-selection problems, the theory of post-stratification is described. Then it is explored if and when post-stratification can reduce a bias caused by these problems.

To be able to carry out post-stratification, one or more qualitative auxiliary variables are needed. The theory is described for one such variable, but the case of more variables is not essentially different. Suppose there is an auxiliary variable $X$ having $L$ categories. So, it divides the population $U$ into $L$ strata $U_1, U_2, ..., U_L$. The number of population elements in stratum $U_h$ is denoted by $N_h$, for $h = 1, 2, ..., L$. So, $N = N_1 + N_2 + \cdots + N_L$. These stratum sizes are supposed to be known. This is the population distribution of the variable $X$.

Assume a simple random sample of size $n$ is selected without replacement from the population. If $n_h$ denotes the number of sample elements in stratum $U_h$ (for $h = 1, 2, ..., L$), then $n = n_1 + n_2 + \cdots + n_L$. Note that the values of the $n_h$ are the result of a random selection process. So, they are random variables.

To get a sample that is representative with respect to the variable $X$, the proportion of elements in stratum $h$ should be equal to $N_h/N$, for $h = 1, 2, ..., L$. However, the proportion of sample elements in stratum $h$ is equal to $n_h/n$. To correct for this, each observed element $i$ in stratum $U_h$ is assigned a correction weight equal to

$$(12.19) \qquad\qquad c_i = \frac{N_h/N}{n_h/n}.$$

If the values of the inclusion weights, $d_i = n/N$, and correction weights (12.19) are substituted in expression (12.16), the result is the *post-stratification estimator*

$$(12.20) \qquad\qquad \bar{y}_{PS} = \frac{1}{N} \sum_{h=1}^{L} N_h \bar{y}^{(h)},$$

where $\bar{y}^{(h)}$ is the mean of the observed elements in stratum $h$. So, the post-stratification estimator is equal to a weighted sum of sample stratum means. Example 12.2 shows a case study about this issue.

**EXAMPLE 12.2 Computing weights by means of post-stratification**

To show the effects of post-stratification, a fictitious population is used. It is the population described in Section 10.3 in Chapter 10. The population consists of all eligible voters of a town. The size of the target population is 30,000 persons.

An election survey is conducted among these voters. A simple random sample of size 1,000 is drawn from this population. Age (in three classes) is used as an auxiliary variable. Table 12.1 contains the population and sample frequencies of this variable.

TABLE 12.1   Computing post-stratification weights

|  | Population | | Sample | | |
|---|---|---|---|---|---|
| Age | Frequency | Percentage | Frequency | Percentage | Weight |
| Young | 11,949 | 39.8300 | 402 | 40.2000 | 0.990796 |
| Middle | 10,582 | 35.2733 | 342 | 34.2000 | 1.031384 |
| Elderly | 7,469 | 24.8967 | 256 | 25.6000 | 0.972526 |
| Total | 30,000 | 100.0000 | 1,000 | 100.0000 | |

Young people are slightly overrepresented. The sample percentage (40.2%) is larger than the population percentage (39.8%). Therefore, post-stratification assigns a weight smaller than 1. The weight for this group is obtained by dividing the population percentage by the sample percentage. The result is 0.990796.

Likewise, middle-aged persons are underrepresented. The sample percentage (34.2%) is smaller than the population percentage (35.3%). Therefore, the weight is larger than 1. It is obtained by dividing 35.2733 by 34.2, resulting in a weight of 1.031384.

The adjustment weights $w_i$ are obtained by multiplying the correction weights $c_i$ by the inclusion weights $d_i$. Here all inclusion weights are equal to $N/n = 30$. Suppose the weights are used to estimate the number of old persons in the population. The weighted estimate would be $0.972526 \times 30 \times 256 = 7,469$, and this is exactly the population frequency. Thus, application of weights to the auxiliary variables results in perfect estimates. If there is a strong relationship between the auxiliary variable and the target variable, also estimates for the target variable will be improved if these weights are used.

There is no simple exact analytical expression for the variance of post-stratification estimator as defined by (12.20). There is, however, a large sample approximation:

$$(12.21) \qquad V\left(\bar{y}_{PS}\right) = \left(\frac{1}{n} - \frac{1}{N}\right) \sum_{h=1}^{L} W_h S_h^2 + \frac{1}{n^2} \sum_{h=1}^{L} (1 - W_h) S_h^2,$$

where $W_h = N_h/N$ is the relative size of stratum $h$ and $S_h^2$ is the (adjusted) population variance of the target variable in stratum $h$. The post-stratification estimator is precise if the strata are homogeneous with respect to the target variable. This implies that variation in the values of the target variable is typically caused by differences in means between strata and not by variation within strata. This is shown in Example 12.3.

■ EXAMPLE 12.3 **The variance of post-stratification estimator**

Suppose one of the aims of the election survey in Example 12.1 is to estimate the percentage of people voting for the National Elderly Party (NEP).
To that end a simple random sample of size 1,000 is drawn from the population if size 30,000.

The variance of the sample percentage turns out to be equal to 1.832. If
post-stratification is carried out with age (in three categories) as auxiliary
variable, the variance of the estimator is equal to 1.294. So post-
stratification reduces the variance of the estimator. Apparently the age strata
are more homogeneous with respect to voting behavior than the population
as a whole.

The *effective sample size* $n_{eff}$ is sometimes used as an indicator of how
effective a sampling design or estimation procedure is. It is the sample size
needed to obtain the same level of precision as the sample mean in simple
random sampling. In case of post-stratification, it is defined as

$$n_{eff} = n\frac{V(\bar{y})}{V(\bar{y}_{PS})}.$$

The effective sample size for the election survey is $1,000 \times 1.832/$
$1.294 = 1,416$. So, the sample mean requires 416 more sample elements
to obtain the same precision as the post-stratification estimator.

Now suppose the sample is affected by *nonresponse*. Then the post-
stratification estimator takes the form

$$(12.22) \qquad \bar{y}_{R,PS} = \frac{1}{N}\sum_{h=1}^{L} N_h\bar{y}_R^{(h)},$$

where $\bar{y}_R^{(h)}$ denotes the mean of the responding elements in stratum $h$. The bias of
this estimator is equal to

$$(12.23) \qquad B(\bar{y}_{R,PS}) = \frac{1}{N}\sum_{h=1}^{L} N_h\frac{R_{\rho Y}^{(h)}S_{\rho}^{(h)}S_Y^{(h)}}{\bar{\rho}^{(h)}},$$

where $R_{\rho Y}^{(h)}$ is the correlation between $Y$ and $\rho$ in stratum $h$; $S_{\rho}^{(h)}$ and $S_Y^{(h)}$ are the
standard errors of $\rho$ and $Y$ in stratum $h$, respectively; and $\bar{\rho}^{(h)}$ is the mean of the
response probabilities in stratum $h$. The bias of the post-stratification estimator is
small if the biases within strata are small. A stratum bias is small in the following
situations:

- There is little or no relationship between the target variable and response behavior within the stratum. Then their correlation is small.

- All response probabilities within a stratum are more or less equal. Then their standard error is small.

- All values of the target variable within a stratum are more or less equal. Then their standard error is small.

These conclusions give some guidance with respect to the construction of strata. Preferably, strata should be used that are homogeneous with respect to the target variable, response probabilities, or both. The more elements resemble each other within strata, the smaller the bias will be, as presented in Example 12.4.

**EXAMPLE 12.4 Using post-stratification for reducing nonresponse bias**

Suppose one of the aims of the election survey in Section 10.3 is to estimate the percentage of people voting for the National Elderly Party (NEP). To that end a simple random sample of size 1,000 is drawn from the population if size 30,000. Three situations are compared. The first one is the ideal situation of a simple random sample from the target population with full response. The selection of the sample is repeated 1,000 times. For each sample the percentage of voters for the NEP is computed. The distribution of these estimates is represented in the upper box plot in Figure 12.1.

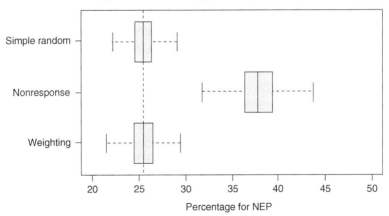

FIGURE 12.1 **Estimating the percentage voters for the NEP in case of nonresponse**

The vertical dotted line denotes the population percentage (25.4%). The box plot is symmetric around this line. It is clear that this estimator is unbiased.

The second situation describes what happens if nonresponse occurs. Each element is assigned a probability of response. This probability is equal to 0.2 for young people, 0.5 for middle-aged people, and 0.8 for the elderly. The box plot in the middle of Figure 12.1 shows the distribution of 1,000 estimates. This estimator is substantially biased.

The expected value of this estimator is equal to 37.8%, which is much higher than 25.4%. This is not surprising. Young people have a low response probability and therefore are underrepresented in the sample. So, elderlies are overrepresented and they typically vote for the NEP.

The third situation examines the effect of post-stratification on the non-response bias of the second situation. Age is used as auxiliary variable. Adjustment weighting is successful here. The bias is completely removed. This could be expected as the response probabilities were equal within the age classes.

Post-stratification is not always successful in reducing the bias. This is shown in a different example. Again, the same population of 30,000 is used, but now the percentage of voters for the New Internet Party (NIP) is estimated. There is different nonresponse mechanism: persons with Internet have a response probability of 0.8, and those without it have a response probability of 0.2. Again, the effect of post-stratification by age is explored. Figure 12.2 displays the results.

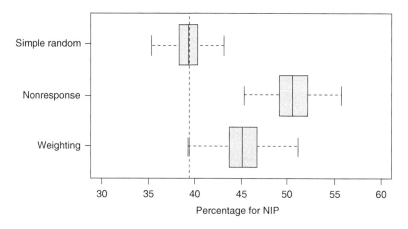

FIGURE 12.2  **Estimating the percentage voters for the NIP in case of nonresponse**

It is clear that a simple random sample results in an unbiased estimator for the population percentage (39.5%). Nonresponse leads to a substantial bias. The expected value of the estimator is now 50.7%. People with Internet are overrepresented in the response, and they are the ones that typically vote for the NIP. Post-stratification by age is not successful here. Only part of the bias is removed. The expected value goes down from 50.7% to 45.2%, but it is still too high. The reason is there is no direct relationship between age and response behavior. Since there is a direct relation between access to Internet and response behavior, and there is some relation between access to Internet and age, part of the bias is removed.

Web surveys may suffer from *under-coverage*. This can happen when some people in the target population do not have access to the Internet. As a result, estimators may be biased. An expression for the bias was given in Section 12.1. The question is whether this bias can be removed or reduced by applying post-stratification. It was shown in Chapter 10 that the bias after post-stratification is equal to

$$(12.24) \qquad B(\bar{y}_{I,PS}) = \sum_{h=1}^{L} W_h \frac{N_{NI,h}}{N_h} \left( \overline{Y}_I^{(h)} - \overline{Y}_{NI}^{(h)} \right),$$

where $N_{NI,h}$ is the number of people in stratum $h$ without Internet, $\overline{Y}_I^{(h)}$ is the mean of $Y$ for those with Internet in stratum $h$, and $\overline{Y}_{NI}^{(h)}$ is the mean for those without Internet in stratum $h$. The bias will be small if there is (on average) no difference between elements with and without Internet within the strata. This is the case if there is a strong relationship between the target variable $Y$ and the stratification variable $X$. The variation in the values of $Y$ will manifest itself in this case between strata but not within strata. In other words, the strata are homogeneous with respect to the target variable. Example 12.5 discusses this issue comparing different samples.

---

**EXAMPLE 12.5 Using post-stratification for reducing under-coverage bias**

Suppose one of the aims of the election survey in Example 10.1 is to estimate the percentage of people voting for the National Elderly Party (NEP). Three situations are considered:

- A simple random sample from the complete population;
- A simple random sample from the Internet population;
- A simple random sample from the Internet population, followed by post-stratification.

In all three situations, the distribution of the estimator is determined by repeating the selection of the sample 1,000 times. The sample size is always 1,000 cases. Figure 12.3 contains the results.

The population was constructed such that Internet access decreases with age. Moreover, Internet access for natives was much higher than for non-natives. Voting for the NEP depended on age only.

A simple random sample from the complete target population results in an unbiased estimator for the population percentage of 25.4%. Just sampling the Internet population leads to an estimator with a substantial bias. The expected value of this estimator is substantially too low: 20.3%.

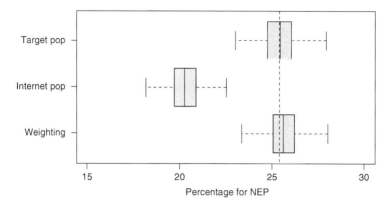

**FIGURE 12.3** Estimating the percentage voters for the NEP in case of under-coverage

This can be explained by the fact that the elderly is underrepresented in the samples because they have less access to Internet. The elderly typically vote for the NEP. Application of post-stratification by age solves the problem. After weighting, the estimator is unbiased. There is a direct relation between voting behavior and age, and there is a direct relation between age and having Internet. So, correcting the age distribution also corrects the estimator.

Figure 12.4 shows the analysis for voting for the New Internet Party (NIP). The same three situations are compared.

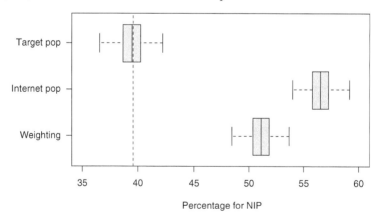

**FIGURE 12.4** Estimating the percentage voters for the NIP in case of under-coverage

Just sampling the Internet leads to a biased estimator. The estimated values are substantially too high. The expected value of the estimator is 56.5%, whereas it should have been 39.5%. Post-stratification is not successful. The expected value of the estimator decreases from 56.5% to 51.1%, but this is still too high. This is not surprising as there is a direct relation between voting for the NIP and having access to Internet.

Web surveys may suffer from *self-selection*. This is the phenomenon that the sample is not selected by means of a probability sample. Instead, it is left to the Internet users themselves to participate in a web survey. The survey questionnaire is simply put on the web. Respondents are those people who happen to have Internet, visit the website, and decide to participate in the survey. The survey researcher is not in control of the selection process. As a result, estimators may be biased. An expression for the bias was given in Section 12.1. The question is whether this bias can be removed or reduced by applying post-stratification. The bias after post-stratification is equal to

$$(12.25) \qquad B\left(\bar{y}_{S,PS}\right) = \sum_{h=1}^{L} W_h \frac{R_{\rho Y}^{(h)} S_{\rho}^{(h)} S_Y^{(h)}}{\overline{\rho}^{(h)}},$$

where the subscript $h$ indicates that the respective quantities are computed just for stratum $h$. $R_{\rho Y}^{(h)}$ is the correlation coefficient between target variable and the response behavior, $S_{\rho}^{(h)}$ is the standard deviation of the participation probabilities, $S_Y^{(h)}$ is the standard deviation of the target variable, and $\overline{\rho}^{(h)}$ is the average participation probability. The bias will be small if:

- The participation probabilities are similar within strata;
- The values of the target variable are similar within strata;
- There is no correlation between participation behavior and the target variable within strata.

These conditions can be realized if there is a strong relationship between the target variable $Y$ and the stratification variable $X$. Then the variation in the values of $Y$ manifests itself between strata and not within strata. In other words, the strata are homogeneous with respect to the target variable. Also, if the strata are homogeneous with respect to the participation probabilities, the bias will be reduced. Example 12.6 compares random sample and two different self-selected samples and discusses results with respect to the bias.

🖥 **EXAMPLE 12.6 Using post-stratification for reducing self-selection bias**

The fictitious population of Section 11.3 is used to illustrate the possible effects of post-stratification on a self-selection bias. This population consists of 100,000 persons. Most persons (99%) are passive Internet users. Active users make up only 1% of the population. Active users have a high participation probability of 0.99. Passive users have a low participation probability (0.01). The percentage of active Internet users decreases with age.

An election survey is conducted. Aim is to estimate the percentage of people voting for the National Elderly Party (NEP). Three situations are considered:

- A simple random sample from the complete population;
- A self-election sample from the population;
- A self-selection sample from the population, followed by post-stratification.

In all three situations, the distribution of the estimator is determined by repeating the selection of the sample 1,000 times. The average participation probability in the population is 0.01971. Therefore, the expected sample size in a self-selection survey is equal to 1,971.

Figure 12.5 contains the results. The upper box plot shows that the estimator is unbiased in case of simple random sampling (of size 1,971) from the target population. The expected value is equal to 25.6%. The middle box plot shows what happens if samples are selected by means of self-selection. The shape of the distribution remains more or less the same, but the distribution as a whole has shifted to the left. All values of the estimator are systematically too low. The expected value of the estimator is only 20.4%. The estimator is biased. The explanation of this bias is simple: relative few elderlies are active Internet users. Therefore, they are underrepresented in the samples. They are typically people who will vote for the NEP. The lower box plot shows the distribution of the estimator in case of post-stratification by age. The bias is removed. This was possible because there is direct relation between participation and the weighting variable age.

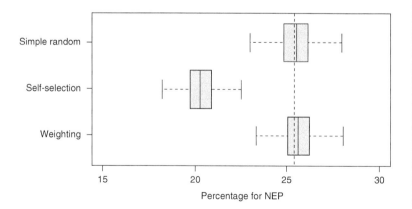

FIGURE 12.5  Estimating the percentage voters for the NEP in case of self-selection

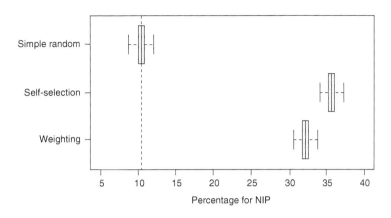

FIGURE 12.6 **Estimating the percentage voters for the NIP in case of self-selection**

Figure 12.6 shows the analysis for voting for the New Internet Party (NIP). The same three situations are compared.

In case of simple random sampling, the estimator is unbiased. The population value of 10.4% is correctly estimated. Self-selection leads to a biased estimator. The estimated values are substantially too high. The expected value of the estimator is 35.6%, whereas it should have been 10.4%. Post-stratification is not successful. The expected value of the estimator decreases from 35.6% to 32.1%, but this is still too high. This is not surprising as there is a direct relation between voting for the NIP and having access to Internet.

Just one auxiliary variable (age) was used for weighting in the examples in this section. It is possible to use more than one variable for weighting. For example, if the auxiliary variables Age and Education are available, they can be crossed. If age has three categories and Education has two categories, this leads to $3 \times 2 = 6$ strata. This weighting model is denoted by

$$\text{Age} \times \text{Education.}$$

The idea of crossing variables can be extended to more than two variables. As long as the table with population frequencies is available, and all response frequencies are greater than 0, weights can be computed. However, if there are no observations in a stratum, the corresponding weight cannot be computed. This leads to incorrect estimates. If the sample frequencies in the strata are very small, say, less than 5, weights can be computed, but estimates will be unstable.

As more variables are included in a weighting model, there will be more strata. This increases the risk of empty strata or strata with too few observations. There are two solutions for this problem. One is not to have so many auxiliary variables in the model, but then a lot of auxiliary information is not used. Another is *collapsing strata*. This means merging a stratum with too few observations with another stratum. It is important to combine strata that resemble each other as much as possible. Collapsing strata is not a simple job, particularly if the number of auxiliary variables and strata is large.

Another problem of a large weighting model is that not all required population information is available. It may happen that the population distribution of the complete crossing of all auxiliary variables can simply not be obtained. A possible cause could be that the population distributions of these variables come from different sources. Without the complete population distribution no weights can be computed, see Example 12.7.

One way to solve this problem is to use less auxiliary variables, but that would mean ignoring all available information with respect to the other variables. What is needed is a weighting technique capable of using partial population information. There are weighting techniques that can do this: generalized regression estimation and raking ratio estimation. These techniques are described in Sections 12.2.3 and 12.2.4.

---

▦ EXAMPLE 12.7 **Incomplete population information**

Suppose the population distributions of the variables Age (with categories Young, Middle, and Old) and Education (with categories Low and High) are known separately, but the distribution in the cross-classification is not known (see Table 12.2). In this case the post-stratification Age × Education cannot be carried out, because weights cannot be computed for the strata in the cross-classification.

TABLE 12.2 **Incomplete population information**

| Population | Low | High | Total | Response | Low | High | Total |
|---|---|---|---|---|---|---|---|
| Young | ? | ? | 19,925 | Young | 111 | 732 | 843 |
| Middle | ? | ? | 15,069 | Middle | 461 | 725 | 1,186 |
| Old | ? | ? | 15,006 | Old | 807 | 923 | 1,730 |
| Total | 29,963 | 20,037 | 50,000 | Total | 1,379 | 2,316 | 3,696 |
| Weights | | | | | | | |
| | Low | High | | | | | |
| Young | ? | ? | | | | | |
| Middle | ? | ? | | | | | |
| Old | ? | ? | | | | | |

## 12.2.3 GENERALIZED REGRESSION ESTIMATION

The *generalized regression estimator* is based on a linear model that attempts to explain a target variable of the survey from one or more auxiliary variables. This estimator is not only capable of producing precise estimates, but it also can reduce the bias. It is shown that regression estimation in fact is a form of weighting. This technique is sometimes also called *linear weighting*.

The weights resulting from generalized regression estimation make the response representative with respect to the auxiliary variables in the model. It is shown that post-stratification is a special case of generalized regression estimation.

In principle, the auxiliary variables in the linear model must be continuous variables, i.e., they measure a size or value. However, it is also possible to use categorical variables. The trick is to replace a categorical variable by a number of dummy variables, where each dummy variable indicates whether or not a person belongs to a specific category.

The theory of generalized regression estimation is described assuming the data has been collected by means of simple random sampling without replacement. The theory can easily be generalized for other sampling designs. For more on this, see, e.g., Bethlehem (1988).

First, the ideal case (no bias) is considered. Suppose there are $p$ (continuous) auxiliary variables available. The $p$-vector of values of these variables for element $k$ is denoted by

$$(12.26) \qquad X_k = \left(X_{k1}, X_{k2}, ..., X_{kp}\right)'.$$

The symbol $'$ denotes transposition of a matrix or vector. Let $Y$ be the $N$-vector of all values of the target variable, and let $X$ be the $N \times p$ matrix of all values of the auxiliary variables. The vector of population means of the $p$ auxiliary variables is defined by

$$(12.27) \qquad \overline{X} = \left(\overline{X}_1, \overline{X}_2, ..., \overline{X}_p\right)'.$$

This vector represents the population information assumed to be available. If the auxiliary variables are correlated with the target variable, then for a suitably chosen vector $B = (B_1, B_2, ..., B_p)'$ of regression coefficients for a best fit of $Y$ on $X$, the residuals $E = (E_1, E_2, ..., E_N)'$, defined by

$$(12.28) \qquad E = Y - XB,$$

will vary less than the values of the target variable itself. In the ideal case of a perfect relation between $Y$ and $X$, all residuals will be 0. Application of ordinary least squares results in

$$(12.29) \qquad B = (X'X)^{-1}XY' = \left(\sum_{k=1}^{N} X_k X_k'\right)^{-1} \left(\sum_{k=1}^{N} X_k Y_k\right).$$

For a simple random sample without replacement, the vector $B$ can be estimated by

$$(12.30) \quad b = \left( \sum_{k=1}^{N} a_k X_k X_k' \right)^{-1} \left( \sum_{k=1}^{N} a_k X_k Y_k \right) = \left( \sum_{i=1}^{n} x_i x_i' \right)^{-1} \left( \sum_{i=1}^{n} x_i y_i \right),$$

where $x_i = (x_{i1}, x_{i2}, \ldots, x_{ip})'$ denotes the $p$-vector of values of the $p$ auxiliary variables for sample element $i$ (for $i = 1, 2, \ldots, n$). The quantity $a_k$ indicates whether or not element $k$ is selected in the sample. The estimator $b$ is an asymptotically design unbiased (ADU) estimator of $B$. It means the bias vanishes for large samples. The *generalized regression estimator* is now defined by

$$(12.31) \qquad \bar{y}_{GR} = \bar{y} + (\overline{X} - \bar{x})' b,$$

where $\bar{x}$ is the vector of sample means of the auxiliary variables.

The generalized regression estimator is an ADU estimator of the population mean of the target variable. If there exists a $p$-vector $c$ of fixed numbers such that $Xc = J$, where $J$ is a $p$-vector consisting of 1's, the generalized regression estimator can also be written as

$$(12.32) \qquad \bar{y}_{GR} = \overline{X}' b.$$

This condition holds if there is a constant term in the regression model. The model also holds if the model contains a set of dummy variables corresponding to all the categories of a categorical variable. It can be shown that the variance of the generalized regression estimator is approximated by

$$(12.33) \qquad V(\bar{y}_{GR}) = \frac{1-f}{n} S_E^2,$$

where is the population variance of the residuals $E_1, E_2, \ldots, E_N$. Expression (12.33) is identical to the expression of the variance of the simple sample mean if the values $Y_k$ are replaced by the residuals $E_k$. This variance will be small if the residual values $E_k$ are small. Hence, use of auxiliary variables that can explain the behavior of the target variable will result in a precise estimator.

Bethlehem and Keller (1987) have shown that the generalized regression estimator (12.31) can be rewritten in the form of the weighted estimator (12.16). The adjustment weight $w_i$ for observed element $i$ is equal to $w_i = v' X_i$, where $v'$ is a vector of *weight coefficients* that is equal to

$$(12.34) \qquad v = n \left( \sum_{i=1}^{n} x_i x_i' \right)^{-1} \overline{X}.$$

Post-stratification is a special case of generalized regression estimation where the auxiliary variables are categorical variables. To show this, categorical auxiliary variables are replaced by sets of dummy variables. Suppose there is one auxiliary variable with $L$ categories. Then $L$ dummy variables $X_1, X_2, \ldots, X_L$ are defined. For an observation in a certain stratum $h$, the corresponding dummy variable $X_h$ is assigned the value 1, and all other dummy variables are set to 0. Consequently, the vector of population means of these dummy variables is equal to

$$(12.35) \qquad \overline{X} = \left( \frac{N_1}{N}, \frac{N_2}{N}, \ldots, \frac{N_L}{N} \right),$$

and $v$ is equal to

$$(12.36) \qquad v = \frac{n}{N} \left( \frac{N_1}{n_1}, \frac{N_2}{n_2}, \ldots, \frac{N_L}{n_L} \right)'.$$

If this form of $v$ is used to compute $w_i = v'X_i$ and the result is substituted in expression (12.16) of the weighted estimator, the post-stratification estimator is obtained. Example 12.8 shows a numerical application.

---

▇  **EXAMPLE 12.8** **Post-stratification as a special case of generalized regression estimation**

This example uses the same data as Example 12.4. Objective of the survey is estimating voting behavior. A self-selection sample has been obtained from the population consisting of 30,000 people. The realized sample size is 3,696 people. There are two auxiliary variables: level of education (in categories Low and High) and age (in categories Young, Middle, and Old). Crossing these two variables produces a table with $2 \times 3 = 6$ cells. A dummy variable is introduced for each cell. So there are six dummy variables $X_1, X_2, \ldots, X_6$.

The possible values of these dummy variables are shown in Table 12.3. For example, $X_4$ is the dummy variable for the stratum consisting of young people with high education. Note that always one dummy variable has the value 1, while all other five dummy variables have the value 0.

The table also contains the vector of population means of the auxiliary variables. These values are equal to the population fractions in the cells of the population table. So, the fraction of young people with high education in the population is equal to 0.161.

TABLE 12.3   **Post-stratification by Education × Age**

| Educ. | Age | $X_1$ | $X_2$ | $X_3$ | $X_4$ | $X_5$ | $X_6$ |
|---|---|---|---|---|---|---|---|
| Low | Young | 1 | 0 | 0 | 0 | 0 | 0 |
| Low | Middle | 0 | 1 | 0 | 0 | 0 | 0 |
| Low | Old | 0 | 0 | 1 | 0 | 0 | 0 |
| High | Young | 0 | 0 | 0 | 1 | 0 | 0 |
| High | Middle | 0 | 0 | 0 | 0 | 1 | 0 |
| High | Old | 0 | 0 | 0 | 0 | 0 | 1 |
| Population means | | 0.238 | 0.181 | 0.181 | 0.161 | 0.121 | 0.119 |
| Sample means | | 0.030 | 0.122 | 0.215 | 0.195 | 0.193 | 0.246 |
| Weight coefficients | | 8.052 | 1.474 | 0.842 | 0.825 | 0.625 | 0.486 |

By comparing the population means with the sample means, it becomes clear that low-educated young people are substantially underrepresented in the sample. Their population fraction is 0.238, whereas their sample fraction is only 0.030.

The weight coefficients in the vector $v$ are given in the bottom row of the table. They have been computed using expression (12.34).

These weight coefficients are used to compute the adjustment weights for the observed elements. The weight of a person is obtained by summing the relevant weight coefficients. In the case of post-stratification, there is always only one relevant weight coefficient. So, the weight is equal to this weight coefficient. For example, the weight for a low-educated young person is equal to 8.052. This implies that every sample person in this stratum counts for eight persons. Note that high-educated old people are overrepresented. This is why they get a weight of only 0.486. Each person in this stratum counts for less than half a person.

In case of under-coverage, generalized regression estimator changes to

$$(12.37) \qquad \bar{y}_{GR,I} = \bar{y}_I + \left( \overline{X} - \overline{x}_I \right)' b_I = \overline{X}' b_I.$$

The subscript $I$ indicates that the corresponding quantities have been computed just using data from the Internet population. The vector of coefficients $b_I$ is defined by

$$(12.38) \qquad b_I = \left( \sum_{k=1}^{N} a_k I_k X_k X_k \right)^{-1} \left( \sum_{k=1}^{N} a_k I_k X_k Y_k \right),$$

in which $a_k$ is the sample indicator and $I_k$ is the Internet indicator. So $b_I$ is the analogue of $b$, but just based on Internet population data. Bethlehem (1988) shows that the bias of estimator (12.37) is approximately equal to

$$(12.39) \qquad B(\bar{y}_{GR,R}) = \overline{X}B_I - \overline{Y} = \overline{X}(B_I - B),$$

where $B_I$ is defined by

$$(12.40) \qquad B_I = \left( \sum_{k=1}^{N} I_k X_k X_k' \right)^{-1} \left( \sum_{k=1}^{N} I_k X_k Y_k \right).$$

The bias of this estimator disappears if $B_I = B$. Thus, the regression estimator will be unbiased if under-coverage does not affect the regression coefficients. Particularly, if relationships are strong (the regression line fits the data well), the risk of finding a wrong relationship is small. By writing

$$(12.41) \qquad B_I = B + \left( \sum_{k=1}^{N} I_k X_k X_k' \right)^{-1} \left( \sum_{k=1}^{N} I_k X_k E_k \right),$$

the conclusion can indeed be drawn that the bias will be small if the residuals are small. This theory shows that using the generalized regression estimator has the potential of reducing the bias caused by under-coverage.

In case of self-selection, the generalized regression estimator changes to

$$(12.42) \qquad \bar{y}_{GR,S} = \bar{y}_S + (\overline{X} - \bar{x}_S)' b_S = \overline{X}' b_S.$$

The subscript $S$ indicates that the corresponding quantities have been computed using data from a self-selection sample. The vector of coefficients $b_S$ is defined by

$$(12.43) \qquad b_S = \left( \sum_{k=1}^{N} R_k X_k X_k' \right)^{-1} \left( \sum_{k=1}^{N} R_k X_k Y_k \right),$$

in which $R_k$ is the response indicator for element $k$. The bias of estimator (12.42) is approximately equal to

$$(12.44) \qquad B(\bar{y}_{GR,S}) = \overline{X}B_S - \overline{Y} = \overline{X}(B_S - B),$$

where $B_S$ is defined by

$$(12.45) \qquad B_S = \left( \sum_{k=1}^{N} \rho_k X_k X_k' \right)^{-1} \left( \sum_{k=1}^{N} \rho_k X_k Y_k \right).$$

The bias of this estimator disappears if $B_S = B$. Thus, the regression estimator will be unbiased if self-selection does not affect the regression coefficients. Particularly, if relationships are strong (the regression line fits the data well), the risk of finding a wrong relationship is small. By writing

$$(12.46) \qquad B_S = B + \left( \sum_{k=1}^{N} \rho_k X_k X_k' \right)^{-1} \left( \sum_{k=1}^{N} \rho_k X_k E_k \right),$$

the conclusion can also here be drawn that the bias will be small if the residuals are small. This theory shows that use of the generalized regression estimator has the potential of reducing the bias due to self-selection.

Generalized regression estimation can address the problem of the lack of sufficient population information. It is possible to include variables in the weighting model without having to know the population frequencies in the cells obtained by cross-tabulating all variables. The trick is to use a different set of dummy variables. Instead of defining one set of dummy variables for the complete crossing of all auxiliary variables, a set of dummy variables is defined for each variable separately or for each crossing of subsets of variables separately.

Suppose there are three auxiliary variables: $X_1$, $X_2$, and $X_3$. Post-stratification would come down to crossing the three variables using one set of dummy variables. If only the marginal population distributions of the three variables can be used, there are three sets of dummy variables, each corresponding to the categories of one auxiliary variable. And if, for example, the population distribution of the crossing of $X_1$ and $X_2$ is available and only the marginal distribution of $X_3$, there are two sets of dummy variables: one for $X_1 \times X_2$ and one for $X_3$. Of course, other combinations and subsets are possible, depending on the available auxiliary information and the number of observations in each cell of each cross-classification. See Example 12.9.

In case of post-stratification, the weight is equal to one of the weight coefficients. If the weighting model contains more than one set of dummy variables, there will also be more weight coefficients contributing to the weight. In fact, each set contributes a weight coefficient, and these weights are added to obtain the weight.

It should be noted that a weighting model containing more than one set of dummy variables will use less information than the model for the complete crossing of all auxiliary variables. Nevertheless, it uses more information than a post-stratification corresponding to one of the subsets.

**EXAMPLE 12.9 Generalized regression estimation using only marginal distributions**

Continuing Example 12.8, it is now shown how to use just only the marginal distributions of education and age. Two sets of dummy variables are introduced: one set of two dummy variables for the categories of education and another set of three dummy variables for the categories of age. Then there are $2 + 3 = 5$ dummy variables. In each set, always one dummy has the value 1, whereas all other dummies are 0. The possible values of the dummy variables are shown in Table 12.4.

TABLE 12.4 **Weighting with the marginal distributions**

| Educ. | Age | $X_1$ | $X_2$ | $X_3$ | $X_4$ | $X_5$ | $X_6$ |
|---|---|---|---|---|---|---|---|
| Low | Young | 1 | 1 | 0 | 1 | 0 | 0 |
| Low | Middle | 1 | 1 | 0 | 0 | 1 | 0 |
| Low | Old | 1 | 1 | 0 | 0 | 0 | 1 |
| High | Young | 1 | 0 | 1 | 1 | 0 | 0 |
| High | Middle | 1 | 0 | 1 | 0 | 1 | 0 |
| High | Old | 1 | 0 | 1 | 0 | 0 | 1 |
| Population means | | 1.000 | 0.599 | 0.401 | 0.399 | 0.301 | 0.300 |
| Weight coefficients | | 1.359 | 0.673 | −0.673 | 0.916 | −0.255 | −0.662 |

The first dummy variable $X_1$ represents the constant term in the regression model. It always has the value 1. The second and third dummy variable relate to the two categories of education (Low and High), and the last three dummies represent the three categories of age (Young, Middle, and Old). The vector of population means is equal to the fractions for all dummy variables separately. Note that in this weighting model always three dummies in a row have the value 1.

The weight for an observed element is now obtained by summing the appropriate elements of this vector. The first value corresponds to the dummy $X_1$, which always has the value 1. So there is always a contribution 1.359 to the weight. The next two values correspond to the categories of education. Note that their sum equals zero. For a low education, an amount 0.673 is added, and for high education, the same amount is subtracted. The final three values correspond to the categories of age. Depending on the age category, a contribution is added or subtracted. For example, the weight for a low-educated young person is equal to $1.359 + 0.673 + 0.916 = 2.948$.

Example 12.9 does not use information about the crossing of Education by Age. Only the marginal distributions are included in the computation of the weights. Therefore, a different notation is introduced. This weighting model is denoted by

Education + Age.

Due to the special structure of the auxiliary variables, the computation of the weight coefficients $v$ cannot be carried out without imposing extra conditions. Here, for every categorical variable, the condition is imposed that the sum of the weight coefficients for the corresponding dummy variables must equal zero.

Example 12.9 uses only two auxiliary variables. More variables can be included in a weighting model. This makes it possible to define various weighting models with these variables. Suppose there are three auxiliary variables: Education, Age, and Gender. If the complete population distribution on the crossing of all three variables is available, then the weighting model

Education × Age × Gender

can be applied. If only the bivariate population distributions of every crossing of two variables are available, the following weighting scheme could be applied:

(Education × Age) + (Age × Gender) + (Education × Gender).

Note that this scheme comes down to doing three post-stratifications simultaneously. If only marginal frequency distributions are available, the model

Education + Age + Gender

could be considered. See Example 12.10. More details about the theory of generalized regression estimation can be found, e.g., in Bethlehem and Keller (1987).

Up to now only generalized regression estimation with categorical auxiliary variables was described. It is also possible to apply this estimation technique with continuous auxiliary variables or a combination of categorical and continuous variables. See Bethlehem (2009) for more details.

■ EXAMPLE 12.10 Using generalized regression estimation for reducing self-selection bias

A fictitious population is used to illustrate the possible effects of generalized regression estimation on a self-selection bias. This population consists of 50,000 eligible voters in a town. The aim of a web survey is to measure whether or not people intend to vote at the next local elections.

Voting depends on age and level of education. Voting increases with age. Voting is higher among high-educated people than among low-educated people. The percentage of voters in the population is 46.1%.

Participation in the web survey depends on age and education: older people are more likely to participate than younger people. The average participation probability in the population was 0.075. Hence the expected response was $0.075 \times 50,000 = 3,750$.

A simulation was carried out in which 1,000 samples were selected from this population. The distribution of the five different estimators was compared:

- The mean of a simple random sample;
- The mean of a self-selection sample;
- Post-stratification by education of the self-selection sample;
- Post-stratification by age of the self-selection sample;
- The generalized regression estimator that uses only the marginal distributions of education and age;
- Post-stratification by education and age of the self-selection sample.

Figure 12.7 contains the results. The upper box plot shows the distribution of the sample mean in case of simple random sampling. It is clear that this estimator is unbiased.

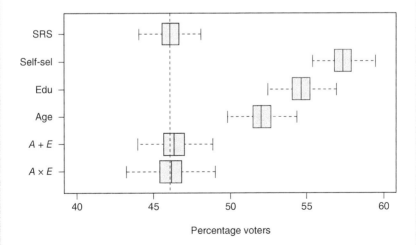

FIGURE 12.7 **Estimating the percentage voters for the NEP in case of selection**

The second box plot shows what happens in case of self-selection. The estimator has a substantial upward bias. This is not surprising as people with a high participation probability also are more inclined to vote.

The third and fourth box plot shows to what extent the bias is reduced if post-stratification carried out with just one auxiliary variable. In both cases the bias is reduced somewhat, but not completely removed. Age seems to be more effective than Education.

The sixth box plot ($A \times E$, at the bottom) displays the distribution of the estimator in case of post-stratification by Age and Education. The bias is completely removed. This can be expected as the participation probabilities are equal within the strata formed by crossing Age and Education.

Finally, the fifth box plot shows the result of using the regression estimator with only the marginal distributions ($A + E$). Apparently, this estimator performs as well as post-stratification with a complete crossing. This can be explained by the fact that there are no special interaction effects between the two auxiliary variables. This effect is often observed in practice. What matters is to have a specific set of auxiliary variables in the model. How they are used (crossed or marginally) is often less important.

## 12.2.4  RAKING RATIO ESTIMATION

If generalized regression estimation is applied, correction weights are obtained by taking the sum of a number of *weight coefficients*. It is also possible to compute correction weights in a different way, namely, as the product of a number of *weight factors*. This weighting technique is usually called *raking ratio estimation, raking, rim weighting*, or *multiplicative weighting*. Here it is denoted by *raking ratio estimation*. Weights are obtained as the product of a number of factors contributed by the various auxiliary variables in the model.

Raking ratio estimation can be applied in the same situations as generalized regression estimation as long as only categorical auxiliary variables are used. It computes correction weights by means of an iterative procedure. The resulting weights are the product of factors contributed by all cross-classifications in the model.

The technique of raking ratio estimation was already described by Deming and Stephan (1940). Skinner (1991) discussed application of this technique in multiple frame surveys. Little and Wu (1991) described the theoretical framework and showed that this technique comes down to fitting a log-linear model for the probabilities of getting observations in strata of the complete cross-classification given the probabilities for marginal distributions.

Raking ratio estimation can be seen as a technique in which a number of post-stratifications are carried out simultaneously. Each post-stratification consists of a single variable or a complete cross-classification of a set of variables. One extreme case is that in which there is just one cross-classification. Then raking ratio estimation reduces to simple post-stratification. For example, if variables $A$, $B$, and $C$ are used, this weighting technique would be denoted by $A \times B \times C$. Another extreme case is that in which each post-stratification consists of just one variable. In case of variables $A$, $B$, and $C$, there would be three post-stratifications: one for variable $A$, one for variable $B$, and one for variable $C$. This is denoted by $A + B + C$. So, only the marginal population distributions are used for computing weights.

To compute the weight factors, the following scheme is carried out:

Step 1: Introduce a weight factor for each stratum in each post-stratification. Set the initial values of all factors to 1.

Step 2: Adjust the weight factors for the first post-stratification so that the weighted sample becomes representative with respect to the cross-classified auxiliary variables.

Step 3: Adjust the weight factors for the next post-stratification so that the weighted sample becomes representative for the cross-classified variables in this post-stratification. Generally, this will disturb representativeness with respect to the other post-stratifications.

Step 4: Repeat this adjustment process until all post-stratifications have been dealt with.

Step 5: Go back to step 2. Repeat steps 2–4 until the weight factors do not change anymore. Then go to step 6.

Step 6: The weight of a respondent is obtained by the multiplying the relevant weight factors in all post-stratifications.

For the computation of a raking ratio estimation see Example 12.11.

### EXAMPLE 12.11 Raking ratio estimation

Raking ratio estimation is illustrated using the same data as in Example 12.9. Two variables are used in the weighting model: Age (three categories) and Education (two categories).

Suppose only the marginal population distributions of Age (three categories) and Education (two categories) are available and not the cross-classification. Table 12.5 contains the starting situation. The upper-left part of the table contains the unweighted relative frequencies in the sample for each combination of Age and Education. The row and column denoted by "Weight factor" contain the initial values of the weight factors (1.000). The values in the row and column denoted by "Weighted sum" are obtained by first computing the weight for each sample cell (by multiplying

TABLE 12.5   **The starting situation**

|  | Low educ. | High educ. | Weight factor | Weighted sum | Population distribution |
|---|---|---|---|---|---|
| Young | 0.030 | 0.195 | 1.000 | 0.225 | 0.399 |
| Middle | 0.122 | 0.193 | 1.000 | 0.315 | 0.301 |
| Old | 0.215 | 0.246 | 1.000 | 0.461 | 0.300 |
| Weight factor | 1.000 | 1.000 |  |  |  |
| Weighted sum | 0.367 | 0.634 |  | 1.000 |  |
| Popul. distr. | 0.599 | 0.401 |  |  | 1.000 |

the relevant row and column factor) and then summing the weighted cell fractions. Since the initial values of all factors are equal to 1, the weighted sums in the table are equal to the unweighted sample sums. The row and column denoted by "Population distribution" contain the fractions in the Age and Education categories in the population.

The iterative process must result in row and column factors with such values that the weighted sums match the population distribution. This is clearly not the case in the starting situation. First, the weight factors for the rows are adjusted. This leads to weight factors 1.773, 0.956, and 0.651 for the categories Young, Middle, and Old (see the Table 12.6). The weighted sums for the rows are now correct, but the weighted sums for the columns are 0.310 and 0.690 and thus still show a discrepancy.

TABLE 12.6   **Situation after adjusting for age**

|  | Low educ. | High educ. | Weight factor | Weighted sum | Population distribution |
|---|---|---|---|---|---|
| Young | 0.030 | 0.195 | 1.773 | 0.399 | 0.399 |
| Middle | 0.122 | 0.193 | 0.956 | 0.301 | 0.301 |
| Old | 0.215 | 0.246 | 0.651 | 0.300 | 0.300 |
| Weight factor | 1.000 | 1.000 |  |  |  |
| Weighted sum | 0.310 | 0.690 |  | 1.000 |  |
| Popul. distr. | 0.599 | 0.401 |  |  | 1.000 |

The next step is to adjust the weight factors for the columns such that the weighted column sums match the corresponding population frequencies. Note that this adjustment for Education will disturb the adjustment for Age. The weighted sums for the age categories no longer match the relative population frequencies. However, the discrepancy is much smaller than in the initial situation. See Table 12.7.

TABLE 12.7 **Situation after adjusting for education**

|  | Low educ. | High educ. | Weight factor | Weighted sum | Population distribution |
|---|---|---|---|---|---|
| Young | 0.030 | 0.195 | 1.773 | 0.304 | 0.399 |
| Middle | 0.122 | 0.193 | 0.956 | 0.333 | 0.301 |
| Old | 0.215 | 0.246 | 0.651 | 0.364 | 0.300 |
| Weight factor | 1.934 | 0.581 |  |  |  |
| Weighted sum | 0.599 | 0.401 |  | 1.000 |  |
| Popul. distr. | 0.599 | 0.401 |  |  | 1.000 |

The process of adjusting for Age and Education is repeated until the weight factors do not change anymore. The final situation is reached after a few iterations. Table 12.8 contains the final results.

The adjustment weight for a specific sample element is now obtained by multiplying the relevant weight factors. For example, the weight for a young male is equal to $2.422 \times 2.080 = 5.037$. Note that for this example the adjustment weights differ from those obtained by the generalized regression estimator in Example 12.9.

TABLE 12.8 **Situation after convergence**

|  | Male | Female | Weight factor | Weighted sum | Population distribution |
|---|---|---|---|---|---|
| Young | 0.030 | 0.195 | 2.422 | 0.399 | 0.399 |
| Middle | 0.122 | 0.193 | 0.848 | 0.301 | 0.301 |
| Old | 0.215 | 0.246 | 0.521 | 0.300 | 0.300 |
| Weight factor | 2.080 | 0.525 |  |  |  |
| Weighted sum | 0.599 | 0.401 |  | 1.000 |  |
| Popul. distr. | 0.599 | 0.401 |  |  | 1.000 |

There are many situations in which both generalized regression estimator and the raking ratio estimator can be applied. This raises the question which estimation method should be preferred. A number of observations may help to take a decision.

In the first place, generalized regression estimation is based on a simple linear model that describes the relationship between a target variable and a number of auxiliary variables. If this model fits well, weighting adjustment will be effective. For raking ratio estimation there is no straightforward model allowing simple interpretation.

In the second place, computations for generalized regression estimation are straightforward. Weights are obtained by application of ordinary least squares. The weights of raking ratio estimation are obtained as the solution of an iterative process. There is no guarantee this process will always converge.

In the third place, for generalized regression estimation, it is possible to derive an analytical expression of the variance of weighted estimates. No simple expressions are available for estimates based on raking ratio estimation.

In the fourth place, weights produced by linear weighting may sometimes turn out to be negative. This seems counterintuitive, but it is simply a consequence of the linear model applied. Negative weights usually indicate that the linear model does not fit to well. A disadvantage of negative weights is that some statistical analysis packages do not accept negative weights. This may prevent weighted analysis of the survey outcomes.

In the fifth place, it has been shown (see Section 12.2.5) that, in many situations, estimators based on linear weights have asymptotically the same properties as those based on multiplicative weights.

## 12.2.5 CALIBRATION ESTIMATION

Deville and Särndal (1992) and Deville, Särndal, and Sautory (1993) have proposed a general framework for weighting of which generalized regression estimation and raking ratio estimation are special cases. Assuming simple random sampling, their starting point is that the correction weights $c_i$ in $w_i = c_i \times d_i$ have to satisfy two conditions:

1. The correction weights $c_i$ have to be as close as possible to 1.
2. The weighted sample distribution of the auxiliary variables has to match the population distribution, i.e.,

$$(12.47) \qquad \bar{x}_W = \frac{1}{N}\sum_{i=1}^{n} w_i x_i = \bar{X}.$$

The first condition sees to it that the resulting estimators are unbiased, or almost unbiased, and the second condition guarantees that the weighted sample is representative with respect to the auxiliary variables used.

Deville and Särndal (1992) introduce general distance measure $D(c_i, 1)$ measuring the difference between $c_i$ and 1. The problem is then to minimize

$$(12.48) \qquad \sum_{i=1}^{n} D(c_i, 1)$$

under the condition (12.47). This problem can be solved by using the method of Lagrange. By choosing the proper distance function, generalized regression estimation and raking ratio estimation are obtained as special cases of this general approach. For generalized regression estimation, the distance function is defined by

$$(12.49) \qquad D(c_i, 1) = (c_i - 1)^2,$$

which is the Euclidean distance, and for raking ratio estimation the distance

$$(12.50) \qquad D(c_i, 1) = c_i \log (c_i) - c_i + 1$$

must be used.

Deville and Särndal (1992) and Deville et al. (1993) only consider the full response situation. They show that estimators based on weights computed within their framework have asymptotically the same properties. This means that for large samples it does not matter which of the two weighting adjustment techniques is applied. Estimators based on both weighting techniques will behave approximately the same. Note that although the estimators behave in the same way, the individual weights computed by means of generalized regression estimation or raking ratio estimation may differ substantially.

Under nonresponse, under-coverage, or self-selection, the situation is different. Then the asymptotic properties of both estimation techniques will generally not be equal. The extent to which the chosen weighting technique is able to reduce the bias depends on how well the corresponding underlying model can be estimated using the observed data. Generalized regression estimation assumes a linear model to hold with the target variable as dependent variable and the auxiliary variables as explanatory variables. Raking ratio estimation assumes a log-linear model for the cell frequencies. An attempt to use a correction technique for which the underlying model does not hold will not help to reduce the bias.

## 12.2.6 CONSTRAINING THE VALUES OF WEIGHTS

There are several reasons why survey researchers may want to have some control over the values of the adjustment weights. One reason is that extremely large weights are generally considered undesirable. Large weights usually correspond to population elements with rare characteristics. Use of such weights may lead to unstable estimates of population parameters. To reduce the impact of large weights on estimators, a weighting method is required that is able to keep adjustment weights within prespecified boundaries and that at the same time enables valid inference.

Another reason to have some control over the values of the adjustment weights is that application of generalized regression estimation may produce negative weights. Although the theory does not require weights to be positive, negative weights should be avoided, since they are counterintuitive, they cause problems in subsequent analyses, and they are an indication that the regression model does not fit the data well.

Negative weights can be avoided by using a better regression model. However, it is not always possible to find such models. Another solution is to use the current model and force weights within certain limits. Several techniques have been proposed for this. A technique developed by Deville et al. (1993) comes down to repeating the regression estimation process a number of times. First a lower bound $L$ and an upper bound $U$ are specified. After the first run, weights smaller than $L$ are set to $L$, and weights larger than $U$ are set to $U$. Then, the weighting process is repeated, but records from the strata with the fixed weights $L$ and $U$ are excluded. Again, weights may be produced not satisfying the conditions. These weights are also set to either the value $L$ or $U$. The weighting process is repeated until all computed weights fall within the specified limits. Convergence of this iterative process is not guaranteed. Particularly, if the lower bound $L$ and upper bound $U$ are not far apart, the algorithm may not converge.

Huang and Fuller (1978) use a different approach. Their algorithm produces weights that are a smooth, continuous, monotone increasing function of the original weights computed from the linear model. The algorithm is iterative. At each step, the weights are checked against a user-supplied criterion value $M$. This value $M$ is the maximum fraction of the mean weight by which any weight may deviate from the mean weight. For example, if $M$ is set to 0.75, then all weights are forced into the interval with lower bound equal to 0.25 times the mean weight and upper bound equal to 1.75 times the mean weight. Setting the value to 1 implies that all weights are forced to be positive. Huang and Fuller (1978) prove that the asymptotic properties of the regression estimator constructed with their algorithm are asymptotically the same as those of the generalized regression estimator. So, restricting the weights has (at least asymptotically) no effect on the properties of population estimates computed with these weights.

## 12.2.7 CORRECTION USING A REFERENCE SURVEY

Post-stratification, generalized regression estimation, and raking ratio estimation can be effective bias reduction techniques provided auxiliary variables are available that have a strong correlation with the target variables of the survey. If such variables cannot be used because their population distribution is not available, one might consider estimating these population distributions in a different survey, a so-called reference survey. This reference survey must be based on a probability sample, where data collection takes place with a mode different from the web, e.g., CAPI (computer-assisted personal interviewing, with laptops) or CATI (computer-assisted telephone interviewing). Preferably, the sample size of this reference survey must be small in order to keep costs within limits. The reference survey approach has been applied by several market research organizations. See, for example, Börsch-Supan et al. (2004) and Duffy et al. (2005).

Under the assumption of no nonresponse, or ignorable nonresponse, this reference survey can produce unbiased estimates of quantities that have also been measured in the web survey. Unbiased estimates for the target variable can be computed, but due to the small sample size, these estimates will have a substantial variance. The question is now whether estimates of population characteristics can be improved by combining the large sample size of the web survey with the unbiasedness of the reference survey.

First, it will be explored whether a reference survey can reduce an undercoverage bias. See also Example 12.12. Then the effect on a self-selection bias is analyzed. Only post-stratification with one auxiliary variable is considered as adjustment method.

It is assumed that one categorical auxiliary variable is observed in both the web survey and the reference survey and that this variable has a strong correlation with the target variable of the survey. Then a form of post-stratification can be applied where the stratum means are estimated using the web survey data and the stratum weights are estimated using the reference survey data. Suppose $m$ is the sample size of the reference survey and $m_h$ is the number of observed elements in stratum $h$. This leads to the post-stratification estimator

$$(12.51) \qquad \bar{y}_{I,RS} = \sum_{h=1}^{L} \frac{m_h}{m} \bar{y}_I^{(h)}$$

where $\bar{y}_I^{(h)}$ is the web survey-based estimate for the mean of stratum $h$ of the Internet population (for $h = 1, 2, \ldots, L$) and $m_h/m$ is the relative sample size in stratum $h$ as estimated in the reference survey sample (for $h = 1, 2, \ldots, L$). Under the conditions described above, the quantity $m_h/m$ is an unbiased estimate of $W_h = N_h/N$.

Let $I$ denote the probability distribution for the web survey and let $P$ be the probability distribution for the reference survey. Then the expected value of the post-stratification estimator is equal to

$$(12.52) \quad E\left(\bar{y}_{I,RS}\right) = E_I E_P\left(\bar{y}_{I,RS} \mid I\right) = E_I\left(\sum_{h=1}^{L} \frac{N_h}{N} \bar{y}_I^{(h)}\right) = \sum_{h=1}^{L} W_h \overline{Y}_I^{(h)} = \widetilde{Y}_I,$$

where $W_h = N_h/N$ is the relative size of stratum $h$ in the target population and $\overline{Y}_I^{(h)}$ is the mean of the target variable of stratum $h$ of the Internet population. The expected value of this estimator is identical to that of the post-stratification estimator (10.20). The bias of this estimator is equal to

$$B\left(\bar{y}_{I,RS}\right) = E\left(\bar{y}_{I,RS}\right) - \overline{Y} = \widetilde{Y}_I - \overline{Y} = \sum_{h=1}^{L} W_h\left(\overline{Y}_I^{(h)} - \overline{Y}^{(h)}\right)$$

(12.53)

$$= \sum_{h=1}^{L} W_h \frac{N_{NI,h}}{N_h}\left(\overline{Y}_I^{(h)} - \overline{Y}_{NI}^{(h)}\right).$$

If a strong relationship exists between the target variable and the auxiliary variable, there is little or no variation of the target variable within the strata. This implies that if the stratum means for the Internet population and for the target population do not differ much, this results in a small bias. So, using a reference survey with the proper auxiliary variables can substantially reduce the bias of web survey estimates.

Note that the expression for the bias of the reference survey estimator is equal to that of the post-stratification estimator. An interesting aspect of the reference survey approach is that any variable can be used for adjustment weighting as long as it is measured in both surveys. For example, some market research organizations use "webographics" or "psychographic" variables that divide the population in "mentality groups." People in the same groups have more or less the same level of motivation and interest to participate in such surveys. Deployment of effective weighting variables resembles the MAR situation. This implies that within weighting strata there is no relationship between participating in a web survey and the target variables of the survey.

Bethlehem (2007) shows that if a reference survey is used, the variance of the post-stratification estimator is equal to

$$V\left(\bar{y}_{I,RS}\right) = \frac{1}{m}\sum_{h=1}^{L} W_h\left(\overline{Y}_I^{(h)} - \widetilde{Y}_I\right)^2 + \frac{1}{m}\sum_{h=1}^{L} W_h(1 - W_h)V\left(\bar{y}_I^{(h)}\right)$$

(12.54)

$$+ \sum_{h=1}^{L} W_h^2 V\left(\bar{y}_I^{(h)}\right).$$

The quantity $\bar{y}_I^{(h)}$ is measured in the web survey. Therefore, its variance $V\left(\bar{y}_I^{(h)}\right)$ will be of the order $1/n$. This means that the first term in the variance of the post-stratification estimator will be of the order $1/m$, the second term of order $1/mn$, and the third term of order $1/n$. Since $n$ will generally be much larger than $m$ in practical situations, the first term in the variance will dominate, i.e., the (small) size of the reference survey will determine the accuracy of the estimates. So, the large number of observations in the web survey does not help to produce accurate estimates. One could say that the reference survey approach reduces the bias of estimates at the cost of a higher variance.

**EXAMPLE    12.12 Using    a    reference    survey    for    reducing under-coverage bias**

Suppose one of the aims of the election survey in Example 12.5 is to estimate the percentage of people voting for the National Elderly Party (NEP). Four situations are considered:

- A simple random sample from the complete population;
- A simple random sample from the Internet population;
- A simple random sample from the Internet population, followed by post-stratification;
- A simple random sample from the Internet population, followed by post-stratification based on a reference survey;

In all four situations, the distribution of the estimator was determined by repeating the selection of the sample 1,000 times. The sample size is always 1,000 cases. Figure 12.8 contains the results.

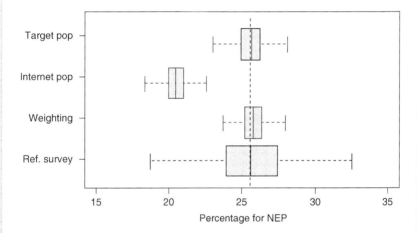

FIGURE 12.8 **Estimating the percentage voters for the NEP in case of under-coverage using a reference survey**

The population was constructed such that Internet access decreases with age. Moreover, Internet access for natives was much higher than for non-natives. Voting for the NEP depended on age only.

A simple random sample from the complete target population results in an unbiased estimator for the population percentage of 25.4%. Just sampling the Internet population leads to an estimator with a substantial bias.

The expected value of this estimator is substantially too low: 20.3%. This can be explained by the fact that the elderlies are underrepresented in the samples because they have less access to Internet. The elderlies typically vote for the NEP.

Application of post-stratification by age solves the problem. After weighting, the estimator is unbiased. There is a direct relation between voting behavior and age, and there is a direct relation between age and having Internet. So, correcting the age distribution also corrects the estimator.

The lower box plot shows the distribution of the estimator if the population distribution of the weighting variable age is estimated in reference survey with a sample size of $m = 100$. The bias is removed, but at the cost of a substantial increase of the variance. This is due to small sample size of the reference survey. Of course, one could consider increasing this sample size, but this also increases the costs. One may even wonder why to conduct a web survey at all, if also a reference survey is carried out.

Figure 12.9 shows the analysis for voting for the New Internet Party (NIP). The same four situations are compared.

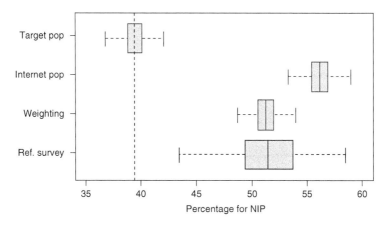

FIGURE 12.9 **Estimating the percentage voters for the NIP in case of under-coverage using a reference survey**

Just sampling the Internet leads to a biased estimator. The estimated values are substantially too high. The expected value of the estimator is 56.5%, whereas it should have been 39.5%. Post-stratification is not successful. The expected value of the estimator decreases from 56.5% to 51.1%, but this is still too high. This is not surprising as there is a direct relation between voting for the NIP and having access to Internet. Post-stratification based on a reference survey will not solve the problem here. The bias remains, and at the same time the variance increases.

Now the case of self-selection is considered. See also Example 12.13. Again, it is assumed that one categorical auxiliary variable is observed in the web survey and the reference survey and that this variable has a strong correlation with the target variable of the survey. Post-stratification is applied where the stratum means are estimated using web survey data and the stratum weights are estimated using the reference survey data. This leads to the post-stratification estimator

$$(12.55) \qquad \bar{y}_{S,RS} = \sum_{h=1}^{L} \frac{m_h}{m} \bar{y}_S^{(h)}$$

where $\bar{y}_S^{(h)}$ is the web survey-based estimate for the mean of stratum $h$ of the target population (for $h = 1, 2, \ldots, L$) and $m_h/m$ is the estimated relative sample size in stratum $h$ using the reference survey (for $h = 1, 2, \ldots, L$). Under the conditions described above, the quantity $m_h/m$ is an unbiased estimate of $W_h = N_h/N$.

Let $I$ denote the probability distribution for the web survey and let $P$ be the probability distribution for the reference survey. Then the expected value of the post-stratification estimator is equal to

$$(12.56) \qquad E\left(\bar{y}_{S,RS}\right) = E_I E_P\left(\bar{y}_{S,RS} \mid I\right) = E_I\left(\sum_{h=1}^{L} \frac{N_h}{N} \bar{y}_h\right) = \sum_{h=1}^{L} W_h \overline{Y}_h^* = \widetilde{Y}^*.$$

So, the expected value of this estimator is identical to that of the post-stratification estimator (11.34). The bias of this estimator is equal to

$$B\left(\bar{y}_{S,RS}\right) = E\left(\bar{y}_{S,RS}\right) - \overline{Y} = \widetilde{Y}^* - \overline{Y} = \sum_{h=1}^{L} W_h\left(\overline{Y}_h^* - \overline{Y}_h\right)$$

$$(12.57)$$

$$= \sum_{h=1}^{L} W_h \frac{R_{\rho Y}^{(h)} S_\rho^{(h)} S_Y^{(h)}}{\overline{\rho}^{(h)}}.$$

If there is a strong relationship between the target variable and the auxiliary variable used for computing the weights, there is little or no variation of the target variable within the strata. Consequently, the correlation between target variable and response behavior will be small, and the same applies to the standard deviation of the target variable. So, using a reference survey with the proper auxiliary variables can substantially reduce the bias of web survey estimates.

Bethlehem (2008) shows that the variance of estimator (12.55) is equal to

$$V\left(\bar{y}_{S,RS}\right) = \frac{1}{m}\sum_{h=1}^{L} W_h\left(\overline{Y}_h^* - \widetilde{Y}^*\right)^2 + \frac{1}{m}\sum_{h=1}^{L} W_h(1 - W_h)V\left(\bar{y}_S^{(h)}\right)$$

$$(12.58)$$

$$+ \sum_{h=1}^{L} W_h^2 V\left(\bar{y}_S^{(h)}\right).$$

The quantity $\bar{y}_S^{(h)}$ is measured in the online survey. Therefore, its variance $V\left(\bar{y}_S^{(h)}\right)$ will be at most of the order $1/E(n_S) = 1/(N\bar{p})$, where $n_S$ is the size of the self-selection sample. This means that the first term in the variance of the post-stratification estimator will be of the order $1/m$, the second term of order $1/(mE(n_S))$, and the third term of order $1/E(n_S)$. Since $E(n_S)$ will generally be much larger than $m$ in practical situations, the first term in the variance will dominate, i.e., the (small) size of the reference survey will determine the accuracy of the estimates.

Moreover, since strata preferably are based on groups of people with the same psychographic characteristics, and target variables may very well be related to the psychographic variables, the stratum means $\bar{Y}_h^*$ may vary substantially. This also contributes to a large value of the first variance component.

The conclusion is that a large number of observations in the web survey do not help to produce accurate estimates. The reference survey approach may reduce the bias of estimates, but it does so at the cost of a higher variance.

The effectiveness of a survey design is sometimes also indicated by means of the *effective sample size*. This is the sample size of a simple random sample of elements that would produce an estimator with the same precision. Use of a reference survey implies that the effective sample size is much lower than the size of the web survey.

■ **EXAMPLE 12.13** Using a reference survey for reducing self-selection bias

The fictitious population of Section 10.3 is used to illustrate the possible effects of post-stratification with a reference survey on a self-selection bias. This population consists of 100,000 persons. Most persons (99%) are passive Internet users. Active users make up only 1% of the population. Active users have a high participation probability of 0.99. Passive users have a low participation probability (0.01). The percentage of active Internet users decreases with age.

An election survey is conducted. Aim is to estimate the percentage of people voting for the National Elderly Party (NEP). Three situations are considered:

- A simple random sample from the population;
- A self-selection sample from the population;
- A self-selection sample from the population, followed by post-stratification;
- A self-selection sample from the Internet population, followed by post-stratification with a reference survey.

In all four situations, the distribution of the estimator determined by repeating the selection of the sample 1,000 times. The average participation probability in the population is 0.01971. Therefore, the expected sample size in a self-selection survey is equal to 1,971.

Figure 12.10 contains the results. The upper box plot shows that the estimator is unbiased in case of simple random sampling (of size 1,971) from the target population. The expected value is equal to 25.6%.

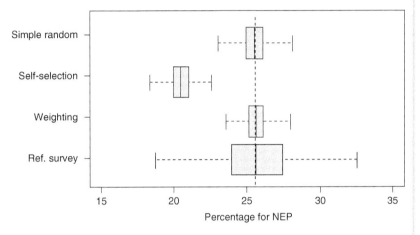

FIGURE 12.10 **Estimating the percentage voters for the NEP in case of self-selection using a reference survey**

The middle box plot shows what happens if samples are selected by means of self-selection. The shape of the distribution remains more or less the same, but the distribution as a whole has shifted to the left. All values of the estimator are systematically too low. The expected value of the estimator is only 20.4%. The estimator is biased. The explanation of this bias is simple: relative few elderlies are active Internet users. Therefore, they are underrepresented in the samples. They are typically people who will vote for the NEP.

The third box plot shows the distribution of the estimator in case of post-stratification by age. The bias is removed. This was possible because there is direct relation between participation and the weighting variable age.

The lower box plot shows the distribution of the estimator if the population distribution of the weighting variable age is estimated in reference survey with a sample size of $m = 100$. The bias is removed, but at the cost of a substantial increase of the variance. This is due to small sample size of the reference survey.

Figure 12.11 shows the analysis for voting for the New Internet Party (NIP). The same four situations are compared. In case of simple random sampling, the estimator is unbiased. The population value of 10.4% is correctly estimated.

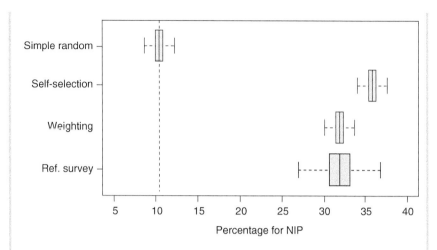

FIGURE 12.11  **Estimating the percentage voters for the NIP in case of self-selection using a reference survey**

Self-selection leads to a biased estimator. The estimated values are substantially too high. The expected value of the estimator is 35.6%, whereas it should have been 10.4%.

Post-stratification is not successful. The expected value of the estimator decreases from 35.6% to 32.1%, but this is still too high. This is not surprising as there is a direct relation between voting for the NIP and having access to Internet.

Post-stratification based on a reference survey does not work here. The bias remains, and at the same time the variance increases.

# 12.3  Application

There are three nationwide public TV channels in the Netherlands. One of these channels ("Nederland 1") has a current affairs program called "EenVandaag." This program maintains a web panel. It is used to measure public opinion with respect to topics that are discussed in the program. The "EenVandaag Opinion Panel" started in 2004. In 2008 it contained approximately 45,000 members.

The panel is a self-selection panel. Participants were recruited among the viewers of the program. For these reasons the panel lacks representativity. It is explored how unbalanced the composition of the panel is and whether estimates can be improved by applying some form of weighting adjustment.

In the period before the start of the Olympic Games in Beijing in August of 2008, there was a lot of discussion in the Netherlands about a possible boycott of the games. Suggestions ranged from not showing up at the opening ceremony to athletes not participating in the games at all. This boycott was considered because of the lack of respect of the Chinese for the human rights of the Tibetan people. One of the waves of the opinion panel was conducted in April 2008 in order to determine the public opinion of the Dutch with respect to this issue. The members of the panel were invited to complete a questionnaire. This questionnaire also contained topics about other issues, like preference for political parties. The questionnaire was completed by 19,392 members of the panel.

The representativity of the response is affected by two phenomena. In the first place, the panel was constructed by means of self-selection. In the second place, not all members of the panel responded to the request to fill in the questionnaire. The response rate was $100 \times 19,392/45,632 = 42.5\%$. Possible deviations from representativity are analyzed in this section. It is also explored to what extent weighting adjustment can improve the situation.

If persons apply for membership of the panel, they have to fill in a basic questionnaire with a number of demographic questions. These demographic variables can be used as auxiliary variables. The following variables were used in the analysis:

- Gender in two categories: male and female;
- Age in five categories: 18–24, 25–39, 40–54, 55–64, and 65+;
- Marital status in four categories: never married, married, divorced, and widowhood;
- Province of residence in 12 categories: Groningen, Friesland, Drenthe, Overijssel, Flevoland, Gelderland, Utrecht, Noord-Holland, Zuid-Holland, Zeeland, Noord-Brabant, and Limburg;
- Ethnic background in three categories: native, first-generation non-native, and second-generation non-native;
- Voting at the 2006 general elections in 12 categories: CDA (Christian democrats), PvdA (social democrats), SP (socialists), VVD (liberals), PVV (right-wing populists), GroenLinks (green party), ChristenUnie (right-wing Christians), D66 (liberal democrats), PvdD (party for the animals), SGP (right-wing Christians), other party, and did not vote. The population distributions were available for all these variables. Most distributions could be found in *StatLine*, the online statistical database of Statistics Netherlands (www.cbs.nl). The distribution of the voting variable came from a different source (the Electoral Council). For this reason, it was not possible to cross this variable with other auxiliary variables.

The first step in the analysis was to compare the response distribution of each variable with its population distribution. Table 12.9 contains the result for the variable Gender. Is a clear that males are substantially overrepresented.

Table 12.10 compares the response distribution of the variable Age with its population distribution. Persons in the age group from 18 to 39 years are

TABLE 12.9    **The distribution by gender in the response and the population**

| Gender | Panel response (%) | Population (%) | Difference (%) |
|---|---|---|---|
| Male | 72.4 | 49.0 | 23.4 |
| Female | 27.6 | 51.0 | −23.4 |
| Total | 100.0 | 100.0 | |

TABLE 12.10    **The distribution by age in the response and the population**

| Age | Panel response (%) | Population (%) | Difference (%) |
|---|---|---|---|
| 18–24 | 5.5 | 10.7 | −5.2 |
| 25–39 | 15.5 | 25.6 | −10.1 |
| 40–54 | 29.0 | 28.8 | 0.3 |
| 55–64 | 33.4 | 16.2 | 17.2 |
| 65+ | 16.5 | 18.8 | −2.3 |
| Total | 100.0 | 100.0 | |

underrepresented. Also the elderly (65+) are somewhat underrepresented. People in the age group from 55 to 64 years are clearly overrepresented.

Table 12.11 contains the comparison for the variable Marital status. People that never married and widowed persons are underrepresented. Married people are overrepresented. The patterns in Tables 12.10 and 12.11 may partly coincide. Young people are typically found in the category Never married. Widowed persons will often belong to the elderly.

Table 12.12 shows the regional distribution of panel response. Response percentages and population percentages are compared for the 12 Dutch provinces. The differences are small. One could conclude that the response is more or less representative with respect to the variable Province.

It should be noted that the pattern is usually different for the response to surveys based on probability sampling. Due to high nonresponse rates in highly

TABLE 12.11    **The distribution of marital status in the response and the population**

| Marital status | Panel response (%) | Population (%) | Difference (%) |
|---|---|---|---|
| Never married | 26.9 | 31.5 | −4.6 |
| Married | 61.4 | 53.6 | 7.7 |
| Divorced | 8.6 | 6.8 | 1.8 |
| Widowed | 3.2 | 8.1 | −4.9 |
| Total | 100.0 | 100.0 | |

TABLE 12.12 **Response and population percentage distribution by province of residence**

| Province | Panel response (%) | Population (%) | Difference (%) |
|---|---|---|---|
| Groningen | 3.6 | 3.6 | 0.0 |
| Friesland | 3.5 | 3.9 | −0.4 |
| Drenthe | 3.0 | 3.0 | 0.1 |
| Overijssel | 6.2 | 6.7 | −0.6 |
| Flevoland | 2.3 | 3.9 | 0.5 |
| Gelderland | 12.4 | 12.0 | 0.4 |
| Utrecht | 8.1 | 7.2 | 0.8 |
| Noord-Holland | 16.9 | 16.1 | 0.7 |
| Zuid-Holland | 21.3 | 21.1 | 0.2 |
| Zeeland | 2.2 | 2.3 | −0.1 |
| Noord-Brabant | 13.8 | 14.8 | −1.1 |
| Limburg | 6.4 | 7.1 | −0.6 |
| Total | 100.0 | 100.0 | |

TABLE 12.13 **The distribution of ethnic background in the response and the population**

| Marital status | Panel response (%) | Population (%) | Difference (%) |
|---|---|---|---|
| Native | 89.4 | 81.2 | 8.2 |
| Non-native (first generation) | 3.4 | 11.7 | −8.3 |
| Non-native (second generation) | 7.2 | 7.1 | 0.1 |
| Total | 100.0 | 100.0 | |

urbanized areas, the response rates are usually low in the provinces of Utrecht, Noord-Holland, and Zuid-Holland. This is not the case for the EenVandaag Opinion Panel.

Table 12.13 compares the response distribution and the population distribution of the variable Ethnic background. There is a problem with the first generation of non-natives. These persons and at least one of their parents are born outside the Netherlands. A language problem is one of the main causes for not participating in surveys. Note that the problem is less severe for second-generation non-natives. They are better integrated in the population.

Finally, Table 12.14 shows the difference between the response distribution and the population distribution for voting behavior in 2006 general elections in the Netherlands.

The more traditional CDA (Christian democrats) voters are underrepresented in the response, and the more activist SP (socialist party) voters are overrepresented. Also notable is the fact that, apparently, voters are overrepresented among

TABLE 12.14    The distribution of voting in 2006 in the response and the population

| Party | Panel response (%) | Population (%) | Difference (%) |
|---|---|---|---|
| CDA (Christian democrats) | 15.3 | 21.3 | −5.9 |
| PvdA (social democrats) | 20.0 | 17.0 | 3.0 |
| SP (socialists) | 20.2 | 13.3 | 6.9 |
| VVD (liberals) | 14.6 | 11.8 | 2.9 |
| PVV (right-wing populists) | 7.1 | 4.7 | 2.4 |
| GroenLinks (green party) | 5.8 | 3.7 | 2.1 |
| ChristenUnie (right-wing Christians) | 4.4 | 3.2 | 1.2 |
| D66 (liberal democrats) | 3.3 | 1.6 | 1.7 |
| PvdD (party for the animals) | 2.1 | 1.5 | 0.6 |
| SGP (right-wing Christians) | 0.3 | 1.3 | −0.9 |
| Other party | 1.6 | 1.0 | 0.6 |
| Did not vote | 5.2 | 19.8 | −14.6 |
| Total | 100.0 | 100.0 | |

the respondents. This phenomenon is also observed in other surveys. There is a relationship between voting and participating in surveys. Voters tend to participate in surveys and nonvoters tend to refuse.

Now, how these auxiliary variables are to be used for weighting adjustment is explored. Select here one target variable. It asks respondents whether they have a paid job for at least 12 hours per week. The response percentage is 48.6%. The population distribution is also available. The population percentage of people with a paid job is 57.6%. The response estimate is significantly too low. The question is whether this estimate is improved by weighting.

As a first step, each auxiliary variable separately is used in post-stratification weighting. The results are presented in Table 12.15.

The effects are small for most variables. There is almost no change in the estimate; there is also no reduction of the standard error. The effect is different for the variable Age. The estimate is adjusted in correct direction, from 48.6 to 50.9. The standard error is also smaller. This is an indication that Age should be included in the weighting model. Note also that the estimate is still far from the population value.

The next step is to explore whether the weighting model can be improved by adding another variable to Age. Table 12.16 contains the results for five generalized regression models with only main effects. Interaction effects of variables are not included.

TABLE 12.15   **Post-stratification weighting with a single variable**

| Weighting model | Estimate | Standard error |
|---|---|---|
| No weighting | 48.55 | 0.36 |
| Age | 50.92 | 0.29 |
| Gender | 47.44 | 0.36 |
| Marital status | 47.58 | 0.36 |
| Province | 48.45 | 0.36 |
| Ethnic background | 48.49 | 0.36 |
| Vote in 2006 | 46.70 | 0.36 |
| Population | 57.62 | |

Objective is estimation of the percentage with a paid job.

It is clear that the estimate cannot be improved by adding another variable to Age. The best weighting model is the one containing Age and Province, but it is not better than the model just containing the variable Age.

Table 12.16 contains also the estimates for the model obtained by crossing the variables Age and Province (Age × Province). There is some improvement with respect to the model Age + Province. This is the best estimate that can be obtained with this set of auxiliary variables. The conclusion can be that weighting adjustment can reduce the bias of the estimate for the percentage of people with a paid job. Unfortunately, the bias cannot be removed. Apparently, the proper auxiliary variables are lacking.

Analysis of another target variable shows that not always the same auxiliary variables are effective in adjustment weighting. This time the target variable measures whether one intends to vote for the Socialist Party in the next elections. The percentage of voters in the response is 15.43%. There is no population value available, but other opinion polls suggest a percentage between 11% and 12%. This

TABLE 12.16   **Generalized regression estimation with two variables**

| Weighting model | Estimate | Standard error |
|---|---|---|
| No weighting | 48.55 | 0.36 |
| Age + Gender | 48.50 | 0.29 |
| Age + Marital status | 50.77 | 0.29 |
| Age + Province | 50.85 | 0.29 |
| Age + Ethnic background | 50.81 | 0.29 |
| Age + Vote in 2006 | 50.66 | 0.29 |
| Age × Province | 51.03 | 0.29 |
| Population | 57.62 | |

Objective is estimation of the percentage with a paid job.

TABLE 12.17   **Post-stratification weighting with a single variable**

| Weighting model | Estimate | Standard error |
| --- | --- | --- |
| No weighting | 15.43 | 0.28 |
| Age | 15.18 | 0.28 |
| Gender | 15.92 | 0.28 |
| Marital status | 15.59 | 0.28 |
| Province | 15.53 | 0.28 |
| Ethnic background | 15.82 | 0.28 |
| Vote in 2006 | 12.43 | 0.21 |
| Vote 2006 + Age | 12.43 | 0.21 |
| Vote 2006 + Gender | 12.30 | 0.21 |
| Vote 2006 + Marital status | 12.43 | 0.21 |
| Vote 2006 + Province | 12.44 | 0.21 |
| Vote 2006 + Ethnic background | 12.67 | 0.21 |

would indicate the response-based estimate is too high. Table 12.17 contains the results of various weighting adjustment attempts.

If post-stratification is applied with just one variable, then there is no effect for five of the six weighting variables. The estimate changes only for the variable Vote 2006. The estimate goes down by 3% from 15.43% to 12.43%. Also, the standard error of the estimate is smaller. This result is not surprising as one can expect there may be a relationship between current and past voting behavior.

Table 12.17 also shows that adding another auxiliary variable to the weighting model does not lead to substantial changes. By adding the variable Gender, the estimate is reduced from 12.43% to 12.30%. One can conclude that this estimate is in line with the results of the other polls, although there is no guarantee that they reflect the true population value.

Note that it is not possible to cross the variable Vote 2006 with other auxiliary variables. This is because Vote 2006 is obtained from different source: Electoral Council of the Netherlands. This council does not record other variables like gender and age.

This application shows that weighting adjustment may help to reduce the bias of web survey-based estimates. However, there is now guarantee the bias is completely removed. The examples in this section also make clear that different target variables may need different weighting adjustment models.

# 12.4  Summary

There can be several reasons to carry some kind of weighting adjustment on the response to a web survey:

- The sample is selected with *unequal probability sampling*. To obtain unbiased estimates, an estimator such as the Horvitz–Thompson must be used. This comes down to weighting adjustment where the weights are equal to the design weights (which are equal to 1 over the first-order inclusion probabilities).

- *Nonresponse* may cause estimators of population characteristics to be biased. This happens when specific groups are over- or underrepresented in the survey response and these groups behave differently with respect to the survey variables.

- If the target population is wider than the Internet population, people without Internet can never be selected for the survey. This is called *under-coverage* and it may lead to biased estimates.

- If the sample is selected by means of *self-selection*, the true selection probabilities are unknown. Assuming equal selection probabilities leads to biased estimates.

Weighting adjustment techniques may help to reduce a bias. These techniques assign *weights* to observed elements, where underrepresented elements get a weight larger than 1 and overrepresented elements get a weight smaller than 1.

To be able to compute adjustment weights, *auxiliary variables* are required. Such variables have to be measured in the survey; also, the population distribution (or complete sample distribution) must be available.

Weighting adjustment will only help to reduce a bias if there is a strong relationship between the survey variables and the auxiliary variables and/or response behavior and auxiliary variables.

*Post-stratification* is the most frequently used weighting adjustment technique. Using auxiliary variables, the population is divided into a number of strata (sub-populations). All observed elements in a stratum are assigned the same weight. Post-stratification reduces a bias if the strata are homogeneous, i.e., all elements within a stratum resemble each other.

Practical limitations may hinder application of post-stratification. If many auxiliary variables are used to form strata, there may be strata without observations. Consequently, it is impossible to compute weights for such strata. It can also happen that insufficient population is available with respect to the distribution of the auxiliary variables.

If it is not possible to carry out post-stratification, there are two alternative weighting methods. One is *generalized regression estimation*. It is based on a linear regression model that predicts the values of the target variable of the survey from a set of auxiliary variables. Such regression models offer more flexibility with respect to the way auxiliary information is used to compute adjustment weights. Another possibility is to use *raking ratio estimation*. This weighting model is based on iterative proportional fitting. It can be shown that estimates based on generalized regression estimation and raking ratio estimation behave approximately the same in many situations.

*Calibration* is a theoretical framework for adjustment weighting. Post-stratification, linear weighting, and multiplicative weighting are special cases for this framework. It has more possibilities, like imposing constraints on the values of the weights.

If auxiliary variables cannot be used because their population distribution is not available, one might consider estimating them with a *reference survey*. This is a survey not affected by the problems of web surveys. It might be a CAPI or CATI survey. This approach can be effective in reducing or removing a bias, but the price to be paid is a substantial increase of the variances of estimates.

## KEY TERMS

**Auxiliary variable**: A variable that has been measured in the survey and for which also the distribution in the population (or the complete sample) is available.

**First order inclusion probability**: The probability that a population element is selected in the sample. The first-order inclusion probability is determined by the sampling design.

**Generalized regression estimation**: A weighting technique that computes weights using a generalized linear regression model that predicts the target variable of the survey from a set of auxiliary variables. This is sometimes also called linear weighting.

**Homogeneous**: A stratum (subpopulation) is called homogeneous if all its elements resemble each other with respect to the target variables of the survey.

**Internet population**: The subpopulation of the target population consisting of only elements that have access to the Internet.

**Nonresponse**: The phenomenon that elements in the selected sample and that are also eligible for the survey and do not provide the requested information or that the provided information is not usable.

**Post-stratification**: A weighting method that divides the population in strata and subsequently assigns the same weight to all observed elements within a stratum.

**Raking ratio estimation**: A weighting technique that computes weights using an iterative fitting procedure that adjusts the weight so that the weighted sample distributions of the auxiliary variables fit their population distributions. This technique is also known as iterative proportional fitting or multiplicative weighting.

**Reference survey**: A survey conducted with the objective to obtain unbiased estimates of the population distributions of auxiliary variables.

**Representative**: The (weighted) survey response is representative with respect to a variable if the (weighted) response distribution is equal to its population distribution.

**Self-selection survey**: A survey for which the sample has been recruited by means of self-selection. It is left to the persons themselves to decide to participate in a survey. No probability sample is selected.

**Stratification**: A division of the population into a number of subpopulations (strata) by cross-classifying a number of auxiliary variables.

**Under-coverage**: The phenomenon that not all elements in the population are represented in the sampling frame.

## EXERCISES

**Exercise 12.1**    Which property of an auxiliary variable makes is useful for including in a weighting adjustment model?

**a.** The response distribution of the variable is approximately equal to its population distribution.

**b.** The sample distribution of the variable is approximately equal to its population distribution.

**c.** The response distribution of the variable differs considerably from its sample distribution.

**d.** The response distribution of the variable is approximately equal to its sample distribution.

**Exercise 12.2**    A large company has 2,500 employees. The management has installed coffee machines everywhere in the building. After a while, the management wants to know whether or not the employees are satisfied with the coffee machines. It is decided to conduct a web survey. A simple random sample without replacement of 500 employees is drawn. It turns out that 380 employees complete the web questionnaire form. Of those, 310 are satisfied with the coffee machines.

**a.** Compute the 95% confidence interval of the percentage of employees in the company who are satisfied with the coffee machines.Only 380 of the 500 selected employees responded. So, there is a nonresponse problem.

**b.** Compute a lower bound and an upper bound for the percentage of employees in the sample who are satisfied with the coffee machines.
Previous research has showed that employees with a higher level of education are less satisfied with the coffee facilities. The management knows the level of education of each employee in the company: 21% has a high education and 79% has a low education. The table below shows the

relationship between coffee machine satisfaction and level of education for the 380 respondents:

|                | Low education | High education | Total |
|----------------|:-------------:|:--------------:|:-----:|
| Satisfied      | 306           | 4              | 310   |
| Not satisfied  | 40            | 30             | 70    |
| Total          | 346           | 34             | 380   |

A weighting adjustment procedure is carried out to reduce the nonresponse bias.

c. Compute weights for low- and high-educated employees.

d. Compute the weighted estimate of the percentage of employees in the company satisfied with the coffee facilities.

**Exercise 12.3**    There are plans in the Netherlands to introduce a system of road pricing. It means car drivers are charged for the roads they use. Such a system could lead to better use of the available road capacity and therefore could reduce traffic congestion. An Automobile Association wants to know what the attitude of the Dutch is toward road pricing. It conducts a web survey. People are asked two questions:

- Are you in favor of road pricing?
- Do you have a car?

The results are summarized below:

| Has a car? | In favor of road pricing? | |
|------------|:-----:|:----:|
|            | Yes   | No   |
| Yes        | 128   | 512  |
| No         | 60    | 40   |

a. Using the available data, and assuming simple random sampling, estimate the percentage in favor of road pricing.

b. From another source it is known that 80% of the target population owns a car and 20% does not have one. Use this additional information to apply weighting adjustment. Compute a weight for car owners and a weight for those without a car.

c. Make a table like the one above, but with weighted frequencies.

d. Compute a weighted estimate for the percentage in favor of road pricing.

e. Explain the difference between the weighted and unweighted estimate.

**Exercise 12.4**   A transport company carries out a web survey to determine how healthy its truck drivers are. Only 21 drivers complete the web questionnaire form. Each respondent was asked whether he has visited a doctor because of medical problems. Also experience of the driver (little, much) and age (young, middle, old) were recorded. The results are in the table below:

| No. | Age | Experience | Doctor visits | No. | Age | Experience | Doctor visits |
|-----|-----|-----------|---------------|-----|-----|-----------|---------------|
| 1 | Young | Much | 2 | 12 | Middle | Little | 6 |
| 2 | Young | Much | 3 | 13 | Middle | Little | 6 |
| 3 | Young | Much | 4 | 14 | Middle | Little | 7 |
| 4 | Young | Little | 3 | 15 | Old | Much | 8 |
| 5 | Young | Little | 4 | 16 | Old | Much | 10 |
| 6 | Young | Little | 4 | 17 | Old | Much | 10 |
| 7 | Young | Little | 5 | 18 | Old | Much | 8 |
| 8 | Middle | Much | 5 | 19 | Old | Little | 8 |
| 9 | Middle | Much | 6 | 20 | Old | Little | 9 |
| 10 | Middle | Much | 7 | 21 | Old | Little | 10 |
| 11 | Middle | Little | 5 | | | | |

**a.** Estimate the average number of doctor visits assuming the response can be seen as a simple random sample.

**b.** Assume the population distributions of experience and age are available for the population of all drivers of the company:

| Experience | Percentage (%) | Age | Percentage (%) |
|-----------|----------------|-----|----------------|
| Much | 48 | Young | 22 |
| Little | 52 | Middle | 30 |
| | | Old | 48 |

Establish whether or not the response is selective. Explain which of these two auxiliary variables should be preferred for computing adjustment weights.

**c.** For each auxiliary variable separately, carry out weighting adjustment. Compute weights for each of the categories of the auxiliary variable.

**d.** Compute for both weighting adjustments a weighted estimate of the average number of doctor visits.

**e.** Compare the outcomes under 12.4.a and 12.4.d. Explain differences and/or similarities.

## REFERENCES

Bethlehem, J. G. (1988), Reduction of Nonresponse Bias through Regression Estimation. *Journal of Official Statistics*, 4, pp. 251–260.

Bethlehem, J. G. (2002), Weighting Nonresponse Adjustments Based on Auxiliary Information. In: Groves, R. M., Dillman, D. A., Eltinge, J. L., & Little, R. J. A. (eds.), *Survey Nonresponse*. Wiley, New York.

Bethlehem, J. G. (2007), *Reducing the Bias of Web Survey Based Estimates*. Discussion Paper 07001. Statistics Netherlands, Voorburg/Heerlen, the Netherlands.

Bethlehem, J. G. (2008), *How accurate are self-selection web surveys? Discussion Paper 08014*. Statistics Netherlands, The Hague/ Heerlen, the Netherlands.

Bethlehem, J. G. (2009), *Applied Survey Methods, A Statistical Perspective*. Wiley, Hoboken, NJ.

Bethlehem, J. G. & Keller, W. J. (1987), Linear Weighting of Sample Survey Data. *Journal of Official Statistics*, 3, pp. 141–154.

Börsch-Supan, A., Elsner, D., Faßbender, H., Kiefer, R., McFadden, D., & Winter, J. (2004), *Correcting the Participation Bias in an Online Survey*. Report, University of Munich, Germany.

Deming, W. E. & Stephan, F. F. (1940), On a Least Squares of Adjustment of a Sampled Frequency Table When the Expected Totals Are Known. *Annals of Mathematical Statistics*, 11, pp. 427–444.

Deville, J. C. & Särndal, C. E. (1992), Calibration Estimation in Survey Sampling. *Journal of the American Statistical Association*, 87, pp. 376–382.

Deville, J. C., Särndal, C. E., & Sautory, O. (1993), Generalized Raking Procedures in Survey Sampling. *Journal of the American Statistical Association*, 88, pp. 1013–1020.

Duffy, B., Smith, K., Terhanian, G., & Bremer, J. (2005), Comparing Data from Online and Face-to-face Surveys. *International Journal of Market Research*, 47, pp. 615–639.

Holt, D. & Smith, T. M. F. (1979), Post Stratification. *Journal of the Royal Statistical Society, Series A*, 142, pp. 33–46.

Horvitz, D. G. & Thompson, D. J. (1952), A Generalization of Sampling Without Replacement from a Finite Universe. *Journal of the American Statistical Association*, 47, pp. 663–685.

Huang, E. T. & Fuller, W. A. (1978), Nonnegative Regression Estimation for Survey Data. *Proceedings of the Social Statistics Section of the American Statistical Association*, Washington, DC, pp. 300–303.

Kruskal, W. & Mosteller, F. (1979a), Representative Sampling, I: Non-scientific Literature. *International Statistical Review*, 47, pp. 13–24.

Kruskal, W. & Mosteller, F. (1979b), Representative Sampling, II: Scientific Literature. Excluding Statistics. *International Statistical Review*, 47, pp. 111–127.

Kruskal, W. & Mosteller, F. (1979c), Representative Sampling, III: The Current Statistical Literature. *International Statistical Review*, 47, pp. 245–265.

Little, R. J. A. & Wu, M. M. (1991), Models for Contingency Tables with Known Margins When Target and Sampled Populations Differ. *Journal of the American Statistical Association*, 86, pp. 87–95.

Skinner, C. J. (1991), On the Efficiency of Raking Ratio Estimates for Multiple Frame Surveys. *Journal of the American Statistical Association*, 86, pp. 779–784.

# CHAPTER THIRTEEN

# Use of Response Propensities

## 13.1 Introduction

Some of the main problems of web surveys are caused by under-coverage, self-selection, and nonresponse. These phenomena may lead to biased estimators of population characteristics, and therefore wrong conclusions are drawn from the web survey data. To avoid this, some kind of correction technique is required. Weighting adjustment is one such technique. This topic is treated in Chapter 12. It is also possible to use response propensities to correct biased estimates. This chapter is about response propensities. It is described what they are, how they can be computed, and what can be done with them.

The problems described above may particularly occur if a general population survey is conducted using Internet data collection. The target population is usually wider than just those with access to Internet. This implies that individuals without Internet cannot be selected for the survey. The sample for a general population web survey is selected from a sampling frame. Such a frame does not contain information about which people have Internet and which do not. It will also not contain e-mail addresses of those having Internet. The sample is usually selected using a different mode. For example, selected individuals are sent a letter with an invitation to complete the web questionnaire. Those without Internet will not respond. For those with Internet, it is up to them whether they will respond or not. In conclusion, the ultimate group of respondents is the result of a selection process with unknown probabilities.

*Handbook of Web Surveys*, Second Edition. Silvia Biffignandi and Jelke Bethlehem.
© 2021 John Wiley & Sons, Inc. Published 2021 by John Wiley & Sons, Inc.

Even if all members of the target population have Internet access, problems are not solved. Selecting a proper probability sample requires a sampling frame containing e-mail addresses of all individuals in the population. Such sampling frames do not exist, and probably will never exist. The way out is, again, to use a different mode to recruit the sample. Selected individuals can be sent an invitation letter, or they can be called by telephone. It is up to them whether they will respond or not. Again, the response may be seen as the result of a probability mechanism with unknown probabilities.

In any case, if the target population is the general population, and only Internet users can complete the survey questionnaire, the question arises whether estimates of population characteristics are biased, and if so, whether these estimates can be corrected so that the bias is reduced or eliminated.

Summing up, web survey-based estimates may be biased due to under-coverage, self-selection, or nonresponse. Although such a selection bias can have various causes, the methodological consequences are similar. Therefore, the treatment of these problems is also more or less the same.

Various methods have been proposed in the literature to deal with selection bias. Adjustment weighting is one of them (see Chapter 10). Use of response propensity is another. Although this approach was developed already many years ago, it witnessed a revival with the emergence of web surveys.

An important problem of causal inference is how to estimate treatment effects in observational studies. This is a situation in (like in an experiment) in which a group of objects is exposed to a well-defined treatment and another group (the control group) does not receive this treatment. However, it is not a controlled experiment. Therefore, observed effects can be biased. They may partially be artifacts caused by the way the treatment group and the control are composed.

So-called *propensity score matching* methods can be used to correct for a sample selection bias due to observable differences between the treatment and the control group. Matching involves pairing objects in the treatment and control group that are similar in terms of their observable characteristics. When the relevant differences between any two objects are captured in observable covariates, which occur when outcomes are independent of assignment to treatment conditional on the covariates, matching methods can yield an unbiased estimate of the treatment effect.

The propensity score matching techniques was developed in the 1980s in the context of biomedical studies (Rosenbaum and Rubin, 1983), and its original conceptual framework is described in Rubin (1974).

The method of propensity score matching can only be applied if certain conditions are satisfied. Nevertheless, the method is widely applied, not only in biomedical analyses, but in many other research fields like labor market research and policy evaluation. See, for example, Dehejia and Wahba (1999). The basic assumptions underlying the method are as follows:

- Selection in the treatment group (or control group) can be explained purely in terms of observable characteristics. If one can control for observable differences in characteristics between the treatment and the control group, the observed effect can be attributed to the treatment. It is a true treatment effect. This assumption is called the *conditional independence assumption* or the *strong ignorability assumption*. If the conditional independence assumption holds, the matching process is analogous to creating an experimental data set where, conditional on observed characteristics, the selection process is random.

- The decision to assign the treatment to on object does not depend on the decision to assign the treatment to other objects.
- The observed outcome of a variable for an object depends only on the object itself and not on the mechanism assigning treatment to objects.

Application of the method of propensity scores requires the individual values off all variables explaining the group differences to be available. Particularly if the two groups are the participants and the nonparticipants, it may be difficult to collect all these values for the nonparticipants. This restriction may prevent application. Moreover, all values must not have been affected by measurement errors.

In this chapter, the method of propensity score matching is applied to web surveys. This is only possible if the conditions stated above are satisfied. In addition, the web survey situation should be put in the proper theoretical framework. This means:

- Identifying what the treatment is;
- Identifying the treatment group and the control group;
- Selecting the observable variables to be included in the matching process;
- Assessment of the conditional independence assumption.

As mentioned before, application of propensity scores to the web survey situation may be hampered by the availability of proper data.

The idea of applying the propensity score method in survey methodology was introduced by Harris Interactive. See, for example, Taylor et al. (2001). Harris Interactive used the propensity score method to solve the problems of undercoverage and self-selection in Internet panels. Terhanian et al. (2001) proposed the use of this technique as a tool for weighting self-selection samples of web respondents. The idea is to define the propensity score as the probability that an object selects itself for the web survey. Consequently, the term *response propensity* will be used in this context. The treatment group consists of all objects in the sample, and the control group consists of all objects not in the sample.

The main idea is now that within a group of objects with the same response propensities, there are no selection effects. So, within-group estimates based on sample elements from this group will be unbiased. By combining the group estimates into a population estimate (considering group sizes), an unbiased population estimate is obtained.

There is a growing literature on experiences with the use of the response propensities method in web surveys (Steinmetz et al., 2014). The approach seems promising. Nevertheless, further theoretical and empirical work is required to prepare it for regular application in web survey and web panel environments. There are still some unanswered questions. There are, for example, studies that criticize the application of propensity matching and show that its application critically depends on assumptions about the nature of the process by which participants select themselves and on the data available to survey researcher. See Smith and Todd (2000) and Heckman et al. (1998). See Example 13.1.

▣ EXAMPLE 13.1 **Response propensities using a reference survey**

Schonlau et al. (2004) and Lee (2006) apply the propensity score method in web surveys. To obtain groups that are similar with respect to the response propensities, they divided the sample in to strata using variables that explain the response behavior. Such variables can be measured in the survey, but their values are not available for individuals not participating in the survey.

To solve this problem, they conduct another survey that does not suffer from under-coverage and self-selection. Moreover, nonresponse in this survey is considered missing completely at random. The idea is to just measure the variables that are required to estimate response propensities. Such a survey is called here a *reference survey*.

This reference survey can be seen as a benchmark for the web survey participants by balancing the distribution of these variables for the web respondents so that it becomes similar to its distribution for the reference survey respondents.

Duffy et al. (2005) uses behavioral, attitudinal, and sociodemographic variables for this purpose. These variables are sometimes called *webographic* or *psychographic variables*. Examples of questions measuring this kind of variables are "How often do you watch TV programs alone?" and "Do you often feel unhappy?"

The response propensity is defined as the conditional probability that a sample element responds in the web survey, given the values of the explanatory variables. The estimated values of the response propensities are used to construct groups with (approximately) the same scores. Stratification based on strata corresponding to these groups will remove the selection bias provided all conditions underlying the group are satisfied.

Bethlehem (2007) shows that use of reference surveys for propensity score matching is not without problems. The bias may be reduced but at the cost of a large increase in variance. In addition, it is not realistic to assume that the reference survey does not experience problems with under-coverage or nonresponse. Furthermore, the reference survey will probably be a CAPI or CATI survey, whereas the main survey is a web survey. So there may be also mode effects. Use of a reference survey is treated in some more detail in Chapter 12.

This chapter discusses various problems and ways to use response propensities to improve estimators in web surveys. A definition of response propensity is given, and models for response propensities are described. Methods for correcting bias using response propensities are presented. Particularly response propensity weighting and stratification are proposed as correction techniques. At the end, an application comparing methods for correcting bias is described.

## 13.2 Theory

Let the target population $U$ of the survey consist of $N$ identifiable elements, which are labeled 1, 2, ..., $N$. Therefore, the target population can be denoted by

$$(13.1) \qquad U = \{1, 2, ..., N\}.$$

Associated with each element $k$ is a value $Y_k$ of the target variable $Y$. The aim of the web survey is assumed to be estimation of the population mean

$$(13.2) \qquad \overline{Y} = \frac{1}{N} \sum_{k=1}^{N} Y_k$$

of the target variable $Y$.

Two cases will be discussed here, in which response propensities can be used in an attempt to reduce a selection bias. The first case is that of a simple random sample in which nonresponse occurs. The second case is that of a self-selection survey.

It will be assumed that every individual in the population has Internet access. This is an ideal situation. In practice, this will usually not be the case, which introduces an extra bias due to under-coverage.

### 13.2.1 A SIMPLE RANDOM SAMPLE WITH NONRESPONSE

Suppose a simple random sample is selected without replacement from the population $U$. It means that each element can appear at most once in the sample. Therefore the sample can be represented by a series of indicators

$$(13.3) \qquad a = a_1, a_2, ..., a_N.$$

The $k$th indicator $a_k$ assumes the value 1 if element $k$ is selected in the sample, and otherwise it assumes the value 0. The expected value, i.e., the mean value over all possible samples, of $a_k$ is denoted by

$$(13.4) \qquad \pi_k = E(a_k).$$

The quantity $\pi_k$ is the *first-order inclusion probability* of element $k$ (for $k = 1$, 2, ..., $N$). In a simple random sample, all first-order inclusion probabilities are equal to $\pi_k = n/N$, where $n$ is the sample size, which can be written as

$$(13.5) \qquad n = \sum_{k=1}^{N} a_k.$$

The Horvitz–Thompson estimator (Horvitz and Thompson, 1952) is defined by

$$(13.6) \qquad \bar{y}_{HT} = \frac{1}{N} \sum_{k=1}^{N} a_k \frac{Y_k}{\pi_k}.$$

This is an unbiased estimator of the population mean. Note that for a simple random sample, the Horvitz–Thompson estimator reduces to the simple sample mean

$$(13.7) \qquad \bar{y} = \frac{1}{n} \sum_{k=1}^{N} a_k Y_k.$$

The Horvitz–Thompson can only be applied if every element in the sample responds. Unfortunately, this is not the case in practical situations. There is always nonresponse. To investigate the consequences of nonresponse, it is assumed that the *random response model* applies. This model assumes every element $k$ in the population to have an (unknown) response probability $\rho_k$. If element $k$ is selected in the sample, a random mechanism is activated that results with probability $\rho_k$ in response and with probability $1 - \rho_k$ in nonresponse. Under this model, a set of response indicators

$$(13.8) \qquad R_1, R_2, ..., R_N$$

is introduced, where $R_k = 1$ if the corresponding element $k$ responds and where $R_k = 0$ otherwise. So, $P(R_k = 1) = \rho_k$, and $P(R_k = 0) = 1 - \rho_k$.

Now suppose that a simple random sample without replacement of size $n$ is selected from this population. The response only consists of those elements $k$ for which $a_k = 1$ and $R_k = 1$. Hence, the number of available cases is equal to

$$(13.9) \qquad n_R = \sum_{k=1}^{N} a_k R_k.$$

Note that this realized sample size is a random variable. The number of nonrespondents is equal to

$$(13.10) \qquad n_{NR} = \sum_{k=1}^{N} a_k(1 - R_k),$$

where $n = n_R + n_{NR}$.

The values of the target variable become only available for the $n_R$ responding elements. The mean of these values is denoted by

$$(13.11) \qquad \bar{y}_R = \frac{1}{n_R} \sum_{k=1}^{N} a_k R_k Y_k.$$

It can be shown (see Bethlehem (2009)) that the expected value of the response mean is approximately equal to

$$(13.12) \qquad E(\bar{y}_R) \approx \tilde{Y},$$

where

$$(13.13) \qquad \tilde{Y} = \frac{1}{N} \sum_{k=1}^{N} \frac{\rho_k}{\bar{\rho}} Y_k$$

and

$$(13.14) \qquad \bar{\rho} = \frac{1}{N} \sum_{k=1}^{N} \rho_k$$

is the mean of all response probabilities in the population. From expression (13.12) it is clear that, generally, the expected value of the response mean is unequal to the population mean to be estimated. Therefore, this estimator is biased. This bias is approximately equal to

$$(13.15) \qquad B(\bar{y}_R) = \tilde{Y} - \bar{Y} = \frac{S_{\rho Y}}{\bar{\rho}} = \frac{R_{\rho Y} S_\rho S_Y}{\bar{\rho}},$$

where $S_{\rho Y}$ is the covariance between the values of the target variable and the response probabilities, $R_{\rho Y}$ is the corresponding correlation coefficient, $S_Y$ is the standard deviation of the variable $Y$, $S_\rho$ is the standard deviation of the response probabilities, and $\bar{\rho}$ is the mean of the response probabilities. The estimator can be repaired by introducing the modified Horvitz–Thompson estimator:

$$(13.16) \qquad \bar{y}_{HT}^* = \frac{1}{N} \sum_{k=1}^{N} a_k R_k \frac{Y_k}{\pi_k \rho_k}.$$

Because $E(a_k R_k) = \pi_k \rho_k$, this is an unbiased estimator. The problem is, however, that this estimator cannot be computed because the response probabilities are unknown. This can be solved by first estimating the response probabilities and then substituting these estimates in expression (13.6). The method of response propensities can be used for this purpose. For selection bias in web surveys, see also Biffignandi and Bethlehem (2011).

## 13.2.2 A SELF-SELECTION SAMPLE

Participation in a self-selection web surveys requires respondents to be aware of the existence of the web survey (they have to accidentally visit the website, or they have to follow up a banner or an e-mail message). They also have to decide to fill in the questionnaire on the Internet. This means that each element $k$ in the Internet population has unknown probability $\rho_k$ of participating in the survey, for $k = 1, 2, \ldots, N$.

The responding elements are denoted by a set

$$(13.17) \qquad R1, R2, \ldots, RN$$

of $N$ indicators, where the $k$th indicator $R_k$ assumes the value 1 if element $k$ participates and otherwise it assumes the value 0, for $k = 1, 2, \ldots, N$. The expected value $\rho_k = E(R_k)$ can be called the *response probability* of element $k$. The realized sample size is denoted by

$$(13.18) \qquad n_S = \sum_{k=1}^{N} R_k.$$

Lacking any knowledge about the values of the response probabilities, survey researchers usually implicitly assume all these probabilities to be equal. In other words, simple random sampling is assumed. Consequently, the sample mean

$$(13.19) \qquad \bar{y}_S = \frac{1}{n_S} \sum_{k=1}^{N} R_k Y_k$$

is used as an estimator for the population mean. The expected value of this estimator is approximately equal to

$$(13.20) \qquad E(\bar{y}_S) \approx \tilde{Y} = \frac{1}{N\bar{\rho}} \sum_{k=1}^{N} \rho_k Y_k$$

where $\bar{\rho}$ is the mean of all response probabilities in the population. This expression was derived by Bethlehem (1988).

It is clear from expression (13.20) that, generally, the expected value of the sample mean is not equal to the population mean of the population. One situation in which the bias vanishes is that in which all response probabilities in the population are equal. In terms of nonresponse correction theory, this comes down to Missing Completely Missing at Random (MCAR). This is the situation in which the cause of missing data is completely independent of all variables measured in the survey. For more information on MCAR and other missing data mechanisms, see Little and Rubin (2002). Indeed, in the case of MCAR, self-selection does not lead to an unrepresentative sample because all elements have the same selection probability. Bethlehem (2002) shows that the bias of the sample mean (13.20) can be written as

$$(13.21) \qquad B(\bar{y}_S) = E(\bar{y}_S) - \overline{Y} \approx \widetilde{Y} - \overline{Y} = \frac{C_{\rho Y}}{\bar{\rho}} = \frac{R_{\rho Y} S_\rho S_Y}{\bar{\rho}},$$

in which

$$(13.22) \qquad C_{\rho Y} = \frac{1}{N} \sum_{k=1}^{N} (\rho_k - \bar{\rho})(Y_k - \overline{Y}).$$

The bias can be repaired by introducing a modified estimator defined by

$$(13.23) \qquad \bar{y}_S^* = \frac{\bar{\rho}}{n_S} \sum_{k=1}^{N} \frac{R_k Y_k}{\rho_k}.$$

This estimator is approximately unbiased. Note that the mean response probability can simply be estimated by $n_S/N$, changing estimator (13.23) into

$$(13.24) \qquad \bar{y}_S^* = \frac{1}{N} \sum_{k=1}^{N} \frac{R_k Y_k}{\rho_k}.$$

Again, the problem is that the response probabilities are unknown. However, they can be estimated using the method of response propensities.

### 13.2.3 THE RESPONSE PROPENSITY DEFINITION

Let $U = \{1, 2, \ldots, N\}$ be the target population of the survey. Associated with each element $k$ is a value $Y_k$ of the target variable $Y$. Furthermore, there is a set of auxiliary variables. For every element $k$ the vector of values of these auxiliary variables is denoted by $X_k = (X_{k1}, X_{k2}, \ldots, X_{kp})'$. It is assumed that these values are known for all elements in the sample (in the case of a random sample) or for all elements in the population (in case of a self-selection survey). Such information can, for instance, be found in a population register.

It is assumed that every element $k$ in the population has a nonzero, unknown response probability, denoted by $\rho_k$. If element $k$ is selected in a sample, a random mechanism is activated that results with probability $\rho_k$ in response and with probability $(1 - \rho_k)$ in nonresponse. In case of a self-selection survey, there is no sample selection. The whole population could be seen as the sample. Also here $\rho_k$ is the probability to respond.

The response probability $\rho$ is a latent variable, but it is not observed. Instead the corresponding response indicator $R$ is observed. The response indicator $R_k$ equals 1 if element $k$ responds, and otherwise it is 0. A vector $R$ of response indicators can be introduced, i.e., $R = (R_1, R_2, \ldots, R_N)'$, where $P(R_k = 1) = \rho_k$, and $P(R_k = 0) = 1 - \rho_k$. These probabilities can be estimated using an appropriate model based on auxiliary information.

The first step is to introduce the *response propensity* $\rho(X_k)$. It is defined by

$$(13.25) \qquad \rho_k(X) = P(R_k = 1 \mid X = X_k).$$

It can be interpreted as the probability of response given the values of the set of auxiliary variables $X$. It is assumed that all auxiliary variables required to explain the response behavior are included in the set $X$. To say it otherwise, given the values of the auxiliary variables, response behavior is independent of the target variables of the surveys. This comes down to the assumption of *Missing at Random* (MAR). This assumption is also known as the *conditional independence* assumption (Lechner, 1999), *selection on observables* (Barnow, Cain, and Goldberger, 1980), the *unconfoundedness* assumption or the *ignorable treatment* assumption (Rosenbaum and Rubin, 1983), and *exogeneity* (Imbens, 2004).

The definition of the concept of response probability is not straightforward. It involves at some stage a decision on how to deal with the dependence of the response probabilities on the circumstances under which the survey is being conducted. They may, for example, depend on the number and timing of contact attempts and the interviewer characteristics. If these circumstances change, it is very likely that the individual response probabilities also change.

In addition, response probabilities may vary over time. However, the more conditions are imposed on the response probabilities, the more fixed they become. The fixed response model arises as a special case of the random response model when the response probabilities are viewed conditional on very detailed circumstances. No variation is left and response becomes deterministic.

## 13.2.4 MODELS FOR RESPONSE PROPENSITIES

To be able to estimate the response propensities, a model must be chosen. The most frequently used one is the *logistic regression model*. It assumes the relationship between response propensity and auxiliary variables can be written as

$$(13.26) \qquad \log \mathrm{it}(\rho_k(X)) = \log \left( \frac{\rho_k(X)}{1 - \rho_k(X)} \right) = \sum_{j=1}^{p} X_{kj} \beta_j,$$

where $\beta = (\beta_1, \beta_2, ..., \beta_p)'$ is a vector of $p$ regression coefficients. The *logit* transformation ensures that estimated response propensies are always in the interval [0, 1].

Another model sometimes used is the *probit model*. It assumes the relationship between response propensity and auxiliary variables can be written as

$$(13.27) \qquad \text{probit}(\rho_k(X)) = \Phi^{-1}(\rho_k(X)) = \sum_{j=1}^{p} X_{kj}\beta_j,$$

in which $\Phi^{-1}$ is the inverse of the standard normal distribution function. Both models are special cases of the generalized linear model (GLM):

$$(13.28) \qquad g(\rho_k(X)) = \sum_{j=1}^{p} X_{kj}\beta_j,$$

where $g$ is called the *link function* that has to be specified. It is also possible to use the *identity* link function. This means the relationship between response propensity and auxiliary variables can be written as

$$(13.29) \qquad \rho_k(X) = \sum_{j=1}^{p} X_{kj}\beta_j.$$

This is a simple linear model. It has advantages and disadvantages. A first advantage of the linear model is that coefficients are much easier to interpret. They simply represent the effects of the auxiliary variables on the response propensity. These effects are "pure" effects. The coefficient of a variable is corrected for the interdependencies of the other auxiliary variables in the model. Interpretation of a logit or probit model is not so straightforward. The logit or probit transform hampers the interpretation of the linear parameters.

A second advantage is that the computations are simpler. Estimates of the coefficients can be obtained ordinary least squares. Estimation of the logit and probit model requires maximum likelihood estimation.

An advantage of the probit and logit model is that estimated response propensities are always in the interval [0, 1]. The linear model does not prevent estimated probabilities to be negative or larger than 1. However, according to Keller, Verbeek, and Bethlehem (1984), the probability of estimates outside the interval [0, 1] vanishes asymptotically if the model is correct and all response probabilities are strictly positive.

It should be noted that the linear model is not necessarily a worse approximation of reality than the probit or logit model. The logit and probit transformations were introduced for convenience only and not because their models were "more likely." Dehejia and Wahba (1999) conclude that the choice of model does not influence the results very strongly.

Figure 13.1 contains the graphs of the logit and the probit function. It can be observed that both functions are more or less linear for values of $p$ between, say, 0.2 and 0.8. So, the linear link functions can be seen as an approximation of the other two link functions.

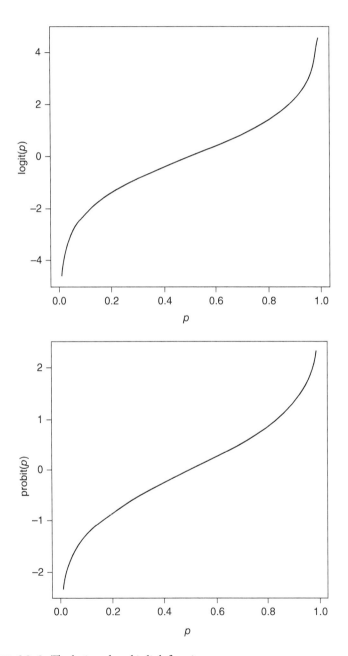

FIGURE 13.1  **The logit and probit link functions**

The logit model is the most commonly used model for estimating response propensities. It will also be applied in this chapter. Rewriting model (13.26) leads to the expression

$$(13.30) \qquad\qquad \rho(X_k) = \frac{\exp\left(X'_k\beta\right)}{1 + \exp\left(X'_k\beta\right)}$$

for the response propensities. The response propensities have to be estimated using the available data. Estimation is only possible if there are both respondents and nonrespondents for each set of values of $X$. This comes down to what is called the *matching assumption*. It states that

$$(13.31) \qquad\qquad 0 < \rho(X_k) < 1.$$

This assumption ensures that for each value of $X$, there are elements that participate in the web surveys and elements that do not participate in the web survey. Note that elements with a response probability equal to 0 or 1 cannot be compared because for these elements there are no counterparts. This is not important for elements $k$ with $\rho(X_k) = 1$, because they are all observed in the sample. It is a problems for elements $k$ with $\rho(X_k) = 0$, because they are never observed and thus may be the cause of a bias.

Estimation also requires the individual values of the auxiliary variables to be known for nonrespondents (in case of a probability sample with nonresponse) or for the nonparticipants (in case of a self-selection survey). This is often not the case. Such values may be available if the sample is selected from a sampling frame containing all relevant variables. See Example 13.2.

### ▣ EXAMPLE 13.2 Estimating response propensities

Response propensities are estimated for a real survey data set. This data set is based on a Dutch survey that has been carried out by Statistics Netherlands. It is called the General Population Survey (GPS). The sampling frame was the population register in the Netherlands. The sampling design was such that each person had the same probability of being selected (a so-called self-weighting sample). The sample of the GPS consisted of 32,019 persons. The number of respondents was 18,792.

To find model for the response propensities, auxiliary variables are required. Statistics Netherlands has an integrated system of social statistics. This system is called the *Social Statistics Database* (SSD). The SSD contains a wide range of characteristics on each individual in the Netherlands. There are data on demography, geography, income, labor, education, health, and social protection. These data are obtained by combining data from registers and other administrative data sources. By linking the sample file of the GPS to the SSD, the values of a large set of auxiliary variable become available for both respondents and nonrespondents.

Table 13.1 contains the subset of variables that turned out to have a significant contribution in the response propensity model.

TABLE 13.1    **Auxiliary variables in response propensity model**

| Variable | Description | Categories |
|---|---|---|
| Gender | Gender | 2 |
| Married | Is married | 2 |
| Age 13 | Age (in 13 age groups) | 13 |
| Ethnic | Type of non-native | 5 |
| HHSize | Size of the household | 5 |
| HHType | Type of household | 5 |
| Phone | Has listed phone number | 2 |
| Hasjob | Has a job | 2 |
| Region | Region of the country | 5 |
| Urban | Degree of urbanization | 5 |

Note that all variables in this table are categorical variables. To include them in model (13.26), they have to be replaced by sets of dummy variables, where there is a dummy variable for each category of each variable. Furthermore, to be able to estimate the model parameters, extra restrictions must be imposed. This is usually accomplished by setting one of the parameters for each set of dummies to 0.

A logit model has been fitted with the variables in Table 13.1 as explanatory variables. Just main effects were included in the model, no interaction effects. The estimated model was used to estimate the response propensity for each sample person. This distribution of these probabilities is displayed in Figure 13.2.

There is a wide variation in these probabilities. The values of the response propensities range roughly between 0.1 and 0.8. The average value is 0.587. This is equal to the response rate. Table 13.2 contains the characteristics of the persons with the lowest and the highest response propensity.

The person with the lowest response propensity is an unmarried middle-aged non-native male. He lives in a big city and has no job. There are two people in the household, but the type of household is unclear. He does not have a listed phone number.

The person with the highest response probability is a native young girl. She is one of the children in a larger household (five persons or more) living in a rural area. She has a job and a listed phone number.

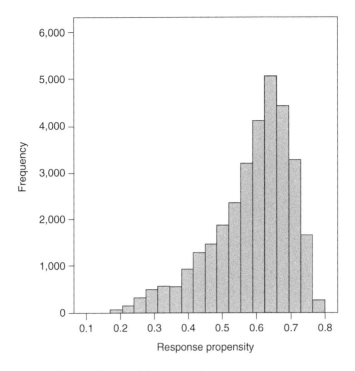

FIGURE 13.2 The distribution of the estimated response propensities

TABLE 13.2 The lowest and highest response propensity

| Variable | Value for lowest propensity | Value for highest propensity |
|---|---|---|
| Gender | Male | Female |
| Is married | No | No |
| Age in 13 age groups | 45–49 | 18–19 |
| Type of non-native | First generation non-western | Native |
| Size of the household | 2 | 5 or more |
| Type of household | Other | Couple with children |
| Has listed phone number | No | Yes |
| Has a job | No | Yes |
| Region of the country | Metropolis | Woodlands |
| Degree of urbanization | Very strong | Not |
| Response propensity | 0.109 | 0.798 |

Another issue is that it is not always clear which variables are required in the model for explaining response behavior. See also Example 13.3. Often one has to do it with the available variables, while proper estimation of response propensities may require more variables. Imbens and Rubin (2015) describe how the model can be built with first and second term interactions of explanatory variables. They also describe how the adequacy of the model specification can be assessed by inspection of the estimated response propensities.

Schouten (2004) shows that just including main terms in the model is often sufficient for estimating response propensities. Adding interaction terms does not improve the explanatory power of the models.

---

█  EXAMPLE 13.3  **Constructing a model for response propensities**

Bethlehem, Cobben, and Schouten (2011) describe how a model for the response propensities can be constructed. First, the bivariate relationship between the auxiliary variables and the response indicator is evaluated using Cramér's $V$. The value of this quantity is always between 0 and 1, where 0 means no relationship and 1 a very strong relationship. The results are displayed in Table 13.3.

TABLE 13.3   **Cramér's $V$ statistic the strength of the relationship between the auxiliary variables and response behavior**

| Auxiliary variable | $V$ |
|---|---|
| Region of the country | 0.163 |
| Degree of urbanization | 0.153 |
| Has listed phone number | 0.150 |
| Percentage non-natives in neighborhood | 0.138 |
| Percentage non-Western non-natives in neighborhood | 0.133 |
| Average house value in neighborhood | 0.115 |
| Type of non-native | 0.112 |
| Type of household | 0.106 |
| Size of the household | 0.099 |
| Marital status | 0.097 |
| Is married | 0.097 |
| Is non-native | 0.087 |
| Has social allowance | 0.077 |
| Age in 13 classes | 0.061 |
| Has an allowance | 0.061 |
| Children in household | 0.056 |
| Has a job | 0.037 |
| Age in 3 classes | 0.030 |
| Has disability allowance | 0.021 |
| Gender | 0.011 |
| Has unemployment allowance | 0.000 |

Apparently, variables related to the degree of urbanization have the strongest relationship. Also the variable indicating whether or not someone has a listed telephone number seems to be important. A variable like gender is less important.

Next, a multivariate response model is constructed by starting with the most significant variable and stepwise including less significant variables until no more significant relationships with the response indicator remain. Table 13.4 contains the resulting model.

TABLE 13.4   **Multivariate model for the response propensities**

| Variable | Wald $\chi^2$ |
|---|---|
| Having a listed phone | 242.0 |
| Region | 164.8 |
| Ethnic background | 93.7 |
| Size of household | 52.6 |
| Age (3 categories) | 11.0 |
| Has a job | 23.7 |
| Marital status | 74.1 |
| Gender | 14.7 |
| Type of household | 23.6 |
| Has a social allowance | 8.6 |
| Average house value | 25.3 |
| Degree of urbanization | 16.3 |
| Pseudo R2 | 0.042 |
| $\chi^2$ | 1,805.62 |
| Df | 40 |

The significance is tested with a Wald test. The $\chi^2$ value for each individual variable and that for the full model are given. Furthermore, the Nagelkerke pseudo $R^2$ (Nagelkerke, 1991) is reported as a measure of the model fit.

## 13.2.5 CORRECTION METHODS BASED ON RESPONSE PROPENSITIES

Once response propensities have been estimated, they can be used to reduce a possible selection bias. There are two general approaches: *response propensity weighting* and *response propensity stratification*. They will be described in the subsequent subsections.

The theory will be restricted to the situation in which first a probability sample has been selected and problems are encountered in obtaining the required information from every sample element (nonresponse).

### 13.2.5.1  Response Propensity Weighting.  *Response propensity weighting* is an approach that recognizes the principle of survey sampling that unbiased estimators can only be constructed and computed if the selection probabilities of the observed elements are known. In case of selection problems (under-coverage, self-selection, and nonresponse), the true selection probability of an element is the product of the selection probability as defined in the sampling design and the response probability. To obtain (estimates of) these true selection probabilities, the (unknown) response probabilities are replaced by estimated response propensities.

In the ideal situation, in which every sample element can be observed, the Horvitz–Thompson estimator defined by

$$(13.32) \qquad \bar{y}_{HT} = \frac{1}{N} \sum_{k=1}^{N} a_k \frac{Y_k}{\pi_k}.$$

This is an unbiased estimator of the population mean. The indicator variable $a_k$ denotes whether element $k$ is selected in the sample ($a_k = 1$) or not ($a_k = 0$), and $\pi_k$ is the first-order inclusion probability of element $k$.

In case of nonresponse, each sample element $k$ has a certain, unknown probability $\rho_k$ of responding. To avoid a possible bias, the Horvitz–Thompson estimator could be modified to include this response probabilities:

$$(13.33) \qquad \bar{y}_{HT,R} = \frac{1}{N} \sum_{k=1}^{N} a_k R_k \frac{Y_k}{\pi_k \rho_k},$$

where $R_k$ indicates whether element $k$ responds or not. This is an unbiased estimator, but it cannot be computed because the values of the $\rho_k$ are unknown. The solution is to replace each $\rho_k$ by its estimated response propensity $\hat{\rho}(X_k)$, resulting in

$$(13.34) \qquad \hat{\bar{y}}_{HT,R} = \frac{1}{N} \sum_{k=1}^{N} a_k R_k \frac{Y_k}{\pi_k \hat{\rho}(X_k)}.$$

Note that it is not so easy to establish the statistical properties of this estimator. Its distribution is not only determined by the sampling design but also by the response behavior mechanism and response propensity model. However, if the proper model is used, this estimator should be approximately unbiased.

Kalton and Flores-Cervantes (2003) note that if only categorical auxiliary variables are used in the logistic regression model, and there are no interactions included in the model, weighting with estimated response propensities is similar to raking ratio estimation. This type of adjustment weighting is described in Chapter 12. There is also a difference: if raking ratio estimation is applied, the weighted marginal distributions of the auxiliary variables are equal to their corresponding population distributions. So representativity with respect to these variables is guaranteed. This is not the case for weighting with response propensities.

The Horvitz–Thompson is only a simple estimator that does not make use of any additional information. If there exists a relationship between the target variable of the survey and a number of auxiliary variables, and also the distribution of these auxiliary variables in the population is known, better estimators can be constructed producing more precise estimates. An example of such an estimator is the *generalized regression estimator*. It was already described in Chapter 4.

Suppose there are $p$ auxiliary variables available. The $p$-vector of values of these variables for element $k$ is denoted by

$$(13.35) \qquad X_k = \left( X_{k1}, X_{k2}, \ldots, X_{kp} \right)'.$$

The vector of population means of the $p$ auxiliary variables is denoted by

$$(13.36) \qquad \overline{X} = \left( \overline{X}_1, \overline{X}_2, \ldots, \overline{X}_p \right)'.$$

This vector is supposed to be known. If the auxiliary variables are correlated with the target variable, then for a suitably chosen vector $B = (B_1, B_2, \ldots, B_p)'$ of regression coefficients for a best fit of $Y$ on $X$, the residuals $E_k$, defined by

$$(13.37) \qquad E_k = Y_k - X_k B$$

vary less than the values $Y_k$ of the target variable itself. Application of ordinary least squares results in

$$(13.38) \qquad B = \left( \sum_{k=1}^{N} X_k X_k' \right)^{-1} \left( \sum_{k=1}^{N} X_k Y_k \right).$$

If all sample elements respond, then for any sampling design, the vector $B$ can be estimated by

$$(13.39) \qquad b = \left( \sum_{k=1}^{N} a_k \frac{X_k X_k'}{\pi_k} \right)^{-1} \left( \sum_{k=1}^{N} a_k \frac{X_k Y_k}{\pi_k} \right).$$

The estimator $b$ is an asymptotically design unbiased (ADU) estimator of $B$. It means the bias vanishes for large samples. Using expression (13.39), the generalized regression estimator is defined by

$$(13.40) \qquad \bar{y}_{GR} = \bar{y}_{HT} + \left(\bar{X} - \bar{x}_{HT}\right)' b,$$

where $\bar{x}_{HT}$ and $\bar{y}_{HT}$ are the Horvitz–Thompson estimators for the population means of $X$ and $Y$, respectively. The generalized regression estimator is an ADU estimator of the population mean of the target variable.

If there are selection problems, each sample element $k$ has a certain, unknown probability $\rho_k$ of responding. To avoid a possible bias, the generalized regression estimator can be modified to include these response probabilities, resulting in

$$(13.41) \qquad b_R = \left(\sum_{k=1}^{N} a_k R_k \frac{X_k X_k'}{\pi_k \rho_k}\right)^{-1} \left(\sum_{k=1}^{N} a_k R_k \frac{X_k Y_k}{\pi_k \rho_k}\right),$$

where $R_k$ indicates whether element $k$ responds or not. This is an approximately unbiased estimator, but it cannot be computed because the values of the $\rho_k$ are unknown. The solution is to replace each $\rho_k$ by its estimated response propensity $\hat{\rho}(X_k)$, resulting in

$$(13.42) \qquad \hat{b}_R = \left(\sum_{k=1}^{N} a_k R_k \frac{X_k X_k'}{\pi_k \hat{\rho}(X_k)}\right)^{-1} \left(\sum_{k=1}^{N} a_k R_k \frac{X_k Y_k}{\pi_k \hat{\rho}(X_k)}\right).$$

Again, computation of the statistical properties of this estimator is not straightforward. Its distribution is determined by the sampling design, the response behavior mechanism, and response propensity model. However, if the proper models for the response model and for the target variable are used, this estimator should be approximately unbiased and be more precise than the Horvitz–Thompson estimator.

It was already mentioned that response propensity weighting does not force the weighted distribution of the auxiliary variables in the logit model to be equal to their population distributions. To perform this kind of calibration, the auxiliary variables can be included in the generalized regression model for the target variable.

More information about response propensity weighting can be found, for example, in Kalton and Flores-Cervantes (2003), Särndal (1981), Särndal and Lundström (2005, 2008), and Little (1986).

### 13.2.5.2 Response Propensity Stratification. *Response propensity stratification* takes advantage of the fact that estimates will not be biased if all response probabilities are equal. In this case, selection problems will only lead to fewer observations, but the composition of the sample is not affected. The idea is to divide the sample in strata in such a way that all elements within a stratum have (approximately) the same response probabilities. Consequently, unbiased estimates can be computed within strata. Next, stratum estimates are combined into a population estimate.

In case of response propensity stratification, the final estimates rely less heavily on the correctness of the model that is used to calculate the response propensities. The reason is that not the exact values are used in the computation. They are just used to construct strata. Hence, the propensity score $\rho(X)$ is smoothed.

Suppose the sample is stratified into $L$ strata based on the response propensities. Cochran (1968) suggests that five strata are sufficient, i.e., $L = 5$. The strata are denoted by $U_1, U_2, \ldots, U_L$. The sample size in stratum $h$ is denoted by $n_h$. These sample sizes are random variables and not fixed numbers. Assuming simple random sampling, the *response propensity estimator* for the population mean of the target variable $Y$ is now defined by

$$(13.43) \qquad \bar{y}_{RPS} = \frac{1}{n} \sum_{h=1}^{L} n_h \bar{y}_R^{(h)},$$

where $\bar{y}_R^{(h)}$ is the mean of the responding elements in stratum $h$, for $h = 1, 2, \ldots, L$.

Note that this post-stratification estimator calibrates the response to the sample level instead of the population level (as the post-stratification estimator in Chapter 4).

There are a number of ways to construct strata. Preferably, the strata should be constructed in such a way that the response propensities vary as little as possible within strata. Starting point is the distribution of the estimated response propensities. Then decisions have to be made about the number of strata and about the width of the strata (in terms of values of the response propensities).

Imbens and Rubin (2015) propose an iterative procedure to determine the number of strata in which the auxiliary variables are balanced between participants and nonparticipants. According to Cochran (1968), five strata are enough for stratification purposes. This is a rule of thumb. However, one should notice that the more strata there are, the less variation there will be within strata and the more the distribution of the strata will resemble a continuous distribution. See Example 13.4.

Once the strata have been constructed, the conditional independence can be checked by testing for a bivariate relationship between the response indicator and the auxiliary variables within each of the strata. This can be done, for instance, by Cramér's $V$ statistic, which is based on the $\chi^2$ test statistics.

■ EXAMPLE 13.4 **Constructing response propensity strata**

Example 13.2 described how the response propensity model was constructed for the General Population Survey (GPS). The estimated response propensities varied between 0.10 and 0.80.

The response propensities were divided in five strata by dividing the interval [0.20, 0.80] in five intervals of equal width. Table 13.5 summarizes this stratification. Note that the number of responding elements increases as the response propensity increases, whereas the number of sample elements in the strata does not necessarily increase.

TABLE 13.5 **The response propensity strata**

| Stratum | Range | | Sample size | Respondents |
|---|---|---|---|---|
| 1 | 0.10 | −0.24 | 303 | 63 |
| 2 | 0.24 | −0.38 | 1,913 | 609 |
| 3 | 0.38 | −0.52 | 5,385 | 2,504 |
| 4 | 0.52 | −0.66 | 14,690 | 8,777 |
| 5 | 0.66 | −0.80 | 9,728 | 6,839 |
| Total | | | 32,019 | 18,792 |

Figure 13.3 uses a kernel density technique to show the distribution of the estimated response propensities. The five bars represent the strata.

In addition to the five strata, a kernel density line has been estimated. This represents the continuous distribution of the response propensity in the sample. It is clear that the histogram is not a very accurate approximation of the distribution of the response propensities.

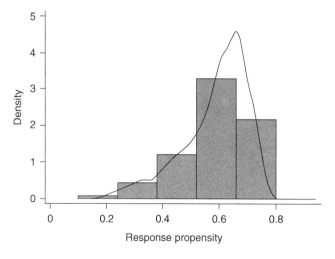

FIGURE 13.3 **Response propensity stratification with 5 strata**

Figure 13.4 shows what happens if 25 strata are constructed instead of 5. The histogram is now much closer to the density function.

One should be careful in constructing too many strata. This will reduce the number of observations per strata and therefore may lead to less stable estimates.

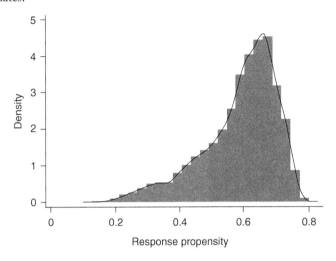

FIGURE 13.4 **Response propensity stratification with 25 strata**

## 13.3 Application

The theory described in the previous sections is now applied to the case of a self-selection in web surveys. Putting the web survey in the context of a medical experiment, there are two possible "treatments": participating in the web survey and not participating in the web survey. In line with Biffignandi and Pratesi (2006), the treatment is participation in the web survey. Note that in a web survey the assignment to the treatment is the result of being a web user in the general population. The untreated units are those who belong to the target population but do not participate into the web survey. For the target variable of the survey, only the values of participants in the web survey become available.

The population is generated for this application. This has the advantage that the properties of the population and relationships between variables are exactly known. Web survey samples are generated for studying the properties of estimators and correction techniques that attempt to reduce the bias using response propensities. It is noted once more that if the values of the response propensities are known, the inference problem could be solved by traditional estimation methods like the Horvitz–Thompson estimator.

It is assumed in this simulation study that the values of auxiliary variables for participating and nonparticipating individuals can be retrieved from a population register. So the reference survey approach is not used. Three different estimation techniques are compared: (1) naïve estimation without any kind of correction, (2) response propensity weighting, and (3) response propensity stratification.

## 13.3.1 GENERATION OF THE POPULATION

A small population has been generated consisting of 260 web users and 1,500 non-web users. There are three auxiliary variables: $X_1$, $X_2$, and $X_3$. The values of these variables have been generated differently for the web users and non-web users.

For the web users, the values of $X_1$ were drawn from a normal distribution, the values of $X_2$ from a gamma distribution, and the values of $X_3$ from a distribution that describes the fact that the probability of response decreases when $t$ (time from first contact) increases:

$$X_1: \text{ Normal}(6, 4)$$

$$X_2: \text{ Gamma}(5)$$

$$X_3: \ e^{-t} + 3$$

For the non-web users, covariates were obtained using the same distributions, but with different parameters:

$$X_1: \text{ Normal}(10, 5)$$

$$X_2: \text{ Gamma}(3)$$

$$X_3: \ e^{-t}$$

Table 13.6 shows the mean and standard deviations of the three variables in the two subpopulations. The table also contains the $p$-values of the test of equality of the means. It is clear that the means differ significantly.

Given the values of the auxiliary variables $X_1$, $X_2$, and $X_3$, the response propensities can be estimated using a model that explains being a web user from

TABLE 13.6  **Characteristics of the generated population (size, mean, and standard deviation)**

|         | Web users     | Non-web users  | $p$-Value |
|---------|---------------|----------------|-----------|
| $N$     | 260           | 1,500          |           |
| $X_1$   | 5.87 ± 3.95   | 9.97 ± 5.00    | <0.005    |
| $X_2$   | 5.13 ± 2.05   | 3.01 ± 1.73    | <0.005    |
| $X_3$   | 4.07 ± 1.12   | 1.01 ± 0.99    | <0.005    |

the auxiliary variables. The target variable $Y$ is constructed as a linear function of the auxiliary variables and a noise term:

$$Y = 0.2X_1 + 0.3X_2 + 0.5X_3 + 10U,$$

where the random variable $U$ has a uniform distribution on the interval $[0, 1]$.

### 13.3.2 GENERATION OF RESPONSE PROBABILITIES

The probabilities of the elements of being web users are generated using a logistic regression model. The auxiliary variables $X_1$, $X_2$, and $X_3$ are used as explanatory variables. The dependent variable is the indicator of being a web user. These probabilities were used as response probabilities of the elements in the self-selection web survey.

It is noted again that the values of the three auxiliary variables are assumed to be available in a population register. Thus the problem is avoided what to do if the auxiliary variables are only available for all elements in the sample. This would raise the question whether or not to use adjustment weights. The literature is not clear about this.

### 13.3.3 GENERATION OF THE SAMPLE

Samples were selected from the population of web users. The response propensities were taken equal to probability of being a web user. So, it was assumed that someone with a high probability of being a web user will also have a high probability to participate in a web survey when confronted with an invitation to do so.

The sample selection mechanism can be seen as a form of Poisson sampling with individual participation probabilities being equal to the generated response probabilities.

### 13.3.4 COMPUTATION OF RESPONSE PROPENSITIES

The information about the responding elements was used to estimate the response propensities. Again, a logistic regression model (13.30) was used with the auxiliary variables $X_1$, $X_2$, $X_3$. It was assumed that the values of these variables for the non-participating elements can be retrieved from a population register. The participation indicator $R$ (with values 0 and 1) is the dependent variable in the model. The expected value of $R_i$ for element $i$ is the response propensity of element $i$. Note that the self-selection mechanism causes the sample size $n_S$ to be a random variable.

### 13.3.5 MATCHING RESPONSE PROPENSITIES

The estimated response propensities were used to find similar elements in the target population. So, participating elements are matched to other (nonparticipating) elements using their response propensities.

The first step was to put elements in the same stratum if their response propensities are equal for the first five decimal digits. The next step was to match the remaining unmatched web survey participants on the basis of four decimal digits. This was continued down to a one-digit match.

Figures 13.5 and 13.6 show the response probability distribution for the original population (separately for web users and non-web users). Figure 13.7 shows the response propensity distribution for the web survey participants.

The set of web survey respondents after matching on the response propensities was reduced to only those participants having matches among the nonparticipants. This ensured that, conditional on the auxiliary variables in the model for the response propensities, the assumption of conditional independence was satisfied.

There is a substantial difference between the range of response probabilities for the web users and non-web users. For the web users, the minimum response probability was 0.088422, and the maximum response probability was 0.999997.

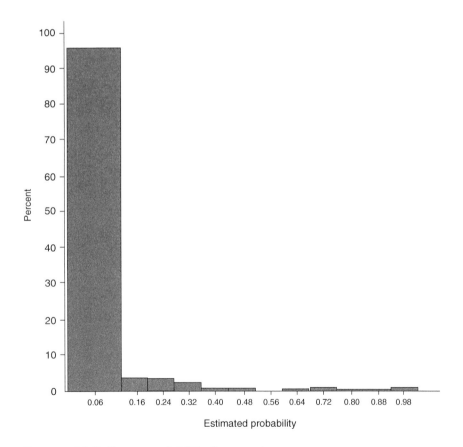

FIGURE 13.5  **Response probabilities for non-web users**

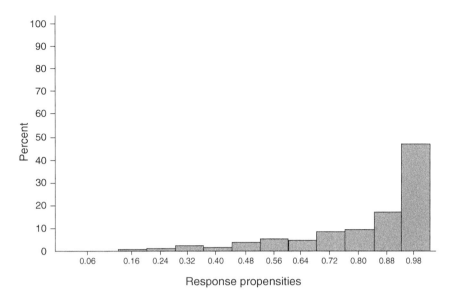

FIGURE 13.6 **Response probabilities for web users**

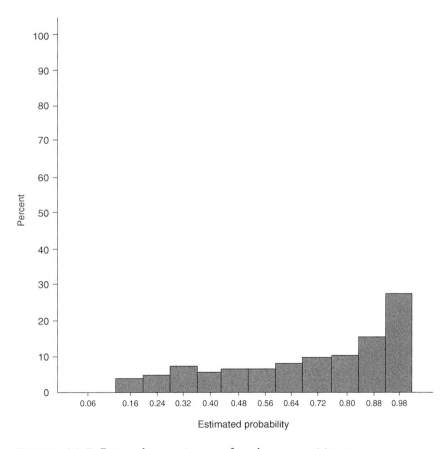

FIGURE 13.7 **Estimated propensity scores for web survey participants**

TABLE 13.7    **Characteristics of response probabilities and response propensities**

| Quantile | Response probabilities for web users | Response probabilities for non-web users | Response propensities for participants |
|---|---|---|---|
| 100% (maximum) | 0.999997 | 0.997206 | 0.999997 |
| 75% (quantile 3) | 0.960129 | 0.011523 | 0.977122 |
| 50% (median) | 0.825558 | 0.002353 | 0.904643 |
| 25% (quantile 1) | 0.591330 | 0.000704 | 0.733165 |
| 0% (minimum) | 0.088422 | 0.000021 | 0.135205 |

For the non-web users, all response probabilities should theoretically be equal to 0. However, the estimation process introduces small deviations from 0. In any case, 90% of the probabilities are equal to 0.

For the generated population of potential web survey participants, the minimum response propensity was 0.135205, and the maximum propensity was 0.999997. Table 13.7 shows some characteristics for each distribution. The web survey participants with the highest propensity and the non-web users with the lowest probability have been excluded.

## 13.3.6 ESTIMATION OF POPULATION CHARACTERISTICS

Weights were assigned to the respondents of the web survey. These weights were the reciprocals of the propensity scores. The weighted data were used to estimate the mean, the total, and the standard deviation of the target variable $Y$. Two approaches were explored. The first approach was *response propensity weighting*. This is an application of the Horvitz–Thompson estimator where the inclusion probabilities are replaced by the estimated response propensities. This approach is described in Section 13.2.5.1. In the case of self-selection, estimator (13.34) is replaced by

$$(13.43) \qquad \hat{\bar{y}}_{HT,SS} = \frac{1}{N} \sum_{k=1}^{N} R_k \frac{Y_k}{\hat{\rho}(X_k)}.$$

Similar expression can be derived for estimating population totals and population standard deviations.

The second approach was to apply *response propensity stratification*. This is a form of post-stratification estimation where strata are constructed on the basis of the values of the estimated response propensities. This approach is described in Section 13.2.5.2. The basic formula is

$$(13.44) \qquad \bar{y}_{PS,SS} = \frac{1}{n_S} \sum_{h=1}^{L} n_{S,h} \bar{y}_R^{(h)},$$

where $n_{S,h}$ is the number of responding elements in stratum $h$ and $\bar{y}_R^{(h)}$ is the mean of target variable of the responding elements in stratum $h$. Similar expressions can be derived for estimating other population characteristics, like population totals and population standard deviations.

### 13.3.7 EVALUATING THE RESULTS

To illustrate the performance of the proposed estimators based on propensity scores relative to other traditional estimation methods (no adjustment and post-stratification using auxiliary variables), estimates of the population mean, total, and standard deviation were compared.

Estimators were compared using the empirical relative bias (RB)

$$(13.45) \qquad RB = \frac{T - \theta}{\sqrt{MSE(T)}},$$

in which $MSE(T)$ is the empirical mean square error of the estimator $T$ and $\theta$ is the parameter of interest. The performance of the response propensity estimation approaches described above were compared for two different situations:

1. Comparing estimators for parameters of the whole target population with estimators for parameters of the population of web users;
2. Comparing response propensity approaches with traditional estimators like no correction (i.e., no weights) and post-stratification with auxiliary variables.

With respect to (1), the RB was lower if the target was the web population. The performance of the proposed estimators was not poor for the whole population, especially for estimating the total ($RB = 0.02$ versus $RB = -0.03$ for propensity weighting, and $RB = 0.01$ versus $RB = -0.02$ for the propensity stratification). Estimation of the standard deviation produced the same level of RB both for the whole and for the web population.

With respect to (2), the results were good for propensity stratification. The RB was smaller compared to traditional estimation methods. The results for propensity weighting were not as good. This was caused by the relatively high values and small variation of the response propensities among the participants, which makes this approach more or less similar to estimation without weighting. Note that in case of post-stratifying with auxiliary variables, only two strata were constructed for each variable: one stratum with individuals having a value below the mean and a stratum with individuals having a value above the mean.

Also note that the response propensity strata have to be constructed so that within the strata the response behavior is homogeneous. Cochran (1968) suggests that it is enough to use five strata.

### 13.3.8 MODEL SENSITIVITY

A final step in this experiment was investigating how sensitive the results were for changes in the model that was used for generating the population. To get an answer to this question, the linear relationship between the study variable and the auxiliary variables was replaced by a nonlinear relationship. Moreover, the shapes of the distributions of the auxiliary variables were made different for web users and non-web users.

Response propensity stratification performed best when the relationship between the target variable and the auxiliary variables was not linear anymore. There was no difference in performance between response propensity weighting and response propensity stratification if the distributions of the auxiliary variables differed for web users and non-web users.

The results of this experiment show that the response propensity matching combined with response propensity stratification is a promising approach for reducing the self-selection bias of web surveys. More research is needed to implement further adjustments to the propensity weighting. In fact, several studies are going on. For instance, Schonlau et al. (2009) investigate whether adjustment using weights or matching on a small set of auxiliary variables makes the distribution of the target variables representative of the population. They extract an Internet sample from the Health and Retirement Study (HRS), which is representative of the U.S. population of 50-year-old or older. Several studies are in progress with respect to selection bias in volunteer web panels, too. For example, Lee and Valliant (2009) and Bethlehem, Cobben, and Schouten (2011) study the effects of a combination of response propensity weighting or stratification with traditional correction techniques.

# 13.4 Summary

In web surveys, selecting a proper probability sample requires a sampling frame containing e-mail addresses of all individuals in the population. Such sampling frames rarely exist. Actually, general population sampling frames do not contain information about which people have Internet and which do not. Thus, one should bear in mind that people not having Internet will not respond to a web questionnaire (if no special data contact and collection strategies are performed). Moreover, people having Internet will also not always participate. Taking these facts into account, it is evident that the ultimate group of respondents is the result of a selection process (mostly self-selected) with unknown probabilities. Even if the target population coincides with all individuals having Internet access, some problems remain due to self-selection.

One possible solution for correcting the bias due to selection problems is using response propensities. The response propensity is the conditional probability that a person responds to the survey request, given the available background characteristics. In order to compute response propensities, auxiliary information for all sample elements is needed.

The response propensities can be used in a direct way for estimation of the target variables by using the response propensities as weights. This is called response propensity weighting. The direct approach attempts to estimate the true selection probabilities by multiplying the first-order inclusion probabilities with the estimated response propensities. Bias reductions will only be successful if the available auxiliary variables are capable of explaining the response behavior.

The response propensities can also be used indirectly by forming strata of elements having (approximately) the same response propensities. This is called response propensity stratification. The final estimates rely less heavily on the accuracy of the model for the response propensities.

Some studies show that response propensity matching combined with response propensity stratification is a promising strategy for the adjustment of the self-selection bias in web surveys. Research is going on to implement further improvements for response propensity weighting.

Propensity score adjustment is a frequently adopted solution to improve representativity of web panels. Note that there is no guarantee this correction techniques are successful. See also Chapter 14.

## KEY TERMS

**Propensity score:** The conditional probability of assignment to a particular treatment given a vector of observed auxiliary variables $X$.

**Propensity score method:** The propensity score method originates from evaluation studies that estimate average treatment effects. In treatment effect studies, there usually are two groups involved: one group that receives the treatment and one group that serves as a control group and does not receive the treatment. The statistic of interest is the effect of the treatment. However, to measure this effect without bias, it is necessary to remove all possible differences in outcome that arise due to a different composition of the treatment and the control group. For this purpose, the propensity score is used to balance the composition of the two groups.

**Response probability:** The probability that a person responds to a survey. This is a theoretical quantity.

**Response propensity:** The response propensity $\rho_k(X)$ is the probability of response of element $k$ given the values of a set of auxiliary variables $X$.

**Response propensity stratification:** A post-stratification method to adjust for nonresponse bias where the strata are based on the response propensities. This method uses response propensities to construct strata.

**Response propensity weighting:** A weighting method to adjust for nonresponse bias where the weights are based on the response propensities and directly used in the estimation of the target variable.

## EXERCISES

**Exercise 13.1** The aim of this exercise is to perform a propensity weighting adjustment as described in Section 13.2.5.1 using the data from the GPS. The data can be found on the website: www.web-survey-handbook.com.

First, estimate response probabilities using a binary logistic regression model and all the available auxiliary information. In SPSS one can use the forward selection strategy based on the Wald criterion to do this. Save the estimated response propensity as *rprob* and label the variable "Response propensity." Compute weights by taking the inverse of the estimated response probabilities *rprob*.

Use **Transform | Compute** to introduce a new variable *propwght*. Its value is obtained for a respondent by *propwgt* = 1/*rprob*.

In order to scale the weights properly, they have to be divided by average response probability of the respondents. Consequently, the weights will average to 1 over the response. First, select the respondents with **Data | Select cases**. Them compute the average weight of the respondents with **Analyze | Descriptive statistics | Descriptives**. Then, use **Transform | Compute** again to adjust the values of *propwgt*.

Weight the response data using **Data | Weight cases**. Be sure that only respondents have been selected. Compute weighted estimates of the target variables "Owns a house" and "PC in household" by **Analyze | Descriptive Statistics | Frequencies**. What are the estimates? Compare them to the response means.

**Exercise 13.2** The aim of this exercise is to perform a response propensity stratification adjustment as described in Section 13.2.5.2 using the data from the GPS. This comes down to post-stratification using strata constructed on the basis of response probabilities.

First, a categorical variable *propclas* is needed. This is obtained by aggregating individuals into five classes based on the estimated response probabilities *rprob*. To determine these classes, look at the distribution of the estimated response probabilities by plotting a histogram of the estimated response probabilities for the complete, unweighted sample. **Graphs | Histogram** can be used for this.

Classes can either be formed to divide the individuals equally over the response probabilities, i.e., every class contains the same number of individuals. Another option is to form classes according to the response probabilities, i.e., every class has the same width of the response probabilities. In this exercise, classes with the same width in response propensities are constructed. The first class comprises sample persons with a response propensity between 0 and 0.2, the second class between 0.2 and 0.4, the third between 0.4 and 0.6, the fourth between 0.6 and 0.8, and the last between 0.8 and 1. The last class, however, is empty because the maximum response propensity is <0.8. So there will be only four classes.

Use **Transform | Recode | Into different variables** to recode the variable *rprob* into the four desired classes. Recode to an output variable *propclas* and assign the label "propensity class." The resulting variable *propclas* will be the stratification variable for the nonresponse adjustment.

First, compute the four weights. As we weight to the sample, use

$$c_i = \frac{n_h/n}{r_h/r}$$

for the weight of an element in class $h$, and use the estimator

$$\bar{y}_{PS} = \frac{1}{n} \sum_{h=1}^{L} n_h \bar{y}^{(h)}$$

to estimate the target variables "Owns a house" and "Has a PC." Here, $n_h$ is the sample size of class $h$, $n$ is the sample size, $r_h$ is the response size in class $h$, $r$ is the response size, and $\bar{y}^{(h)}$ is the response mean of the target variable in class $h$. In order to be able to compute the post-stratification estimator, these quantities need to be computed first. To this end, complete the table below:

| $h$ | $n_h$ | $r_h$ | $w_h$ |
|-----|-------|-------|-------|
| 1 | | | |
| 2 | | | |
| 3 | | | |
| 4 | | | |
| Total | 32,019 | 18,792 | |

Derive the class sizes $n_h$ and $r_h$ by **Analyze | Descriptive Statistics | Crosstabs** by crossing the response indicator *response* with the variable *propclas*. Compute the weights by hand and import them in SPSS by constructing a new weight variable *propwgt2* using **Transform | Compute** (you have to do it four times for every category of *propclas*).

Finally, estimate the target variables by activating the weights with **Data | Weight cases**, respondent selection in **Data | Select cases**, and weighted estimation in **Analyze | Descriptive Statistics | Frequencies**. What are the estimates? Are they different from the first approach? Also compare them to the response means.

## REFERENCES

Barnow, B. S., Cain, G. G., & Goldberger, A. S. (1980), Issues in the Analysis of Selectivity Bias. In: Stromsdorfer, E. & Farkas, G. (eds.), *Evaluation Studies, Vol. 5*. Sage, San Francisco, CA, pp. 42–59.

Bethlehem, J. G. (1988), Reduction of Nonresponse Bias through Regression Estimation. *Journal of Official Statistics*, 4, pp. 251–260.

Bethlehem, J. G. (2002), Weighting Nonresponse Adjustments Based on Auxiliary Information. In: Groves, R. M., Dillman, D. A., Eltinge, J. L., & Little, R. J. A. (eds.), *Survey Nonresponse*. Wiley, New York.

Bethlehem, J. G. (2007), *Reducing the Bias of Web Survey-based Estimates*. Discussion paper 07001. Statistics Netherlands, Voorburg/Heelen, the Netherlands.

Bethlehem, J. G. (2009), *Applied Survey Methods: A Statistical Perspective*. Wiley, Hoboken, NJ.

Bethlehem, J. G., Cobben, F., & Schouten, B. (2011), *Handbook on Nonresponse in Household Surveys*. Wiley, Hoboken, NJ.

Biffignandi, S. & Bethlehem, J. G. (2011), Web Surveys: Methodological Problems and Research Perspectives. In: Di Ciaccio, A., Coli, M., & Angulo Ibanez, J. M. (eds.), *Advanced Statistical Methods for the Analysis of Large Data-Sets*. Springer, Belin.

Biffignandi, S. & Pratesi, M. (2006), Indagini WEB: propensity score matching e inferenza. Un'analisi empirica ed uno studio di simulazione (Web Surveys: Propensity Scores Matching and Inference. An Empirical Analysis and a Simulation Study). In: Falorsi, P., Pallara, A., & Russo, A. (eds.), *L'Integrazione di dati di fonti diverse. Tecniche e applicazioni del Record Linkage e metodi di stima basati sull'uso congiunto di fonti statistiche e amministrative*. F. Angeli, Milano, Italy.

Cochran, W. (1968), The Effectiveness of Adjustment by Subclassification in Removing Bias in Observational Studies. *Biometrics*, 24, pp. 205–213.

Dehejia, R. & Wahba, S. (1999), Causal Effects in Non-Experimental Studies: Re-Evaluating the Evaluation of Training Programs. *Journal of the American Statistical Association*, 94, pp. 1053–1062.

Duffy, B., Smith, K., Terhanian, G., & Bremer, J. (2005), Comparing Data from Online and Face-to-Face Surveys. *International Journal of Market Research*, 47, pp. 615–639.

Heckman, J., Ichimura, H., Smith, J., & Todd, P. (1998), Characterizing Selection Bias Using Experimental Data. *Econometrica*, 66, pp. 1017–1098.

Horvitz, D. G. & Thompson, D. J. (1952), A Generalization of Sampling Without Replacement from a Finite Universe. *Journal of the American Statistical Association*, 47, pp. 663–685.

Imbens, G. (2004), Nonparametric Estimation of Average Treatment Effects under Exogeneity: A Review. *The Review of Economics and Statistics*, 86, pp. 4–29.

Imbens, G. & Rubin, D. B. (2015), *Causal Inference for Statistics, Social and Biomedical Science*. Cambridge University Press, Cambridge.

Kalton, G. & Flores-Cervantes, I. (2003), Weighting Methods. *Journal of Official Statistics*, 19, pp. 81–97.

Keller, W. J., Verbeek, A., & Bethlehem, J. G. (1984), *ANOTA: Analysis of Tables*. Report 5766-84-M1-3. Statistics Netherlands, Department for Statistical Methods, Voorburg, the Netherlands.

Lechner, M. (1999), *Identification and Estimation of Causal Effects of Multiple Treatments under the Conditional Independence Assumption*. IZA Discussion Papers 91, Institute for the Study of Labor (IZA), Bonn, Germany.

Lee, S. (2006), Propensity Score Adjustment as a Weighting Scheme for Volunteer Panel Web Surveys. *Journal of Official Statistics*, 22, pp. 329–349.

Lee, S. & Valliant, R. (2009), Estimation for Volunteer Panel Web Surveys Using Propensity Score Adjustment and Calibration Adjustment. *Sociological Methods & Research*, 37, pp. 319–343.

Little, R. (1986), Survey Nonresponse Adjustments for Estimates of Means. *International Statistical Review*, 54, pp. 139–157.

Little, R. & Rubin, D. (2002), *Statistical Analysis with Missing Data, 2nd edition*. Wiley, New York.

Nagelkerke, N. J. D. (1991), Miscellanea, A Note on a General Definition of the Coefficient of Determination. *Biometrika*, 78, pp. 691–692.

Rosenbaum, P. R. & Rubin, D. B. (1983), The Central Role of the Propensity Score in Observational Studies for Causal Effects. *Biometrica*, 70, pp. 41–55.

Rubin, D. B. (1974), Estimating Causal Effects of Treatments in Randomized and Nonrandomized Studies. *Journal of Educational Psychology*, 66, pp. 688–701.

Särndal, C. E. (1981), Frameworks for Inference in Survey Sampling with Application to Small Area Estimation and Adjustment for Non-Response. *Bulletin of the International Statistical Institute*, 49, pp. 494–513.

Särndal, C. E. & Lundström, S. (2005), *Estimation in Surveys with Nonresponse*. Wiley, Chichester, UK.

Särndal, C. E. & Lundström, S. (2008), Assessing Auxiliary Vectors for Control of Nonresponse Bias in the Calibration Estimator. *Journal of Official Statistics*, 24, pp. 167–191.

Schonlau, M., Van Soest, A., Kapteyn, A., & Couper, M. P. (2009), Selection Bias in Web Surveys and the Use of Propensity Scores. *Sociological Methods & Research*, 37, pp. 291–318.

Schonlau, M., Zapert, K., Simon, L. P., Sanstad, K. H., Marcus, S. M., Adams, J., Spranca, M., Kan, H., Turner, R., & Berry, S. H. (2004), A Comparison between Responses from a Propensity-weighted Web Survey and an Identical RDD Survey. *Social Science Computer Review*, 22, pp. 128–138.

Schouten, B. (2004), *Adjustment for Bias in the Integrated Survey on Living Conditions (POLS) 1998*. Discussion paper 04001. Statistics Netherlands, Voorburg, the Netherlands.

Smith, J. & Todd, P. (2000), *Does Matching Overcome Lalonde's Critique of Nonexperimental Estimators?* Paper Presented at the University of North Carolina, Chapel Hill, NC.

Steinmetz, S., Bianchi, A., Tijdens, K., & Biffignandi, S. (2014), Improving Web Survey Quality—Potentials and Constraints of Propensity Score Weighting. In: Callegaro, M., Baker, R., Bethlehem, J., Göritz, A., Krosnick, J. A., & Lavrakas, P. J. (eds.), *Online Panel Research: A Data Quality Perspective*. Wiley, Chichester, UK, pp. 273–298.

Taylor, H., Bremer, J., Overmeyer, C., Siegel, J. W., & Terhanian, G. (2001), The Record of Internet-Based Opinion Polls in Predicting the Results of 72 Races in the November 2000 U.S. Elections. *International Journal of Market Research*, 43, pp. 127–136.

Terhanian, G., Smith, R., Bremer J., & Thomas, R. K. (2001), *Exploiting Analytical Advances: Minimizing the Biases Associated with Internet–Based Surveys of Non–Random Samples*. ARF/ESOMAR: Worldwide Online Measurement, ESOMAR Publication Services, 248, pp. 247–272.

# CHAPTER FOURTEEN

# Web Panels

## 14.1 Introduction

There is a growing demand for information. Surveys are an important means to collect this information. The way in which surveys are conducted continuously changes over time. New developments in computer technology have caused changes in survey methodology, as did new challenges in society like increasing nonresponse rates and decreasing budgets.

It is the task of researchers to produce reliable and valid statistics. Traditionally, they conducted face-to-face or telephone surveys to collect the data for these statistics. This is an expensive way of survey data collection, but experience has shown that such surveys produce high-quality data. Nowadays, budget constraints cause researchers in many countries to look for less expensive ways of data collection while maintaining data quality as much as possible.

A web panel (also called an *online panel*, *Internet panel*, or *access panel*) seems a promising alternative approach. It is a survey system in which the same individuals are interviewed via the web at different points in time. Information is therefore collected in longitudinal way using the same group of individuals.

Web panel data collection has become increasingly popular, particularly in the world of market research. This is not surprising as it is a simple, fast, and inexpensive way to collect a large amount of data. The largest advantage of a web panel is the speed with which data can be collected over the Internet, combined with the longitudinal aspect of a panel design.

*Handbook of Web Surveys*, Second Edition. Silvia Biffignandi and Jelke Bethlehem.
© 2021 John Wiley & Sons, Inc. Published 2021 by John Wiley & Sons, Inc.

A web panel can be used in two different ways:

1. For *longitudinal research*. The same set of variables is measured for the same group of individuals at subsequent points in time. In principle, all members are re-approached each time. Focus of these studies is on measuring change. It can, for example, be interesting to follow specific concepts over time, such as unemployment, health status, and financial behavior. Measurements at different points in time on the same set of persons are correlated, thereby reducing the variance of estimates of change.

2. For *cross-sectional research*. The panel is used as a sampling frame for specific surveys that may address different topics and thus measure different variables. Also, samples may be selected from different groups, like the elderly or the high-educated. The focus of this chapter is on the cross-sectional use of web panels for the general population.

For longitudinal panels, there is only one phase of sampling. It takes place during recruitment. Individuals are selected from the target population, and these individuals are invited to become a member of the panel. For cross-sectional panels, there can be two phases of sampling. The first phase is the same as for longitudinal panels: sampling of panel members in the recruitment phase. The researcher may decide to introduce a second phase of sampling by selecting a sample from the panel for a specific cross-sectional survey. One reason may be that a small sample is sufficient for the purpose of the researcher. Another reason could be to spread the response burden by using different samples for different surveys. And a third reason could be to select the sample from a special group in the population (for example, voters for a specific party, smokers, freelancers, and schoolteachers). Characteristics that were measured in previous waves can be used for this purpose, thus allowing for samples of persons with very specific characteristics that otherwise would not have been available for sampling.

Of course, the two different types of panel use can be combined in one panel. For example, a set of questions measuring change in behavior is asked in every specific survey, and in addition new questions are added. Note there always is a risk of order effects. To avoid these effects, the longitudinal questions should be placed at the beginning of the questionnaire. The type of panel sets restrictions on the design. For example, when the objective is to monitor changes in behavior, it is important that panelists participate in every wave. This becomes less important when the web panel only serves as a sampling frame for a specific survey.

Why a web panel, and not a single web survey? The answer is that web panels have some attractive properties:

- Once a web panel is in place, it is an ideal sampling frame from which random samples can be selected in a quick, fast, and cheap way.

- Since an e-mail address is available for each panel member, the procedure of sending invitations and reminders for participation in a specific survey is very simple.

- Little time is needed to prepare a survey from the panel. Particularly opinion polls are sometimes set up, carried out, analyzed, and published in just one day.

- There is a lot of information available about the panel members, because they participated in earlier surveys. Such information can, for example, be used to correct for nonresponse in a specific survey.

Web panels are increasingly used in market research and more particularly for public opinion polls. See, for example, the overviews given by Postoaca (2006), Comley (2007), and Callegaro et al. (2014). Use of web panels is now being recognized as one of the most important market research survey tools. In 2006, ESOMAR (the European Society for Opinion and Marketing Research) organized a Conference on Panel Research, with a special focus on web panels. Note that ESOMAR's mission is to promote the use of opinion and market research for improving decision making in business and society worldwide.

 **EXAMPLE 14.1 Examples of web panels**

There are various web panels in the United States. The *Gallup Panel* is one of the largest. It contains around 100,000 members. The *KnowledgePanel* is also a large panel. It was developed in 1999 by Knowledge Networks, a GfK company. It contains approximately 55,000 members. A much smaller panel is the *American Life Panel*. It was set up by Rand, and it contains around 6,000 members. The members of the three panels were all recruited by means of probability sampling (Couper and Dominitz, 2007).

The *LISS Panel* (Long-term Internet Study for the Social Sciences) in the Netherlands is a web panel consisting of 5,000 Dutch households. It is maintained by the Tilburg University. Recruitment is based on a probability sample from the population register. Surveys can be carried out free of charge by academic social science researchers. For a short description of this panel, see Chapter 4 in this book, and for more details, see Scherpenzeel (2008).

Germany has the *GESIS Panel*. GESIS stands for German Social Science Infrastructure Services. Recruitment is based on a random sample from municipal population registers. There are around 4,900 members. Only noncommercial surveys projects can be carried out with this panel. Surveys can be conducted free of charge by social science academics.

*ELIPSS* (Étude Longitudinal par Internet pour les Sciences Sociales) is a web panel that is representative of the population living in metropolitan France (France without overseas regions). The panel has approximately 3,500 members. They were randomly selected by the French National

Institute of Statistics and Economic Studies (INSEE). All panel members are provided with a touchscreen tablet and a mobile Internet subscription. Surveys are conducted monthly.

The *PAADEL Project* (Agri-food and Demographic Panel for Lombardy) is managed by the Center for Statistical Analyses and Interviewing (CASI) of the University of Bergamo in Italy. It is supported by a grant of the Lombardy region in Italy. This project manages two probability-based panels with the aim of continuously collecting data related to company innovation in the agricultural sector and food consumer behavior. One panel contains companies in the agricultural sector, and the other contains households (viewed as consumers).

There are many web panels that were not recruited by means of probability sampling. Particularly, market research companies often use self-selection web panels. As an example, there were eleven major opinion polling companies active during the campaign for the general elections in the United Kingdom on May 7, 2015. Seven of the eleven pollsters (*BMG, Opinium, Panelbase, Populus, Survation, TNS,* and *YouGov*) used self-selection web panels. There were serious problems with the polls selected from these panels. See Section 14.3.2.

All major opinion polling organizations in the Netherlands (*Ipsos, Maurice de Hond, TNS NIPO, I&O Research, GfK*) use web panels that are at least partly based on self-selection. Their estimates of election outcomes show significant differences.

The validity of the outcomes of web panel surveys is to a large extent determined by the way the panel was set up. Preferably, panel recruitment should be based on random sampling, and the impact of recruitment nonresponse should be small. If this is the case, a random sample from the panel is representative. Unfortunately, recruitment of many web panels is based on self-selection. This leads to panels that lack representativity. Consequently, random samples from these panels also lack representativity.

### EXAMPLE 14.2 The Dutch online panel study

Vonk, Van Ossenbruggen, and Willems (2006) describe how web panels became the primary means of data collection for market research in the Netherlands. One reason was the rapidly improving Internet technology and the acceptance of online data collection by consumers and users of market research. Another reason was the substantial increase of the costs of traditional data collection by interviewers. And a third reason was the drop-ping response rates for telephone surveys.

Around 2006 there were approximately 30 commercial web panels in the Netherlands. Together they had 1,650,000 members. Note that at this time there were 12,752,000 people of age 18 and over in the Netherlands. This suggest that approximately 13% of the population was a member of a panel.

Vonk, Van Ossenbruggen, and Willems (2006) conducted a comparison study across 19 of these panels together containing 90% of all panel members. It turned out that the 1,650,000 members were not unique. In fact, there were only 900,000 unique panel members. Of those, 700,000 participated in only one panel. The other 200,000 individuals were member of more panels. On average, there were a member of 4.7 panels.

Vonk, Van Ossenbruggen, and Willems (2006) also investigated the representativity of these self-selection panels by taking a sample of 1,000 members from each of the 19 panels. It turned out that these panels were representative with respect to basic demographic variables, like age, gender, level of education, and region of the country. However, there was a lack of representativity with respect to other variables. The panel contained too few non-Western non-natives and too many heavy Internet users. There were too few churchgoers. Also, voters for the Christian Democrats were under-represented, and voters for the Socialist Party overrepresented.

As indicated, web panels are mainly applied in commercial market research. To a lesser extent, this mode of data collection is also used in academic research. For example, Tortora (2009) and Göritz (2007) describe how many sociological and psychological phenomena are being investigated with web panels. Couper (2007) shows that web panels are also not uncommon in medical research.

### EXAMPLE 14.3 Sources of quality guidelines

Many organizations and associations have provided guidelines on the use of web panels. Some sources for these guidelines are as follows:

- The Interactive Marketing Research Organization (IMRO) released a document called "Guidelines for best practices in online sample and panel management." See Interactive Marketing Research Organization (2006).

- The American Association for Public Opinion Research (AAPOR) established a task force on online panels in 2010. See Baker et al. (2010) for the report of this task force.

- ESOMAR issued a document on "Conducting market and opinion research using the Internet." See European Society for Opinion and Marketing Research (2005). A more recent publication is "26 Questions to help research buyers of online samples." See European Society for Opinion and Marketing Research (2008).

- The European Federation of Associations of Market Research Organizations (EFAMRO) drafted a document on "Quality standards for access panels." See European Federation of Associations of Market Research Organizations (2004). This document is for a large part based on the European Society for Opinion and Marketing Research (ESOMAR) guidelines for online surveys; a consortium of German market and social research institutes released their "standards for quality assurance for online surveys." See Arbeitskreis Deutscher Markt (2001).

At first sight, a web panel is just another mode of data collection. A sample is selected from a panel, and questions are not asked face-to-face or by telephone, but over the Internet. There are some methodological challenges, however. To be able to obtain reliable and valid results, a number of issues have to be dealt with:

- *Under-coverage.* Not everyone in the target population may have access to the Internet. If this is the case, portions of the target population are excluded from the panel. Therefore, estimates of characteristics of the target population can be biased.

- *Recruitment.* Ideally, recruiting people for a panel should be based on a random sample from the target population. Only then can a random sample from the panel be seen as a random sample from the population. Often however, recruitment of a web panel is based on self-selection of its members. Then the panel will not be representative. Moreover, random samples from the panel will also not be representative. Hence estimates can be biased.

- *Nonresponse.* Nonresponse is an important problem in web panels. Nonresponse occurs in two phases of online panel research: (1) during the recruitment phase and (2) in the specific surveys taken from the panel. Recruitment nonresponse may be high because participating in a panel requires substantial commitment and effort of respondents. Nonresponse in a specific survey is often low as the invitation to participate is a consequence of agreeing to be a panel member.

- *Measurement errors.* Interviewer-assisted surveys like CAPI and CATI surveys produce high-quality data. However, interviewer assistance is missing for surveys taken from a web panel. This may lead to measurement errors.

- *Maintenance.* Panel members should remain active in the panel. If not, representativity may be affected. Nevertheless, after participating in a number of surveys, people may lose interest and drop out. This is called *attrition.* So, the size of the panel is reduced over time. Therefore, it may become necessary to add new members to the panel after some time. This is called *panel refreshment.* Also, *panel conditioning* may occur. People may get more and more experienced in completing the questionnaires. They may even learn tricks to follow the shortest route through the questionnaire. If such phenomena occur, it may be necessary to refresh the panel after a while.

This chapter gives an overview of various aspects of web panels. It describes its advantages and disadvantages. Also, some examples of existing web panels are given.

## **14.2** Theory

### 14.2.1 UNDER-COVERAGE

Under-coverage is the phenomenon that not every member of the target population can be selected in the sample. Under-coverage occurs in a web panel if not every person in the target population has access to the Internet, i.e., the target population is wider than just those with Internet access.

According to data of Eurostat, the statistical office of the European Union (EU), 90% of the households in the EU had access to the Internet in 2019. There were large variations between countries. The countries with the highest percentages of Internet access were the Netherlands, Iceland and Norway (98%), Sweden, United Kingdom (96%) and Luxembourg, Germany and Denmark (95%). Internet access was lowest in Bulgaria (75%), Greece (79%), source Eurostat (isoc_ci_in_h). Most recent data about Internet access in the United States are from 2013. According to File and Ryan (2014), the level of Internet access in the United States was 74% at the time.

Under-coverage would not be a problem in web panels if there were no differences between those with and without Internet access. Then a random sample from the population with Internet access would still be representative of the whole population. Unfortunately, there are differences. Internet access is not equally spread across the population. For example, in several countries, the elderly, the low-educated, people living in rural area and ethnic minority groups are less well represented among those with Internet access.

If recruitment for a web panel is based on a random sample, and also the surveys taken from the web panel are random samples, then the bias due to under-coverage of the sample mean $\bar{y}_I$ as an estimator of the population mean $\overline{Y}$ of the target variable $Y$ is equal to

$$(14.1) \qquad B(\bar{y}_I) = \frac{N_{NI}}{N}(\overline{Y}_I - \overline{Y}_{NI}),$$

where $N_{NI}$ is the number of people in the subpopulation without Internet access (the non-Internet population), $\overline{Y}_I$ is the mean of the target variable for people with Internet access (the Internet population), and $\overline{Y}_{NI}$ is the mean of $Y$ for those without Internet access. The magnitude of this bias is determined by two factors:

- The relative size $N_{NI}/N$ of the non-Internet population. The more people have access to the Internet, the smaller the bias will be.
- The contrast $\overline{Y}_I - \overline{Y}_{NI}$ between the people with and without Internet. It is the difference between the population means of the two subpopulations. The more the mean of the target variable differs for these two subpopulations, the larger the bias will be.

The relative size of the non-Internet population cannot be neglected in many countries. Moreover, there are substantial differences between those with and without Internet. Specific groups are underrepresented in the Internet population. So, the conclusion is that generally a random sample from the Internet population will lead to biased estimates for the parameters of the target population.

It is to be expected that Internet coverage will increase over time. Hence, the factor $N_{NI}/N$ will become smaller, and this will reduce the bias. It is unclear, however, whether the contrast will also decrease over time. It is even possible that it increases, as the non-Internet population may differ more and more from the Internet population. So, the combined effect of an increased Internet coverage and a larger contrast need not necessarily lead to a smaller bias.

One way to solve the under-coverage problem is to provide free Internet access to sample persons without it. This approach was implemented in a number of web panels. An example is the LISS Panel in the Netherlands. Households without Internet access, and those who worried that filling in a questionnaire on the Internet would be too complicated for them, were offered a simple-to-operate computer with Internet access that could be installed and used for the duration of the LISS project. Leenheer and Scherpenzeel (2013) describe how this approach improved the representativity of the LISS Panel. The surveyed population was closer to the target population.

The KnowledgePanel in the United States has implemented a similar approach. Non-Internet households that are selected in the sample are provided a web-enabled computer and free access so that they can participate as online panel members.

The ELIPSS Panel in France has implemented an even more drastic solution. All panel members were provided with a touchscreen tablet (a Samsung Galaxy Tab) and a 3G Internet connection so that everybody could participate in monthly surveys. This had the advantage that all people in the panel used the same device (no device effects) and no computers had to be installed in their homes.

This approach of giving everybody Internet access also raises new questions. Is this approach still feasible if the under-coverage is substantial? And what about the people who have no experience at all with the Internet? Will this lead to incomplete interviews or measurement errors?

A different solution of the under-coverage problem is to extend the web panel to a mixed-mode panel. This implies there will be a group of panel members without Internet access. This group is approached in a different mode than web (mail,

CATI, or CAPI). There is risk that this will cause mode effects, i.e., the same question will be answered differently in a different mode.

The German GESIS Panel is an example of a mixed-mode panel. The members of this panel were selected by means of a random sample from municipal population registers. All people in the sample were first interviewed at home (CAPI) and asked to participate in the panel. Mail questionnaires were sent to those participants who were not able or did not want to participate in web surveys.

Traditional face-to-face panel are moving to web by the adoption of a mixed-mode approach with web component. The U.K. Household Longitudinal Study (UKHLS) is a major longitudinal household panel survey started in 2009 and is the largest study of its kind, with around 40,000 households interviewed, face-to-face, at Wave 1. The study collects data from household members aged 10 and above on an annual basis. It is commissioned by the Economic and Social Research Council (ESRC) and led by the Institute for Social and Economic Research (ISER) at the University of Essex. Since Wave 8, the sample was separated into two issue modes: "CAWI first" (40% of the sample) and "CAPI first" (60% of the sample). Fieldwork for the CAWI first sample followed a sequential mixed-mode design. Households were initially invited to take part online. At the end of the initial web fieldwork period, any individuals or whole households that had not taken part online were issued to a face-to-face interviewer. From this point on, the majority of interviewing was completed face-to-face although the web survey remained available for sample members to complete that way.

Solving under-coverage problems costs money. Either the researcher has to provide the panel members with Internet access, or another mode of data collection has to be implemented.

## 14.2.2 RECRUITMENT

To set up a web panel allowing for valid statistical inference about a target population, a probability sample must be selected, after which selected people are invited to become a member of the panel.

Often, the following steps are distinguished in the recruitment phase:

1. Contact and invite potential panel members. For probability-based panels, the recruitment mode can be e-mail, mail, telephone, or face-to-face. For self-selection panels, the researcher depends on the spontaneous contact attempts by respondents. Candidates for the panel must always be checked for eligibility.

2. Conduct a *profile survey* (sometimes called a *welcome survey*) in order to collect basic demographic information about the panel members.

3. Once the panel is ready, a specific survey can be conducted. The researcher has the following options of selecting respondents for this survey:

   - Select all panel members.
   - Select a random sample.

- Select a subset of panel members using the values of one or more variables measured in the profile survey. For example, the subset could consist of all panel members of age 65 and older.
- Select a subset of panel members by asking a screening question (if this question was not asked in the profile survey).

The best recruitment approach is to select a random sample with equal probabil ities. But how to do this in practice?

To select a random sample, a sampling frame is required. The ideal sampling frame would be a list of e-mail addresses of all people in the target population. Then it is easy to draw a random sample, after which all people in the sample can be sent an e-mail with an invitation to become a panel member. To do this, the e-mail must contain a link to the Internet page where they can register. Usually a profile survey is conducted, in which new members are asked to answer some basic demographic questions. The values of the profile variables are available for all panel members. Therefore, they can be used to select samples from specific subpopulations (for example, a particular age group or people in a certain region). They can also be used as auxiliary variables in a weighting adjustment procedure.

Sampling frames with e-mails only exist in some situations. One example is a target population consisting of students of a specific university, all of them having a university-supplied e-mail address. Another example is a target population consisting of the employees of a company, who all have a company e-mail address. Unfortunately, there are no e-mail sampling frames for general population surveys. Moreover, not everyone has an e-mail address. It is also not possible to generate random e-mail address in a fashion similar to random digit dialing (RDD) for telephone surveys. So, e-mail addresses cannot be used to recruit people for a web panel. Another means of communication has to be used. Some approaches are mentioned here.

If a population register is available, a random sample of people can be selected from this register. Selected people are sent a letter by ordinary mail. The letter contains an invitation to become a member of the panel and the Internet address of the registration page. As part of the registration process, the panel member will be asked to answer a number of profile questions. Of course, new panel members must also provide their e-mail address.

Instead of recruitment by mail, also the telephone can be considered for contacting potential panel members. If a population register is available, a random sample of people can be selected from this register. Then an attempt can be made to link selected persons to telephone numbers. This is often not completely successful as many people have unlisted fixed-line telephones or they only have a mobile phone. People with a listed telephone number are called and asked to complete a short profile questionnaire, after which they are invited to become a panel member.

Many countries do not have a population register. Then an alternative could be to use an address list. For example, there are several countries in which there is a Postal Address File (PAF). This is a list of all locations in the country where post can be delivered. A random sample of addresses can be selected. Then invitation letters are sent by ordinary mail to these addresses. Usually, one person is randomly drawn at each selected address. This person is asked to become a panel member and therefore must provide his/her e-mail address.

Note that a PAF may suffer from over-coverage if it also contains non-household addresses. Sometimes, the status of each address is known so that one can filter out these addresses. If the status of the addresses is unknown, a sample will also contain addresses of companies. These addresses are not eligible and should be ignored. Also note that the selection probabilities are not equal if a random person is drawn at each selected address. People in large households have a smaller selection probability than people in small households. A disadvantage of using an address list is that the names of people living on an address are unknown. The salutation of the letter will be impersonal ("Dear Resident"), and this does not help to realize a high response rate.

An address list can also be used for recruitment by telephone. This requires telephone numbers to be available for the selected addresses. Again, it may be a problem to find a telephone number for everybody in the sample, as there are many unlisted telephone numbers. Ignoring this problem may result in a panel that is not representative for the population.

Note that in many cases there is one household per address. So, addresses correspond to households. This makes an address list suitable for setting up a web panel of households. A simple random sample of addresses corresponds to a simple random sample of households. It was already mentioned that for a panel of individual persons, these persons do not have equal selection probabilities. The selection probabilities of persons in a large household are smaller than those of persons in a small one. To obtain valid estimates, it is therefore necessary to correct for these unequal selection probabilities.

 EXAMPLE 14.4 **Panel recruitment**

### Recruitment for the LISS Panel

The LISS Panel was established in 2006 in the Netherlands. The idea was to set up a panel consisting of approximately 5,000 households. LISS stands for Longitudinal Internet Studies for the Social Sciences. Universities were offered the possibility of conducting web surveys for their research projects free of charge with this panel. See Scherpenzeel (2008) for more details about the LISS Panel.

Recruitment for this panel was based on probability sampling. The population register of the Netherlands was used to draw a simple random sample of households. An initial sample of 10,150 addresses was selected. Unusable addresses were removed. These were, among others, nonexisting addresses, non-inhabited addresses, business addresses, and addresses with people unable to participate (due to long-time illness or language problems). After cleaning the initial sample, 9,844 addresses remained.

For each sample address an attempt was made to find a corresponding fixed-line telephone number. This was successful in 70% of the cases. These addresses were approached by telephone (CATI) for a recruitment interview. The other 30% of the cases were approached face-to-face (CAPI) for recruitment.

The recruitment interview was a short interview. It took approximately 10 minutes. Some basic profile questions were asked: demographic questions, questions about an Internet connection at home, social integration, political interest, leisure activities, survey attitudes, loneliness, and personality. At the end of this profile interview, the respondent was invited to become a member of the web panel.

As a result of the recruitment process, 48% of the invited households actually participated in the web panel. So, the recruitment rate was 48%. This comes down to approximately 4,800 households.

### Recruitment for the KnowledgePanel

The KnowledgePanel is an U.S. panel that was set up by GfK. The panel members are recruited by means of probability sampling. Until April 2009 telephone recruitment was used. Random fixed-line telephone numbers were selected using a form of RDD (list-assisted RDD sampling). Of each selected number, an attempt was made to find a corresponding address.

This was successful in 60–70% of the cases. The households for which there was an address available received an advance mailing, typically seven to nine days before the recruitment telephone call. The letter informed them that they had been selected for the panel. Following the advance letter, the telephone recruitment process began for all sampled telephone numbers. Cases sent to telephone interviewers were dialed for up to 90 days, with at least 10 dial attempts when no one answered the phone and when phone numbers were known to be associated with households. Extensive refusal conversion was also performed. Experienced interviewers conducted all recruitment interviews. The growing number of mobile-only households caused GfK in 2009 to change to address-based sampling. Mobile telephone numbers could not be sampled with the RDD system. So, representativity was affected due to the absence of the growing group of mobile-only people. The address list used is a Postal Address File. It is the Postal Service's Delivery Sequence File of the U.S. Postal Service. It contains approximately 97% of the household addresses in the United States. Randomly sampled addresses are invited to join KnowledgePanel through a series of mailings, including an initial invitation letter, a reminder postcard, and a subsequent follow-up letter. Approximately 45% of the sample addresses can be matched to a corresponding telephone number. About five weeks after the initial mailing, telephone refusal conversion calls are made to households for whom a telephone number is available.

The KnowledgePanel contained approximately 55,000 members in 2016. About 40% of its members were recruited through RDD, and 60% through address-based sampling. According to GfK (in 2010), 14 out of 100 eligible households responded positively to mailed invitation by indicating an interest in joining KnowledgePanel. Of those, three out of four follow to become participating panel members. This comes down to a recruitment rate of approximately 10%.

To set up a web panel that allows for valid statistical inference about a general population, a probability sample must be selected. Ideally, an e-mail sampling frame is available, but this often not the case. A solution can be to use a different mode of recruitment, such as sending a letter or calling people by telephone. Such approaches affect some of the advantages of web surveys. It is more cumbersome, it increases the costs of the survey, and it is not so fast anymore. Therefore, many web panels rely on some form of self-selection. Self-selection (also called opt-in) means that it is completely left to people to select themselves for the web panel or not. Panel members are those who happened to have access to Internet, encountered an invitation to become a member, visited the appropriate website, and spontaneously decided to participate.

Recruitment for self-selection web panels can take several forms. Here are some examples of how one can become a member of a panel:

- Participants go to the specific panel recruitment portal themselves. They could, for example, know about the web panel via an advertisement in the media.

- Participants are redirected through banners or pop-ups on a different web page.

- Some websites are designed to offer people the possibility of registering for several web panels at the same time. People do this because they can earn money by filling in survey questionnaires. Examples for such websites are www.surveymonster.net and www.yellowsurveys.com.

- At the end of another (possibly offline) survey, participants are asked to become a member of a web panel. In that case, the panel is populated with a subset of the respondents of this survey.

In case of self-selection, the survey researcher is not in control of the selection process. Each person has an unknown participation probability, which makes it impossible to construct unbiased estimators. Another problem is that also people from outside the target population can become panel member. Moreover, a person can have multiple memberships by becoming a member several times (under different identities). Another problem is that people may try to manipulate the outcome of the survey.

■ EXAMPLE 14.5 **Panel manipulation**

Bronzwaer (2012) describes an attempt to manipulate a panel. The panel is a Dutch panel, called *Peil.nl*. It is a self-selection panel containing approximately 50,000 members. Every week, a political poll is conducted focusing on party preference.

During the campaign for the parliamentary elections in 2012, there was a group of 2,500 people who applied for membership of the web panel. Their plan was to first behave as voters for the Christian Democrats Party (CDA). After a while they would gradually step over to a different party: the old people party 50PLUS. Their hope was that this move would also trigger other voters to change to the 50PLUS.

Unfortunately for them, their plan was detected when polling organization discovered that suddenly a lot of people at the same time wanted to become a panel member. So, this manipulation attempt failed. Nevertheless, this anecdote shows that manipulation of a self-selection web panel is technically possible.

To show the effects of self-selection on estimators, it is assumed that each person $k$ in the Internet population has an unknown probability $\tau_k$ of becoming a panel member. This probability is called the participation probability. A naive researcher assuming every element in the Internet population to have the same probability of being selected in the web panel will use the sample mean as estimator. It is shown in Chapter 11 that the bias of this estimator is approximately equal to (14.2)

$$(14.2) \qquad\qquad B(\bar{y}_S) = \frac{R_{\tau Y} S_{\tau} S_Y}{\bar{\tau}}.$$

This bias is determined by three factors:

- The *average participation probability* $\bar{\tau}$. If people are more likely to become a panel member, the average participation probability will be higher, which leads to a smaller bias.
- *The standard deviation* $S_{\tau}$ of the participation probabilities. The more these probabilities vary, the larger the bias will be.
- The *correlation* $R_{\tau Y}$ between the participation probabilities and the target variable. The stronger the correlation, the larger the bias.

The average participation probability can be very small in a self-selection web panel. For example, there are web panels in the Netherlands with 100,000 members. The target population (all Dutch from the age of 18) consists of 12.8 million people. This means the average participation probability is equal to 100,000/ 12,800,000 = 0.008.

The bias of the estimator vanishes if all participation probabilities are equal. In this case, the recruitment process is comparable to a simple random sample. The bias also vanishes if participation does not depend on the value of the target variable. Then the correlation is 0.

A self-selection panel should be considered out of the question for compiling accurate statistics about the general population. Indeed, a special task force of the American Association for Public Opinion Research (AAPOR) concluded that "Researchers should avoid nonprobability online panels when one of the researcher objectives is to accurately estimate population values." See Baker et al. (2010).

### 14.2.3 NONRESPONSE

Nonresponse is an important issue for web panels. It occurs in two phases of the process of using a web panel for data collection: (1) during the recruitment phase and (2) in the specific surveys taken from the panel. Recruitment nonresponse may be high because participating in a panel requires substantial commitment and effort of respondents. The specific survey nonresponse is often low as the invitation to participate in the survey is a consequence of agreeing to be a panel member.

Nonresponse can have various causes. These causes also depend on the recruitment approach used to set up the panel. If a sampling frame with e-mail addresses is available, nonresponse can be caused by noncontact (wrong e-mail address, message is intercepted by a spam filter, message is no read), refusal (after reading the e-mail message), or not-able (browser not working properly). If recruitment is by mail or telephone, the causes of nonresponse are similar to those for mail or telephone surveys. Note that for mail and telephone recruitment not having access to the Internet must be recorded as a cause for nonresponse (not-able) and not as under-coverage.

*Attrition* is a specific type of nonresponse that may occur in the surveys taken from the panel. It is the phenomenon that panel members drop out of the panel in the course of time. People can get tired of completing the specific survey questionnaires, and therefore they may decide to stop their cooperation. In fact, they cease to be a panel member. Once they stop, they will never start again.

The problem of nonresponse is that it may be selective. To show what the effects of nonresponse can be, it is assumed that every person $k$ in the population has an unknown response probability $\rho_k$. Then, according to Bethlehem, Cobben, and Schouten (2011), the bias of the response mean $\bar{y}_R$ is equal to

$$(14.3) \qquad B\left(\bar{y}_R\right) \approx \frac{R_{\rho Y} S_\rho S_Y}{\bar{\rho}},$$

in which $R_{\rho Y}$ is the correlation between the response probabilities and the values of the target variable. Furthermore, $S_\rho$ is the standard deviation of the response probabilities, $S_Y$ is the standard deviation of the target variable, and $\bar{\rho}$ is the mean of all response probabilities. The bias of the response mean (as an estimator of the population mean) is determined by three factors:

- The *average response probability*. The lower the response rate, the larger the bias will be.

- The *variation* (standard deviation) of the response probabilities. The more these probabilities vary in magnitude, the larger the bias will be.
- The *correlation* between the target variable and the response probabilities. A strong correlation between the values of the target variable and the response probabilities will lead to a large bias.

The bias vanishes if all response probabilities are equal. Then response can be seen as a simple random sample. The bias also vanishes if response probabilities do not depend on the value of the target variable.

The recruitment mode has an impact on the response rate in the recruitment phase. Typically, interviewer-assisted surveys (CAPI, CATI) have higher response rates than self-administered surveys (mail, web). From the point of view of representativity, interviewer-assisted recruitment should therefore be preferred. The other side of the coin is, however, that this increases recruitment costs considerably.

This book stresses the importance of applying the principles of probability sampling: samples must be selected by means of probability sampling, and the selection probabilities must be known. Self-selection was rejected as a scientifically sound sampling technique, because selection probabilities are unknown. Moreover, these probabilities may be 0 for some groups. But what about a proper probability sample that is affected by a substantial amount of nonresponse? Does this not almost resemble a self-selection survey? To compare both survey situations, an upper bound of the bias is computed. Note that the expression (14.2) for the self-selection bias is similar to expression (14.3) for the nonresponse bias. The only difference is that the participation probabilities $\tau_k$ are replaced by the response probabilities $\rho_k$. Given the mean response probability $\bar{\rho}$, there is a maximum value the standard deviation $S_\rho$ of the response probabilities cannot exceed:

$$(14.4) \qquad\qquad S(\rho) \leq \sqrt{\bar{\rho}(1-\bar{\rho})}.$$

This implies that in the worst case $S_\rho$ assumes its maximum value and the correlation coefficient $R_{\rho Y}$ is equal to either $+1$ or $-1$. Hence, the maximum of absolute value of the bias will be

$$(14.5) \qquad\qquad B_{MAX} = S_Y \sqrt{\frac{1}{\bar{\rho}} - 1}.$$

As an example, the recruitment response rate of the probability sampling-based LISS Panel was 54%; the maximum absolute bias is therefore equal to $0.923 \times S_Y$. In a similar fashion, it can be shown that in case of self-selection, the worst bias case is equal to

$$(14.6) \qquad\qquad B_{MAX} = S_Y \sqrt{\frac{1}{\bar{\tau}} - 1}.$$

There are several self-selection web panels in the Netherlands with a size of around 100,000 members. Assuming the target population to be all Dutch from the age of 18, the size of the target population is approximately 12,800,00. Since the average participation probability is $100,000/12,800,000 = 0.0078$, the maximum absolute bias is $11.271 \times S_Y$. This is more than 12 times larger than the worst bias case of a probability-based survey with nonresponse. So, the bias of the outcomes of a self-selection panel can be much larger than the bias of the outcomes of probability sampling panel.

The response rate is not the only factor in expression (14.3) for the nonresponse bias. Another important factor is the variation of the response probabilities. This is measured by the standard deviation of the response probabilities. If all response probabilities are equal, the standard deviation is 0, and there is no bias. The more the response probabilities vary, the larger the bias. Schouten, Cobben, and Bethlehem (2009) propose to compute the *R-indicator* as a measure of representativity. This indicator is defined as $R = 1 - 2S_\rho$, where $S_\rho$ is the standard deviation of the response probabilities. $R$ assumes a value in the interval $[0, 1]$. $R$ is equal to 1 if all response probabilities are the same. This is the case of complete representativity. The closer the value of $R$ is to 0, the larger the lack of representativity is.

> **EXAMPLE 14.6 Nonresponse in the LISS Panel**
>
> Table 14.1 gives an example of nonresponse in a web panel. The data relate to the LISS Panel and are taken from Scherpenzeel and Schouten (2011). The table shows the response rates in the subsequent phases of the process. Percentages are with respect to the initial sample.
>
> Contact could be established with 91% of the sample households. In 75% of the cases, households agreed to do a short recruitment interview. 54% of the households decided to participate in the panel, but only 48% became active in the panel. So, the response rate of the recruitment phase was 54%. The table also shows the effect of attrition. The response rate decreases over time. After four years, only one out of three original sample households is still active in the panel.
>
> TABLE 14.1    **Response rates and R-indicator for the LISS Panel**
>
> | Phase | Response (%) | R-indicator |
> |---|---|---|
> | Recruitment contact | 91 | 0.85 |
> | Recruitment interview | 75 | 0.80 |
> | Agree to participate in panel | 54 | 0.71 |
> | Active in panel in 2007 | 48 | 0.67 |
> | Active in panel in 2008 | 41 | 0.70 |
> | Active in panel in 2009 | 36 | 0.75 |
> | Active in panel in 2010 | 33 | 0.78 |

> The R-indicator is high in the contact phase (0.85), but it decreases in the course of the recruitment process. The R-indicator is only 0.67 when the panel starts in 2007. Therefore, the panel is not so representative. Surprisingly, the R-indicator increases again over the years. Apparently, attrition causes the composition of the panel to become more balanced.

The response rate is an important indicator for the quality of single cross-sectional surveys, because low response rates present a serious risk of biased estimates. See, for example, Bethlehem, Cobben, and Schouten (2011). Interpretation of response rates in the context of panels is more complex. A high response rate in a specific survey taken from a panel does not guarantee of small bias if the recruitment response was low. So, a good sample from a bad panel does not necessarily produce accurate estimates.

There is ample literature about the question of the usefulness and meaning of response rates as quality indicators. See, for example, AAPOR (2006) and Eysenbach (2004). AAPOR states that "Response percentage does not indicate sample or panel quality. It reflects a panel business strategy. The response rate is an indication of the level of efficiency of the panel provider." The literature shows that response rate alone is not a good indicator of the nonresponse bias and thus of web survey and web panel quality. See Bethlehem (2011), Groves (2006), and Groves and Peytcheva (2008).

A basic aspect of self-selection recruitment for a panel is that everyone can volunteer to participate. There is no sampling frame. Although a target population may have been defined, it is not always clear whether or not each respondent belongs to this target population. Therefore, it is not possible to compute exact response rates for these panels. The problems is discussed by Fricker and Schonlau (2002); Schonlau, Fricker and Elliott (2002); and AAPOR (2006). Only completion rates can be calculated for self-selection panels.

For panels based on probability sample recruitment, a *cumulative* or *multiplicative* response rate can be used. This computation considers the response rate of the recruitment phase and the response rates of the subsequent surveys. See Huggins and Eyerman (2001); Schlengen (2002); Tourangeau (2003); the Office of Management and Budget (2006); Schouten et al. (2009); and Couper (2007).

No rigorous and well-accepted terminology and definition of the response rate in web panels exists. An example of a definition of the response rate and the completion rate is given by Interactive Marketing Research Organization (2006):

- The *response rate* is "based on the people who have accepted the invitation to the survey and have started to complete the survey. Even if they are disqualified during screening, the attempt qualifies as a response."

- The *completion rate* "is calculated as the proportion of those who have started, qualified, and then completed the survey."

Another example of an alternative evaluation criterion is the *initial response rate,* i.e., the percentage of people who initially agree to become a member of the panel and complete a second in-house profile interview. This definition has been proposed by Saris (1991, 1998). He applied it in the analysis of the probability-based Dutch *Telepanel.*

The terminology used by different companies and organizations varies, and often the same term is used with a different meaning. Alternative concepts, evaluation criteria, and rates are also provided within different contexts. Each of the guidelines quoted in this section uses its own terminology. Here are a number of indicators that can be used in evaluation criteria:

- Percentage response based on the total amount of invited individuals;
- Percentage of questionnaires opened;
- Percentage of questionnaires completed;
- Percentage in the target group (quota sampling).

One has to bear in mind that although a variety of criteria and rate definitions exists, it is of vital importance to use standardized concepts and definitions when comparing the performance and quality of different panels. Before introducing and describing a number of indicators, the following points are stressed:

- There are different stages in building a web panel. Each stage has different kinds of response rates and indicators.
- Cumulative response rates over subsequent stages can be used to evaluate the performance and quality of studies using probability-based panels.
- Computation of indicators should be restricted to only the active part of the panel. For the computation of response rates over time, the availability of panel members at a given point in time must be considered.

Callegaro and DiSogra (2008) propose a systematic framework of concepts and indicators related to response in panels. The basic idea is that the computation of response rates for a probability-based panel has to consider all steps in the recruitment and maintenance. At each step, different response rates can be computed, and each of them provides insight in different aspects of the quality and success of the survey. An example of the response rates for each of the different stages of recruitment in the Dutch CentER Panel can be found in Sikkel and Hoogendoorn (2008). Below, some important indicators are described in detail. More indicators can be found in Callegaro and DiSogra (2008).

### 14.2.3.1 The Recruitment Rate.

Recruitment is the first step in setting up a panel. Recruitment may also be required for maintaining the panel. Recruitment is only well defined for probability-based panels. It comes down to selecting a sample from the target population and inviting the selected individuals to become a member of the panel. The recruitment rate is an indicator of the success of this activity.

If a sample of households is selected, the researcher has a choice to just invite one (randomly selected) member per household or to invite all eligible members of the household. Different choices for selection procedures may lead to different selection probabilities of individuals. For example, if just one person per household is selected, persons in larger households have a smaller probability of selection. The proper selection probabilities should be considered when computing estimates.

The recruitment rate cannot be computed for self-selection panels. The reason is that the target population is undefined. It is also unclear how many people are invited to participate as it is unknown how many people see the invitation.

The *recruitment rate RECR* is defined by

$$(14.7) \qquad \text{RECR} = \frac{\text{IC}}{\text{IC} + (R + \text{NC} + O) + e \times (\text{UH} + \text{UO})},$$

where

- IC = number of initial consent cases
- $R$ = number of cases directly and actively refusing
- NC = number of noncontacts
- $O$ = number of other cases
- UH = number of cases for which it is unknown whether household is occupied
- UO = number of unknown other cases
- $e$ = estimated proportion of cases of unknown eligibility that are eligible

The proportion of eligible cases among those of unknown eligibility cannot be computed in practice, because information with respect to UH and UO is usually not available. Therefore, an estimate has to be used. If panel members are recruited during the first contact with the household, there are no separate recruitment rates. Instead, both steps are combined, and one indicator is used (see Couper (2007)).

If first contact and recruitment for the panel are separate steps, and thus take place at two different points in time (see, e.g., Arens and Miller-Steiger (2006)), then the RECR represents only the second step, i.e., the consent to join the web panel.

If all eligible members of the household are recruited, the RECR can be computed at either the household or person level. Note that:

- *At the household level,* the total number of eligible households is in the denominator. Each household must have at least one potentially eligible member to be recruited.

- *At person level,* the total number of eligible persons across all households needs to be known. The denominator describes all eligible persons, and the numerator refers to all recruited persons. The factor *e* would then be a multiplier that gives the estimated number of eligible persons expected from the number of "unknown" or "other" households.

The *RECR* computed at a household level is the same as RECR computed at person level if a within-household selection at the recruitment stage is applied where only one member per household is recruited for the web panel.

When several members per household have been recruited, but a member-level sample for a given study is drawn in which only one random member per household is selected (among all eligible members, if there is an eligibility criterion) and no substitutions are allowed, a household-level RECR measure can be used since it is similar to recruiting only one person per household.

### ▓ EXAMPLE 14.7 Computing recruitment rate

Suppose that in a web panel 11,420 households are eligible for panel selection. Table 14.2 contains the results that were obtained during the recruitment.

TABLE 14.2   **Household-level recruitment data for a fictitious web panel**

| Household level | |
| --- | --- |
| IC | 10,200 |
| R | 400 |
| NC | 370 |
| O | 200 |
| UH | 150 |
| UO | 100 |
| Total eligible | 11,420 |
| Estimated proportion of unknown eligibility that are eligible | 0.01 |

The recruitment rate at the household level is computed as follows:

$$RECR = \frac{10,200}{10,200 + (400 + 370 + 200) + 0.01 \times (150 + 200)} = 0.91.$$

Thus, the recruitment rate is 91% of the eligible population.

**14.2.3.2 The Profile Rate.** The second step of the recruitment process for a web panel is often a profile survey. All those who have agreed to become a member are asked to complete the profile survey questionnaire. Alternatively, the survey organization can redirect recruited respondents to an online registration page that functions as the profile survey. By responding to this short survey, respondents become active panel members.

In the case of probability-based panels, there can be a difference between the number of initially recruited panel members and the number of active panel members. The cause is that recruited candidates could choose not to complete their profile survey and therefore drop out before being registered as panel members. To correctly account for this initial dropout effect, an indicator based on the active web panel should be computed. This indicator is called the profile rate *PROR*. No unknown eligibility or ineligible cases exist at this stage, because these cases will have previously been removed. The *profile rate* is defined by

$$(14.8) \qquad \text{PROR} = \frac{(I + P)}{(I + P) + (R + \text{NC} + O)},$$

where

- $I$ = number of complete cases in the profile survey
- $P$ = number of partially complete cases in the profile survey
- $R$ = refusals (direct or active)
- NC = noncontacts (including passive refusals)
- $O$ = other cases

Noncontact (NC) occurs if there is no reply from a respondent in the profile survey (Couper et al., 2007). Because these people were contacted at the recruitment stage, they could be contacted again to record their reasons for not completing the profile survey or, if relevant, to confirm that the e-mail invitation for the profile survey actually reached them. Note that noncontacts can be a form of passive refusal behavior. This phenomenon, if included in the NC, could contribute to membership bias.

As with the RECR, the PROR can be computed either at the household level or at the person level. If the sample for a given study is limited to selecting only one random panel member per household, the household-level PROR is equal to the person-level PROR.

For self-selection panels, the profile rate has a different meaning, since determining the number of refusals and noncontacts is not as straightforward as it is for probability-based panels. If people applying for the panel immediately become a member (single opt-in), there is no profile rate. If applicants are asked to confirm that they want to join the panel (double opt-in), a profile rate can be computed by comparing those indicating they want to join the panel with those confirming they really want to be a panel member.

**EXAMPLE 14.8 Computing the profile rate**

In Example 14.7, 10,200 households had given their initial consent. Suppose they all respond in the profile survey, either completely or partially. Unfortunately, this almost never happens in practice. Assume there was a complete response in 7,000 cases and a partial response in 3,200 cases. The profile rate is now

$$
\text{PROR} = \frac{(7,000 + 3,200)}{(7,000 + 3,200) + (400 + 370 + 200)} = 0.91.
$$

The profile rate is here equal to the recruitment rate. In practice, however, some households that initially agree to become a member of the panel do not respond in the profile survey. Suppose that only 7,800 households complete the survey (5,500 complete responses and 2,300 partial responses). Then the profile rate is

$$
\text{PROR} = \frac{(5,500 + 2,300)}{(5,500 + 2,300) + (400 + 370 + 200)} = 0.89.
$$

**14.2.3.3 The Absorption Rate.** Once the panel is in operation, members of the panel are invited to participate in specific surveys. This is done by means of e-mail. Not every selected member will respond. Pratesi et al. (2004) distinguish four steps in the web survey participation process. Nonresponse may occur in each step:

1. *Sending the invitation by e-mail.* In this contact step, some e-mails may not reach the members due to technical problems. This may happen in around 10% of the cases. For example, an e-mail invitation might end up in a spam filter and will subsequently be deleted or moved to a spam map. In this case, the e-mail appears not to be "absorbed" by the panel member. There is no feedback indicating that e-mail was treated as spam. This prevents the researcher successfully contacting an invited person. This is a problem. The researcher has no means of measuring the relevance or magnitude of this problem. Other examples of e-mails not reaching the selected person are a wrong e-mail address, a full mailbox, or a network error.

2. *Access to the introductory page of the web survey.* The contacted person accesses the questionnaire, but this action does not yet imply survey participation.

3. *Start of questionnaire completion.* The respondent starts answering questions. This means at least partial completion. There is always a risk that some questions are skipped (if the route through the questionnaire is not forced) or that completion of the questionnaire is interrupted.

**4.** *Completion of the survey questionnaire.* At this step, participation is completed, although some answers may have been skipped in the process.

In order to take possible absorption effects into account, Lozar Manfreda and Vehovar (2002) propose to compute the so-called absorption rate (ABSR). According to Callegaro and DiSogra (2008), the *ABSR* is defined as

$$(14.9) \qquad \qquad ABSR = \frac{EI - BB - NET}{EI},$$

where

- EI = number of e-mail invitations sent
- BB = number of undeliverable e-mail invitations (bounce back)
- NET = network error – undeliverable e-mails

The absorption rate is 1 (i.e., all e-mails are absorbed) if every selected member actually receives the e-mail. Therefore, this rate can be considered an indicator of the quality of the e-mail list of the sampled web panel members. It is a proxy indicator of how many objects receive the e-mail, because, as stated before, it is impossible to check whether or not e-mails get lost in the system. In other words, it is impossible to determine exactly the number of selected members that received an initial e-mail.

**14.2.3.4 The Completion Rate.** The completion rate is the proportion of selected and invited eligible members who completed a specific web survey. Thus, this rate reflects the success of a specific study. The *completion rate* is defined by

$$(14.10) \qquad \qquad COMR = \frac{(I + P)}{(I + P) + (R + NC + O)},$$

where

- $I$ = number of completed cases in the survey
- $P$ = number of partially completed cases in the survey
- $R$ = refusals (direct or active)
- NC = noncontacts (including passive refusals)
- $O$ = other cases

The variables are the same as those found in the profile rate. The difference is that these variables are not defined for the profile survey, but rather for a specific survey.

Note that the completion rate can also be computed for self-selection panels, but with some modifications.

### 14.2.3.5 The Break-Off Rate.

In expressions (14.8) and (14.10), the partially completed interviews ($P$) are seen as successful cases. However, in some situations, an incomplete questionnaire is not considered to be an acceptable result. To quantify the extent of this phenomenon, the concept of "break-off" can be introduced. Break-off means that the questionnaire was started but not finished. The break-off rate (BOR) is defined by

$$(14.11) \qquad \mathrm{BOR} = \frac{\mathrm{BO}}{I + P + \mathrm{BO}},$$

where

- $I$ = number of complete cases in the survey
- $P$ = number of partially complete cases that are considered successful
- BO = number of break-offs, i.e., the number of unsuccessful partially completed cases (according to the criteria of the researcher)

### 14.2.3.6 The Screening Completion Rate and the Study-Specific Rate.

For a specific study, the target population need not coincide with the panel. Therefore, the researcher has to select eligible respondents. This is called *screening*. There are two options:

1. The screening process can be based on previously collected information (in the profile survey or in another specific survey).
2. If no screening variables are available, an e-mail invitation can be sent to all (or a large subset of) panel members. The first questions of the survey must aim at assessing eligibility of the respondents. Routing instructions will see to it that only eligible respondents answer the remaining questions.

Bearing in mind the abovementioned options, two indicators can be computed: the screening completion rate (S_COMP) and the study-specific eligibility rate (S_ELIG). The *screening completion rate* is defined by

$$(14.12) \qquad S_{\mathrm{COMP}} = \frac{\mathrm{SCQ} + \mathrm{SCNQ}}{\mathrm{INV}},$$

where

- SCQ = the number of people who were successfully screened and qualified for the study
- SCNQ = the number of people who were successfully screened and did not qualify for the study
- INV = the number of survey invitations sent

A potential problem is that nonresponse may hamper correct interpretation of this rate. If selected persons do not answer the screening questions, the researcher cannot establish whether or not they qualify for the survey.

The *study-specific eligibility rate* is defined by

$$(14.13) \qquad S_{\text{ELIG}} = \frac{\text{SCQ}}{\text{SCQ} + \text{SCNQ}},$$

where

- SCQ = the number of people who were successfully screened and qualified for the study
- SCNQ = the number of people who were successfully screened and did not qualify for the study

### 14.2.3.7 Cumulative Response Rates.

In case of probability-based panels, cumulative response rates can be computed. These indicators consider what happens in the different steps of the survey process, from panel recruitment to response in a specific study. These cumulative response rates are obtained by multiplying rates that have been obtained for each step in the process.

Note that an approach often adopted to panel maintenance is to add new cohorts of members to the current panel. Consequently, response rates should be computed separately for each cohort. As a next step, response rates for the panel as a whole are obtained by taking a weighted average of the cohort response rates. Here focus is on computing the response rate for a single cohort. Two types of cumulative response rates can be calculated. They are called the cumulative response rate 1 and the cumulative response rate 2.

The *cumulative response rate 1* (CUMRR1) is defined by

$$(14.14) \qquad \text{CUMRR1} = \text{RECR} \times \text{PROR} \times \text{COMR}.$$

This response rate reflects the percentage of cases left over after nonresponse in the recruitment phase, nonresponse in the profile survey, and nonresponse in the specific web survey.

The cumulative response rate 2 introduces a fourth component: the retention rate. The *retention rate* (RETR) is the proportion of an original cohort that remains in the active panel at the time the sample for the specific survey is drawn. Therefore, cumulative response rate 2 is defined with reference to a specific cohort. For a given cohort, this indicator is obtained by multiplying the cumulative response rate 1 by the retention rate, and *cumulative response rate 2* is defined by

$$(14.15) \qquad \text{CUMRR2} = \text{RECR} \times \text{PROR} \times \text{RETR} \times \text{COMR}.$$

**14.2.3.8 The Attrition Rate.** Attrition is the phenomenon that people, after participating in a number of surveys, lose interest and drop out of the panel. After dropping out, they will not become active again. The *attrition rate* is defined as the proportion of active panel members that drop out of the panel in a specific time period. Attrition has an effect on cumulative response rates. The overall representativeness of the web panel can also be affected by differential attrition rates. For example, subgroups with higher attrition rates than other subgroups will become underrepresented in the panel in the course of time. Therefore, these groups will also be underrepresented in specific surveys.

DiSogra et al. (2007) and Sayles and Arens (2007) stress the importance of studying differential attrition rates for subgroups of the population. The attrition rate of group $a$ from month $t$ to month $t + 1$ is defined by

$$(14.16) \qquad \text{ATTR}_{M_t} = \frac{\text{Cohort}_a@\text{Time}_t - \text{Cohort}_a@\text{Time}_{t+1}}{\text{Cohort}_a@\text{Time}_t}$$

where

- $\text{Cohort}_a@\text{Time}_t$ = the size of the specific cohort at time (month) $t$
- $\text{Cohort}_a@\text{Time}_{t+1}$ = the size of the specific cohort at time (month) $t + 1$

Roughly speaking, attrition is measured by counting how many recruits stay in the web panel month after month (Clinton, 2001).

The web panel attrition rate is an important indicator, because a high value could mean the specific surveys are too long or have a poor questionnaire design (European Society for Opinion and Marketing Research, 2008). Monitoring attrition is also crucial for assessing the representativeness of any panel, particularly because attrition is rarely equal across all demographic subgroups.

Several indicators for the quality of probability-based web panels were presented in this section. All these indicators may help to judge the reliability and validity of the outcomes of a web panel. It is important to include indicators like the recruitment rate, the profile rate, the study-specific survey completion rate, and the final cumulative response rate (CUMRR1) in the survey report. This helps users of web panel data to assess its usefulness.

There are also other factors that may play a role in the quality of the outcomes. Some of them are:

- The field period (starting and closing dates and length of the fieldwork period length);
- Number of reminders sent and follow-up mode (e-mail, letter, telephone call);
- Use of incentives.

For example, the use of incentives generally increases the response rates, but some literature suggests it may not help to improve the composition of the response, resulting in a larger bias. Another example is shortening the fieldwork period. It may be attractive to have timely information, but if this leads to an overrepresentation of frequent e-mail users and early respondents, this may also be the cause of a bias.

An obvious indicator for self-selection panels is the completion rate (COMR). This indicator can be interpreted as the respondent's interest in the survey and/or the ability of the survey organization to maximize cooperation. Note that it is possible to increase the survey completion rate by just selecting the most cooperative web panel members. Unfortunately, this can seriously affect the composition of the sample. For example, Vonk (2006) conducted an experiment in which the same survey was carried out at the same time by 19 different self-selection web panels in the Netherlands. The completion rate turned out to vary between 8% and 77%. Thus, completion rate is an extremely volatile indicator.

The ABSR is an interesting indicator for measuring ability in managing and updating the web panel database, whereas the BOR could suggest that problems exist either with respect to the design of the questionnaire (for example, too long or boring) or to technical aspects during the survey administration (e.g., streaming media or animations that may "break" a survey at some point).

The ABSR and the BOR should be reported for both probability-based and volunteer panel research.

Web panel research is often claimed to have high response rates. Such claims are based on the high response rates of samples selected from the panel for specific surveys. This ignores the fact that the response rate in the recruitment phase of the panel may have been very low, and as a result, the estimates using the survey data may be substantially biased.

If recruitment is based on self-selection, or if recruitment nonresponse is high, high response rates in a specific survey may hide a wide range of other survey errors. For an overview of survey errors, see Bethlehem (2009).

It is always possible in a self-selection panel to generate a high response in a specific survey by selecting only those members who always participate when they are invited ("the low-hanging fruit"). This increases the sample size, but it is unlikely to improve survey estimates, as this special group might produce highly biased results. This shows again that high response rates in a panel will generally not be able to compensate for low initial cooperation or an unknown selection bias. Computing the response rate only using willing respondents or boosting response rates by only using the most cooperative panel members makes response rates difficult to compare with those of probability samples. For this reason, response rates in non-probability samples do not have the same meaning as those in probability samples.

If self-selection is used for web panel recruitment, one should follow the advice of Fowler (2002) and be transparent. If a researcher decides to use a non-probability sample, readers should be told how the sample was drawn, the fact that it is likely to be biased in the direction of availability and willingness to be interviewed, and that the normal assumptions for calculating sampling errors do not apply. This may help to avoid the results of non-probability samples to be seriously misrepresented, thereby contributing to a loss of credibility of social science research.

### 14.2.4 REPRESENTATIVITY

Many web panels are large. This sometimes leads to the claims that therefore they are representative, and thus the quality of the survey results is high. These claims are often not justified. The concept of representativity and its relationship to web panels are discussed in this section.

The concept of *representativity* plays a crucial role in the discussion about the foundations of survey sampling. This concept is often used in survey research, but usually it is not clear what it means. Kruskal and Mosteller (1979a, 1979b, 1979c) present an extensive overview of what representative is supposed to mean. They found the following meanings for "representative sampling":

- General acclaim for data;
- Absence of selective forces;
- Miniature of the population;
- Typical or ideal case(s);
- Coverage of the population;
- A vague term, to be made precise;
- Representative sampling as a specific sampling method;
- As permitting good estimation;
- Good enough for a particular purpose.

Kruskal and Mosteller (1979b) recommend not using the word representative, unless one explains what it means. The problem is that both for probability sampling and other forms of sampling claims are made that samples are representative, often with quite different meanings, and sometimes with no concrete meaning at all besides conveying a vague sense of good quality.

In this chapter, the term "representative" is used for indicating that a sample is *representative with respect to a variable*. It means that the distribution of the variable in the sample is the same as its distribution in the population. The idea is that if a sample is representative with respect to many auxiliary variables, the hope is it will also be representative with respect to the target variables of the survey, therefore allowing unbiased estimation of population characteristics. In case a sample is not representative with respect to a number of auxiliary variables, weighting adjustment can be applied in attempt to improve representativity (Schonlau et al., 2007; Schonlau et al., 2009).

Bethlehem and Stoop (2007) discuss the frequent misunderstanding about online research that large numbers make the sample better. Couper (2000) comments on the claims of a self-selected online survey in the United States: "We received more than 50,000 responses – twice the minimum required for scientific validity," whereas survey was not based on a random sample, and the selection probabilities were unknown. Not surprisingly, despite the large number of respondents, the sample did not resemble the U.S. population on a number of key indicators.

Dillman and Bowker (2001) express a similar opinion about online surveys: "Conductors of such surveys have in effect been seduced by the hope that large numbers, a traditional indicator of a high survey quality (because of low sampling error), will compensate in some undefined way for whatever coverage and nonresponse problems that might exist. Large numbers of volunteer respondents, by themselves, have no meaning. Ignoring the need to define survey populations, select probability samples, and obtain high response rates, together provide a major threat to the validity of web surveys."

Couper (2001) also pointed to the misguided assumption that large samples necessarily mean more valid outcomes. This only works for probability samples where a larger sample size leads to a smaller variance of estimates and thus to an increase of precision. For other sampling techniques, no inference about the underlying population is possible, and larger samples do not necessarily produce better estimates than smaller samples. Generally, the bias is unaffected by the sample size. So, the bias remains, even for a very large sample.

Large web panels and web surveys have the advantage that specific subgroups of the population can be identified. Information about such groups may be difficult to obtain in traditional surveys, because only a few people belong to these groups, they are hard to identify, or they are unlikely to participate in surveys. The underlying assumption is that the elderly single women, low-educated, ethnic minorities, or other usually underrepresented groups who participate in a web survey are similar to people with the same characteristics but who do not participate. In some cases, this might be not an unlikely assumption, but in others definitely not, and, in most cases, it will be difficult to test.

An additional caveat is that self-selection in web panels may require heavy weighting adjustment because of vastly varying participation probabilities. Because of large weights, the effective sample size is likely to be much smaller than the number of participants in a survey (Duffy et al., 2005).

When analyzing the data collected by means of a web panel, it is implicitly assumed that the panel is a small copy (a miniature) of the population it came from. So, relationships found in the panel are not an artifact of the sample but are identical to relationships that exist in the whole population. However, even if there are a large number of respondents, they might not reflect the population structure. For example, Faas and Schoen (2006) have studied whether participants in self-selection web surveys represent Internet users in general. They conclude this is not the case. Hence, the conclusion must be that self-selection web surveys do not yield results representative of Internet users in general (in terms of marginal distributions of variables or in terms of relationships between variables). If the aim is to have survey results that are representative of Internet users, one must carefully select a sample of such users.

In considering the effects of the lack of representativity and sample size, it is important to distinguish the various types of sample selection mechanisms:

- *Probability sampling*. The sample is selected from the population using some kind of random selection mechanism. Each element in the population must have a nonzero probability of selection, and all selection probabilities must be known. The simplest form is a simple random sample, in which all elements have the same selection probability. Correct inference from the sample to the population is possible. Unbiased estimates can be computed. The accuracy of estimates increases as the sample size increases.

- *Non-probability sampling*. This comes in many forms. See, for example, Kalton (1987), Couper (2000) and Schonlau, Fricker, and Elliott (2002). Some forms are:

  ○ *Convenience sampling*. Elements are drawn for such a sample because of their convenient accessibility or proximity to the researcher. Convenience sampling is fast, simple, and cheap. Self-selection samples can be seen as a form of convenience sampling.

  ○ *Purposive sampling*. The researcher picks units that are "representative" in a subjective way. The sample is selected such that its characteristics resemble to the characteristics of the population. Purposive sampling is similar to the "representative method" proposed by Kiaer (1895). See also Chapter 1. He constructed rather large samples that were a miniature of the population with respect to the regional distribution (rural areas and cities, well-to-do streets and not well-to-do streets). Since selection was not based on probability sampling, the accuracy of the estimates could not be computed.

  ○ *Quota sampling*. The population is divided into groups. The size of the groups is supposed to be known. A sample of predefined size (a quotum) is selected from each group. It is left to the judgment of interviewers to pick elements in a group. The basic assumption is that the probability of being available for an interview is the same for each element within a group. See also Sudman (1966). The problem is that often this assumption is not satisfied. For example, people with a higher probability of being at home may differ from those that are frequently not at home. Also, for quota sampling, it is not possible to compute selection probabilities.

Because of the problems of non-probability sampling described above, it is not possible to apply probability theory. This prohibits making proper inference to the target population of the panel or survey. It is also not possible to compute indicators for survey data quality like response rates.

## 14.2.5  WEIGHTING ADJUSTMENT FOR PANELS

The representativity of specific surveys taken from web panels may be affected by under-coverage, self-selection, or nonresponse. To repair the lack of representativity, one could consider applying some kind of adjustment weighting. Adjustment weighting in surveys was already described in detail in Chapter 12.

Representativity problems can occur in two phases of the selection process: during recruitment and during the subsequent specific surveys taken from the panel. This implies that also two corrections are required. A first approach could be to ignore the two phases. This implies that weights are obtained by directly confronting response distributions for auxiliary variables in the survey with their population distributions. This may be not the most effective way to conduct adjustment weighting. Weighting in two steps may be preferred.

In the first place, recruitment nonresponse may be a different phenomenon than survey nonresponse. Therefore, it may require a different model containing different variables. In the second place, there are more auxiliary variables available to correct for the specific survey nonresponse. For many web panels, new members conduct a profile survey. Then these variables can be used to improve the representativity of the set of panel members. Once people are panel members, they participate in specific surveys. Therefore, many more variables become available for weighting the response of these surveys.

To summarize, weighting adjustment in a web panel is a two-step process:

1. Compute weights for all panel members in such a way that the panel becomes representative with respect to the target population.

2. For each specific survey, compute weights in such a way that the it becomes representative with respect to the panel.

The final weights are obtained by multiplying the recruitment weights by the survey weights.

Various techniques are available to compute adjustment weights. A well-known and frequently used weighting adjustment technique is *post-stratification*. Strata (subpopulations) are constructed by crossing qualitative auxiliary variables. All respondents within a stratum are assigned the same weight. Post-stratification is effective if the strata are homogeneous, i.e., persons within a stratum resemble each other in terms of response behavior and target variables.

If there are many auxiliary variables, it may not be possible to apply post-stratification. For example, there could be strata without observations preventing computations of weights. Also, the population distribution of the crossing of the variables may not be available. Other, more general, weighting techniques can be applied in these situations, such as *generalized regression estimation (linear weighting*) or *raking ratio estimation* (*multiplicative weighting*). See Chapter 12 for more information.

A different approach to weighting is called *propensity weighting*. Participation in the survey is modeled by means of a logistic regression model. This comes down to predicting the probability of participation from a set of auxiliary variables. The estimated participation probabilities are called *response propensities*. A next step could be to carry out post-stratification where strata are constructed by grouping respondents with (approximately) the same response propensity. A drawback of propensity weighting is that the individual values of the auxiliary variables for the nonparticipating persons are required. Such information is often not available. Note that response propensities can also be used in other ways to reduce the bias (see Chapter 13).

Weighting adjustment is only effective if two conditions are satisfied:

1. The set of auxiliary variables must be able to completely explain participation behavior. This implies that auxiliary variables must be correlated with participation behavior.
2. The auxiliary variables must be able to completely explain the target variables. This imply they must be correlated with the target variables.

If these two conditions are not completely fulfilled, the bias of estimators will only partly be reduced. The availability of effective auxiliary variables is often a problem. Usually, there are not many variables that have a known population distribution and that satisfy the two conditions.

If proper auxiliary variables are not available, one might consider conducting a *reference survey*. The objective of this survey is unbiased estimation of the population distribution of relevant auxiliary variables. Such a reference survey could be based on a small probability sample, where data collection takes place with a mode different from the web, e.g., CAPI or CATI. Under the assumption of no nonresponse, or ignorable nonresponse, this reference survey will produce unbiased estimates of the population distribution of auxiliary variables.

The reference survey approach also has disadvantages. In the first place, it is expensive to conduct an extra survey. However, it should be noted this survey need not be very large as it is just used for estimating the population distribution of auxiliary variables. And the information can be used for more than one web survey. In the second place, Bethlehem (2010) shows that the variance of the post-stratification estimator is, for a substantial part, determined by the size of the (small) reference survey. So, the large number of observations in the web survey does not guarantee precise estimates. The reference survey approach reduces the bias of estimates at the cost of a higher variance.

The conclusion can be that some form of weighting adjustment must certainly be applied in order to reduce the bias of estimator based on a specific survey taken from a web panel. However, success is not guaranteed. The ingredients for effective bias reduction may not always be available.

### EXAMPLE 14.9 Comparing web panels

Example 14.2 shows the experiment of Vonk, Van Ossenbruggen, and Willems (2006) on 19 Dutch panels and the evaluation of the representativity of these web panels. It turned out that there was representativity for some variables and lack of representativity with respect to other variables.

Brüggen, Van den Brakel and Krosnick (2016) compared 18 of these self-selection panels with the LISS Panel (based on probability sampling), two CAPI surveys (also based on probability sampling), and data from the Dutch population register (Municipal Basic Administration). They conclude that the self-selection web panels behave in an inconsistent way. Panels that perform well for register variable estimates do not necessarily perform well for nonregister variable estimates. Some variables are estimated incorrectly by only a subset of panels (for example, gender), whereas other variables (for example, level of education or employment) are estimated incorrectly by most panels. Some panels simply perform poorly for most variables.

They find that weighting does not reduce the selection bias of the estimates for variables in the panels. The available auxiliary information used to construct adjustment weights often fail to explain the participation behavior of the panels.

Moreover, there are also problems with estimating relationships between variables. Panels differ significantly from each other with respect to the strength and even sign of relationships. Unfortunately, weighting adjustment does not improve the situation.

The conclusion of the authors is that researchers should be very careful when using data that are obtained by a survey taken from a self-selection web panel. According to the Code of Ethics of the American Association for Public Opinion Research (AAPOR), the survey report must at least clearly document the methods used to recruit the panel members. See www.aapor.org.

## 14.2.6 PANEL MAINTENANCE

As discussed in Section 14.1, there are two types of panel research: longitudinal research and cross-sectional research. In case of longitudinal research, the same set of questions is asked at regular time intervals. So behavior of panel members can be monitored over time. In case of cross-sectional research, a sample is selected from the panel members for a single, unique survey. Characteristics of panel members that have been measured in earlier surveys can be used to screen panel members. So samples can be selected from specific subpopulations.

In case of longitudinal research, it is important that panel members, once recruited, remain active in the panel and participate in (almost) every wave. If not, it becomes difficult explain changes over time. Can they be attributed to real changes, or are they caused by changes in the composition of the panel? To avoid this problem, it is important to keep people in the panel. In case of cross-sectional research, it is less important to keep the same people in the panel. It is more important to keep the panel representative. If the panel is used for both longitudinal research and cross-sectional research, maintenance of the panel should be guided by the most restrictive type of research, and this is longitudinal research.

In this section, three important aspects of web panel maintenance discussed: frequency of the surveys, panel refreshment, and maximum stay of people in the panel.

### 14.2.6.1 Frequency of Surveys.

In a panel study, repeated measurements can be made on the same group of persons at different points in time. This allows for longitudinal analysis such as measurement of change. One could, for example, be interested to follow specific variables over time, such as unemployment, health status, financial situation, and voting behavior. Measurements at different points in time on the same set of persons are correlated, thereby reducing the variance of estimates of change.

The accuracy of estimates for surveys from a panel is related to the frequency of these surveys. The higher the frequency of the surveys, the higher the perceived response burden, and thus the higher the nonresponse (attrition). The higher nonresponse rate will increase the magnitude of the bias. The reduced number of observations will also increase the variance of estimates.

It will be clear that a high survey frequency will have a negative impact on the quality of the outcomes of these surveys. Seen from this perspective, it is better to avoid a high frequency of surveys. But what is a reasonable frequency? It should be noted that the perceived response burden is determined not only by the frequency of the surveys but also by aspects like the topic of the survey and the length of the survey questionnaire.

The survey frequency should also not be too low. If panel members are asked only now and then to complete a survey questionnaire, they may lose interest and stop participating. Their engagement will disappear.

For longitudinal research, subject-matter experts usually determine the survey frequency. For slowly changing characteristics (for example, the employment situation), a survey once a year may be sufficient. Quickly changing characteristics (like voting intention) should be monitored more frequently. One could think of conducting a survey each month.

To reduce the response burden, and hence to control errors due to nonresponse and attrition, a rotating design can be chosen. This implies that the panel consists of a number of groups or cohorts. At regular intervals, for example, each month, a new group of panel members is recruited, and another group reaches the maximum survey frequency and drops out of the panel. This affects the accuracy of estimates of change.

The KnowledgePanel invites panel members three to four times per month to do a survey. The surveys are kept short, with durations of 5–20 minutes at most. For each survey, panel members earn points with a cash value of $1. Surveys taking longer than 16 minutes are rewarded with an additional small incentive. This is done to minimize panel attrition. In addition, the KnowledgePanel has a so-called Panel Relations program to keep panel members involved. The members of the LISS panel are sent one survey each month, which takes about half an hour to complete. Like for the KnowledgePanel, they receive a small amount of money for each survey they complete.

According to Baker et al. (2010), attrition is most likely among the newest members. Once people agree to participate in the panel and have completed some survey questionnaires, they are likely to continue doing so. This is because people like to behave in a way that is consistent with their own previous behavior. So most of the effort should be directed at recruiting panel members and stimulating them to participate with the first few surveys.

### 14.2.6.2 Panel Refreshment.
Panel refreshment means that a fresh sample is selected and recruited for the panel to compensate for attrition. If it is assumed that the panel after recruitment is representative for the target population, finding replacement for the attrition could prevent the panel to become less representative. A refreshment sample can be a simple random sample. This is, however, not a solution to keep the panel representative. It would be wise to replace those vanished from the panel in a more selective way, which comes down to drawing a selective random sample from a group that resembles the attrition group. Then the panel remains as representative as it was before the attrition. The question arises how this can be done in such a way that it is still possible to calculate the inclusion probabilities and therefore unbiased estimates.

This approach has some drawbacks. For selective selection of a refreshment sample, only characteristics can be used that are available for the entire target population (or sample). To really compensate for selective attrition, the Missing at Random (MAR) assumption needs to hold for the characteristics that are not available for sampling. The refreshment sample has to be recruited, leading to (recruitment) nonresponse that is probably also selective. And to fine-tune the selective sampling to the selective attrition, it is not possible to recruit a refreshment sample before the actual attrition occurred. Therefore, from wave to wave, there will always be some selectivity due to attrition. This can be dealt with by weighting adjustment methods (see Chapter 12).

Another approach could be recruiting possible future panel members and to use not only variables in the sampling frame but also variables that were measured during the recruitment interview. This way, the MAR assumption is more likely to hold. However, this also introduces an additional source of nonresponse.

Selective panel refreshment increases the quality of the web panel and should be implemented in practice. The R-indicator can be used to identify groups that contribute most to the selectivity of the panel composition (Schouten et al., 2009).

**14.2.6.3 Maximum Stay.** Frequent participation may influence behavior and opinions of panel members and therefore introduce a bias in their outcomes. For example, persons learn how to take the shortest route through a questionnaire. This effect is known as *panel conditioning*.

To avoid panel conditioning, it is common to set a maximum to the time that respondents can be a member of the panel. What the maximum duration should be depends on the frequency of the specific surveys, the length of the surveys, and the variation in survey topics. Research has shown that in probability-based panels, the effect of panel conditioning can be controlled by varying the topic of the panel surveys from wave to wave.

A comparison of 19 Dutch online panels, already quoted in Example 14.2 (Vonk, Van Ossenbruggen, and Willems 2006), showed that most attrition took place in the first few months of the panel membership. It should be noted that all these panels were non-probability, self-selection panels. After six months, the response stabilized. In the first year of the panel membership, the response was highest. Vonk, Van Ossenbruggen, and Willems (2006) compared online panels that were recruited in different ways. The number of survey requests per month was 1–2 for 60% of the panels; for 35% of the panels, this was even more. The average time needed to complete a survey was somewhere between 10 and 15 minutes.

# 14.3 Applications

Two applications of web panels are described in this section. The first one is the account of a test carried out by Statistics Netherlands, the national statistical institute of the Netherlands. Objective was to find out whether web panels can be a replacement for the more expensive and more time-consuming CAPI and CATI surveys.

The second application describes the use of self-selection web panels for opinion polls during the campaign of the general elections in the United Kingdom on May 7, 2015. The polls did so bad in predicting the election result that some people referred to it as the U.K. polling disaster.

## 14.3.1 APPLICATION 1: THE WEB PANEL PILOT OF STATISTICS NETHERLANDS

Statistics Netherlands carried out a pilot with a web panel in 2012. The main objective of this project was to get some first experiences with building web panels and to keep up-to-date with the newest developments. There was little time available and resources were limited. For this reason it was decided to recruit panel members from the respondents of an existing survey of Statistics Netherlands: OViN (Onderzoek Verplaatsingen in Nederland). This is a mobility survey. The sample for this survey was randomly selected from the population register. At the end of the survey questionnaire, all respondents were asked whether they would be willing to participate in other Statistics Netherlands surveys. Those with a positive response (and having access to Internet) were invited by letter to become a member of the web panel for a period of one year. They were informed that they were expected to complete one questionnaire each month.

Statistics Netherlands could link the OViN sample to several registrations. Hence, a large set of auxiliary variables became available. The values of these variables were measured for both panel members and people not in the panel. Therefore, these auxiliary variables allow for analysis of the representativity of the panel.

It should be noted that the sample for web panel was selected from the respondents of the OViN. This implies the target population of the web panel is the set of respondents of OViN and not the general population of the Netherlands (as represented in the population register). This should be taken into account when investigating the representativity of the web panel.

The recruitment process for the web panel is summarized in Table 14.3. The original sample for OViN consisted of 12,406 persons selected from the population register. There were 6,928 respondents, which comes down to a response rate of 57.5%. Of those respondents, 4,251 (35.3%) agreed to participate in the panel. Only 1,231 willing people really registered for the panel, and only 1,134 willing people really completed the questionnaire forms. Consequently, in the end, only 9.4% of the initial OViN sample became active panel members. This is a low score, which raised concerns about representativity.

As was shown in Section 14.2.3, the magnitude of the nonresponse bias is determined by several factors. The response rate is one of them. Another one is the variation of the response probabilities. This variation can be measured by means of the R-indicator. Its value is 1 in case of complete representativity (all response probabilities are equal) and approaches 0 as the lack of representativity increases. Table 14.4 contains the value of the R-indicator for the data in the several steps of the recruitment process. The OViN response lacks some representativity as the R-indicator is 0.78. Surprisingly, the R-indicator increases to 0.88 in the subsequent steps. The remaining group of people becomes more representative as more people drop off. The final composition of the panel is apparently more balanced than the initial group of OViN respondents.

TABLE 14.3   **Response rates in the panel recruitment process**

| Phase | Size | Percent of sample | Percent of previous |
|---|---|---|---|
| OViN sample | 12,406 | | |
| OViN response | 6,928 | 57.5 | 57.5 |
| Willing to participate in panel | 4,251 | 35.3 | 61.4 |
| Selected for panel | 4,227 | 35.1 | 99.4 |
| Registered as panel member | 1,231 | 10.2 | 29.1 |
| Participated in panel | 1,134 | 9.4 | 92.1 |

TABLE 14.4   **The R-indicator in the steps of the recruitment process**

| Phase | Size | R-indicator |
|---|---|---|
| OViN sample | 12,406 | |
| OViN response | 6,928 | 0.78 |
| Willing to participate in panel | 4,251 | 0.84 |
| Selected for panel | 4,227 | 0.84 |
| Registered as panel member | 1,231 | 0.88 |

TABLE 14.5   **The relation between recruitment response and OViN data collection mode**

| OViN data collection mode | OViN response | Willing (% of response) | In panel (% of willing) |
|---|---|---|---|
| Web | 2,370 | 55.4 | 55.4 |
| CATI | 2,946 | 59.9 | 16.9 |
| CAPI | 1,612 | 72.8 | 17.5 |
| Sample | 6,928 | 61.4 | 29.0 |

OViN was a mixed-mode survey. Sample persons were first asked to complete the questionnaire on the Internet. If they did not respond, they were re-approached by CATI (if a listed telephone number was available), and otherwise they were re-approached by CAPI. The second column in Table 14.5 shows how many people responded in each mode. There are substantial differences in the percentages of respondents willing to participate in the panel. Of the OViN web respondents, only 55.4% agrees to become a panel. This percentage is slightly higher for CATI respondents (59.9%), but much higher for CAPI respondents (72.8%). One possible explanation is that CAPI respondents gave more socially desirable answers. This phenomenon typically occurs when interviewers ask the questions. Of the willing web respondents, 55.4% actually became panel members. The percentages are much lower for the CATI respondents (16.9%) and the CAPI respondents (17.5%). Again, this pattern may indicate socially desirable answers for the question inviting people to participate in the panel.

The representativity of the web panel was explored by analyzing two target variables: level of education and main activity. The values of these two variables were available because the OViN sample could be linked to administrative sources. The distributions in the panel of these two variables were compared with their corresponding distributions in the OViN sample (the target population of this pilot). Furthermore, it was investigated whether weighting adjustment could improve the web panel-based estimates of the distributions.

Table 14.6 shows the results for level of education. Low-educated people seem to be underrepresented in the panel. The percentage of low-educated in the panel is only 2.6%, whereas it should be more than double as high (5.5%). High-educated people seem to be overrepresented in the panel. The percentage of high-educated in the panel is 45.5%, whereas it should be only 33.6%.

Can weighting adjustment improve the situation? A weighting model was constructed containing the variables age, household income, and socioeconomic status. The selection of the auxiliary variables was based on an analysis of the relationship between available auxiliary variables on the one hand and the target variables and response behavior on the other. Table 14.6 shows that weighting adjustment improves the estimates somewhat. The weighted panel estimates are closer to the true OViN values. The bias is smaller for all categories of level of education. Still, substantial biases remain.

Table 14.7 shows the results for the target variable main activity. Also, here, there are differences between the panel estimates and the true population values. Pensioners and the employed are overrepresented, and students are underrepresented. The effects of weighting are mixed. The estimates for some categories (housewife/man, school/student, disabled, unemployed) have improved. The weighted estimates are closer to the true value. The opposite effect can be observed for the pensioners and the employed: the weighted estimate is further away from the true value.

TABLE 14.6    **Estimates (unweighted and weighted) for level of education (%)**

| Level of education | Panel | Weighted | OViN |
|---|---|---|---|
| Primary | 2.6 | 4.3 | 5.5 |
| Lower secondary | 15.2 | 16.5 | 21.0 |
| Higher secondary | 34.4 | 35.8 | 37.6 |
| Bachelor/master | 45.5 | 40.6 | 33.6 |

TABLE 14.7    **Estimates (unweighted and weighted) for main activity (%)**

| Main activity | Panel | Weighted | OViN |
|---|---|---|---|
| Housewife/man | 11.9 | 12.2 | 12.5 |
| Pensioner | 16.8 | 17.8 | 14.7 |
| At school/student | 6.1 | 10.6 | 9.8 |
| Disabled | 2.4 | 2.8 | 2.8 |
| Unemployed | 1.9 | 2.1 | 2.3 |
| Employed | 59.2 | 52.4 | 56.1 |

Another conclusion of this web panel pilot was that Statistics Netherlands could use this mode of data collection only for a limited number of surveys. Not every CAPI survey can be moved to the web. For example, some survey questionnaires are simply too long. Moreover, recruiting panel members from the respondents of previous CAPI and CATI surveys leads to panels with a serious lack of representativity. Also the costs are substantial. Particularly, recruitment costs are high. There are also other costs, like yearly maintenance costs (for example, for keeping the panel representative) and costs per survey. This leads to the decision to postpone the introduction of a web panel.

## 14.3.2 APPLICATION 2: THE U.K. POLLING DISASTER

Pre-election polling for the general election in the United Kingdom on May 7, 2015, turned out to be a disaster. Many polls were wrong with their prediction of the final election result. Moreover, they were consistently wrong. All polls predicted a neck-and-neck race between the Conservative Party and the Labour Party, likely leading Britain to a "hung parliament." The real election result was, however, completely different. The Conservative Party got much more votes than the Labour Party. The difference was 6.5%. So there was a comfortable majority for the Conservatives.

Table 14.8 contains an overview of the 11 most important polls that were conducted just before the election. The first seven polls all drew their samples from self-selection web panel. The last three polls were telephone poll. They selected samples using some form of RDD or a database with information about consumers. The final column in Table 14.8 contains the predicted differences (in percentage points) between the Conservatives and Labour. It is clear that these differences are very small.

TABLE 14.8   **Estimates (unweighted and weighted) for main activity (%)**

| Poll | Mode | Sample | Difference (Con-Lab) (%) |
| --- | --- | --- | --- |
| Populus | Web panel | 3,917 | 0 |
| YouGov | Web panel | 10,307 | 0 |
| Survation | Web panel | 4,088 | 0 |
| Panelbase | Web panel | 3,019 | −2 |
| Opinium | Web panel | 2,916 | 1 |
| TNS | Web panel | 1,185 | 1 |
| BMG | Web panel | 1,009 | 0 |
| Ipsos MORI | Telephone | 1,186 | 1 |
| ComRes | Telephone | 1,007 | 1 |
| ICM | Telephone | 2,023 | −1 |
| Lord Ashcroft | Telephone | 3,028 | 0 |

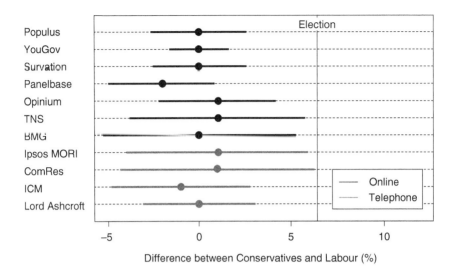

FIGURE 14.1  **The polling disaster in the United Kingdom on 7 May, 2015**

The performance of the polls is summarized in a different way in Figure 14.1. The dot plot compares the predictions of the polling companies with the true election result. The big dots represent the predicted differences between the Conservatives and Labour. All differences are clearly close to 0. The vertical line represents the election result (6.5%).

The horizontal line segments represent the margins of error (under the assumption that all samples are simple random samples). Note that the election result is outside these margins, which indicates that all poll results differ significantly from the value to be estimated. This would mean that all polls are systematically wrong.

The poll disaster caused a lot of concern about the reliability of the election polls carried out in the United Kingdom. Therefore, the British Polling Council (BPC), an association of polling organizations that publish polls, decided to set up an independent inquiry to look into the possible causes of the problems and to make recommendations for future polling.

Main ingredient of the inquiry was a comparison of the polls with two benchmark surveys. The first one was the British Election Survey (BES), and the second was British Social Attitudes Survey (BSA). Both surveys were face-to-face surveys based on multistage, stratified probability samples of addresses. At each selected address, one random eligible person was drawn. The sampling frame was the PAF. The response rates were 56% for the BES and 51% for the BSA. These are reasonably high response rates when compared to the opinion polls. The BES estimate of the difference between the Conservatives and Labour is 7%, and the BSA estimate is 6%. Both estimates were very close to the true difference of 6.5 percentage points. It is clear that these two surveys could be used as "gold standards" for the analysis of the polls.

The Inquiry panel concluded that the primary cause of the failure was the lack of representativity of the samples selected by the polling companies. Labour voters were systematically overrepresented, and Conservative voters were systematically underrepresented. Moreover, it turned out that weighting adjustment was not effective for reducing the bias in the predictions. See Sturgis et al. (2016) for more details.

The Inquiry panel ruled out other possible causes. One possible cause was differential misreporting of voting intentions. This phenomenon was sometimes also called the "Shy Tory Factor." It is the effect that more Conservative voters than Labour voters said in the polls they were not going to vote. This effect would lead to an underrepresentation of Conservatives in the polls. The Inquiry panel concluded that there was no substantial "Shy Tory Factor."

There was also no evidence of a so-called late swing. This is the phenomenon that people change their mind between the final polls and the day of the election. This would mean that ultimately people voted for a different party than indicated in the polls.

It is remarkable that the polls were consistently wrong. They all had more or less the same bias. This suggest "herding" as a possible cause of the problems. Herding means that polling organizations make design decisions that cause their polls to be closer to the results of other polling organizations. Consequently, they avoid the risk of being the only polling organization with a wrong result. The Inquiry panel could not completely rule out herding.

Table 14.8 shows that two different modes of data collection were used in the polls. The first seven polling organizations did online polls. They all had web panels, and people in these panels were all recruited by means of self-selection. So, there is no guarantee that these panels are representative. This also means that random samples from these online panels lack representativity. Of course, the polling organizations tried to remove or reduce the bias in their estimators by carrying out some kind of weighting adjustment. Typical weighting variables were age, sex, region, social grade, and working status. Most companies also used vote at the previous election as weighting variable. Unfortunately, the investigation of the Inquiry panel showed the polling companies did not succeed in removing the bias of estimates.

Four of the polling organizations in Table 14.8 used telephone polls. This mode of data collection is not without problems. The telephone directory cannot be used as a sampling frame. It is far from complete. Many landline telephone numbers are not in this directory. And it contains almost no mobile telephone numbers. So there is a substantial under-coverage problem. The way out for these polling organizations is to use RDD. They generate and call randomly generated telephone numbers (both landline and mobile). Unfortunately, response rates are very low. They often do not exceed 20% and are sometimes (in urban areas) even below 10%. This means these polls can have a large nonresponse bias. Therefore, one can conclude that telephone polls are not an alternative for self-selection online polls.

# 14.4 Summary

A *web panel* (also called an *online panel*, *Internet panel*, or *access panel*) is a mode of data collection in which the same group of individuals is interviewed via the web at different points in time. A web panel can be used for longitudinal research, in which the same individuals are interviewed repeatedly. The focus is on measuring change. A web panel can also be used for cross-sectional research if it is used as sampling frame for specific surveys, where a new sample is selected from the panel members for each survey.

Before a web panel can be used, it has to be built up. In this recruitment phase, individuals are invited to join the panel. There are two ways to do this:

- Draw a probability sample from a sampling frame representing the target population, and ask the selected individuals to become a member of the web panel. This can be done, for example, by means of a face-to-face, telephone, or mail survey.

- Apply self-selection. Let individuals select themselves for the web panel. Invitations can take the form of banners or pop-up windows on websites or advertisements in other media (radio, TV, newspapers).

Probability-based web panels have the advantage that they allow for proper statistical inference from the panel to the population. The lack of representativity of self-selection web panels may prevent reliable inference. Weighting adjustment may be applied to improve representativity, but there is no guarantee that this will be successful.

It is more expensive and time consuming to develop a probability-based panel than a self-selection panel. A well-designed recruitment campaign may cause many people to become a member of a self-selection panel. Such a large panel has the advantage that it will also contain people belonging to special groups in the population. The inference problems caused by the lack of representativity of self-selection panels are not solved by a large panel size.

During recruitment, new members usually complete a profile questionnaire. The collected data can be used for weighting adjustment and to select special groups from the panel for specific surveys.

For a specific survey, all panel members could be asked to complete the questionnaire. Often only a random sample is selected from the panel, or a special group is selected using the available variables.

The response rates of specific surveys are often high. This is not surprising as the people in the panel have already agreed in the recruitment phase to regularly complete questionnaires. This response rate is not a good indicator of the quality of the response in a specific survey. Also the response rate in the recruitment phase should be taken into account. This results in so-called cumulative response rates.

It is not possible to compute response rates for self-selection panels as the target population is not defined and also no sampling frame is used. There are other indicators, like the completion rate, that provide some insight in the quality of the response.

## KEY TERMS

**Access panel**: See Self-selection panel.

**Attrition**: The phenomenon that panel members drop out of the panel.

**Attrition rate**: The proportion of members who drop out of the panel in a defined time period.

**Completion rate**: The proportion of qualified panel members who started the survey questionnaire and completed the questionnaire.

**Cumulative response rate**: The response rate obtained by multiplying the response rate in the recruitment phase by the response rates in the subsequent waves or specific surveys.

**Internet population**: The subpopulation of the target population consisting of only those elements having access to the Internet.

**Opt-in panel**: See Self-selection panel.

**Probability-based panel**: A panel for which members are recruited by means of probability sampling.

**Reference survey**: A survey conducted with the objective of obtaining unbiased estimates of the population distributions of auxiliary variables.

**Representative**: The (weighted) survey response is representative with respect to a variable if the (weighted) response distribution is equal to its population distribution.

**Self-selection panel**: A web panel for which people select themselves spontaneously in response to a banner, pop-up window, or advertisements in other media (radio, TV, newspapers).

**Volunteer panel**: See Self-selection panel.

## EXERCISES

**Exercise 14.1**    The table below contains the results of the recruitment phase for a panel of persons. Compute the recruitment rate.

| Household level | |
|---|---:|
| IC | 23,460 |
| R | 1,200 |
| NC | 800 |
| O | 500 |
| UH | 380 |
| UO | 150 |
| Total eligible | 26,490 |
| Estimated proportion of unknown eligibility that are eligible | 0.06 |

**Exercise 14.2** A list of eligible households has been contacted with the invitation to become a member of a web panel. 7,000 households are on the list and have been sent e-mail invitations. 200 e-mails have bounced back as undeliverable, and 250 e-mails can be characterized as network error-undeliverable e-mails. Compute the absorption rate.

**Exercise 14.3** In web panel recruitment of the KnowledgePanel, the following rates have been calculated:

- Profile rate (PROR) = 0.568.
- Completion rate (COMR) = 0.845.
- Break-off rate (BOR) = 0.0056.
- Retention rate (RETR) = 0.390.

Compute cumulative response rate 1 and cumulative response rate 2.

**Exercise 14.4** A large self-selection web panel is made representative with respect to gender and age by removing members of overrepresented groups from the panel until the gender by age distribution in the panel is the same as this distribution in the population. The researcher claims that the resulting panel is representative. Is this correct?

**a.** Yes, surveys from the panel can now be treated as equal probability samples.
**b.** Yes, but weighting adjustment by gender and age should be applied to specific surveys from the panel.
**c.** No, but it can be made representative by repeating this for other auxiliary variables.
**d.** No, the panel will never be completely representative because a specific part of the population is always missing.

**Exercise 14.5** Why is it not possible to compute the response rate for the recruitment phase of a self-selection web panel?

**a.** There are only respondents, no nonrespondents.
**b.** It is not possible to distinguish nonresponse from over-coverage.
**c.** Recruitment is continuous activity.
**d.** No initial sample has been selected.

**Exercise 14.6** What is an advantage of a self-selection panel over a probability-based panel?

**a.** It is less expensive and less time consuming to construct a web panel.

**b.** A much wider population can be covered, because no sampling frame is used.

**c.** Only people become a member who are really interested.

**d.** More surveys per month can be offered.

**Exercise 14.7**   In which situation is it wise to use a reference survey for adjustment weighting?

**a.** To improve the accuracy of estimates after weighting adjustment.

**b.** If "webographic" variables are unrelated to the target variable of the survey.

**c.** If the specific survey lacks representativity and effective weighting variables cannot be retrieved from another source.

**d.** Only if the panel was recruited by means of probability sampling.

**Exercise 14.8**   The recruitment sample for a longitudinal study is obtained by means of a probability sampling. The response rate is 50%. There are three waves of interviewing after recruitment. In each wave, 10% of the participants decide to stop. What is the cumulative response rate in the last wave?

## REFERENCES

American Association for Public Opinion Research (AAPOR). (2006), *Final Dispositions of Case Codes and Outcomes Rates for Surveys, 4th Edition*. AAPOR, Lenexa, KS.

Arbeitskreis Deutscher Markt. (2001), *Standards for Quality Assurance for Online Surveys*. ADM, Bonn, Germany.

Arens, Z. & Miller-Steiger, D. (2006), Time in Sample: Searching for Conditioning in a Consumer Panel. *Public Opinion Pros*, August, www.publicopinionpros.norc.org.

Baker, R., Blumberg, S. J., Brick, J. M., Couper, M. P., Courtright, M., Dennis, J. M., Dillman, D., Frankel, M. R., Garland, P., Groves, R. M., Kennedy, C., Krosnick, J., Lavrakas, P. J., Lee, S., Link, M., Piekarski, L., Rao, K., Thomas, R. K., & Zahs, D. (2010), Research Synthesis: AAPOR Report on Online Panels. *Public Opinion Quarterly*, 74, pp. 711–781.

Bethlehem, J. G. (2009), *Applied Survey Methods: A Statistical Perspective*. Wiley, Hoboken, NJ.

Bethlehem, J. G. (2010), Selection Bias in Web Surveys. *International Statistical Review*, 78, pp. 161–188.

Bethlehem, J. G., Cobben, F., & Schouten, B. (2011), *Handbook of Nonresponse in Household Surveys*. Wiley, Hoboken, NJ.

Bethlehem, J. G. & Stoop, I. A. L. (2007), *Online Panels – A Theft of Paradigm? The Challenges of a Changing World*, Proceedings of the Fifth International Conference of the Association of Survey Computing, Southampton, pp. 113–132.

Bronzwaer, S. (2012), Infiltranten probeerden de peilingen van Maurice de Hond te ani-puleren. *NRC*, September 13.

Brüggen, E., Van Den Brakel, J., & Krosnick, J. (2016), *Establishing the Accuracy of Online Panels for Survey Research*. Report 2016|04. Statistics Netherlands, The Hague, the Netherlands.

Callegaro, M., Baker, R., Bethlehem, J. G., Göritz, A. S., Krosnick, J. A., & Lavrakas, P. (2014), *Online Panel Research, A Data Quality Perspective*. Wiley, Chichester, U.K.

Callegaro, M. & DiSogra, C. (2008), Computing Response Metrics for Online Panels, *Public Opinion Quarterly*, 72, pp. 1008–1032.

Clinton, J. D. (2001), *Panel Bias from Attrition and Conditioning: A Case Study of the Knowledge Networks Panel*. Paper presented at 56th Annual Conference of the American Association for Public Opinion Research, May, Montreal, Canada.

Comley, P. (2007), Online Market Research. In: ESOMAR (ed.), *Market Research Handbook*, Wiley, Hoboken, NJ, pp. 401–420.

Couper, M. P. (2000), Web Surveys. A Review of Issues and Approaches. *Public Opinion Quarterly*, 64, pp. 464–494.

Couper, M. P. (2001), The Promises and Perils of Web Surveys. In: Westlake, A., Sykes, W., Manners, T., & Riggs, M. (eds.), *The challenge of the Internet*. Association for Survey Computing, London, pp. 35–56.

Couper, M. P. (2007), Issues of Representation in Health Research (with a Focus on Web Surveys). *American Journal of Preventive Medicine*, 32, pp. 83–89.

Couper, M. P. & Dominitz, J. (2007), *Using an RDD Survey to Recruit Online Panel Members*. Paper presented at 2007 Biennial Conference of the European Survey Research Association, June 25–29, Prague.

Couper, M. P., Kapteyn, A., Schonlau, M., & Winter, J. (2007), Noncoverage and Non-response in an Internet Survey. *Social Science Research*, 36, pp. 131–148.

Dillman, D. A. & Bowker, D. K. (2001), The Web Questionnaire Challenge to Survey Methodologists, In: Reips, U. D. & Bosnjak, M. (eds.), *Dimensions of Internet Science*, Pabst Science Publishers, Lengerich, Germany.

DiSogra, C., Slotwiner, D., Clinton, S., Chan, E., Hendarwan, E., & Zheng, W. (2007), *Nonresponse Bias in Two Methods of Panel Recruitment*. Paper presented at Joint Statistical Meetings (JSM), July 29–August 2, Salt Lake City, UT.

Duffy, B., Smith, K., Terhanian, G., & Bremer, J. (2005), Comparing Data from Online and Face-to-Face Surveys. *International Journal of Market Research*, 47, pp. 615–639.

European Federation of Associations of Market Research Organizations (EFAMRO). (2004), *Quality Standards for Access Panel (QSAP)*. www.efamro.com/short-print2.html.

European Society for Opinion and Marketing Research (ESOMAR). (2005), *Conducting Market and Opinion Research Using the Internet*. www.esomar.org/uploads/pdf/.

European Society for Opinion and Marketing Research (ESOMAR). (2008), *26 Questions to Help Research Buyers of Online Samples ESOMAR Codes & Guidelines Conducting Research Using Internet*. www.esomar.org/uploads/pdf/professionalstandards/26ques-tions.pdf.

Eysenbach, G. (2004), Improving the Quality of Web Surveys: The Checklist for Report-ing Results from Internet E-Surveys (Cherries). *Journal of Medical Internet Research*, 6, e34.

Faas, T. & Schoen, H. (2006), Putting a Questionnaire on the Web Is Not Enough: A Comparison of Online and Offline Surveys Conducted in the Context of the German Federal Election 2002. *Journal of Official Statistics*, 22, pp. 177–190.

File, T. & Ryan, C. (2014), *Computer and Internet Use in the United States: 2013.* American Community Survey Reports, U.S. Census Bureau, Washington DC.

Fowler, F. J. Jr. (2002), *Survey Research Methods, 3rd Edition.* Sage, Thousand Oaks, CA.

Fricker, R. D. & Schonlau, M. (2002), Advantages and Disadvantages of Internet Research Surveys: Evidence from the Literature. *Social Science Computer Review*, 14, pp. 347–367.

Göritz, A. S. (2007), Using Online Panels in Psychological Research. In: Joinson, A. N., McKenna, K. Y. A., Postmes, T., & Reips, U. D. (eds.), *The Oxford Handbook of Internet Psychology.* Oxford University Press, New York, pp. 473–485.

Groves, R. M. (2006), Nonresponse Rates and Nonresponse Bias in Household Surveys. *Public Opinion Quarterly*, 70, pp. 646–675.

Groves, R. M., & Peytcheva, E. (2008), The Impact of Nonresponse Rates on Nonresponse Bias. A Meta-Analysis. *Public Opinion Quarterly*, 72, pp. 167–189.

Huggins, V. & Eyerman, J. (2001), *Probability Based Internet Surveys: A Synopsis of Early Methods and Survey Research Results.* Paper presented at Federal Committee on Statistical Methodology Research Conference, November 14–16, Arlington, VA.

Interactive Marketing Research Organization (IMRO). (2006), *IMRO Guidelines for Best Practices in Online Sample and Panel Management.* www.imro.org/pdf/.

Kalton, G. (1987), *Introduction to Survey Sampling.* Sage, Thousand Oaks, CA.

Kiaer, A. N. (1895), Observations et Expériences Concernant des Dénombrements Représentatives. *Bulletin of the International Statistical Institute*, IX, Book 2, pp. 176–183.

Kruskal, W. & Mosteller, F. (1979a), Representative Sampling, I: Non-scientific Literature. *International Statistical Review*, 47, pp. 13–24.

Kruskal, W. & Mosteller, F. (1979b), Representative Sampling, II: Scientific Literature. Excluding Statistics. *International Statistical Review*, 47, pp. 111–127.

Kruskal, W. & Mosteller, F. (1979c). Representative Sampling, III: The Current Statistical Literature. *International Statistical Review*, 47, pp. 245–265.

Leenheer, J. & Scherpenzeel, A. C. (2013), Does It Pay Off to Include Non-internet Households in an Internet Panel? *International Journal of Internet Science*, 8, 1, pp. 17–29.

Lozar Manfreda, K. & Vehovar, V. (2002), *Survey Design Features Influencing Response Rates in Web Surveys.* Paper presented at International Conference on Improving Surveys, August 25–28, Copenhagen, Denmark.

Office of Management and Budget (OMB). (2006), *Questions and Answers When Designing Surveys for Information Collections.* Office of Management and Budget, Washington, DC.

Postoaca, A. (2006), *The Anonymous Elect. Market Research through Online Access Panels.* Springer, Berlin, Germany.

Pratesi, M., Lozar Manfreda, K., Biffignandi, S., & Vehovar, V. (2004), List-Based Web Surveys: Quality, Timeliness, and Nonresponse in the Steps of the Participation Flow. *Journal of Official Statistics*, 20, pp. 451–465.

Saris, W. E. (1991), *Computer Assisted Interviewing.* Sage, Newbury Park, CA.

Saris, W. E. (1998), Ten Years of Interviewing Without Interviewers: The Telepanel. In: Couper, M. P., Baker, R. P., Bethlehem, J. G., Clark, C. Z. F., Martin, J., Nicholls, W. L., & O'Reilly, J. M. (eds.), *Computer Assisted Survey Information Collection*, Wiley, New York, pp. 409–429.

Sayles, H. & Arens, Z. (2007), *A Study of Panel Member Attrition in the Gallup Panel.* Paper presented at 62nd Annual Conference of the American Association for Public Opinion Research, May 17–20, Anaheim, CA.

Scherpenzeel, A. (2008), An Online Panel as a Platform for Multi-Disciplinary Research. In: Stoop, I. & Wittenberg, M. (eds.), *Access Panels and Online Research, Panacea or Pitfall?* Aksant, Amsterdam, the Netherlands, pp. 101–106.

Scherpenzeel, A, & Schouten, B. (2011), *LISS Panel R-indicator: Representativity in Different Stages of Recruitment and Participation of an Internet Panel.* Paper presented at the 22nd International Workshop on Household Survey Nonresponse, Bilbao, Spain.

Schlengen, W. E., Caddell, J. M., Ebert, L., Jordan, K. B., Rourke, K. M., Wilson, D., Thalji, T., Dennis, M. J., Fairbank, J. A., & Kulka, R. (2002), Psychological Reactions to Terrorist Attacks. Findings from the National Study of American's Reactions to September 11. *Journal of the American Medical Association*, 288, pp. 581–588.

Schonlau, M., Fricker, R. D., & Elliott, M. N. (2002), *Conducting Research Surveys Via E-Mail and the Web.* RAND, Santa Monica, CA.

Schonlau, M., Van Soest, A., Kaypten, A., & Couper, M. P. (2007), Are "Webographic" or Attitudinal Questions Useful for Adjusting Estimates from Web Surveys Using Propensity Scoring? *Survey Research Methods*, 1, pp. 155–156.

Schonlau, M., Van Soest, A, Kapteyn, A., & Couper, M. (2009), Selection Bias in Web Surveys and the Use of Propensity Scores. *Sociological Methods and Research*, 37, pp. 291–318.

Schouten, B., Cobben, F., & Bethlehem, J. (2009), Indicators for the Representativeness of Survey Response. *Survey Methodology*, 35, 1, pp. 101–113.

Sikkel, D. & Hoogendoorn, A. (2008), Panel Surveys. In: de Leeuw, E., Hox, J., & Dillman, D. A. (eds.), *International Handbook of Survey Methodology*, Lawrence Erlbaum. New York, pp. 479–499.

Sturgis, P., Baker, N., Callegaro, M., Fisher, S., Green, J., Jennings, W., Kuha, J., Lauderdale, B., & Smith, P. (2016), *Report of the Inquiry into the 2015 British General Election Opinion Polls.* Market Research Society and British Polling Council, London, U.K.

Sudman, S. (1966), Probability Sampling With Quotas. *Journal of the American Statistical Association*, 61, pp. 749–771.

Tortora, R. (2009), Attrition in Consumer Panels. In: Lynn, P. (ed.), *Methodology of Longitudinal Surveys*, Wiley, Hoboken, NJ, pp. 235–249.

Tourangeau, R. (2003), Web-Based Data Collection. In: Cork, D. L., Cohen, M. L., Groves, R., & Kalsbeek, W. (eds.), *Survey Automation, Report and Workshop Proceedings*, National Academies Press, Washington, DC, pp. 183–198.

Vonk, T., Van Ossenbruggen, R., and Willems, P. (2006), The Effect of Panel Recruitment and Management on Research Results. A Study Across 19 Online Panels. *Panel Research 2006*, ESOMAR Publications Series, Amsterdam, the Netherlands.

# Index

*Handbook of Web Surveys*, Second Edition. Silvia Biffignandi and Jelke Bethlehem.
© 2021 John Wiley & Sons, Inc. Published 2021 by John Wiley & Sons, Inc.